Functions of a Complex Variable

SIAM's Classics in Applied Mathematics series consists of books that were previously allowed to go out of print. These books are republished by SIAM as a professional service because they continue to be important resources for mathematical scientists.

Classics in Applied Mathematics

C. C. Lin and L. A. Segel, *Mathematics Applied to Deterministic Problems in the Natural Sciences*
Johan G. F. Belinfante and Bernard Kolman, *A Survey of Lie Groups and Lie Algebras with Applications and Computational Methods*
James M. Ortega, *Numerical Analysis: A Second Course*
Anthony V. Fiacco and Garth P. McCormick, *Nonlinear Programming: Sequential Unconstrained Minimization Techniques*
F. H. Clarke, *Optimization and Nonsmooth Analysis*
George F. Carrier and Carl E. Pearson, *Ordinary Differential Equations*
Leo Breiman, *Probability*
R. Bellman and G. M. Wing, *An Introduction to Invariant Imbedding*
Abraham Berman and Robert J. Plemmons, *Nonnegative Matrices in the Mathematical Sciences*
Olvi L. Mangasarian, *Nonlinear Programming*
*Carl Friedrich Gauss, *Theory of the Combination of Observations Least Subject to Errors: Part One, Part Two, Supplement.* Translated by G. W. Stewart
Richard Bellman, *Introduction to Matrix Analysis*
U. M. Ascher, R. M. M. Mattheij, and R. D. Russell, *Numerical Solution of Boundary Value Problems for Ordinary Differential Equations*
K. E. Brenan, S. L. Campbell, and L. R. Petzold, *Numerical Solution of Initial-Value Problems in Differential-Algebraic Equations*
Charles L. Lawson and Richard J. Hanson, *Solving Least Squares Problems*
J. E. Dennis, Jr. and Robert B. Schnabel, *Numerical Methods for Unconstrained Optimization and Nonlinear Equations*
Richard E. Barlow and Frank Proschan, *Mathematical Theory of Reliability*
Cornelius Lanczos, *Linear Differential Operators*
Richard Bellman, *Introduction to Matrix Analysis, Second Edition*
Beresford N. Parlett, *The Symmetric Eigenvalue Problem*
Richard Haberman, *Mathematical Models: Mechanical Vibrations, Population Dynamics, and Traffic Flow*

*First time in print.

Classics in Applied Mathematics (continued)

Peter W. M. John, *Statistical Design and Analysis of Experiments*
Tamer Başar and Geert Jan Olsder, *Dynamic Noncooperative Game Theory, Second Edition*
Emanuel Parzen, *Stochastic Processes*
Petar Kokotović, Hassan K. Khalil, and John O'Reilly, *Singular Perturbation Methods in Control: Analysis and Design*
Jean Dickinson Gibbons, Ingram Olkin, and Milton Sobel, *Selecting and Ordering Populations: A New Statistical Methodology*
James A. Murdock, *Perturbations: Theory and Methods*
Ivar Ekeland and Roger Témam, *Convex Analysis and Variational Problems*
Ivar Stakgold, *Boundary Value Problems of Mathematical Physics, Volumes I and II*
J. M. Ortega and W. C. Rheinboldt, *Iterative Solution of Nonlinear Equations in Several Variables*
David Kinderlehrer and Guido Stampacchia, *An Introduction to Variational Inequalities and Their Applications*
F. Natterer, *The Mathematics of Computerized Tomography*
Avinash C. Kak and Malcolm Slaney, *Principles of Computerized Tomographic Imaging*
R. Wong, *Asymptotic Approximations of Integrals*
O. Axelsson and V. A. Barker, *Finite Element Solution of Boundary Value Problems: Theory and Computation*
David R. Brillinger, *Time Series: Data Analysis and Theory*
Joel N. Franklin, *Methods of Mathematical Economics: Linear and Nonlinear Programming, Fixed-Point Theorems*
Philip Hartman, *Ordinary Differential Equations, Second Edition*
Michael D. Intriligator, *Mathematical Optimization and Economic Theory*
Philippe G. Ciarlet, *The Finite Element Method for Elliptic Problems*
Jane K. Cullum and Ralph A. Willoughby, *Lanczos Algorithms for Large Symmetric Eigenvalue Computations, Vol. I: Theory*
M. Vidyasagar, *Nonlinear Systems Analysis, Second Edition*
Robert Mattheij and Jaap Molenaar, *Ordinary Differential Equations in Theory and Practice*
Shanti S. Gupta and S. Panchapakesan, *Multiple Decision Procedures: Theory and Methodology of Selecting and Ranking Populations*
Eugene L. Allgower and Kurt Georg, *Introduction to Numerical Continuation Methods*
Leah Edelstein-Keshet, *Mathematical Models in Biology*
Heinz-Otto Kreiss and Jens Lorenz, *Initial-Boundary Value Problems and the Navier-Stokes Equations*
J. L. Hodges, Jr. and E. L. Lehmann, *Basic Concepts of Probability and Statistics, Second Edition*
George F. Carrier, Max Krook, and Carl E. Pearson, *Functions of a Complex Variable: Theory and Technique*

Functions of a Complex Variable

Theory and Technique

George F. Carrier
Max Krook
Carl E. Pearson

Society for Industrial and Applied Mathematics
Philadelphia

This SIAM edition is an unabridged republication of the work first published by McGraw-Hill, New York, 1966.

10 9 8 7 6 5 4 3 2 1

Library of Congress Control Number: 2005925865
ISBN 0-89871-595-4

contents

4 *conformal mapping* 111

5 *special functions* 183

6 *asymptotic methods* 241

7 *transform methods* 301

8 *special techniques* 376

preface

In addition to being a rewarding branch of mathematics in its own right, the theory of functions of a complex variable underlies a large number of enormously powerful techniques which find their application not only in other branches of mathematics but also in the sciences and in engineering. Chapters 1, 2, and 5 of this book provide concisely but honestly the classical aspects of the theory of functions of a complex variable; the rest of the book is devoted to a detailed account of various techniques and the ideas from which they evolve. Many of the illustrative examples are phrased in terms of the physical contexts in which they might arise; however, we have tried to be consistent in including a mathematical statement of each problem to be discussed.

For the acquisition of skill in the use of these techniques, practice is even more important than instruction. Accordingly, we have inserted many exercises, including some which are essential parts of the text. The reader who fails to carry out a substantial number of these exercises will have missed much of the value of this book.

The individual chapters segregate specific topical items, but many readers will find it profitable to study selected parts of Chaps. 3 to 7 as they encounter the related underlying theory in Chap. 2.

We hope that through our presentation the reader will be able to discern the fascination of complex-function theory, recognize its power, and acquire skill in its use.

George F. Carrier
Max Krook
Carl E. Pearson

erratum

On page 340, Exercise 8 should read as follows:
8.(a) Let $g(x, y, z, t; \xi, \eta, \zeta, \tau)$ satisfy the wave equation

$$g_{xx} + g_{yy} + g_{zz} - \frac{1}{c^2} g_{tt} = \delta(x - \xi)\delta(y - \eta)\delta(z - \zeta)\delta(t - \tau).$$

That is, in two places, the symbol ξ should be replaced by ζ.

1

complex numbers and their elementary properties

1-1 Origin and Definition

Complex numbers originated from the desire for a symbolic representation for the solution of such equations as $x^2 + 1 = 0$. Continued usage for this purpose gradually endowed them with some degree of conceptual existence; however, they were not generally accepted as providing a legitimate extension of real numbers until experience showed that their use gave a completeness and insight that had previously been lacking. It was found, for example, that if complex numbers were allowed, then every polynomial had a zero; moreover, such results as the previously mysterious divergence at $x = 1$ of the power series for $(1 + x^2)^{-1}$ became more explicable.

Complex numbers are now so widely used in applied mathematics (for example, in evaluation of integrals, series representations of functions, and solutions of ordinary and partial differential equations) that they tend to be accepted without question. It was not always so; in forcing their way into mathematics, these "imaginary" numbers experienced as much opposition as negative numbers had in their earlier turn. It is easy to appreciate the remark of Leibniz that complex numbers are a "wonderful refuge of the divine spirit . . . an amphibian between being and non-being" or the succinct statement of Gauss that the "true metaphysics" of $(-1)^{1/2}$ is hard.

The quantity $(-1)^{1/2}$ will be denoted by i, and the general form of a complex number will be taken to be

$$c = a + ib$$

where the real numbers a and b are referred to as the *real* and *imaginary* parts of c, written Re c and Im c, respectively. If b is zero, then c is an ordinary real number; if a is zero, then c is a pure imaginary. The number c may be depicted geometrically (in what is called the *Argand diagram*, or z plane) as a point in the xy plane having the coordinates (a,b). The *modulus* of any complex number $z = x + iy$, written $|z|$, is a positive real number defined by

$$|z| = (x^2 + y^2)^{1/2}$$

and so equals the distance of the representative point z from the origin (see Fig. 1-1). The angle θ measured from the positive x axis to the radius vector drawn from the origin to z is called the *argument* of z and is denoted by arg z; it satisfies

$$\tan\theta = \frac{y}{x} \qquad \cos\theta = \frac{x}{|z|} \qquad \sin\theta = \frac{y}{|z|}$$

The counterclockwise direction of measurement of θ is taken as positive. The angle θ defined by these relations is determined except for an arbitrary multiple of 2π radians; that value of θ which satisfies $-\pi < \theta \leq \pi$ is called the *principal value* of the argument. It is clear that

$$z = |z|(\cos\theta + i\sin\theta) \tag{1-1}$$

and also that a complex number is uniquely determined by a knowledge of its modulus and argument.

By definition, two complex numbers are equal if their real and imaginary parts are respectively equal, so that their representative points coincide.

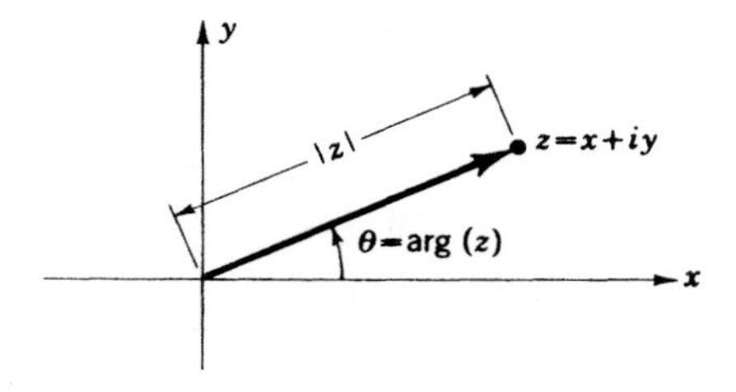

Fig. 1-1 Argand diagram.

The sum or difference z_3 of the two numbers $z_1 = x_1 + iy_1$ and $z_2 = x_2 + iy_2$ is defined by

$$z_3 = z_1 \pm z_2 = (x_1 \pm x_2) + i(y_1 \pm y_2)$$

This is geometrically equivalent to the parallelogram law for composition of vectors. The triangle inequality of geometry gives at once

$$|(|z_1| - |z_2|)| \leq |z_1 + z_2| \leq |z_1| + |z_2| \tag{1-2}$$

The product z_3 of z_1 and z_2 is defined to be the result obtained by a formal application of the distributive law, so that

$$z_3 = z_1 z_2 = (x_1 + iy_1)(x_2 + iy_2) = (x_1 x_2 - y_1 y_2) + i(x_1 y_2 + x_2 y_1)$$

A simple calculation shows that the modulus of this product is the product of the original moduli (so that a product can vanish only if one of the constituent factors is zero) and that within an integral multiple of 2π the argument is the sum of the original arguments. By induction, this statement holds for the product of any number of complex numbers. Note that the geometrical equivalent of the multiplication of z by i is the rotation, without magnification, of the radius vector through a counterclockwise angle of $\pi/2$.

In particular, the result of multiplying z by itself n times (n a positive integer) is written z^n. By induction and use of Eq. (1-1) it is easily shown either geometrically or analytically that

$$z^n = |z|^n(\cos n\theta + i \sin n\theta) \tag{1-3}$$

Conversely, if a number ζ can be found such that $\zeta^n = z$, then ζ is said to be an nth *root* of z.

The above definitions of addition and multiplication imply that the usual commutative, associative, and distributive laws hold, so that, for example,

$$\begin{aligned} z_1 + z_2 &= z_2 + z_1 \\ z_1 z_2 &= z_2 z_1 \\ (z_1 + z_2) + z_3 &= z_1 + (z_2 + z_3) \\ (z_1 z_2) z_3 &= z_1(z_2 z_3) \\ z_1(z_2 + z_3) &= z_1 z_2 + z_1 z_3 \end{aligned}$$

Given any number z_1 and any nonzero number z_2, we define the quotient of z_1 by z_2, written z_1/z_2, as the unique number z_3 satisfying

$$\begin{aligned} |z_3| &= \frac{|z_1|}{|z_2|} \\ \arg z_3 &= \arg z_1 - \arg z_2 \pm 2\pi n \qquad n \text{ integral} \end{aligned}$$

Note that $z_1 = z_2 z_3$, so that this division operation is inverse to multiplication.

The complex conjugate of $z = x + iy$ is defined by $z^* = x - iy$ and so corresponds geometrically to a reflection of z in the real axis. It may be checked that the complex conjugate of a sum, product, or quotient is the sum, product, or quotient of the individual complex conjugates. Since the product zz^* is always real, a convenient way of calculating z_1/z_2 is to multiply both numerator and denominator by z_2^*; thus

$$\frac{z_1}{z_2} = \frac{x_1 + iy_1}{x_2 + iy_2} = \frac{(x_1 + iy_1)(x_2 - iy_2)}{x_2^2 + y_2^2} = \frac{x_1x_2 + y_1y_2}{x_2^2 + y_2^2} + i\frac{y_1x_2 - x_1y_2}{x_2^2 + y_2^2}$$

The resulting formula, which uses only real numbers, could be used as the definition of the process of division for complex numbers.

More generally, it is worth pointing out that all operations on complex numbers involve only manipulations of real numbers. For any complex number $z = x + iy$ can be thought of as the ordered pair (x,y) of real numbers, and the result of adding or multiplying two such pairs may be defined to be a new ordered pair given by $(x_1 + x_2, y_1 + y_2)$ or $(x_1x_2 - y_1y_2, x_1y_2 + x_2y_1)$, respectively. The advantage of such an ordered-pair interpretation of z is that one need not introduce i as the mysterious $(-1)^{1/2}$, but merely as a label which distinguishes the order of the pair of real numbers. The rule for multiplication then implies that $0 + i1$, or $(0,1)$, is that number whose square is -1. On the other hand, the ordered-pair interpretation does involve some loss of compactness and motivation. Incidentally, it is possible to generalize the ordered-pair concept by considering ordered triplets or even n-tuplets of real numbers, with similar rules for the formation of sums or products. A historically famous such *linear algebra* is that of quaternions.

Finally, if there is some rule whereby, to each choice of a complex number z in a given region of the z plane, there corresponds another complex number w, we say that w is a (*single-valued*) function of z and write $w = f(z)$. Denoting the value of w at z_0 by w_0, we say that such a function is continuous at z_0 if w approaches w_0 as z approaches z_0. More precisely, the condition is that, corresponding to any preassigned real value of $\epsilon > 0$, there exists a real number $\rho > 0$ such that $|w - w_0| < \epsilon$ for all points satisfying $|z - z_0| < \rho$.

In the sequel, we shall also encounter situations in which two or more complex numbers are associated with a single choice of z; $f(z)$ is then said to be *multiple-valued.*

Exercises

1. In terms of the complex numbers z_1, z_2, z_3, forming the vertices of a triangle, find a formula for its centroid. Obtain an analogous result for the mass center of a number of masses in a plane.

2. Prove the triangle inequality algebraically. Show also that $|z_1 + z_2 + \cdots + z_n| \leq |z_1| + |z_2| + \cdots + |z_n|$. What is the condition for equality?

3. If n is a positive integer, show that there are exactly n essentially different nth roots of unity (i.e., roots of the equation $\zeta^n = 1$) and that geometrically these roots form the vertices of a regular n-gon.

4. Show that the equation of the straight line through z_1 and z_2 is $z(z_1 - z_2)^* + z_1(z_2 - z)^* + z_2(z - z_1)^* = 0$. Obtain a similar formula for the general equation of a conic, and give a criterion for type of conic in terms of its complex coefficients.

5. Use Eq. (1-3) and the binomial expansion to obtain a formula for $\cos n\theta$ in terms of powers of $\cos\theta$ and $\sin\theta$, and show that the result is a polynomial in $\cos\theta$.

6. Show that, if r is a real number,

$$1 + r\cos\theta + r^2\cos 2\theta + \cdots + r^n \cos n\theta$$

can be found as the real part of $(1 - z^{n+1})/(1 - z)$, where

$$z = r(\cos\theta + i\sin\theta)$$

Obtain this real part explicitly.

7. In the theory of real functions, it is shown that the sum, product, or quotient (if the divisor is nonzero) of continuous functions is also continuous. Extend this result to complex functions.

1-2 *Sequences and Series*

Just as in the case of real variables, the sequence $\{z_n\}$ of complex numbers

$$z_1, z_2, z_3, \ldots, z_n, \ldots$$

will be said to approach the complex number L as a limit (we write $z_n \to L$) if $|L - z_n| \to 0$ as $n \to \infty$. Formally, we require that, corresponding to any positive ϵ however small, there exists an integer N such that $|z_n - L| < \epsilon$ for all $n > N$. Note that the number L need not be a member of the sequence $\{z_n\}$. Geometrically, this definition of convergence requires the points z_i ultimately to cluster more and more tightly about the point L.

It is easy to show that, if $z_n \to L$ as $n \to \infty$, then $(z_1 + z_2 + \cdots + z_n)/n \to L$; further, if no term in the sequence vanishes and if the appropriate root is taken, $(z_1 z_2 \cdots z_n)^{1/n}$ also $\to L$. For example,

$$n^{1/n} = \left(\frac{1}{1}\frac{2}{1}\frac{3}{2}\frac{4}{3} \cdots \frac{n}{n-1}\right)^{1/n}$$

so that $n^{1/n} \to 1$ as $n \to \infty$. Similarly the reader may show by considering the sequence $2/1$, $(3/2)^2$, $(4/3)^3$, $\ldots$ that $(1/n)(n!)^{1/n} \to 1/e$ as $n \to \infty$.

The familiar Cauchy test for the convergence of real sequences holds also for complex sequences: the sequence $\{z_n\}$ converges if and only if, given any $\epsilon > 0$, an N exists such that $|z_{n+p} - z_n| < \epsilon$ for all $n > N$ and for all $p > 0$. The necessity of this condition is almost obvious, particularly if visualized geometrically; to prove sufficiency, note that the Cauchy condition requires all members of the infinite sequence to lie in some bounded region of the plane, so that the Bolzano-Weierstrass theorem[1] ensures the existence of at least one limit point—and clearly there cannot be more than one. (*Note:* A limit point of a set is defined formally as a point in every neighborhood of which, however small, there are always an infinite number of points of the set. The limit point itself need not be a member of the set.)

Consider now the infinite series of complex numbers

$$\Sigma z_i = z_1 + z_2 + \cdots + z_n + \cdots$$

Define the nth partial sum s_n to be the sum of the first n terms; then the series is said to converge if the sequence $\{s_n\}$ converges. The limit L of $\{s_n\}$ is called the *sum* of the series. Note that convergence of a series of complex numbers is equivalent to simultaneous convergence of the two real series obtained by considering separately the real and imaginary parts of the terms z_i, so that conventional convergence criteria for real series may be used. Clearly, a necessary (but by no means sufficient) condition for convergence is that $z_n \to 0$ as $n \to \infty$.

If the series obtained by replacing each term by its modulus is convergent, the series is said to be *absolutely convergent.* Use of the triangle inequality shows that absolute convergence implies convergence; the converse, of course, does not hold (consider $1 - \frac{1}{2} + \frac{1}{3} - \frac{1}{4} + \cdots$). Complex series which are absolutely convergent have a number of valuable properties which the reader may prove as exercises. First, the two real series obtained as above by taking real and imaginary parts of an absolutely convergent series are absolutely convergent. Second, an arbitrary rearrangement of the terms alters neither the property of absolute convergence nor the sum of the series (every term of the original series must of course appear somewhere in the new series). Third, if two absolutely convergent series are multiplied term by term in any manner which results in every possible product occurring exactly once, the result is again an absolutely convergent series and has a sum equal to the product of the original sums.

Convergence Tests

In testing a series for absolute convergence, we are concerned with a series of positive real terms. Although such a series obviously converges if the sequence of partial sums is bounded, this criterion is not usually

[1] The Bolzano-Weierstrass theorem states that every bounded infinite set of points must possess at least one limit point (which may or may not, however, be a point of the set). To see how the proof goes, consider the special case of an infinite number of points lying in the interval (0,1) of the real axis. Divide this interval into tenths. At least one such tenth interval must contain an infinite number of points; so subdivide that tenth again into tenths, and so on. In this way, the decimal representation of a certain number is constructed, and this number is clearly a limit point. The proof for a plane region is entirely analogous—merely use both horizontal and vertical subdivision.

employed directly. More commonly, an equivalent comparison test is used. If each term is less than the corresponding term of a known convergent series, then the partial sums are bounded, so that the given series is also convergent; conversely, if the given series "dominates" a series which diverges, then the given series also diverges. In such comparison testing, two useful series of positive terms are the geometric series

$$1 + r + r^2 + \cdots + r^n + \cdots$$

which converges for $r < 1$ and diverges for $r \geq 1$, and the harmonic series

$$1 + \frac{1}{2^p} + \frac{1}{3^p} + \cdots + \frac{1}{n^p} + \cdots$$

which converges for $p > 1$ and diverges for $p \leq 1$.

Application of such a comparison test shows at once, for example, that the series of positive terms

$$x_1 + x_2 + \cdots + x_n + \cdots$$

is convergent if there exists a real number r with $0 < r < 1$ such that from some n onward, one of the two conditions

$$(x_n)^{1/n} < r$$

$$\frac{x_{n+1}}{x_n} < r$$

holds. More delicate tests of this general character, associated with the names of Gauss and Raabe, are described in books on infinite series.[1]

Another useful test, applicable to a series of positive monotonely decreasing terms, is the Cauchy integral test. It may be exemplified by observing (visualize geometrically) that the series whose general term is $(n \ln n)^{-1}$ for $n > 2$ converges or diverges together with

$$\int_2^n \frac{dx}{x \ln x} = \ln \frac{\ln n}{\ln 2}$$

and so diverges.

Uniform Convergence

Returning now to sequences $\{s_n\}$ of complex numbers, let us consider the case where each term of such a sequence is a function of a complex variable z; that is, $s_n = s_n(z)$. Suppose that, for any fixed choice of z within a certain region, the sequence converges. Denote the limit function by $f(z)$. If the rate of convergence does not depend on the particular choice of z, then the sequence is said to be uniformly convergent in that region; more precisely, the requirement is that, given any ϵ, there exists an N independent of z such that $|s_n(z) - f(z)| < \epsilon$ for all $n > N$, and for all z in the region.

[1] For example, Konrad Knopp, "Theory and Application of Infinite Series," chap. 9, Hafner Publishing Company, Inc., New York, 1947.

The example $s_n = (1 + nx)^{-1}$, x real and greater than zero, shows that a convergent sequence need not be uniformly convergent. Here $f(x) = 0$, but $|s_n - f| = |(1 + nx)^{-1}|$, and irrespective of N there is always some x for which this quantity is greater than any preassigned small ϵ. On the other hand, this sequence is uniformly convergent in any interval bounded away from the origin.

A series $\Sigma u_i(z)$ is said to be uniformly convergent if its sequence of nth partial sums is uniformly convergent. An almost obvious test for uniform convergence (called the *Weierstrass M test*) is the following: If there is a convergent series of positive real constants, ΣM_n, such that $|u_n| < M_n$ for all n and for all z in the given region, then the series is uniformly convergent in that region.

The reader may now prove the important result that, if a sequence of continuous functions is uniformly convergent in a region of the complex plane, then the limit function is also continuous in that region.

EXERCISES

1. Find the limit function of $\epsilon/(\epsilon^2 + x^2)$ as $\epsilon \to 0$, where x and ϵ are real. For what region is the convergence uniform?

2. Examine for absolute and uniform convergence the infinite series whose nth terms are $n^n z^n$, $z^n/n!$, $(-1)^n n^2/(z - 1)^n$, and $z^n/(z^2 + n^2)$. Omit discussion of the behavior on boundaries between regions of convergence and divergence.

3. Prove (Kummer) that the series of positive terms Σx_i is convergent if there exist a fixed r and a sequence of positive a_i such that, for n sufficiently large,

$$a_n \frac{x_n}{x_{n+1}} - a_{n+1} > r > 0$$

(setting $a_n = n$ would give Raabe's test).

4. Use an argument of the Cauchy-integral-test kind to show that

$$1 + \frac{1}{2} + \frac{1}{3} + \cdots + \frac{1}{n} - \ln n$$

approaches a constant (Euler's constant γ) as $n \to \infty$.

5. Devise a series of positive terms x_n which converges despite the fact that there is no N such that x_{n+1}/x_n is bounded for $n > N$.

1-3 Power Series

Let $\{a_i\}$ be a sequence of fixed complex numbers. Then the infinite series

$$a_0 + a_1 z + a_2 z^2 + \cdots + a_n z^n + \cdots \tag{1-4}$$

is said to be a power series in the complex variable z. A superficially more general power series would be

$$a_0 + a_1(z - \zeta) + a_2(z - \zeta)^2 + \cdots + a_n(z - \zeta)^n + \cdots$$

where ζ is some fixed complex number; however, by defining a new variable $z' = z - \zeta$, the form becomes the same as that of Eq. (1-4).

A power series has the remarkable property of converging—if it converges at all—for all values of z lying within some circle in the complex plane. For let the series converge at some point z_0. Then $|a_n z_0{}^n|$ is bounded by some constant A for all n, so that

$$|a_n z^n| = |a_n z_0{}^n| \left|\frac{z}{z_0}\right|^n < A \left|\frac{z}{z_0}\right|^n \tag{1-5}$$

Consequently this comparison test shows that the series converges absolutely for all z satisfying $|z| < |z_0|$. Suppose second that the series diverges at some z_1. Then it could not converge for any z satisfying $|z| > |z_1|$, for the first result would then imply convergence at z_1. These two results together imply that there exists some circle in the z plane such that the series converges absolutely for all z inside the circle and diverges for all z outside the circle. As special cases, the radius of the circle may be zero or infinite. The behavior of the series for points on the boundary of the circle cannot be inferred from this theorem. Use of the Weierstrass M test of Sec. 1-2 together with an inequality of the form (1-5) shows that a power series converges uniformly in any region that can be enclosed in a circle which in turn lies wholly in the interior of the circle of convergence; it then follows from the continuity of any power of z that a power series converges to a continuous function within the circle of convergence.

Explicit formulas for the radius R of this circle of convergence may be obtained by direct use of the convergence tests of Sec. 1-2. First and most simply, if the limit as $n \to \infty$ of $|a_n/a_{n+1}|$ exists, then R equals this limit. Second, R equals the reciprocal of

$$\lim_{n\to\infty} [\text{l.u.b.}(|a_n|^{1/n}, |a_{n+1}|^{1/(n+1)}, |a_{n+2}|^{1/(n+2)}, \ldots)]$$

Since the least upper bound (l.u.b.) either decreases or remains constant as n increases, this limit always exists—although its value may of course be infinite.

Exercises

1. Find the radii of convergence of the power series whose coefficients are $1/n$, $(-1)^n n^4$, $\ln n$, $i^n/n!$, $1/n^2 + n^2[1 + (-1)^n]$.

2. Given a power series with a finite nonzero radius of convergence, find a change in variable which will make this radius unity. Describe also the effect of the change in variable defined by $\zeta = 1/z$.

In determining the behavior of a power series on the boundary of its circle of convergence, more general tests than those applicable to series of positive real terms must be applied. A number of such tests are described in texts on infinite series; we shall here give only the useful general condition of Weierstrass,[1] which applies to a series arranged to have unit radius of convergence (cf. Exercise 2 above).

In the series $\Sigma a_n z^n$, let

$$\frac{a_{n+1}}{a_n} = 1 - \frac{\alpha}{n} - \frac{A(n)}{n^\lambda}$$

where α is a complex constant, λ is a real constant with $\lambda > 1$, and $A(n)$ is some bounded complex function of n. Then, for points on the circle of convergence $|z| = 1$, the series (1) will converge absolutely if $\operatorname{Re} \alpha > 1$, (2) will diverge if $\operatorname{Re} \alpha \leq 0$, and (3) will converge (but not absolutely) if $0 < \operatorname{Re} \alpha \leq 1$, except at $z = 1$, where it will diverge.

As an example, the series $\Sigma(1/n)z^n$ converges everywhere on the unit circle except at the point $z = 1$.

Continuity, Uniqueness, and Other Properties

Power series have a number of important general properties. Suppose first that, on each of a sequence of distinct points approaching the origin as a limit point, two series $\Sigma a_n z^n$ and $\Sigma b_n z^n$ have the same sum; then it can be deduced that $a_n = b_n$ for all n. For the first series may be written

$$a_0 + z(a_1 + a_2 z + a_3 z^2 + \cdots)$$

and the second series similarly; in each case, the quantity in parentheses is a power series with the same convergence properties as the parent series. Since the parenthetical function is continuous and therefore bounded within a circle around the origin (lying inside both circles of convergence), the two series have the form

$$a_0 + zM_1 \qquad b_0 + zM_2$$

where M_1 and M_2 are bounded. As z is allowed to approach zero, corresponding to a progression along the sequence of points on which the two series have the same value, the second terms above become negligible and it can be concluded that $a_0 = b_0$. It now follows that

$$a_1 z + a_2 z^2 + \cdots = b_1 z + b_2 z^2 + \cdots$$

on each of the points of the sequence; upon dividing by z, the above argument may be repeated to show that $a_1 = b_1$, and similarly $a_n = b_n$ for any n. As a corollary, a power series which vanishes on the above sequence of points is everywhere zero.

Second, let

$$f(w) = a_0 + a_1 w + a_2 w^2 + \cdots$$

[1] See *ibid.*, p. 401.

have radius of convergence R, and let the complex variable w be in its turn a function of z via

$$w(z) = b_0 + b_1 z + b_2 z^2 + \cdots$$

Consider next the series

$$a_0 + a_1(b_0 + b_1 z + b_2 z^2 + \cdots) \\ + a_2(b_0 + b_1 z + b_2 z^2 + \cdots)^2 + \cdots \qquad (1\text{-}6)$$

If z is such that the $w(z)$ series converges to a value w satisfying $|w| < R$, then the series (1-6) clearly converges to $f(w)$ and so may be represented by $f[w(z)]$. The question now arises as to whether the terms in (1-6) can be rearranged (e.g., the powers of z collected). If the value of z in question is such that

$$|b_0| + |b_1| \cdot |z| + |b_2| \cdot |z|^2 + \cdots$$

converges to a value less than R, then this can certainly be done—for the series obtained from (1-6) by replacing each term by its absolute value converges, from which it follows that a rearrangement of terms in the original series (1-6) does not alter its value. A more general result will be obtained subsequently (Exercise 8 of Sec. 2-3).

Consider next the problem of calculating the reciprocal of a power series. Assuming $a_0 \neq 0$, write

$$(a_0 + a_1 z + a_2 z^2 + \cdots)^{-1} = a_0^{-1}(1 + \alpha)^{-1}$$

where $$\alpha = z\left(\frac{a_1}{a_0} + \frac{a_2}{a_0} z + \frac{a_3}{a_0} z^2 + \cdots\right)$$

and so approaches zero as z approaches zero. Now, for α sufficiently small in absolute value,

$$(1 + \alpha)^{-1} = 1 - \alpha + \alpha^2 - \alpha^3 + \cdots$$

(proved by multiplying both sides by $1 + \alpha$), so that for small z

$$(a_0 + a_1 z + a_2 z^2 + \cdots)^{-1} = a_0^{-1}\left[1 - \left(\frac{a_1}{a_0} z + \frac{a_2}{a_0} z^2 + \cdots\right) \right. \\ \left. + \left(\frac{a_1}{a_0} z + \frac{a_2}{a_0} z^2 + \cdots\right)^2 - \cdots\right]$$

and by the preceding paragraph this series may be rearranged at will. It thus follows that the reciprocal of any power series with nonzero constant term is itself a power series. In practice, the coefficients of this reciprocal series are most easily found by denoting them by b_n, writing

$$(a_0 + a_1 z + a_2 z^2 + \cdots)(b_0 + b_1 z + b_2 z^2 + \cdots) = 1$$

and requiring all coefficients after the first in the expansion of the left-hand side to vanish. Even so, the calculation of the coefficients is usually time-consuming. There are two cases which often occur in practice and for which many of the coefficients have been tabulated. These are

$$\begin{aligned} \left(1+\frac{z}{2!}+\frac{z^2}{3!}+\frac{z^3}{4!}+\cdots\right)^{-1} &= \frac{B_0}{0!}+\frac{B_1}{1!}z+\frac{B_2}{2!}z^2+\cdots \\ \left(1+\frac{z^2}{2!}+\frac{z^4}{4!}+\frac{z^6}{6!}+\cdots\right)^{-1} &= \frac{E_0}{0!}+\frac{E_1}{1!}z+\frac{E_2}{2!}z^2+\cdots \end{aligned} \tag{1-7}$$

(*Note:* $0! = 1$.) The coefficients B_n and E_n are termed the *Bernoulli numbers* and *Euler numbers*, respectively. Direct calculation gives the first few of these numbers (those which turn out to be zero being omitted) as

$$\begin{array}{lllllll} B_0 = 1 & B_1 = -\tfrac{1}{2} & B_2 = \tfrac{1}{6} & B_4 = -\tfrac{1}{30} & B_6 = \tfrac{1}{42} & B_8 = -\tfrac{1}{30} & B_{10} = \tfrac{5}{66} \\ E_0 = 1 & E_2 = -1 & E_4 = 5 & E_6 = -61 & E_8 = 1{,}385 & E_{10} = -50{,}521 & \end{array}$$

The reader may easily show that symbolic recursion formulas are

$$(B+1)^n - B^n = 0 \qquad (E+1)^n + (E-1)^n = 0$$

where after the binomial expansion we replace B^n by B_n and E^n by E_n. We shall subsequently find that the left-hand sides of Eq. (1-7) represent the expansions for $z/(e^z - 1)$ and sech z and that the right-hand sides converge for $|z| < 2\pi$, $|z| < \pi/2$, respectively.

EXERCISES

3. Find the radius of convergence of

$$1 + \tfrac{1}{2}z + \frac{\tfrac{1}{2}(\tfrac{1}{2}-1)}{2!}z^2 + \frac{\tfrac{1}{2}(\tfrac{1}{2}-1)(\tfrac{1}{2}-2)}{3!}z^3 + \cdots$$

and prove that the square of this series equals $1 + z$.

4. In terms of the Bernoulli numbers, find the reciprocal of

$$1 - \frac{z}{2!} - \frac{2z^2}{3!} - \frac{3z^3}{4!} - \frac{4z^4}{5!} - \cdots$$

[*Hint:* Multiply by $(1 - z)^{-1}$ the previous expression defining the B_i.]

5. Assuming that the variables z and w are related by the convergent power series

$$\begin{aligned} w &= z + a_2z^2 + a_3z^3 + \cdots \\ z &= w + b_2w^2 + b_3w^3 + \cdots \end{aligned}$$

where the a_n and b_n are known constants, find the power series for z in terms of a new variable y defined by

$$y = z - a_2z^2 + a_3z^3 - a_4z^4 + \cdots + (-1)^{n+1}a_nz^n + \cdots$$

1-4 Powers and Logarithms

Although a born mathematician has been defined as one to whom the value of $e^{i\pi}$ is obviously -1, it is perhaps better to proceed more rationally and define complex powers in such a way that they obey the same rules of manipulation satisfied by real powers and, moreover, provide a simple generalization of real powers which reduces to the latter when all numbers involved are real. A useful way to proceed is via logarithms. For this purpose, consider the power series

$$E(z) = 1 + z + \frac{z^2}{2!} + \frac{z^3}{3!} + \cdots \tag{1-8}$$

This power series converges for all finite z; it is the complex-variable generalization of the Taylor-series expansion of e^x, and we can expect it to lead us to a suitable definition of e^z.

By using Eq. (1-8), it is easily checked that, for any two numbers z_1 and z_2,

$$E(z_1 + z_2) = E(z_1)E(z_2) \tag{1-9}$$

Let us first make use of this addition theorem to prove in a way other than by use of real-variable Taylor series that $E(x) = e^x$ for real x. For real x and integral positive n, repeated application of Eq. (1-9) gives

$$E(nx) = [E(x)]^n$$

Define the number e by $e = E(1)$ [and it may be shown that

$$e = \lim_{p\to\infty}\left(1 + \frac{1}{p}\right)^p$$

—see any text on infinite series]; then $E(n) = e^n$ for all integral $n > 0$. If m is any other positive integer, we also have

$$e^m = E(m) = E\left(n\frac{m}{n}\right) = \left[E\left(\frac{m}{n}\right)\right]^n$$

so that $E(m/n) = e^{m/n}$. Consequently, $E(x) = e^x$ for real positive rational x, and, regarding an irrational number as the limit of a sequence of rational numbers, we can extend this result to all real positive x. Finally, Eq. (1-9) gives

$$1 = E(0) = E(x - x) = E(x)E(-x) = e^xE(-x)$$

so that $E(-x) = 1/e^x = e^{-x}$. Therefore, for all real x, $E(x) = e^x$.

Consider next $E(ix)$, where x is real. Use of Eq. (1-8) and separation of real and imaginary parts give

$$E(ix) = \cos x + i \sin x$$

where the usual Taylor-series expressions for sine and cosine have been used. Thus Eq. (1-9) gives

$$E(z) = E(x + iy) = e^x(\cos y + i \sin y) \tag{1-10}$$

which gives us a remarkably easy method of evaluating $E(z)$. Note that $|E(z)| = e^x$. In particular, we see that values of x and y may be found so that $E(z)$ has any arbitrary (complex) value other than zero, and this result, together with the exponential interpretation for the real-variable case, leads us to define the natural logarithm (written ln) of a nonzero number w as that number z satisfying $w = E(z)$. When w is real, this definition coincides with the usual one. Note, however, that the use of complex numbers has made the logarithm multiple-valued in that, if $z = \ln w$, then $z \pm 2\pi ni$ is also equal to $\ln w$ for any integral n. As examples,

$$\begin{aligned}
\ln 1 &= 0 \pm 2\pi ni \\
\ln(-1) &= i\pi \pm 2\pi ni \\
\ln i &= i\frac{\pi}{2} \pm 2\pi ni \\
\ln(-i) &= -\frac{i\pi}{2} \pm 2\pi ni \\
\ln \frac{e + ie}{2^{1/2}} &= \left(1 + i\frac{\pi}{4}\right) \pm 2\pi ni
\end{aligned}$$

It is convenient to define the *principal value* of $\ln z$, written $\ln_p z$, as that particular value of $\ln z$ for which

$$-\pi < \text{Im}(\ln z) \leq \pi$$

The other possible values of $\ln z$ can differ from this only by an integral multiple of $2\pi i$. Thus the principal values of the logarithms in the foregoing examples are 0, $i\pi$, $i\pi/2$, $-i\pi/2$, and $1 + i\pi/4$, respectively.

Since, by Eq. (1-9),

$$w_1 w_2 = E(\ln w_1)E(\ln w_2) = E(\ln w_1 + \ln w_2)$$

it follows that

$$\ln w_1 w_2 = \ln w_1 + \ln w_2 \pm 2\pi ni$$

Note that we cannot always avoid this multiple-valuedness by claiming that

$$\ln_p w_1 w_2 = \ln_p w_1 + \ln_p w_2$$

since, for example, $w_1 = w_2 = -i$ would violate such an equation.

Natural logarithms of complex numbers having been defined, a very reasonable way of defining complex powers is to write

$$w^z = E(z \ln w)$$

where w and z may each be complex. Again, the result for real w and z agrees with the usual real-variable definition of a power. However, the present, more general case introduces some multiple-valuedness because of the multiple-valuedness of $\ln w$. For example, the possible values of 1^i are $E[i(0 \pm 2\pi ni)]$, that is, $\ldots e^{-6\pi}, e^{-4\pi}, e^{-2\pi}, 1, e^{2\pi}, e^{4\pi}, e^{6\pi}, \ldots$. Of these, the value corresponding to use of the principal value of the logarithm is called the *principal value of the power;* thus the principal value of 1^i is 1.

EXERCISES

1. Show that the principal values of $e^{i\pi}$, $e^{-i\pi}$, 2^{1+i}, i^i, and $(1+i)^{1+i}$ are -1, -1, $2[\cos(\ln_p 2) + i \sin(\ln_p 2)]$, $E(-\pi/2)$, and $2^{1/2} E(-\pi/4)[\cos(\pi/4 + \frac{1}{2}\ln_p 2) + i \sin(\pi/4 + \frac{1}{2}\ln_p 2)]$, respectively.

2. Prove that $\ln w^z = z(\ln_p w \pm 2\pi ni) \pm 2\pi Ni$. Is it true that $\ln_p w^z = z \ln_p w$ without any multiplicity in $2\pi i$?

3. Prove that the principal value of $w^{z_1+z_2}$ is equal to the product of the principal values of w^{z_1} and w^{z_2}. Discuss the relationship between the principal values of $w_1{}^z$, $w_2{}^z$, and $(w_1w_2)^z$.

4. Prove that z^n, n a real integer (positive or negative), is unique.

5. Discuss the possible equivalence of w^z and $1/w^{-z}$.

6. Find the fallacy in the following argument:

$$\ln(-1) = \ln\frac{1}{-1} = \ln 1 - \ln(-1) = -\ln(-1)$$

and thus

$$\ln(-1) = 0$$

7. Find the most general relationship between $\ln z$ and $\log z$.

Since $\ln 0$ does not exist, 0^z is undefined. For $\operatorname{Re} z > 0$, we define it arbitrarily as 0.

Consider now the expression e^z, where z is complex. If we apply the general rule for computation of a power to this expression, we would have to interpret it as $E[z(\ln e)]$, and $\ln e$ can have any of the values $1 \pm 2\pi ni$. Thus, in general, only one of the possible values of e^z would coincide with $E(z)$. However, in this special case, it has become conventional to interpret e^z as meaning only its principal value, unless otherwise stated. We shall adopt this convention throughout the book, so that always $e^z = E(z)$. In future, then, e^z will always be taken as the single-valued quantity

$$e^z = 1 + z + \frac{z^2}{2!} + \frac{z^3}{3!} + \cdots$$

No such convention applies to the other numbers raised to the power z. Thus $(2.7)^z$ has many possible values; $(2.71828 \cdots)^z$ has only one value. Sometimes, for notational convenience, we shall write $\exp(z)$ instead of e^z.

Trigonometric and Inverse Functions

For real x, the familiar Taylor series gives

$$\begin{aligned}\sin x &= x - \frac{x^3}{3!} + \frac{x^5}{5!} - \frac{x^7}{7!} + \cdots \\ &= \frac{1}{2i}(e^{ix} - e^{-ix})\end{aligned}$$

It is natural to define $\sin z$ for complex z by simply replacing x by z in this series. The resulting series converges inside every circle around the origin (in such situations we shall say that the radius of convergence is infinity). Similarly, define

$$\begin{aligned}\cos z &= 1 - \frac{z^2}{2!} + \frac{z^4}{4!} - \frac{z^6}{6!} + \cdots = \frac{1}{2}(e^{iz} + e^{-iz}) \\ \sinh z &= z + \frac{z^3}{3!} + \frac{z^5}{5!} + \frac{z^7}{7!} + \cdots = \frac{1}{2}(e^z - e^{-z}) \\ \cosh z &= 1 + \frac{z^2}{2!} + \frac{z^4}{4!} + \frac{z^6}{6!} + \cdots = \frac{1}{2}(e^z + e^{-z})\end{aligned}$$

$$\begin{aligned}\tan z &= \frac{\sin z}{\cos z} & \tanh z &= \frac{\sinh z}{\cosh z} \\ \cot z &= \frac{\cos z}{\sin z} & \coth z &= \frac{\cosh z}{\sinh z} \\ \sec z &= \frac{1}{\cos z} & \operatorname{sech} z &= \frac{1}{\cosh z} \\ \csc z &= \frac{1}{\sin z} & \operatorname{csch} z &= \frac{1}{\sinh z}\end{aligned}$$

The reader may verify that all the usual identities are fulfilled—for example, $\sin^2 z + \cos^2 z = 1$, $\sin(z_1 + z_2) = \sin z_1 \cos z_2 + \cos z_1 \sin z_2$, and $\cosh^2 z - \sinh^2 z = 1$. Also, $\sinh iz = i \sin z$, etc.

As in the case of real variables, the quantity whose sine is z will be denoted by $\arcsin z$, and the other inverse trigonometric and hyperbolic functions are defined similarly. Now that complex variables are being used, how much indeterminacy is there in $\arcsin z$? Suppose that both w_1 and w_2 equal $\arcsin z$. Then $z = \sin w_1 = \sin w_2$, so that

$$\frac{\exp(iw_1) - \exp(-iw_1)}{2i} = \frac{\exp(iw_2) - \exp(-iw_2)}{2i}$$

from which it follows that either

$$\exp(iw_1) = \exp(iw_2)$$

or

$$\exp(iw_1) = -\exp(-iw_2)$$

In the first case, $w_1 = w_2 \pm 2\pi n$; in the second, $w_1 = (\pi - w_2) \pm 2\pi n$. Consequently, the possible relations between w_1 and w_2 are the same here as they are when w_1 and w_2 are real variables. Similar calculations may be made for the other inverse functions.

Power series for the inverse exponential and trigonometric series are easily found by replacing x in the real-variable Taylor series by z and then verifying that the resultant series satisfies the inverse property. For example, if x is real, then we know that

$$Q(x) = x - \frac{x^2}{2} + \frac{x^3}{3} - \frac{x^4}{4} \cdots$$

is the Taylor series for $\ln(1 + x)$ and thus we know, that when this series is substituted into the Taylor series for $e^{Q(x)}$, the result is $1 + x$. If now x is replaced formally by the complex variable z, the power series

$$Q(z) = z - \frac{z^2}{2} + \frac{z^3}{3} - \frac{z^4}{4} \cdots$$

(which is absolutely convergent for $|z| < 1$) must yield $1 + z$ when substituted into the power series for $e^{Q(z)}$, since the algebra of coefficients is exactly the same as in the real variable case. Consequently,

$$\ln(1 + z) = z - \frac{z^2}{2} + \frac{z^3}{3} - \cdots \tag{1-11}$$

At the point $z = 0$, this series gives the result zero, which is the principal value of $\ln 1$; by the continuity property of a power series, therefore, it follows that this series converges to the principal value of $\ln(1 + z)$ within its circle of convergence.

Using this same method, we can easily obtain the power series for $\arcsin z$, $\operatorname{arcsinh} z$, etc. (where z is complex), as well as a generalized version of the binomial theorem. In particular, we have

$$\arcsin z = z + \frac{1}{2}\frac{z^3}{3} + \frac{1 \cdot 3}{2 \cdot 4}\frac{z^5}{5} + \frac{1 \cdot 3 \cdot 5}{2 \cdot 4 \cdot 6}\frac{z^7}{7} + \cdots \tag{1-12}$$

$$(1 + z)^s = 1 + sz + \frac{s(s-1)}{2!} z^2 + \frac{s(s-1)(s-2)}{3!} z^3 + \cdots \tag{1-13}$$

In Eq. (1-13), s is an arbitrary complex number. Each series converges for $|z| < 1$.

EXERCISES

8. Use the above definition of a power to show that $z^{1/n}$, n integral, has exactly n distinct values.

9. Find all possible values of arctanh 1.

10. If θ and φ are real numbers, use the fact that $e^{i(\theta+\phi)} = e^{i\theta}e^{i\phi}$ to devise a simple system for remembering the trigonometric identities for $\sin(\theta + \varphi)$ and $\cos(\theta + \varphi)$.

11. Show that the binomial series (1-13) converges for $|z| < 1$, and for all s, to the principal value of the left-hand side. Using the criterion of Sec. 1-3 (*Hint:* Set $z = -\zeta$), show that the series converges for all z satisfying $|z| = 1$ if $\operatorname{Re} s > 0$; that it diverges for all such z if

Re $s \leq -1$; and that it diverges at $z = -1$ but converges at all other z satisfying $|z| = 1$ if $-1 < \text{Re } s \leq 0$.

Infinite Products

The logarithm function is useful in examining the convergence of infinite products. Consider a sequence of complex numbers, $\{u_j\}$, and define

$$\Pi_n = (1 + u_1)(1 + u_2) \cdots (1 + u_n)$$

If $\Pi_n \to L$ as $n \to \infty$, where L is finite and nonzero, then the infinite product

$$\prod_{j=1}^{\infty} (1 + u_j) = (1 + u_1)(1 + u_2) \cdots (1 + u_n) \cdots$$

is said to *converge* to L. A necessary condition for convergence is clearly that $u_j \to 0$ as $j \to \infty$ (if $L = 0$ were allowed, this would not be so). The infinite product must converge or diverge together with $\Sigma \ln (1 + u_j)$ (note, however, that even if principal values are used, this logarithmic series need not converge to the *principal* value of $\ln L$).

If $\Pi(1 + |u_j|)$ converges, the infinite product is said to converge *absolutely*. The reader may show that the convergence of any one of the following expressions implies that of the others:

$$\Pi(1 + |u_j|) \qquad \Sigma \ln (1 + |u_j|) \qquad \Sigma|\ln (1 + u_j)| \qquad \Sigma|u_j|$$

[As a help in the proof, note that for j sufficiently large, $|u_j^{-1} \ln (1 + u_j) - 1| < \frac{1}{2}$, which provides upper and lower bounds on $|u_j|^{-1} |\ln (1 + u_j)|$.] Note that for infinite products, just as for infinite series, absolute convergence implies convergence.

1-5 *Geometric Properties of Simple Functions*

Let $w = f(z)$ denote a functional relationship between the complex variables w and z. Depending on whether only one or more than one value of w corresponds to a particular choice of z in a certain domain, w is said to be a single-valued or a multiple-valued function of z in that domain. An example of the first kind of function is $w = z^2$; an example of the second is $w = z^{1/2}$. Suppose now that two complex planes are envisioned, one for z and one for w. Each point z in the z plane possesses, by virtue of $w = f(z)$, one or more image points in the w plane; the relationship can be thought of as a transformation or mapping of the z plane into the w plane.

A particularly simple transformation is the general linear one $w = az + b$, where a and b are fixed (complex) constants. To obtain the image point w of a given point z, it is clear that one must rotate the point z around the origin by an amount arg a, increase its distance from the origin in the ratio $|a|/1$, translate it to the right by an amount Re b, and translate it upward by an amount Im b. Thus this transformation corresponds to a uniform rotation, linear stretching, and uniform translation of the z plane.

Consider next the transformation $w = 1/z$. Here w is also a single-valued function of z, satisfying $|w| = 1/|z|$ and $\arg w = -\arg z$. Each of two points lying on the same straight line through the origin and distant r, r' from it is said to be the other's inverse in the unit circle if $rr' = 1$; consequently, the transformation $w = 1/z$ corresponds to an inversion in the unit circle followed by a reflection in the real axis. The more general transformation $w = a/z + b$ would merely add a final stretching, rotation, and translation.

The transformation $w = 1/z$ does not provide an image point for the origin. However, as $z \to 0$, $w \to \infty$, so that it is conventional to say that the image of the origin is the point at infinity. There is of course no single point at infinity, but the reason for this nomenclature arises from the familiar stereographic projection of a sphere onto a plane. Place a sphere of unit diameter on the z plane so that it touches the origin at its south pole. Connect any particular point z to the north pole by a straight line; the point where this line cuts the sphere is then the image of the z point. As $z \to \infty$, the image point approaches the north pole, so that the north pole itself—a single point—may be thought of as the image of the "point at infinity." Incidentally, the stereographic projection has the interesting property of mapping any circle on the sphere (not passing through the north pole) into a circle on the z plane and any circle passing through the north pole into a straight line; the converse result also holds. If a straight line in the plane is regarded as a circle passing through infinity, then the statement of this result need mention circles only.

A combination of transformations of the foregoing kinds is called the general *bilinear*, or *fractional*, transformation; it has the form

$$w = \frac{az + b}{cz + d}$$

where $a/c \neq b/d$ (otherwise w is a constant). The transformations previously discussed have the property of always mapping circles into circles (again, a straight line is a degenerate circle), and since this bilinear transformation may be written as a sequence of such transformations, it has the same property. Note that the expression for z in terms of w is also linear fractional and that if the points w_1, w_2, w_3, and w_4 correspond to z_1, z_2, z_3, and z_4, respectively, then the cross ratios are equal; i.e.,

$$\frac{(w_1 - w_2)(w_3 - w_4)}{(w_1 - w_3)(w_2 - w_4)} = \frac{(z_1 - z_2)(z_3 - z_4)}{(z_1 - z_3)(z_2 - z_4)}$$

EXERCISES

1. Use the cross-ratio equality to obtain a mapping which transforms the upper-half z plane into the interior of the unit circle in the

w plane. (*Hint:* Replace z_1, w_1 by z, w; set z_2, z_3, z_4 equal to $-1, 0, 1$, etc. Or use point at infinity.) Sketch the w images of various points and curves in the z plane, and vice versa.

2. Show that, if two curves in the z plane intersect at a certain angle, then their w images as obtained by a bilinear transformation intersect at the same angle. (The stereographic projection also has this property of "isogonality.")

3. Show that the net result of two consecutive bilinear transformations is again a bilinear transformation.

4. How many points can coincide with their images in a (non-identity) bilinear transformation?

5. Can a bilinear transformation be found which will map two non-intersecting circles into concentric circles?

A somewhat different kind of transformation is exemplified by $w = z^2$. Although this transformation is single-valued, in that to each value of z corresponds only one value of w, the entire w plane has been covered by the time that z has been allowed to attain all values in the upper half z plane; as z now ranges through the lower half-plane, the entire w plane is covered again. For example, let $z = re^{i\theta}$, where r is fixed and θ varies. Then $w = Re^{i\varphi} = r^2e^{2i\theta}$, so that, as θ goes from 0 to π, φ goes from 0 to 2π. Thus the transformation $w = z^2$ does the following: (1) it expands the upper half z-plane sector so as to cover a total angle of 2π and simultaneously squares the distance of each point from the origin; (2) it does exactly the same for the lower half z plane, so that the entire w plane is covered twice for one covering of the z plane.

Given $w = f(z)$, it is often useful to start at any point z_0 and traverse a simple closed curve ending up again at z_0, recording meanwhile the argument of w. For example, let $w = Az$, where A is a complex constant. Then $\arg w = \arg A + \arg z$; if the z curve circles the origin once counterclockwise, and if we insist that the argument is to change continuously (assume that the curve does not pass through the origin, where the argument is not defined), then $\arg w$ will be found to have increased by 2π (or, if clockwise, by -2π). If the curve does not enclose the origin, there is no change in $\arg w$. Similarly, if $w = Az^2$, then the change in $\arg w$ for each counterclockwise circuit of the origin is 4π (or -4π, if clockwise); again, there is no change in $\arg w$ for a curve not enclosing the origin. In either example, w is a single-valued function of z so that any change in $\arg w$ has to be a multiple of 2π.

EXERCISE

6. Describe the way in which $\arg w$ changes for various circuits for the case $w = A(z - z_1)(z - z_2)^2$, where z_1 and z_2 are fixed points.

Let us now use the considerations of the preceding paragraph to show that any polynomial $w = a_0 + a_1z + a_2z^2 + \cdots + a_nz^n$ vanishes for some value of z. [This is the crucial point in the proof that the polynomial may be written in the form $a_n(z - z_1)(z - z_2) \cdots (z - z_n)$, for if it is shown that there exists a z_1 such that $w(z_1) = 0$, then the usual process of algebraic division by $z - z_1$ gives as the remaining factor a polynomial of order $n - 1$, which by the theorem must vanish for some value of z, etc.] To prove the theorem, we assume the contrary and derive a contradiction. If there is no point for which $w = 0$, then arg w may be required to vary continuously along any closed curve; since w is a single-valued function of z, the change in arg w around any such curve must be a multiple of 2π. Begin with a very large square enclosing the origin—so large that w behaves effectively like a_nz^n. Then the change in $\arg w = n(2\pi) \neq 0$ for one counterclockwise traversal. Subdivide the large square into four smaller squares, and note that the sum of the changes in arg w around each of the smaller squares must equal the change around the larger square, so that for at least one of these smaller squares we have the change in $\arg w \neq 0$. Proceed in this way by repeated subdivision, and it is clear that we must eventually contradict the continuity property of w, which requires that around a sufficiently small circuit the change in $\arg w = 0$.

EXERCISES

7. In the above proof, at exactly which places was the assumption used that $w = 0$ nowhere?

8. Show that, if w is any polynomial, then the change in arg w observed during the counterclockwise traversal of any simple closed curve is 2π times the number of zeros of the polynomial lying within the curve. Here it is assumed that w does not vanish at any point on the curve itself; also, a zero of multiplicity m is counted as m zeros.

9. Use the result of Exercise 3 to locate the general areas in which the roots of $1 + z^3 + 2iz^4 = 0$ lie.

Exponential Function

The mappings corresponding to the exponential functions are easily delineated. First, if $w = e^z$, then w is a single-valued function of z satisfying $w = e^x(e^{iy})$ so that $|w| = e^x$ and $\arg w = y \pm 2n\pi$. Consequently, as z covers any horizontal strip of width 2π (including one horizontal boundary and not the other), the entire w plane is covered once, with the exception of the origin. A horizontal line in the z plane is mapped into a straight semi-line from the origin to infinity in the w plane; an infinite vertical line in the z plane repeatedly covers a particular circle in the w plane. The mapping $w = \ln z$ is the inverse of this and forms an example of a multiple-valued function in that the entire z plane (except the origin) is mapped into each of the previous horizontal strips.

Consider, second, the mapping $w = \sin z$. Set $z = x + iy$; then

$$w = u + iv = (\sin x \cosh y) + i(\cos x \sinh y)$$

A vertical line therefore maps into a hyperbola, a horizontal line into an ellipse. The reader should check that the details of the mapping are as follows:

1. Consider the strip $0 \leq x \leq \pi/2$ in the z plane. The left-hand border maps into the v axis of the w plane, and the right-hand border maps into that part of the u axis satisfying $u \geq 1$. Any other vertical line in the strip maps into a hyperbola intersecting the u axis between 0 and 1, the hyperbola lying in the right half-plane. The v axis and the part of the u axis to the right of $u = 1$ are the two limiting members of this family of hyperbolas. The upper half of the strip thus covers the first quadrant, and the lower half, the fourth quadrant, of the w plane.

2. The strip $\pi/2 < x < \pi$ also maps into the right half w plane in a similar manner, with, however, the difference that the upper and lower parts of the strip now map into the fourth and first w quadrants, respectively.

3. The strip $\pi < x < 3\pi/2$ maps into the left half w plane, with the upper half strip going into the third quadrant, and so on for the other strips in the z plane of width $\pi/2$. For example, the strip $-\pi/2 < x < \pi/2$ would map into the entire w plane.

4. That portion of the horizontal straight line $y = c$ intercepted by the strip $-\pi/2 < x < \pi/2$ maps into the upper or lower half of an ellipse depending on whether c is positive or negative; this ellipse has foci ± 1 and intersects orthogonally the previous (confocal) hyperbolas. If $c = 0$, the ellipse degenerates into the straight line joining the foci. The interior and boundary of a w ellipse with foci at ± 1 is the image of a closed rectangle [1] in the z plane and, in fact, of an infinite number of rectangles.

Exercise

10. Find easy ways to use the properties of the sin z mapping so as to determine the natures of the mappings $w = \cos z$, $w = \sinh z$, $w = \arcsin z$.

Branch Cuts and Riemann Surfaces

Finally, let us define the terms *branch point*, *branch line*, and *Riemann surface* by considering the example $w = z^{1/2}$. Let $z = re^{i\theta}$; then there are exactly two different values of w corresponding to any choice of r, θ given by

$$w_1 = r^{1/2}e^{i(\theta/2)}$$
$$w_2 = r^{1/2}e^{i(\theta/2+\pi)} = -w_1$$

[1] The images of the vertical sides of the rectangle are horizontal cut lines joining the foci of the ellipse to its boundary.

where, for definiteness, $r^{1/2}$ is taken to be the positive square root of r. The two functions w_1 and w_2 defined in this manner are referred to as *branches* of w. Suppose that we start at some point z_0 and choose one of the two possible values of w—say w_1. Traverse now a simple closed curve, ending again at z_0, and requiring w to vary continuously along this curve. It is easily seen that if the curve does not enclose the origin then w returns to the value $w_1(z_0)$, but if the curve does enclose the origin, then w attains the value $w_2(z_0)$. In this latter case, the branches have become interchanged; a second circuit about the origin would change w back to $w_1(z_0)$. The origin is here the only point (apart from the point at ∞) possessing this property that a circuit about it interchanges the branches; such a point is called a *branch point*. In this paragraph, it has been assumed implicitly that the various curves did not pass through the origin, for one could there change arbitrarily from branch to branch without violating the condition of continuous variation of w.

The multiplicity of values of w in such a case as $w = z^{1/2}$ can be awkward, and it may be useful to prevent z from being able to circle the origin. One way of doing this is to "cut" the z plane along the negative real axis and require that no paths in the z plane intersect this barrier. If we then choose the value of w at z_0 to be $w_1(z_0)$ and require that the value of w at any other point z_1 be obtained by the condition of continuous variation of w along any curve joining z_0 and z_1, then the fact that circuits around the origin are prevented means that w is uniquely defined. Any curve in the z plane originating at the origin and proceeding to infinity would serve equally well as a barrier. Such a barrier curve, constructed for the purpose of ensuring uniqueness, is called a *branch line*, which in this case may be said to be terminated by two points—the origin and the point at infinity.

Let us now allow the point z to move arbitrarily in the z plane, requiring as before that the corresponding value of w vary continuously. Starting with one branch of w, each circuit around the origin will interchange the branches. It may be convenient to think of the z plane as two superposed planes such that, as z circles the origin on one of these planes and starts to repeat its previous curve, the point z now shifts to the other plane (where w has its other branch value); another circuit will again interchange the two planes. For aid in visualization, suppose that the second plane is thought of as a cone with apex angle just less than 180°, with its apex touching the origin and one generator lying along the positive real axis. If we start on the first plane and circle the origin, then z is required to move onto the cone as it crosses the positive real axis; a continued circuit of the origin on the cone surface will again meet the positive real axis, and this time z is required to move back onto the plane. Then for any point z, the corresponding value of w is uniquely defined—one branch value if z is on the lower plane, and the other if z is on the cone. The cone is of course now thought of as being flattened out so as to coincide with the first plane; the use of these two *sheets* for the z plane is essentially a topological device for making w single-valued. The resulting two-sheeted range of values for z is called a *Riemann surface* for the function $z^{1/2}$. The reader not familiar with these concepts should pay particular attention to the following exercises.

EXERCISES

11. Describe the branch-point and branch-line situation for $w = z^{1/3}$. How many sheets does the Riemann surface have? Describe its topological character. Repeat for $w = \ln z$ and for $w = [z(z-1)]^{1/2}$. (In this last case, write the formula for w as $(r_1 r_2)^{1/2} \exp [i(\varphi_1/2 + \varphi_2/2)]$, where r_1 and r_2 are the distances of z from the two branch points and φ_1, φ_2 are the angles from the horizontal made by the vectors from the branch points to z; note that a circuit around both branch points does not alter the branch of w, so that a suitable branch line is any curve joining the two branch points. Alternatively, each branch line emanating from these two branch points may take any path to the point at infinity.) Repeat for $w = [z + (z^2 - 1)^{1/2}]^{1/3}$.

12. Verify that a suitably defined branch of

$$f(z) = \ln\left(5 + \sqrt{\frac{z+1}{z-1}}\right)$$

is single-valued in the z plane outside of a line joining the points $z = 1$ and $z = -1$. Show, however, that if one enters another sheet of the Riemann surface by crossing this line, there will be a branch point at $z = 13/12$.

13. Discuss the branch-cut and Riemann-surface situation for each of the following functions:

$$(a)\quad g(z) = \sqrt{1 + \sqrt{z}}$$
$$(b)\quad h(z) = \ln [1 + \sqrt{z^2 + 1}]$$
$$(c)\quad m(z) = \frac{1}{\sqrt{z^2 + \alpha^2}} \ln \frac{z + \sqrt{z^2 + \alpha^2}}{z - \sqrt{z^2 + \alpha^2}} \tag{1-14}$$

where α is a constant with $\operatorname{Re} \alpha = k > 0$. In particular, verify that $m(z)$ can define a function which has no branch points in $\operatorname{Im} z < k$, that it can also define a function which has no branch points in $\operatorname{Im} z > -k$, and that one linear combination of the two functions so defined is $1/\sqrt{z^2 + \alpha^2}$. [It may be helpful to trace the values of $m(z)$ for $z = i\alpha \sin\theta$, as θ varies from $-\pi/2$ to $\pi/2$].

14. Find all complex numbers α satisfying the equation

$$\cos \alpha \cosh \alpha = 1$$

2
analytic functions

2-1 *Differentiation in the Complex Plane*

A function $f(z)$ of a complex variable z is said to have a *derivative* (be differentiable) at the point z if

$$\lim_{\xi \to z} \frac{f(\xi) - f(z)}{\xi - z} \tag{2-1}$$

exists and has the same value *for any mode of approach* of ξ to z. This limit is called the derivative, and it is denoted by $f'(z)$ or df/dz. The formalism of (2-1) is the same as in the real-variable case, but the added requirement that the limit be independent of the way in which $\xi \to z$ is a very restrictive one, and it is not surprising that differentiable functions turn out to have a large number of interesting properties. What is surprising is that so many of the common functions are differentiable. For example, if $f(z) = z^2$, then

$$\lim_{\xi \to z} \frac{\xi^2 - z^2}{\xi - z} = 2z$$

independent of the manner of approach of ξ to z. Thus

$(d/dz)(z^2)$ does exist and equals $2z$. On the other hand, it is easy to find a function $f(z)$ for which $f'(z)$ does not exist. As an example, let $f(z) = zz^*$. Then

$$\lim_{\xi \to z} \frac{\xi\xi^* - zz^*}{\xi - z} = \lim_{\xi \to z} \left(\xi^* + z \frac{\xi^* - z^*}{\xi - z} \right)$$

and if we allow ξ to approach z from a direction making an angle θ with the real axis, so that $\xi - z = re^{i\theta}$ with θ fixed and $r \to 0$, this limit becomes

$$z^* + ze^{-2i\theta}$$

This clearly depends on θ, so that $(d/dz)(zz^*)$ does not exist (except at the origin).

If $f(z)$ has a derivative at a point z_0 and also at each point in some neighborhood of z_0, then $f(z)$ is said to be *analytic* at z_0. The terms *holomorphic*, *monogenic*, and *regular* are also used.

Since the definition of the derivative $f'(z)$ is formally the same as that for a function of a real variable, the rules for differentiating composite expressions in the real-variable domain must also hold for complex variables. Thus, if f and g are each differentiable in a region, so is each of the following combinations and

$$(af + bg)' = af' + bg' \qquad a, b \text{ complex constants}$$
$$(fg)' = fg' + f'g$$
$$\left(\frac{f}{g}\right)' = \frac{f'g - fg'}{g^2} \qquad \text{if } g \neq 0$$
$$\frac{d}{dz}\{f[g(z)]\} = f'[g(z)]g'(z)$$

where in the last equation $f(\xi)$ is assumed to be differentiable for those values of its argument $\xi = g(z)$ that correspond to points z lying in the given region.

The formula for differentiating a product may be used inductively to show that

$$\frac{d}{dz}(z^n) = nz^{n-1}$$

where n is any integer. For the exponential function,

$$\begin{aligned}\frac{d}{dz}(e^z) &= \lim_{\xi \to z} \frac{e^\xi - e^z}{\xi - z} = \lim_{\xi \to z} e^z \frac{e^{\xi - z} - 1}{\xi - z} \\ &= e^z \lim_{\xi \to z} \left[1 + \frac{\xi - z}{2!} + \frac{(\xi - z)^2}{3!} + \cdots \right] \\ &= e^z\end{aligned}$$

since the limit is independent of the mode of approach. Similarly,

$$\frac{d}{dz}(\sin z) = \lim_{\xi \to z} \frac{\sin[z + (\xi - z)] - \sin z}{\xi - z} = \cos z$$

and in fact all the usual formulas for differentiation of the trigonometric and hyperbolic functions are found to hold. To differentiate the ln function let $y = \ln z$; then $z = e^y$ so that $1 = e^y(dy/dz)$, from which

$$\frac{d}{dz}(\ln z) = \frac{1}{z}$$

A similar method may be used for the inverse trigonometric and hyperbolic functions, and again the familiar results are found to hold; e.g.,

$$\frac{d}{dz}(\arctan z) = \frac{1}{1 + z^2}$$

Multiple-Valuedness

A multiple-valued expression sometimes results from the use of the conventional differentiation formulas, and it is then necessary to choose the right branch. This situation usually arises when the function to be differentiated is itself multiple-valued; an important exception is the function $w = \ln z$, which, as we have seen, has the single-valued derivative $w' = 1/z$ irrespective of the branch of w. Consider, however, $w = z^s$, where s is any complex number (including as a special case any real fractional number). Then, if $\ln_1 z$ is any chosen branch of the ln function, the function w in the neighborhood of any point z must be interpreted as one of the functions

$$w_n = \exp\left[s(\ln_1 z + 2\pi i n)\right] \qquad n = 0, \pm 1, \pm 2, \ldots$$

Then, using the previous formulas,

$$\begin{aligned} \frac{dw_n}{dz} &= \exp\left[s(\ln_1 z + 2\pi i n)\right]\left(\frac{s}{z}\right) \\ &= s \exp\left[(s - 1)(\ln_1 z + 2\pi i n)\right] \end{aligned}$$

This expression may be written

$$\frac{dw}{dz} = sz^{s-1}$$

provided that the same branch of the ln function is used to compute z^{s-1} as is used to compute z^s.

As a second example, let $w = \arcsin z$. As before, this function is multiple-valued, and we must complete its definition by specifying the desired branch. Solving

$$\frac{e^{iw} - e^{-iw}}{2i} = z$$

as a quadratic equation in e^{iw}, we have

$$w = -i \ln\left[iz + (1 - z^2)^{1/2}\right]$$

which incidentally exemplifies the fact that any of the inverse trigonometric or hyperbolic functions can be expressed in terms of logarithms. The argument of the ln function is here double-valued because of $(1 - z^2)^{1/2}$; the product of the two possible values of the argument is -1. Upon denoting a chosen one of the branches of $(1 - z^2)^{1/2}$ by $[1 - z^2]_1^{1/2}$ and a chosen one of the branches of ln by $\ln_1$, the possible values of w are then

$$\begin{aligned} w_{1m} &= -i \ln_1 \{iz + [1 - z^2]_1^{1/2}\} + 2\pi m, \qquad & m &= 0, \pm 1, \pm 2, \ldots \\ w_{2n} &= i \ln_1 \{iz + [1 - z^2]_1^{1/2}\} + (2n + 1)\pi & n &= 0, \pm 1, \pm 2, \ldots \end{aligned}$$

(Notice that, within a multiple of 2π, $w_{1m} + w_{2n} = \pi$, in agreement with the familiar real-variable case.) Differentiation gives

$$\begin{aligned} \frac{dw_{1m}}{dz} &= \frac{-i}{iz + [1 - z^2]_1^{1/2}} \left\{ i + \frac{-2z}{2[1 - z^2]_1^{1/2}} \right\} \\ &= \frac{1}{[1 - z^2]_1^{1/2}} \\ \frac{dw_{2n}}{dz} &= \frac{i}{iz + [1 - z^2]_1^{1/2}} \left\{ i + \frac{-2z}{2[1 - z^2]_1^{1/2}} \right\} \\ &= \frac{-1}{[1 - z^2]_1^{1/2}} \end{aligned}$$

Thus, despite the multiplicity of values of arcsin z, there are only two possible values for its derivative. According to whether w is a member of the w_{1m} sequence or the w_{2n} sequence, its derivative is

$$\frac{1}{[1 - z^2]_1^{1/2}} \qquad \text{or} \qquad \frac{-1}{[1 - z^2]_1^{1/2}}$$

respectively.

Cauchy-Riemann Equations

To return now to the general situation, the condition that $f(z)$ have a derivative can be expressed in terms of a pair of equations linking partial derivatives of its real and imaginary parts. Let $f(z) = u(x,y) + iv(x,y)$, where $z = x + iy$ and where u, v are real functions. Then, at the point $z_0 = z_0 + iy_0$,

$$\lim_{z \to z_0} \frac{f(z) - f(z_0)}{z - z_0} = \lim_{x \to x_0, y \to y_0} \frac{[u(x,y) - u(x_0,y_0)] + i[v(x,y) - v(x_0,y_0)]}{(x - x_0) + i(y - y_0)} \tag{2-2}$$

If now $f(z)$ does have a derivative at z_0, then this limit exists for any mode of approach of z to z_0. One such mode is that in which $y = y_0$; for that mode,

$$f'(z_0) = \lim_{x \to x_0} \frac{[u(x,y_0) - u(x_0,y_0)] + i[v(x,y_0) - y(x_0,y_0)]}{x - x_0}$$

We know this limit must exist; therefore, the partial derivatives u_x and v_x must exist, and we have

$$f'(z_0) = u_x(x_0,y_0) + iv_x(x_0,y_0)$$

Similarly, using that mode of approach for which $x = x_0$, we obtain

$$f'(z_0) = \frac{1}{i}\,[u_y(x_0,y_0) + iv_y(x_0,y_0)]$$

Comparing these two results shows that the *Cauchy-Riemann* equations

$$u_x = v_y \qquad u_y = -v_x \tag{2-3}$$

are a necessary condition for the existence of a derivative. They are also sufficient (at least if the partial derivatives are continuous functions of x and y), as may easily be shown by writing

$$u(x,y) - u(x_0,y_0) = u_x(x_0,y_0)(x - x_0) + u_y(x_0,y_0)(y - y_0) + \text{higher-order terms}$$

inserting this and a similar expression for $v(x,y) - v(x_0,y_0)$ in Eq. (2-2), and then using Eq. (2-3).

If the quantities occurring in Eq. (2-3) are further differentiable, then we note that each of u and v satisfies Laplace's equation—that is,

$$u_{xx} + u_{yy} = v_{xx} + v_{yy} = 0$$

Since a real function satisfying Laplace's equation is said to be *harmonic*, we thus have the result that *the real and imaginary parts of an analytic function are harmonic.*[1] This is one illustration of the consequences that result from the existence of a derivative; it also shows that, among all possible functions $f(z) = u(x,y) + iv(x,y)$, where u and v are arbitrarily chosen continuously differentiable real functions, only a very narrow selection can be analytic.

Exercises

1. Show that no purely real function can be analytic, unless it is a constant.

2. Verify the Cauchy-Riemann equations for z^2, e^z, $\ln z$, and $(1 - z^2)^{1/2}$. Prove in an easy way that

$$(x^2 + y^2)^{1/4} \cos\left(\frac{1}{2}\arctan\frac{y}{x}\right)$$

is harmonic. At what points do these various functions have singularities?

3. Show that, if $f(z)$ has a derivative at z_0, then it is also continuous there.

4. Even if a given $f(z)$ is not analytic, the limit (2-1) may still exist for each fixed choice of θ in $\xi - z = re^{i\theta}$, as $r \to 0$. Denote such a directional derivative by $f_\theta(z)$. Show that the set of complex numbers $f_\theta(z)$, obtained by choosing all possible values of θ, lies on a circle

[1] Provided that u and v are indeed sufficiently differentiable; we shall show later that they always are.

when plotted in the complex plane. Obtain the center and radius of the circle, and comment on the condition for degeneracy of the circle into a single point.

5. If a is a complex constant, differentiate a^z, z^z, and arccosh z. Explain how to determine the branch of the result.

6. Prove that $\ln |f|$ is harmonic, if $f(z)$ is analytic. Show that a necessary condition that the curves $F(x,y) = c$, c a real parameter, coincide with the contour lines of the modulus of some analytic function is that

$$\frac{F_{xx} + F_{yy}}{F_x{}^2 + F_y{}^2}$$

be some function of F.

7. If u and v are expressed in terms of polar coordinates (r,θ), show that the Cauchy-Riemann equations can be written

$$u_r = \frac{1}{r} v_\theta \qquad \frac{1}{r} u_\theta = -v_r$$

Why should this result be almost obvious?

2-2 *Integration in the Complex Plane*

The integral of a real-valued function is defined as the limit of a certain sum; we carry over the same formal definition to the complex-variable case. Let $f(z)$ be any complex function, analytic or not. Consider any path C in the plane of the complex variable z, beginning at a point A and terminating at a point B. Choose any ordered sequence $\{z_j\}$ of $n + 1$ points on C, with $z_0 = A$ and $z_n = B$. Denote $z_j - z_{j-1}$ by Δz_j, and form the sum

$$\sum_{j=1}^{n} f(\xi_j)\, \Delta z_j$$

where ξ_j is any point on C between z_{j-1} and z_j. If this sum approaches a limit as $n \to \infty$ (with the largest $|\Delta z_j| \to 0$), and if the value of the limit does not depend on the way in which the z_j and ξ_j are chosen, then the value of this limit is called the *integral* of $f(z)$ along C. Thus

$$\int_C f(z)\, dz = \lim_{\substack{n \to \infty \\ |\Delta z_j| \to 0}} \sum_{j=1}^{n} f(\xi_j)\, \Delta z_j \tag{2-4}$$

Since $f(z)$ and z can be written in terms of real variables, the integral of a complex function does not differ conceptually from that of a real function. In particular, if $z = x + iy$ and $f(z) = u(x,y) + iv(x,y)$,

$$\int_C f(z)\, dz = \int_C (u\, dx - v\, dy) + i \int_C (u\, dy + v\, dx) \tag{2-5}$$

This fact allows us to borrow, from real-variable theory, conditions which can be imposed on $f(z)$ and C in order to ensure the existence of the limit (2-4); the functions we deal with will normally be sufficiently well behaved (e.g., continuous, bounded) that such conditions are indeed fulfilled.

In principle, the definition (2-4) can be used for the actual calculation of an integral. If, for example, we want to evaluate $\int_C z\,dz$, we can write ξ_j in the form $\frac{1}{2}(z_j + z_{j-1}) + \epsilon_j$, where the complex number $\epsilon_j \to 0$ as $\Delta z_j \to 0$; it then follows easily that

$$\int_C z\,dz = \tfrac{1}{2}(B^2 - A^2)$$

where A and B are again the (complex) end points of the curve. In this case the result is independent of C. We shall subsequently find that, whenever $f(z)$ is analytic, the value of $\int_C f(z)\,dz$ depends only on the end points A, B and not on C itself;[1] we shall also be able to use the usual integration formulas (inverse differentiation) to find the value of the integral. First, however, let us use Eq. (2-4) to obtain some general properties of integrals.

One immediate implication of Eq. (2-4) is that the usual rules for manipulation of integrals hold. For example, integration in the reverse direction along C gives the negative of the previous integral. Also, for any two functions $f(z)$ and $g(z)$, and for any two complex constants a and b,

$$\int_C [af(z) + bg(z)]\,dz = a\int_C f(z)\,dz + b\int_C g(z)\,dz$$

If $f(z)$ is bounded on C, say $|f| \leqq M$, then the triangle inequality may be applied to Eq. (2-4) to give

$$\left|\int_C f(z)\,dz\right| \leqq \int_C |f(z)||dz| \leqq ML \tag{2-6}$$

where L is the length of C. Finally, suppose that we have a uniformly convergent series on C,

$$f(z) = f_1(z) + f_2(z) + \cdots$$

Let $s_n(z)$ denote the sum of the first n terms of the series. Then the condition of uniform convergence implies that, given any small $\epsilon > 0$, we can choose n sufficiently large that $|f(z) - s_n(z)| < \epsilon$ everywhere on C. Consequently,

$$\left|\int_C f(z)\,dz - \int_C s_n(z)\,dz\right| < \epsilon L$$

by Eq. (2-6), so that $\int_C s_n(z)\,dz$ (which since n is finite is the sum of the

[1] Provided that we restrict ourselves to those curves C, having common end points A, B, which can be deformed into one another without passing out of the region of analyticity.

integrals of the first n terms) approaches $\int_C f(z)\,dz$ as $n \to \infty$. We have therefore proved that *a uniformly convergent series can be integrated term by term to give the integral of its sum function.*

Cauchy's Theorem

We turn now to one of the most important and useful theorems of complex-function theory. Let $f(z)$ be analytic over some region[1] R of the complex plane, and consider any simple closed curve C which together with its interior is completely contained in R. Then *Cauchy's theorem* (1825) states that

$$\int_C f(z)\,dz = 0 \tag{2-7}$$

Clearly, this result is equivalent to the statement that the integral between any two points is independent of the path linking those points, provided that all paths to be considered can be deformed into one another without leaving R.

One way of proving this theorem is to use Eq. (2-5) and to express each of the real contour integrals in terms of integrals over the area enclosed by C, by use of Green's theorem. This latter theorem—a two-dimensional form of Gauss's divergence theorem—states that for any two functions $\Phi(x,y)$ and $\psi(x,y)$

$$\int_C (\Phi\,dx + \psi\,dy) = \int_A (\psi_x - \Phi_y)\,dA$$

The left-hand integral must traverse the closed contour C in the counterclockwise direction; the domain of integration of the right-hand integral (in which subscripts denote partial derivatives) is that area enclosed by C. Applying this integral identity to the right-hand integrals of Eq. (2-5) gives two area integrals, each of whose integrands vanishes because of the Cauchy-Riemann equations; consequently Eq. (2-7) is proved.[2]

[1] By the word *region* (or *domain*) we shall always mean an open, connected set of points in the plane; i.e., a set in which every point of the set is surrounded by some circular neighborhood of points all of which belong to the set, and in which any two points may be joined by a curve lying entirely in the set.

[2] Strictly speaking, Green's theorem is usually proved in books on calculus for situations in which the partial derivatives of Φ and ψ are continuous; this is here equivalent to requiring that $f'(z)$ be continuous. Such proofs also require that C be a rather well-behaved curve.

A method for the direct proof of Cauchy's theorem, which does not require the assumption that $f'(z)$ be continuous, was devised by Goursat. The method consists in ruling the interior of C with lines parallel to the X and Y axes, thus dividing the interior region into a large number of rectangles (or, at the boundary, part rectangles). The integral around C of $f(z)$ is then equal to the sum of the integrals around each of these rectangular (or part-rectangular) contours, since each inner side of a rectangle is then covered twice, and in opposite directions. For each such rectangle, the value of $f(z)$ on its boundary

The most immediate consequence of Cauchy's theorem is that integrals of analytic functions may be computed by reversion of the differentiation process, just as in the real-variable case. For let $f(z)$ be analytic, and define

$$F(z) = \int_{z_0}^{z} f(z)\, dz \tag{2-8}$$

where the integration is along any path joining a fixed initial point z_0 to a variable end point z. Because of Cauchy's theorem, we need not specify the path itself, but only its end points (provided, again, that any two such paths can be deformed into one another without leaving the region of analyticity). In Eq. (2-8), $F(z)$ is thus a function of the terminal point z alone, since z_0 is considered fixed. To see whether or not $F'(z)$ exists, we compute

$$\lim_{\xi \to z} \frac{\int_{z_0}^{\xi} f(z)\, dz - \int_{z_0}^{z} f(z)\, dz}{\xi - z} = \lim_{\xi \to z} \frac{\int_{z}^{\xi} f(z)\, dz}{\xi - z}$$

Upon using the definition of an integral this expression is simply $f(z)$, irrespective of the mode of approach of ξ to z. Consequently $F'(z)$ exists, hence $F(z)$ is analytic, and

$$F'(z) = f(z) \tag{2-9}$$

Now let $G(z)$ be *any* analytic function satisfying Eq. (2-9); i.e., let

$$G'(z) = f(z)$$

Then

$$\frac{d}{dz}[F(z) - G(z)] = 0$$

so that

$$F(z) = G(z) + K$$

where K is some complex constant (since the only analytic function with zero derivative throughout a region is a constant—as is easily seen from the

---

can be related to the value of $f(z)$ at some point z_1 interior to that rectangle by

$$f(z) = f(z_1) + (z - z_1)f'(z_1) + (z - z_1)m(z)$$

where $m(z) \to 0$ as the mesh is made finer [and, moreover, given any bound ϵ for $m(z)$, a finite mesh can be found so that $|m| < \epsilon$ for *all* rectangles]. Upon using next the easily proved results that $\int dz = \int z\, dz = 0$ around each rectangle, the remainder of the integral around each rectangle is easily seen to be proportional to the area of that rectangle multiplied by ϵ; the sum of all the integrals is then proportional to ϵ, and since ϵ can be made arbitrarily small, the theorem follows.

The condition that f be analytic on and inside C can also be weakened; it is necessary only to require f to be analytic inside C and continuous onto the bounding curve C. Finally, the curve C need be only continuous [e.g., defined by $z = z(t)$, where z is a continuous function of the parameter t] and rectifiable.

Detailed proofs along these lines will be found in G. N. Watson, Complex Integration and Cauchy's Theorem, *Cambridge Tracts* 15, 1914; in K. Knopp, "Theory of Functions," Dover Publications, Inc., New York, 1943; and in E. Hille, "Analytic Function Theory," Ginn and Company, Boston, 1959.

fact that the partial derivatives of the real and imaginary parts of such a function must vanish). Now $F(z_0) = 0$ from Eq. (2-8), so that

$$K = -G(z_0)$$

and Eq. (2-8) now becomes

$$\int_{z_0}^{z} f(z)\, dz = G(z) - G(z_0)$$

Thus the integral can be evaluated in terms of any function $G(z)$ whose derivative equals $f(z)$; just as in the case of real variables, it is conventional to call such a function $G(z)$ an *indefinite* integral and to write

$$\int f(z)\, dz = G(z) + \text{arbitrary const}$$

Since integration is inverse to differentiation, the previously mentioned precautions concerning appropriate branches of multiple-valued functions must be observed. Thus, as in Sec. 2-1, if we write the indefinite integral of z^s (s an arbitrary complex constant) as

$$\frac{1}{s+1} z^{s+1}$$

then the same branch of the logarithm of z must be used in evaluating each of the two quantities z^s and z^{s+1}.

Multiply Connected Regions

So far, we have restricted ourselves to regions of analyticity that are *simply connected*—i.e., for which any closed curve drawn in the region encloses only points belonging to the region. Suppose next that the region has "holes" in it, as in Fig. 2-1, where the shaded region R is that over which

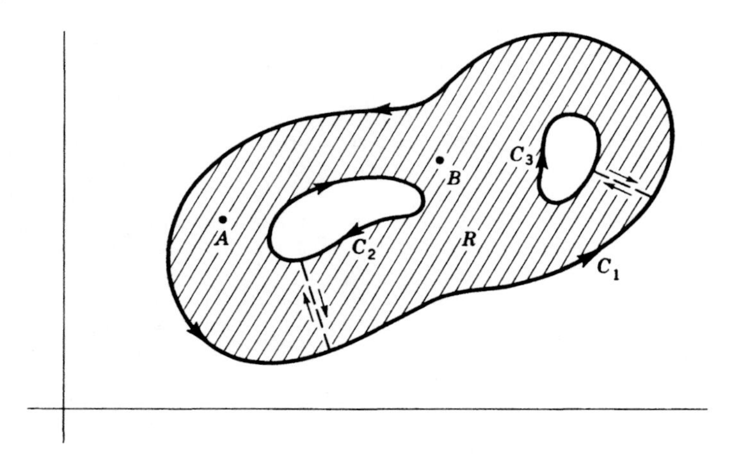

Fig. 2-1 Cauchy's theorem for multiply connected regions.

a single-valued $f(z)$ is analytic. Then it is no longer necessarily true that the integral from A to B is independent of the path, for the two paths may lie on opposite sides of one of the holes, in which case $f(z)$ is not analytic over the entire region between the two paths, and Cauchy's theorem cannot be applied. (Of course, integration along each of two paths lying on the same side of the hole will give the same result.)

Cauchy's theorem may be generalized to apply to a single-valued function analytic in a multiply connected region. In Fig. 2-1, make "cuts" in the region (dotted lines) which join the holes to the outer boundary. If these cuts are treated as barriers which may not be crossed by paths of integration, the resulting region is simply connected and Cauchy's theorem may be applied to the complete boundary, traversed in the way indicated by the arrows. Note that each cut is traversed twice, and in opposite directions, so that the net result of integration along the dotted lines is zero. We therefore have

$$\int_{C_1} f(z)\,dz + \int_{C_2} f(z)\,dz + \int_{C_3} f(z)\,dz = 0$$

More generally, if there are $n - 1$ holes in the region, and if the directions of integrations are always taken clockwise along the hole boundaries and counterclockwise on the outside boundary C_1 (so that a man walking on any of these contours would always have the shaded region to his left), then

$$\sum_{j=1}^{n} \int_{C_j} f(z)\,dz = 0$$

and this is the desired generalization of Cauchy's theorem to multiply connected regions.

This result can also be expressed in another way. Suppose that we have two simple closed curves C_1 and C_2, each lying in R, and each enclosing the same hole (or holes). Then, although it is no longer true in general that $\int f(z) = 0$ around either C_1 or C_2 alone, it is still true that

$$\int_{C_1} f(z)\,dz = \int_{C_2} f(z)\,dz$$

provided that both directions of integration are clockwise or both counterclockwise. The fact that multiply connected domains can usually be made simply connected by the use of appropriate cuts, as in Fig. 2-1, means that it is generally sufficient to prove theorems for simply connected domains.

Exercises

1. Evaluate

$$\int_{1+i}^{3-2i} \sin z\,dz$$

in two ways: (1) by choosing any path between the two end points and using real integrals as in Eq. (2-5), and (2) by use of an indefinite integral. Show that the inequality (2-6) is satisfied.

2. Show that

$$\int \frac{dz}{z} = 2\pi i$$

for any counterclockwise closed curve enclosing the origin. Do this in two ways, first by direct computation (note that, if $z = re^{i\theta}$, $dz = dr\, e^{i\theta} + re^{i\theta}i\, d\theta$), and second by use of the generalized Cauchy theorem to replace the curve by a circle around the origin, prior to such direct computation.

3. Show in an easy way that the integral of each of the following expressions around the circle $|z| = \frac{1}{2}$ vanishes:

$$\frac{z+1}{z^2+z+1} \qquad \exp z^2 \ln (1+z) \qquad \sin^{-1} z$$

4. Let $u(x,y)$ and $v(x,y)$ be two well-behaved functions of x and y. Denote the combination $u + iv$ by F (not necessarily an analytic function), and replace x by $\frac{1}{2}(z + z^*)$, y by $(-i/2)(z - z^*)$ so that $F = F(z,z^*)$. Show that, if C is any closed contour containing an area A,

$$\int_C F(z,z^*)\, dz = 2i \int_A \frac{\partial F}{\partial z^*}\, dA$$

5. Evaluate

$$\int \frac{dz}{(1+z^2)^{\frac{1}{2}}}$$

from $z = -2$ to $z = +2$, around the semicircle in the upper half plane satisfying $|z| = 2$. We complete the definition of the function $(1 + z^2)^{\frac{1}{2}}$ by choosing as branch line that part of the imaginary axis satisfying $|y| \leq 1$; also, at $z = -1$, $(1 + z^2)^{\frac{1}{2}}$ is chosen as the negative square root of 2.

6. Explain why it would be valid to obtain the power series in z for arcsin z by integrating the expansion of $(1 - z^2)^{-\frac{1}{2}}$ term by term. Discuss the choice of branch.

7. Carry out the change in variable $z = 1/\xi$ in order to evaluate

$$\int_\Gamma \frac{dz}{1+z^3}$$

where Γ is the counterclockwise path around the circle $|z| = 2$. Is it legitimate to write $dz = -(1/\xi^2)\, d\xi$? What is the new contour, what are the properties of the integrand within it, and is the new sense of integration clockwise or counterclockwise? Obtain a similar result for the integral of $1/P_n(z)$ around any contour containing all the zeros of the nth-order polynomial $P_n(z)$, if $n \geqq 2$.

8. (*a*) Show that formal, term-by-term differentiation, or integration, of a power series yields a new power series with the same radius of convergence.

(*b*) The uniform-convergence property of a power series implies that term-by-term integration yields the integral of the sum function. Show that the integrated sum function is single-valued and analytic within the circle of convergence.

(*c*) Show that a power series converges to an analytic function within its circle of convergence.

2-3 Cauchy's Integral Formula

In Sec. 2-2 we proved *Cauchy's theorem*, which states that the integral of a function around a closed curve vanishes if that function is analytic at each point within the curve. We now derive a consequence of this theorem which allows us to express the value of such a function at any point *inside* the contour in terms of its values *on* the contour. This result—called *Cauchy's integral formula*—has no analog in real-variable theory; it again illustrates the special nature of the property of differentiability in the complex plane.

Henceforth, we shall require that a closed contour not otherwise specified is to be *simple* (which means that it does not intersect itself and so in particular makes only one complete circuit) and that it is to be traversed in the mathematically *positive* direction, which is defined to be *counterclockwise*. When a multiply connected region such as that of Fig. 2-1 is being considered, the positive direction of traversal for an inner contour is taken as clockwise. Thus a man walking along a contour in the positive direction always has the area of interest to his left, whether the region is simply connected or not.

Let $f(z)$ be analytic in a simply connected region, and let C be any closed contour in that region. Choose any fixed point inside C; then Cauchy's integral formula states that

$$f(z) = \frac{1}{2\pi i} \int_C \frac{f(\zeta)}{\zeta - z}\, d\zeta \tag{2-10}$$

To prove this result, note that the integrand when considered as a function of ζ is analytic everywhere except at $\zeta = z$, so that by Sec. 2-2 the path of integration may be replaced by a small circle of center z and radius r. Moreover, given any $\epsilon > 0$, the radius of the circle may be chosen so small that $|f(\zeta) - f(z)| < \epsilon$ for all points ζ on that circle, because of the continuity of $f(z)$ (cf. Exercise 3 of Sec. 2-1). Upon denoting this circular contour by C', the right-hand side of Eq. (2-10) becomes

$$\frac{1}{2\pi i}\int_{C'}\frac{f(z)}{\zeta - z}\,d\zeta + \frac{1}{2\pi i}\int_{C'}\frac{f(\zeta) - f(z)}{\zeta - z}\,d\zeta$$

The first of these equals $f(z)$ (cf Exercise 2 of Sec. 2-2). The second is clearly bounded in absolute value by

$$\frac{1}{2\pi}\int_{C'}\left|\frac{f(\zeta) - f(\zeta)}{\zeta - z}\right|\,|d\zeta| \leq \frac{1}{2\pi}\int_0^{2\pi}\frac{\epsilon}{r}\,r\,d\theta = \epsilon$$

which can be made arbitrarily small. Hence the second integral must be zero, and the validity of Eq. (2-10) is established.

If a single-valued function $f(z)$ is analytic in a multiply connected region, such as that of Fig. 2-1, and if z is a point in the shaded region of analyticity, then an argument of the kind used in Sec. 2-2 shows that Eq. (2-10) still holds, provided that the integration path consists of all contours. Denoting the composite contour by C, we still have then

$$f(z) = \frac{1}{2\pi i}\int_C \frac{f(\zeta)}{\zeta - z}\,d\zeta = \frac{1}{2\pi i}\int_{C_1}\frac{f(\zeta)}{\zeta - z}\,d\zeta + \frac{1}{2\pi i}\int_{C_2}\frac{f(\zeta)}{\zeta - z}\,d\zeta + \cdots$$

Point z Not Inside Contour

In the foregoing, z has been a point inside the region surrounded by C (e.g., inside the shaded region of Fig. 2-1). If z is outside C, then the integrand in Eq. (2-10) is analytic inside C, so that the integral vanishes. Thus, in either the simply or the multiply connected case,

$$\frac{1}{2\pi i}\int_C \frac{f(\zeta)}{\zeta - z}\,d\zeta = 0 \qquad \text{for } z \text{ outside } C \tag{2-11}$$

If z is *on* C, then the integral does not exist, since the integrand becomes infinite at $\zeta = z$. However, there are various ways of interpreting the integral in this case so as to avoid the singularity; any such interpretation is of course merely a matter of definition. One way of redefining the integral, for the case in which z is on C, is to evaluate the integral for a point z_1, either inside or outside the contour, and then to take the limit as $z_1 \to z$. This clearly gives $f(z)$ or 0, respectively, and so is not very interesting. Another way is to adopt from real-variable theory the idea of the principal value of an integral. Here this corresponds to omitting from C a portion centered on z and of length 2ϵ, evaluating the integral over the remainder of C, and then taking the result as $\epsilon \to 0$. Suppose first that z is not a corner point of the contour. Then, for any ϵ, we can consider a new contour C' consisting of all of C except for that portion near z which is to be omitted, and also a semicircle (at least in the limit) inside the region of analyticity and centered on z, as in Fig. 2-2. Then, since z is outside C',

$$\frac{1}{2\pi i}\int_{C'}\frac{f(\zeta)}{\zeta - z}\,d\zeta = 0$$

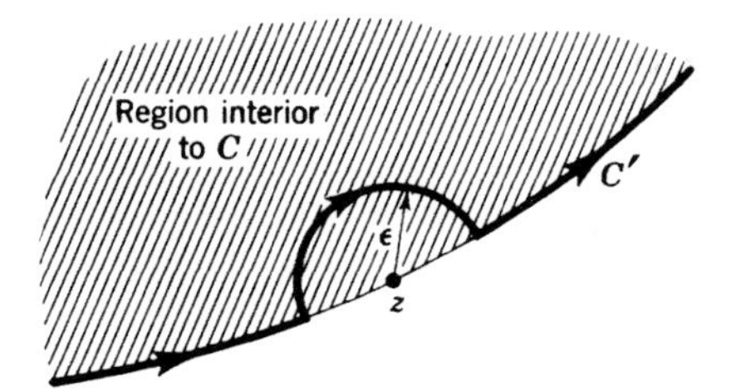

Fig. 2-2 Contour C' for principal-value evaluation of Cauchy integral formula.

As $\epsilon \to 0$, the integral over the part of C' which does not include the semicircle becomes the desired principal value of the original integral. The remainder clearly becomes (with $\zeta = z + \epsilon e^{i\theta}$)

$$\frac{1}{2\pi i} \int_{\pi}^{0} \frac{f(z + \epsilon e^{i\theta})}{\epsilon e^{i\theta}} \epsilon e^{i\theta} i \, d\theta \to -\tfrac{1}{2} f(z)$$

Using the abbreviation PV to denote principal value, we therefore have the result that the sum of the desired principal value and $-\frac{1}{2}f(z)$ vanishes, or that

$$\text{PV} \frac{1}{2\pi i} \int_C \frac{f(\zeta)}{\zeta - z} \, d\zeta = \tfrac{1}{2} f(z) \qquad \text{for } z \text{ on } C \tag{2-12}$$

If z is a corner point, enclosing an interior angle of α radians, then the same kind of argument shows that the factor $\frac{1}{2}$ must be replaced by $\alpha/2\pi$.

Formulas for First and Higher Derivatives

Equation (2-10) expresses the value of an analytic function $f(z)$ at a point interior to a contour in terms of the values of $f(z)$ on the boundary. It is natural to ask if the value of $f'(z)$ at such an interior point [whose existence is assured by the analyticity of $f(z)$] is obtained by simply differentiating under the integral sign. Consider

$$\begin{aligned}
\frac{f(\xi) - f(z)}{\xi - z} &= \frac{1}{2\pi i} \int_C \left[\frac{f(\zeta)}{\zeta - \xi} - \frac{f(\zeta)}{\zeta - z} \right] \frac{d\zeta}{\xi - z} \\
&= \frac{1}{2\pi i} \int_C \frac{f(\zeta) \, d\zeta}{(\zeta - \xi)(\zeta - z)} \\
&= \frac{1}{2\pi i} \int_C \frac{f(\zeta)}{(\zeta - z)^2} \left(1 + \frac{\xi - z}{\zeta - \xi} \right) d\zeta
\end{aligned}$$

The limit of the left-hand side as $\xi \to z$ is $f'(z)$. Now, in the term $(\xi - z)/(\zeta - \xi)$ in the right-hand integrand, ζ is on C and ξ is approaching the interior point z, so that $|\zeta - \xi|$ eventually becomes and remains greater than, say, half the shortest distance from z to C. On the other hand, the numerator of this term approaches zero. Consequently, only the first term remains, and we have

$$f'(z) = \frac{1}{2\pi i} \int_C \frac{f(\zeta)}{(\zeta - z)^2} \, d\zeta \tag{2-13}$$

But we can now apply an exactly similar argument to the expression

$$\frac{f'(\xi) - f'(z)}{\xi - z} = \frac{1}{2\pi i} \int_C \left[\frac{f(\zeta)}{(\zeta - \xi)^2} - \frac{f(\zeta)}{(\zeta - z)^2} \right] \frac{d\zeta}{\xi - z}, \text{ etc.}$$

and we find that the limit of this expression as $\xi \to z$ is given by

$$\frac{2}{2\pi i} \int_C \frac{f(\zeta)}{(\zeta - z)^3} \, d\zeta$$

Thus $f''(z)$ exists, and

$$f''(z) = \frac{2}{2\pi i} \int_C \frac{f(\zeta)}{(\zeta - z)^3} \, d\zeta$$

Consequently, if $f(z)$ is analytic in a region, then not only does $f'(z)$ exist, but $f''(z)$ also exists in that region. But then $f'(z)$ is a new analytic function, so that its second derivative $f'''(z)$ must also exist. Proceeding in this way, we see that *any function which is analytic in a region R has derivatives of all orders in R.* Moreover, by a procedure similar to that used to find integral formulas for $f'(z)$ and $f''(z)$, we obtain for the nth derivative, $f^{(n)}(z)$, the result

$$f^{(n)}(z) = \frac{n!}{2\pi i} \int_C \frac{f(\zeta)}{(\zeta - z)^{n+1}} \, d\zeta \tag{2-14}$$

Further, since the existence of a derivative implies the continuity of the function, it follows that all derivatives of $f(z)$ are continuous in R.

The fact that derivatives of all orders of an analytic function exist and are continuous in R implies that all partial derivatives of the real and imaginary parts of an analytic function also exist and are continuous in R.

EXERCISES

1. Use Cauchy's integral formula to evaluate the integral around the unit circle ($|z| = 1$) of

$$\frac{\sin z}{2z + i} \qquad \frac{\ln (z + 2)}{z + 2} \qquad \frac{z^3 + \operatorname{arcsinh} (z/2)}{z^2 + iz + i} \qquad \cot z$$

2. If $\Phi(z)$ is analytic in a simply connected region in which a closed contour C is drawn, obtain all possible values of

$$\int_C \frac{\Phi(\zeta)}{\zeta^2 - z^2} \, d\zeta$$

where z is not a point on C itself.

3. If n is an integer, positive or negative, and if C is a closed contour around the origin, use Eq. (2-14) to show that

$$\int_C \frac{dz}{z^n} = 0 \qquad \text{unless } n = 1$$

4. Show that the principal-value calculation associated with Fig. 2-2 could have been carried out just as easily if the semicircle went *outside* z rather than inside, provided that $f(\zeta)$ were analytic in some larger region containing that part of C near z.

5. Obtain a version of Cauchy's integral theorem valid for the situation in which $f(z)$ is analytic everywhere *outside* some closed curve C, with $f(z) \to$ some constant α as $z \to \infty$. Relate this result to the transformation $\zeta = 1/z$.

6. Show, by the method used to derive Eq. 2-13, that, if $\Phi(z)$ is any function continuous on C (but not necessarily analytic there or anywhere else), and if a function $f(z)$ is defined for z inside C by

$$f(z) = \frac{1}{2\pi i} \int_C \frac{\Phi(\zeta)}{\zeta - z}\, d\zeta$$

then $f(z)$ is analytic inside C. Invent an example to illustrate the fact that, as an interior point z approaches a boundary point z_0, $f(z)$ need not approach $\Phi(z_0)$.

7. A function $\Phi(z)$ is known to be continuous throughout a simply connected region and to have the property that

$$\int_C \Phi(z)\, dz = 0$$

for any closed contour lying in that region. Show that $\Phi(z)$ must be analytic in that region. This converse of Cauchy's theorem is known as *Morera's theorem*.

8. In Exercise 8 of Sec. 2-2, a proof that a power series is analytic within its circle of convergence was required. Use Morera's theorem to obtain this same result. Show more generally that any series of analytic functions Σf_j, uniformly convergent over a common region, converges to an analytic function f within that region. Show also that all derivatives of f may be obtained by

$$f^{(k)} = \sum_{j=1}^{\infty} f_j^{(k)}$$

Finally prove that if each f_j has the form

$$f_j = \sum_{n=0}^{\infty} a_{nj}(z - z_0)^n$$

and if
$$f = \sum_{n=0}^{\infty} a_n (z - z_0)^n$$

then
$$a_n = \sum_{j=1}^{\infty} a_{nj}$$

This result is called *Weierstrass's double-series theorem;* it provides a generalization of the result in Sec. 1-3 concerning the substitution of one series into another.

2-4 Maximum Modulus Theorem

For the special case in which the contour C of Eq. (2-10) is a circle, center z and radius r, we have $\zeta - z = re^{i\theta}$ and $d\zeta = re^{i\theta} i\, d\theta$, so that Eq. (2-10) becomes

$$f(z) = \frac{1}{2\pi} \int_0^{2\pi} f(z + re^{i\theta})\, d\theta \tag{2-15}$$

which is sometimes called the *mean-value theorem* for analytic functions. It states that the value of an analytic function at any chosen point is equal to the mean value of the function on the circumference of a circle centered on that point, provided of course that the region of analyticity extends out to that circle. Note that a separation into real and imaginary parts shows that the real and imaginary parts of the function satisfy this mean-value theorem individually. Note also that, since Eq. (2-15) is independent of r, we could multiply both sides by $r\, dr$ and integrate with respect to r from 0 to R to obtain

$$f(z) \int_0^R r\, dr = \frac{1}{2\pi} \int_0^R \int_0^{2\pi} f(z + re^{i\theta}) r\, dr\, d\theta = \frac{1}{2\pi} \int_A f(z + re^{i\theta})\, dA$$

where the integration is over the area of the circle of radius R and center z. Thus

$$f(z) = \frac{1}{\pi R^2} \int_A f(z + re^{i\theta})\, dA \tag{2-16}$$

which means that the value of an analytic function at a point is also equal to its mean value over the *area* of a circle centered on that point. Again, the same result holds for the real and imaginary parts individually.

Equation (2-16) may be used to prove the *maximum modulus theorem,* which states that if $f(z)$ is analytic in a region, and if $|f(z)|$ attains a maximum value at some point z in the interior of the region, then $f(z)$ is identically constant. For let z be a point interior to the region such that $|f(\zeta)| \leq |f(z)|$ for all points ζ in the region. Choose any circle, center z and radius R, such that the circle in its entirety lies inside the region. Then Eq. (2-16) requires that

$$|f(z)| \leqq \frac{1}{\pi R^2} \int_A |f(\zeta)| \, dA \tag{2-17}$$

where the area integration is over the interior of the circle. But for all such ζ, $|f(\zeta)| \leqq |f(z)|$, so that Eq. (2-17) cannot possibly hold unless $|f(\zeta)| \equiv |f(z)|$ for all ζ inside the circle; use of the Cauchy-Riemann equations then shows in turn that $|f(\zeta)|$ cannot be constant inside the circle unless $f(\zeta)$ itself is constant. Thus the maximum value of $|f(\zeta)|$ is attained on the boundary of the circle also, and we now construct new circles centered at various boundary points and repeat the argument; the result is clearly that $f(\zeta)$ is constant everywhere in the original region.

Note that the "region" in the above theorem can be taken to be any small neighborhood surrounding a given point; it thus follows that a nonconstant analytic function cannot attain even a *local* maximum at any point inside its region of analyticity.

Another way of stating the above maximum-modulus theorem is that, if $f(z)$ is analytic inside a closed contour, then $|f(z)|$ attains its maximum value on the contour itself. Moreover, if $f(z)$ does not vanish at any point inside the contour, then $|f(z)|$ also attains its minimum value on the boundary [merely consider $1/f(z)$].

EXERCISES

1. Find the maximum for $|z| \leqq 1$ of $|z^2 + 2z + i|$, $|\sin z|$, $|\arcsin (z/2)|$.

2. In what sense does the maximum-modulus theorem hold for a multiply connected region?

3. Let the real number $M > 0$ be a lower bound for $|f(z)|$ on some contour C within which $f(z)$ is analytic. Let z_0 be a point within C, and suppose that $|f(z_0)| < M$. Show that $f(z)$ must have at least one zero inside C. Use this result to prove that every nonconstant polynomial must have a zero.

Liouville's Theorem

In Sec. 2-3 we obtained expressions for the derivatives of an analytic function in terms of the values of the function on an enclosing contour. Consider Eq. (2-14), and as before let C be a circle of center z and radius r. Then

$$f^n(z) = \frac{n!}{2\pi} \int_0^{2\pi} \frac{f(z + re^{i\theta})}{r^n e^{in\theta}} \, d\theta$$

so that, if $M(r)$ is the maximum of $|f(\zeta)|$ on $|\zeta - z| = r$, we have

$$|f^{(n)}(z)| \leq n! \frac{M(r)}{r^n} \tag{2-18}$$

a form of what is called *Cauchy's inequality.*

Define now an *entire* function to be one which is analytic for all (finite) values of z. Examples of entire functions are $z^3 + 2z$, $\sin z$, e^z. The entire function $f(z)$ is said to be bounded if there exists a real constant M such that $|f(z)| < M$ for all z. From Eq. (2-18), we can now deduce *Liouville's theorem*, which states that *a bounded entire function is necessarily a constant.* For choose any value of z, and set $n = 1$ in Eq. (2-18):

$$|f'(z)| < \frac{M}{r}$$

Since M is independent of r, and since r can be taken as large as we wish, this inequality shows that $f'(z)$ must be zero, for any value of z. Consequently $f(z)$ is indeed a constant. An equivalent statement of Liouville's theorem is that outside every circle a nonconstant entire function attains arbitrarily large absolute values.

EXERCISES

4. Use Liouville's theorem to show that any nth-degree polynomial $P_n(z)$, $n \geqq 1$, has a zero. (Consider $1/P_n$.)

5. A particular entire function has its real part everywhere nonnegative. What can be deduced concerning the function?

6. Let $f(z)$ be analytic and single-valued in the ring-shaped region $r_1 \leqq |z| \leqq r_3$. Let $|z| = r_2$ be a circle satisfying $r_1 < r_2 < r_3$. Denote the maximum of $|f(z)|$ on r_j by $M(r_j)$, $j = 1, 2, 3$. Then prove *Hadamard's three-circles theorem*, which states that $\ln M(r)$ is a convex function of $\ln r$, in the sense that

$$\ln M(r_2) \leqq \frac{\ln r_3 - \ln r_2}{\ln r_3 - \ln r_1} \ln M(r_1) + \frac{\ln r_2 - \ln r_1}{\ln r_3 - \ln r_1} \ln M(r_3)$$

[*Hint:* Consider $w(z) = z^s f(z)$, where s is real. Since $|z^s|$ is single-valued, the maximum-modulus theorem is easily seen to apply to $w(z)$. Then choose s appropriately.] When would equality occur?

2-5 *Harmonic Functions*

The real function $u(x,y)$ is said to be *harmonic* in a region R if the second partial derivatives u_{xx} and u_{yy} exist in R, are continuous, and satisfy Laplace's equation,

$$u_{xx} + u_{yy} = 0 \tag{2-19}$$

The theory of analytic functions of a complex variable is closely related to the theory of such harmonic functions. In fact, we have already seen from the Cauchy-Riemann equations that the real and imaginary parts of an analytic function are harmonic; we shall now show that, conversely, given any harmonic function, we can find an analytic function having that function as its real (or imaginary) part. We shall then summarize some of the more basic properties of harmonic functions, partly in order to give further insight into the nature of analytic functions, and partly in preparation for future applications.

We shall first prove that, if $u(x,y)$ is any harmonic function, then a *conjugate* harmonic function $v(x,y)$ exists such that $u(x,y) + iv(x,y)$ is an analytic function of $z = x + iy$. (Of course, the roles of real and imaginary parts may be interchanged by multiplying by i.) Consider first a simply connected region, and let a function $u(x,y)$ satisfying Eq. (2-19) in this region be given. Then, if a conjugate function $v(x,y)$ exists, it must satisfy the Cauchy-Riemann conditions; so let us try to construct such a function by arbitrarily setting $v(x_0,y_0) = v_0$ at some reference point (x_0,y_0), and then defining its value at any other point (x,y) by

$$v(x,y) = \int_{(x_0,y_0)}^{(x,y)} -u_y\,dx + u_x\,dy \tag{2-20}$$

If the value of this integral does *not* depend on the path linking (x_0,y_0) to (x,y), then $v(x,y)$ is a well-behaved function whose partial derivatives are given by

$$v_x = -u_y \qquad v_y = u_x$$

Since the satisfaction of these Cauchy-Riemann equations is sufficient to ensure the analyticity of the combination $u + iv$, we need therefore prove only that the right-hand side of Eq. (2-20) is indeed path-independent, or, equivalently, that the integral around any closed circuit

$$\int_C -u_y\,dx + u_x\,dy$$

vanishes. But Green's theorem (Sec. 2-2) may be used to change this into an area integral with the integrand $u_{xx} + u_{yy}$, which vanishes by hypothesis, so that the proof is complete.

Thus a conjugate function always exists, and an arbitrary harmonic function can always be exhibited as the real (or imaginary) part of an analytic function. Within an arbitrary additive constant, the conjugate function is unique.

If the region is multiply connected, then the same result holds; however, the conjugate function can now be multiple-valued, in that integration paths taken on opposite sides of a "hole" may give different results. A simple example is $u(x,y) = \frac{1}{2}\ln_p(x^2 + y^2)$, defined in some region surrounding but excluding by a circle centered on the origin. Use of Eq. (2-20) gives the multiple-valued conjugate

$$v(x,y) = \arctan\frac{y}{x} \pm 2n\pi + \text{const} \qquad n = 1, 2, 3, \ldots$$

Having now linked up harmonic and analytic functions, let us deduce from the properties of analytic functions some useful properties of harmonic functions. Since derivatives of all orders of analytic functions exist and are continuous, any harmonic function must possess continuous partial derivatives of all orders—a rather remarkable result. (Incidentally, any partial derivative $\partial^{n+m}\varphi/\partial x^n\,\partial y^m$ of a harmonic function φ is clearly again a harmonic function.) Since the real and imaginary parts of an analytic function must separately satisfy Eq. (2-15), it follows that *any* harmonic function satisfies the mean-value theorem; again, the mean value may be computed either for the circumference or for the area of the surrounding circle. By an argument similar to that used in Sec. 2-4, the mean-value theorem can be used to show that the maximum—and also the minimum—of any real-valued function harmonic in a region must occur on the boundary of that region.

Dirichlet and Neumann Problems

The simplest practical problems involving harmonic functions are the Dirichlet and Neumann problems. In either problem, one is asked to find a function which is harmonic in a certain region; in the first case, the function itself is prescribed on the boundary, whereas, in the second case, the normal derivative is prescribed on the boundary. Either one of these problems may be transformed into the other, by merely putting it in terms of the conjugate function. For use of the Cauchy-Riemann equations shows that, on the boundary,

$$\frac{\partial u}{\partial s} = -\frac{\partial v}{\partial n} \tag{2-21}$$

where $\partial u/\partial s$ is the tangential derivative of the function u in the positive direction along the boundary and $\partial v/\partial n$ is the outward normal derivative of the conjugate function.

Since the integral of $\partial u/\partial s$ around the complete circuit C must vanish, it follows that the Neumann problem for a function $v(x,y)$ cannot be solved unless

$$\int_C \frac{\partial v}{\partial n}\,ds = 0 \tag{2-22}$$

For a multiply connected region the contour of Eq. (2-22) must include all internal contours as well as the external one; the reason why Eq. (2-22) need not hold for each individual contour is that the conjugate function u may be multiple-valued so that the integral of $\partial u/\partial s$ around any one contour need not vanish.

From Cauchy's integral formula, an explicit solution for the Dirichlet problem for a circular region can be obtained. Without loss of generality, the circle can be taken to be of unit radius and centered at the origin. Let

$z = re^{i\alpha}$, $\zeta = e^{i\theta}$, $r < 1$. Then since $d\zeta = i\zeta\, d\theta$, Eq. (2-10) gives

$$f(z) = \frac{1}{2\pi}\int_0^{2\pi} \frac{f(\zeta)\zeta}{\zeta - z}\, d\theta$$

But since z is inside the unit circle, $1/z^*$ is outside it, so that

$$0 = \frac{1}{2\pi}\int_0^{2\pi} \frac{f(\zeta)\zeta}{\zeta - 1/z^*}\, d\theta$$

Consequently (using $\zeta = 1/\zeta^*$),

$$f(z) = \frac{1}{2\pi}\int_0^{2\pi} f(\zeta)\left(\frac{\zeta}{\zeta - z} \pm \frac{z^*}{\zeta^* - z^*}\right) d\theta \tag{2-23}$$

Taking first the positive sign in Eq. (2-23), we obtain

$$f(z) = \frac{1}{2\pi}\int_0^{2\pi} f(\zeta)\, \frac{1 - |z|^2}{|\zeta - z|^2}\, d\theta$$

But the factor multiplying $f(\zeta)$ is purely real, so that the process of taking the real part gives

$$\begin{aligned} u(r,\alpha) &= \frac{1}{2\pi}\int_0^{2\pi} u(\theta)\, \frac{1 - r^2}{|e^{i\theta} - re^{i\alpha}|^2}\, d\theta \\ &= \frac{1}{2\pi}\int_0^{2\pi} u(\theta)\, \frac{1 - r^2}{1 - 2r\cos(\alpha - \theta) + r^2}\, d\theta \end{aligned} \tag{2-24}$$

where $u(\theta)$ is the value of the harmonic function u on the boundary. This result, which solves the Dirichlet problem for the circle, is known as *Poisson's formula*.

If in Eq. (2-23) we take the negative sign, we shall obtain the conjugate function v in terms of $u(\theta)$,

$$\begin{aligned} f(z) &= \frac{1}{2\pi}\int_0^{2\pi} f(\zeta)\, \frac{1 - 2\zeta z^* + |z|^2}{|\zeta - z|^2}\, d\theta \\ &= \frac{1}{2\pi}\int_0^{2\pi} f(\zeta)\left[1 + \frac{2i\,\mathrm{Im}\,(z\zeta^*)}{|\zeta - z|^2}\right] d\theta \end{aligned}$$

Consequently,

$$v(r,\alpha) = v(0) + \frac{1}{\pi}\int_0^{2\pi} u(\theta)\, \frac{r\sin(\alpha - \theta)}{1 - 2r\cos(\alpha - \theta) + r^2}\, d\theta \tag{2-25}$$

Finally, we can also obtain $f(z)$ in terms of $u(\theta)$, by combining the two results for u and v,

$$f(z) = iv(0) + \frac{1}{2\pi}\int_0^{2\pi} u(\theta)\, \frac{\zeta + z}{\zeta - z}\, d\theta \tag{2-26}$$

Equation (2-26) exemplifies the fact that we cannot arbitrarily specify both the real and the imaginary parts of an analytic function on the unit circle. The quantity multiplying $u(\theta)$ in the integrand of Eq. (2-24) is called the *Poisson kernel.* We note that it can be exhibited in any of the forms

$$\frac{1 - r^2}{1 - 2r\cos(\alpha - \theta) + r^2} = \frac{1 - |z|^2}{|\zeta - z|^2} = \mathrm{Re}\left(\frac{\zeta + z}{\zeta - z}\right)$$

What we have actually shown so far is that, if a function is known to be harmonic over a region containing the unit circle, then its values on the unit circle determine those inside, via Eq. (2-24). It may be shown more generally (by direct differentiation) that, if $u(\theta)$ is any continuous function with period 2π, then the right-hand side of Eq. (2-24) defines a function which is harmonic inside the circle and which approaches the value $u(\theta)$ as $re^{i\alpha} \to e^{i\theta}$.

EXERCISES

1. Show that the contour curves of two harmonic functions conjugate to one another intersect orthogonally.

2. Prove that the solution of any Dirichlet problem is unique and that the solution of any properly posed Neumann problem [cf. Eq. (2-22)] is unique within an additive constant.

3. Extend the Poisson formula so as to hold for a circle of arbitrary radius R.

4. Show that, if a continuous real function $u(x,y)$ satisfies the mean-value theorem for every circle in a region, then that function is harmonic in the region (*Koebe's theorem*).

5. Let $u(x,y)$ be harmonic and nonnegative inside a circle centered on the origin and of radius R. Show that, for any concentric circle of radius $r < R$, *Harnack's inequality* holds,

$$u(0)\,\frac{R - r}{R + r} \leqq u(r,\theta) \leqq u(0)\,\frac{R + r}{R - r}$$

6. Let the real function $u(x,y) \equiv u(z)$ be continuous in a region and satisfy the inequality

$$u(z_0) \leqq \frac{1}{2\pi}\int_0^{2\pi} u(z_0 + re^{i\theta})\,d\theta$$

for any z_0 in the region and for circles of any radius r lying wholly in the region. Such a function is said to be *subharmonic,* for obvious reasons. Show (*a*) that the maximum of a subharmonic function always occurs on the boundary of a region and (*b*) that, if $u(x,y)$ is subharmonic, then, if $w(x,y)$ is any harmonic function which satisfies $w \geqq u$ on the boundary of a region, that same inequality must be satisfied throughout the region. Show that the modulus of any analytic function is subharmonic.

7. Let u, v be any two real functions of (x,y). Using Green's theorem, obtain the first and second *Green's identities*, respectively,

$$\int_C u \frac{\partial v}{\partial n}\, ds = \int_A (u_x v_x + u_y v_y)\, dA + \int_A u(\Delta v)\, dA$$

$$\int_C \left(u \frac{\partial v}{\partial n} - v \frac{\partial u}{\partial n} \right) ds = \int_A [u(\Delta v) - v(\Delta u)]\, dA$$

Here C is a closed contour of area A; the symbol Δ represents the laplacian $\partial^2/\partial x^2 + \partial^2/\partial y^2$, and $\partial/\partial n$ represents the outward normal derivative. Use these identities to show (*a*) that $\int_C \partial u/\partial n\, ds = 0$ for any function harmonic over the region inside C and (*b*) that a differentiable real function v is subharmonic if and only if $\Delta v \geqq 0$. [*Hint for* (*b*): Set $u = 1$, and let C be a circle.]

8. Show that the solution of a Dirichlet problem may be characterized as that function u which satisfies the boundary condition and also the condition that

$$\int_A (u_x^2 + u_y^2)\, dA$$

be a minimum. Interpret this condition geometrically.

2-6 *Taylor Series*

The fact that an analytic function has continuous derivatives of all orders suggests the existence of power-series representations for such a function. Let $f(z)$ be analytic in a region R, and let z_0 be a point inside this region. With center z_0, construct any circle C lying wholly in R. We shall show that within this circle $f(z)$ may be expressed as the sum of the convergent power series

$$f(z) = a_0 + a_1(z - z_0) + a_2(z - z_0)^2 + \cdots a_n(z - z_0)^n + \cdots \tag{2-27}$$

in which the constants a_n are given by

$$a_n = \frac{f^{(n)}(z_0)}{n!} \tag{2-28}$$

The proof is a straightforward application of Cauchy's integral formula. Let ζ be a point on the circle C; then, for any point z inside C,

$$f(z) = \frac{1}{2\pi i} \int_C \frac{f(\zeta)\, d\zeta}{\zeta - z} = \frac{1}{2\pi i} \int_C \frac{f(\zeta)}{\zeta - z_0} \left(1 - \frac{z - z_0}{\zeta - z_0}\right)^{-1} d\zeta$$

Since $|z - z_0| < |\zeta - z_0|$, the series

$$\left(1 - \frac{z - z_0}{\zeta - z_0}\right)^{-1} = 1 + \frac{z - z_0}{\zeta - z_0} + \left(\frac{z - z_0}{\zeta - z_0}\right)^2 + \cdots \tag{2-29}$$

is uniformly convergent for all points ζ on C, so that we may integrate term by term to give

$$\begin{aligned} f(z) &= \frac{1}{2\pi i}\int_C \frac{f(\zeta)\,d\zeta}{\zeta - z_0} + \frac{z - z_0}{2\pi i}\int_C \frac{f(\zeta)\,d\zeta}{(\zeta - z_0)^2} + \frac{(z - z_0)^2}{2\pi i}\int_C \frac{f(\zeta)\,d\zeta}{(\zeta - z_0)^3} + \cdots \\ &= f(z_0) + \frac{f'(z_0)}{1!}(z - z_0) + \frac{f''(z_0)}{2!}(z - z_0)^2 + \\ &\qquad \cdots + \frac{f^{(n)}(z_0)}{n!}(z - z_0)^n + \cdots \end{aligned}$$

by use of Eq. (2-14). The expansion converges wherever the series (2-29) does, i.e., within C. Note that the coefficients depend only on the derivatives of $f(z)$ at z_0. By analogy with the Taylor-Maclaurin power-series expansion for real functions, an expansion of the form (2-27) is called a *Taylor series.*

We have previously seen (Exercise 8 of Sec. 2-2 or Exercise 9 of Sec. 2-3) that any power series represents an analytic function inside its circle of convergence; we have now proved the converse result that any analytic function can be expanded in a convergent power series.

Some Examples

Equation (2-28) provides a direct method for calculating the coefficients of the power-series expansion of a given function. For example, the expansion of e^z around the point $i\pi$ is obtained by setting

$$a_n = \frac{1}{n!}\left[\frac{d^n}{dz^n}(e^z)\right]_{z=i\pi} = \frac{e^{i\pi}}{n!} = \frac{-1}{n!}$$

and thus

$$e^z = -\left[1 + (z - i\pi) + \frac{(z - i\pi)^2}{2!} + \cdots\right]$$

The function e^z can of course also be expanded around the point $i\pi$ by writing

$$e^z = (e^{z-i\pi})e^{i\pi} = -e^{z-i\pi}$$

which gives the same result. This is an illustration of the fact that the power-series expansion of an analytic function must be unique, because of the identity theorem for power series proved in Sec. 1-3.

In expanding the function $f(z) = \sin(\sin z)$ around the origin, Eq. (2-28) can again be used, though it leads to tedious computations. Alternatively, the known expansion for $\sin z$ can be employed,

$$f(z) = \left(z - \frac{z^3}{3!} + \frac{z^5}{5!} \cdots \right) - \frac{1}{3!}\left(z - \frac{z^3}{3!} + \frac{z^5}{5!} \cdots \right)^3 + \frac{1}{5!}\left(z - \frac{z^3}{3!} + \frac{z^5}{5!} \cdots \right)^5 - \cdots$$

and terms of the same order collected. Notice that in an example like this it is useful to have the lowest order of terms increase from one expression to the next, so that only a finite number of terms need be used to find the coefficient of each power of z. Thus, for $f(z) = \sin(\cos z)$, it would be advantageous to write

$$\begin{aligned} \sin(\cos z) &= \sin[(\cos z - 1) + 1] \\ &= \sin(\cos z - 1)\cos 1 + \cos(\cos z - 1)\sin 1 \end{aligned}$$

before using an expansion similar to that for $\sin(\sin z)$.

It may also be useful to differentiate or integrate a given function before power-series expansion; thus, if $f(z) = \arctan z$, $f'(z) = (1 + z^2)^{-1}$, which is easily expanded, and the result can then be integrated term by term to give the expansion for $f(z)$ itself.

If $f(z)$ is analytic at z_0 but is multiple-valued, then each of its branches is a locally single-valued analytic function which may be expanded about z_0. For example, to expand $f(z) = \ln z$ about $z_0 = 1$, we write

$$f' = \frac{1}{z} = \frac{1}{(z-1)+1} = 1 - (z-1) + (z-1)^2 - \cdots$$

so that $$f(z) = (z-1) - \frac{(z-1)^2}{2} + \frac{(z-1)^3}{3} \cdots + \text{const}$$

Setting the constant equal to zero, we obtain the expansion for the principal value of $\ln z$ (since then $\ln 1 = 0$); that for any other branch is obtained by setting the constant equal to an appropriate multiple of $2\pi i$. It would not have been possible to obtain an expansion for $\ln z$ about the origin, since no branch of $\ln z$ is analytic there.

Further Properties Deduced From Taylor Series

If $f(z)$ is analytic at z_0 (and so also in a neighborhood of z_0, by the definition of analyticity), and if $f(z_0) = 0$, then the power-series expansion (2-27) will have $a_0 = 0$. More generally, it may happen that all of $a_0, a_1, \ldots, a_{p-1}$ vanish, but a_p does not; in that case, all of $f(z_0), f'(z_0), \ldots, f^{(p-1)}(z_0)$ vanish, and $f(z)$ is said to have a pth-order zero at z_0. If $f(z)$ has a zero of order p at z_0, then, given any positive integer $m \leqq p$, the function

$$g(z) = \begin{cases} \dfrac{f(z)}{(z-z_0)^m} & z \neq z_0 \\ a_m & z = z_0 \end{cases}$$

is analytic at z_0. Note that the usual L'Hospital rule holds, in that, if two functions $w(z)$ and $h(z)$ both vanish at z_0,

$$\lim_{z \to z_0} \frac{w(z)}{h(z)} = \frac{w'(z_0)}{h'(z_0)}$$

if this quantity exists; if still indeterminate, second derivatives may be taken, and so on.

Next, let $f(z)$ be analytic and nonconstant in a region R, and consider any closed curve C lying wholly in R. Then, inside C, $f(z)$ can attain any given complex value α at most a finite number of times, for otherwise there would be an infinite number of such points, having a limit point z_1, say. But an expansion of $f(z)$ in a power series about z_1 would have to agree at each of the α points with the power series $g(z) = \alpha + 0$ and so by the identity theorem would have everywhere the value α, which is contrary to hypothesis. A consequence of this result is that if $f(z)$ is analytic at a point z_0, and nonconstant, then it is possible to find a small circle around z_0 such that inside that circle $f(z)$ does not take on the value $f(z_0)$ at any point other than z_0.

There is an interesting Parseval-type formula relating the values of the coefficients a_n in a Taylor-series expansion of $f(z)$ to the values of $f(z)$ on a circle surrounding z_0. Let $f(z)$ be analytic in the circle $|z - z_0| \leqq r$; then, setting $z - z_0 = re^{i\theta}$ on the boundary of the circle gives

$$\begin{aligned} f(z) &= a_0 + a_1 re^{i\theta} + \cdots + a_n r^n e^{in\theta} + \cdots \\ f^*(z) &= a_0^* + a_1^* re^{-i\theta} + \cdots + a_n^* r^n e^{-in\theta} + \cdots \end{aligned}$$

so that

$$\frac{1}{2\pi}\int_0^{2\pi} |f(z_0 + re^{i\theta})|^2 \, d\theta = |a_0|^2 + |a_1|^2 r^2 + \cdots + |a_n|^2 r^{2n} + \cdots \tag{2-30}$$

We already have, from Eqs. (2-14) and (2-28), the inequality

$$|a_n| \leqq \frac{M(r)}{r^n}$$

where $M(r)$ is the maximum value of $|f|$ on the circle $|z - z_0| = r$. Equation (2-30) clearly allows us to sharpen this inequality, to the extent to which $f(z_0)$, $f'(z_0)$, $f''(z_0)$, etc., are known.

The existence of a power-series expansion for an analytic function implies that an arbitrary real harmonic function has a similar property. For let $u(x,y)$ be harmonic in a neighborhood of (x_0,y_0), and let $f(z)$ be that analytic function whose real part it is. Then, if the expansion of $f(z)$ is given by Eq. (2-27), where $z = x + iy$, and if $a_j = \alpha_j + i\beta_j$, where α_j and β_j are real, we have

$$\begin{aligned} u(x,y) = \alpha_0 + \alpha_1(x - x_0) - \beta_1(y - y_0) + \alpha_2(x - x_0)^2 \\ - 2\beta_2(x - x_0)(y - y_0) - \alpha_2(y - y_0)^2 + \cdots \end{aligned}$$

The coefficients can be obtained either from the expansion of $f(z)$, as above, or by computing the various partial derivatives of u at the point (x_0,y_0). If we choose polar coordinates (r,θ) so that $z = z_0 + re^{i\theta}$, the expansion of a harmonic function is a little simpler. For then write $a_j = r_j \exp(i\theta_j)$, and we have

$$f(z) = r_0e^{i\theta} + r_1r \exp[i(\theta + \theta_1)] + r_2r^2 \exp[i(2\theta + \theta_2)] + \cdots$$

so that

$$u(r,\theta) = u(0,0) + r_1r\cos(\theta + \theta_1) + r_2r^2\cos(2\theta + \theta_2) + \cdots$$

This is of course a Fourier series for the function $u(r,\theta)$, where the Fourier expansion is with respect to θ. It has, however, the interesting property that the dependence of the coefficients on r is very simple.

EXERCISES

1. Obtain the power-series expansions for the following functions around the indicated points. Where the function is multiple-valued, give the results for all possible branches.

(*a*) $(1 + z + z^2)^{-1}$; $z_0 = 0, 1 + i$
(*b*) $\sin^2 z$; $z_0 = 0, -1$
(*c*) $z^{1/2}$; $z_0 = 1, i\pi$
(*d*) $(z^2 - 1)^{1/2}$; $z_0 = 0$
(*e*) $(z - \pi)/(\sin z)$; $z_0 = \pi$
(*f*) $\ln[iz + (1 - z^2)^{1/2}]$; $z_0 = 0, i$
(*g*) $\arctan z$; $z_0 = 1$

2. In terms of the Bernoulli numbers B_n of Eq. (1-7), show that

(*a*) $$\frac{z}{e^z - 1} = \sum_{n=0}^{\infty} \frac{B_n}{n!} z^n$$

(*b*) $$\cot z = \frac{1}{z} \sum_{n=0}^{\infty} (-1)^n \frac{2^{2n}B_{2n}}{(2n)!} z^{2n}$$

(*c*) $$\tanh z = \sum_{n=1}^{\infty} \frac{2^{2n}(2^{2n} - 1)B_{2n}}{(2n)!} z^{2n-1}$$

[*Hint:* $\tanh z$ may be expressed as a linear combination of $\coth z$ and $\coth 2z$].

3. Let $f(z)$ be analytic inside the unit circle, with $f(0) = 0$, and with $|f(z)| \leqq 1$ inside and on the circle. Show that $|f(z)| \leqq |z|$ for any point z inside the circle and that if equality holds at any interior point then $f(z) = e^{i\alpha}z$ everywhere, with α some real constant (*Schwarz's lemma*).

4. Obtain a series solution for the Dirichlet problem for the interior of the circle $|z| = 2$, if $u(2e^{i\theta}) = \min(\theta, 2\pi - \theta)$ for $0 \leqq \theta \leqq 2\pi$.

5. The *hypergeometric series* is defined by

$$F(a,b;c;z) = 1 + \frac{a \cdot b}{1 \cdot c} z + \frac{a(a+1)b(b+1)}{1 \cdot 2 \cdot c(c+1)} z^2 + \frac{a(a+1)(a+2)b(b+1)(b+2)}{1 \cdot 2 \cdot 3 \cdot c(c+1)(c+2)} z^3 + \cdots$$

Here, a, b, and c are arbitrary complex numbers, except that c is not to be a negative integer. Show that the series converges inside the unit circle and diverges outside. Using the criterion of Sec. 1-3, show that for $|z| = 1$ it converges absolutely if $\text{Re}\,(a + b - c) < 0$. Prove the following results:

(*a*) $z(1 - z)F'' + [c - (a + b + 1)z]F' - abF = 0$ (*the hypergeometric equation*)
(*b*) $(1 + z)^a = F(-a,b;b;-z)$
(*c*) $\arcsin z = zF(\frac{1}{2},\frac{1}{2};\frac{3}{2};z^2)$
(*d*) $\ln(1 + z) = zF(1,1;2;-z)$

6. Show that:

(*a*) The power-series expansion about the origin of an even analytic function contains only even powers of z.
(*b*) If $f(z)$ is entire, with $|f|/|z|^n$ bounded as $|z| \to \infty$, where n is some positive integer, then $f(z)$ must be a polynomial of degree $\leqq n$.

2-7 *Laurent Series*

If $f(z)$ is analytic inside some circle centered on z_0, then it possesses a Taylor-series expansion about that point. It may happen, however, that $f(z)$ is analytic outside rather than inside the circle, or, more generally, that $f(z)$ is analytic only in some annular region contained between two concentric circles centered on z_0. If $f(z)$ is analytic and single-valued in such an annular region, then a generalization of the Taylor series, called a *Laurent series,* may be used to obtain a series expansion of $f(z)$ about z_0. The generalization consists in permitting negative as well as positive powers of $z - z_0$ to appear in the expansion.

Let $f(z)$ be analytic and single-valued in the annular region R defined by

$$r_1 \leqq |z - z_0| \leqq r_2$$

where r_1 and r_2 are the radii of two circles C_1 and C_2 centered on z_0, with of course $r_1 < r_2$. By Cauchy's integral formula, the value of $f(z)$ at any point z in R is given by

$$f(z) = \frac{1}{2\pi i}\int_{C_2}\frac{f(\zeta)\,d\zeta}{\zeta - z} + \frac{1}{2\pi i}\int_{C_1}\frac{f(\zeta)\,d\zeta}{\zeta - z}$$

Here the first integral is taken counterclockwise, the second clockwise, in accordance with our standard convention for multiply connected regions. Rewriting the equation as

$$f(z) = \frac{1}{2\pi i}\int_{C_2}\frac{f(\zeta)}{\zeta - z_0}\left(1 - \frac{z - z_0}{\zeta - z_0}\right)^{-1} d\zeta - \frac{1}{2\pi i}\int_{C_1}\frac{f(\zeta)}{z - z_0}\left(1 - \frac{\zeta - z_0}{z - z_0}\right)^{-1} d\zeta$$

expanding the integrands by use of the binomial theorem, and integrating term by term (justified by uniform convergence; note that $|\zeta - z_0| < |z - z_0|$ in the second integrand), we obtain

$$\begin{aligned} f(z) &= \cdots + \frac{a_{-2}}{(z - z_0)^2} + \frac{a_{-1}}{z - z_0} + a_0 + a_1(z - z_0) + a_2(z - z_0)^2 + \cdots \\ &= \sum_{n=-\infty}^{\infty} a_n(z - z_0)^n \end{aligned} \tag{2-31}$$

where
$$a_n = \frac{1}{2\pi i}\int_C \frac{f(\zeta)\,d\zeta}{(\zeta - z_0)^{n+1}}$$

Here C can be *any* curve in R enclosing the inner circle and traversed counterclockwise. The inner circle may be contracted and the outer expanded, until they are respectively as small or as large as they can be without encountering any singularities of $f(z)$; the *Laurent series* (2-31) is clearly valid in this maximal region. Moreover, the expansion is unique, for if there were two series yielding $f(z)$, we would have

$$\sum_{n=-\infty}^{\infty} a_n(z - z_0)^n = \sum_{n=-\infty}^{\infty} b_n(z - z_0)^n$$

throughout R. Multiply this equation by $(z - z_0)^k$, where k is any positive or negative integer, and integrate around any closed curve C enclosing C_1; then since

$$\frac{1}{2\pi i}\int_C (z - z_0)^{n+k}\,dz = \begin{cases} 0 & \text{if } n + k \neq -1 \\ 1 & \text{if } n + k = -1 \end{cases}$$

we obtain $a_{-k-1} = b_{-k-1}$ for all k, thus proving the identity of the two Laurent series.

The Laurent series divides naturally into two series, the first of which contains all the positive powers (including zero) and the second of which contains all the negative powers. The first series converges everywhere inside the outer circle, and the second series converges everywhere outside the inner circle. If $f(z)$ is analytic everywhere inside the outer circle, then

Cauchy's theorem shows that $a_n = 0$ for n negative so that the Laurent series reduces to a Taylor series for this case.

As an example, the function $f(z) = [(z - 1)(z + 2)]^{-1}$ has three different expansions about the origin, each of which is most easily obtained by separation into partial fractions and use of the binomial theorem,

$$(a)\ |z| < 1: \qquad f(z) = -\frac{1}{6}\sum_{n=0}^{\infty} [2^{n+1} + (-1)^n]\left(\frac{z}{2}\right)^n$$

$$(b)\ 1 < |z| < 2: \quad f(z) = \frac{1}{3}\sum_{n=1}^{\infty} z^{-n} + \frac{1}{6}\sum_{n=0}^{\infty} (-1)^{n+1}\left(\frac{z}{2}\right)^n$$

$$(c)\ |z| > 2: \qquad f(z) = \frac{1}{6}\sum_{n=0}^{\infty} [2^{-n} - (-1)^n]\left(\frac{2}{z}\right)^{n+1}$$

If $f(z)$ is multiple-valued, it may still happen that a particular branch is single-valued in the annular region of interest; a Laurent series may then be obtained for that branch. This is possible, for example, with each branch of $[(z - 1)(z + 2)]^{1/2}$ in the region $|z| > 2$; the appropriate expansion in powers of z is easily found by use of the binomial theorem. On the other hand, $\ln z$ does not have a single-valued branch in any annular region centered on the origin, so that it has no Laurent-series expansion in any such region.

EXERCISES

1. Expand $1/\sin z$ in powers of z for $0 < |z| < \pi$ and also for $\pi < |z| < 2\pi$.

2. In what annular regions centered on $-i$ could expansions in powers of $z + i$ be obtained for each of $(z^2 - 1)^{1/2}$, $(z^2 + 1)^{1/2}$, $\ln [(z + 1)/(z - 1)]$, $[z(z^2 - 1)]^{1/2}$, $[z(z - i)(z^2 - 1)]^{1/2}$? Obtain an expansion for the last of these, valid for large z.

3. Expand in powers of z the function $\sin (z + 1/z)$ in whatever annular region is closest to the origin. Express the coefficients as simple trigonometric integrals.

4. What can be deduced from the Laurent-series theorem concerning the possibility of series expansions in annular regions for certain harmonic functions?

Isolated Singularity of a Single-valued Function

Suppose now that $f(z)$ is analytic and single-valued (e.g., some chosen branch of a multiple-valued function) everywhere in some neighborhood $|z - z_0| < r$ of a point z_0, except at z_0 itself. Then $f(z)$ is said to have an *isolated singularity* at z_0. For example, each of $1/z$ and $\sin (1/z)$ has an isolated singularity at the origin. Each branch of the function

$[(z-1)^{1/2}-1]^{-1}$ is single-valued near $z = 2$, but only one branch has an isolated singularity at that point.

The simplest kind of isolated singularity at z_0 is one which could be removed by simply giving $f(z_0)$ an appropriate new value. For example, the function

$$f(z) = \begin{cases} e^z & z \neq 0 \\ 0 & z = 0 \end{cases}$$

can be made analytic at its isolated singularity by merely altering $f(0)$ from 0 to 1. Such a singularity is said to be *removable*. Other kinds of isolated singularity at z_0 may be less simple in nature; an obvious tool for use in characterizing such singularities is the Laurent-series expansion of the function about z_0. If a (single-valued) function $f(z)$ has an isolated singularity at z_0, then it is analytic in some circular region from which the central point z_0 has been deleted, i.e., for z satisfying $0 < |z - z_0| < \epsilon$, and may consequently be expanded in a Laurent series in this region. Let this series be

$$f(z) = \sum_{n=-\infty}^{\infty} a_n(z-z_0)^n \qquad a_n = \frac{1}{2\pi i}\int_C \frac{f(\zeta)\,d\zeta}{(\zeta - z_0)^{n+1}} \tag{2-32}$$

where C is any contour around z_0, contained in the region of analyticity. There are now three distinct cases to consider, distinguished from one another by the character of the series of negative powers in this expansion. This negative power series is called the *principal part* of $f(z)$ at z_0; the coefficient a_{-1} is called the *residue* of $f(z)$ at z_0, the name arising from the easily checked fact that

$$\frac{1}{2\pi i}\int_C f(\zeta)\,d\zeta = a_{-1}$$

We digress for a moment to remark that if $f(z)$ is analytic and single-valued everywhere inside a contour C, except at a number of isolated singularities, then as an immediate generalization of this last formula we obtain the *residue theorem*

$$\frac{1}{2\pi i}\int_C f(\zeta)\,d\zeta = \text{sum of residues inside } C \tag{2-33}$$

For example, let C_1 be the unit circle, and let $f(z) = [z^2(z - \frac{1}{2})(z^2 + 4)]^{-1}$. Then there are isolated singularities at $z = 0$, $z = \frac{1}{2}$ inside the circle, and the reader may verify that the sum of the residues is $-\frac{1}{17}$, so that

$$\frac{1}{2\pi i}\int_{C_1} \frac{d\zeta}{\zeta^2(\zeta - \frac{1}{2})(\zeta^2 + 4)} = -\frac{1}{17}$$

The reader will obtain an extensive acquaintance with Eq. (2-33) in Chap. 3.

Returning now to our general discussion of isolated singularities, we consider three cases.

Case 1. *No principal part.* Suppose that $f(z)$ has an isolated singularity at z_0 but that its Laurent series about z_0 is found to contain no negative powers of $z - z_0$. Then the series for $f(z)$ in $0 < |z - z_0| < \epsilon$ is simply a Taylor series. The only way for $f(z)$ to have a Taylor-series representation for $0 < |z - z_0| < \epsilon$, and yet not be analytic at z_0, is if $f(z_0) \neq a_0$. However, merely redefining $f(z_0)$ to be equal to a_0 gives an analytic function, so that the singularity in question must be *removable.* Moreover, it is clear in this case that $f(z)$ is *bounded* in absolute value in the neighborhood of z_0.

Conversely, if the singularity is removable, it is clear that the Laurent series has no principal part at z_0. Also, if $f(z)$ is bounded near z_0, say $|f(z)| < M$ for all z in $0 < |z - z_0| < \epsilon$, and if the contour C in Eq. (2-32) is chosen to be a small circle of radius r and center z_0, then Eq. (2-32) gives

$$|a_n| < \frac{M}{r^n}$$

and, letting $r \to 0$, we see that a_n must vanish for all negative values of n. Consequently each of the three conditions, (1) boundedness near z_0, (2) no principal part at z_0, and (3) removable singularity at z_0, implies the other two.

Case 2. *Principal part has finite number of terms.* Let the principal part of $f(z)$ at z_0 end with the term $a_{-m}(z - z_0)^{-m}$ where m is a positive integer; then $f(z)$ is said to have a *pole of order m* at z_0. To avoid case 1, m must be $\geqq 1$. If $m = 1, 2, \ldots,$ the pole is said to be *simple, double,* . . . . If $f(z)$ has an mth-order pole at z_0, then $(z - z_0)^m f(z)$ is analytic at z_0; conversely, if $(z - z_0)^m f(z)$ is analytic at z_0, $f(z)$ cannot have a singularity worse than an mth-order pole at z_0. If $g(z)$ is analytic at z_0, and $g(z)$ has a zero of order m at z_0, then $f(z) = 1/g(z)$ has a pole of order m at z_0.

Let $f(z)$ have a pole of order m at z_0. Then since $f(z)$ can be written as

$$\begin{aligned} f(z) &= \frac{1}{(z - z_0)^m} [a_{-m} + a_{-m+1}(z - z_0) + \cdots + a_0(z - z_0)^m \\ &\qquad\qquad + a_1(z - z_0)^{m+1} + \cdots] \\ &= \frac{1}{(z - z_0)^m} h(z) \end{aligned}$$

where $h(z)$ is analytic at z_0, it follows that $|f(z)| \to \infty$ as $z \to z_0$. In fact, given any M, a neighborhood of z_0 can be found such that $|f(z)| > M$ for all z in that neighborhood.

Case 3. *Principal part has infinite number of terms.* In this case, $f(z)$ is said to have an *isolated essential singularity* at z_0. There is a remarkable—but easily proved—theorem due to Weierstrass which says that, in every neighborhood (no matter how small) of such an essential singularity, $f(z)$ comes arbitrarily close to every possible complex value. To prove this, let us show that the contrary supposition leads to a contradiction.

If $f(z)$ has an isolated essential singularity at z_0 but if there is some circular neighborhood of z_0 within which $f(z)$ does *not* come arbitrarily close to attaining some particular complex value w, then $f(z) - w$ is bounded away from zero in this neighborhood so that

$$g(z) = \frac{1}{f(z) - w}$$

is bounded. Consequently $g(z)$ has at worst a removable singularity at z_0 and can be written (at least for $z \neq z_0$) as

$$g(z) = a_0 + a_1(z - z_0) + a_2(z - z_0)^2 + \cdots$$

If $a_0 \neq 0$, then $f(z) \to w + (1/a_0)$ as $z \to z_0$, so that $f(z)$ itself is bounded near z_0, which contradicts the existence of a principal part at z_0. If on the other hand $a_0 = 0$, then the series for $g(z)$ has some first nonvanishing term, say $a_m(z - z_0)^m$, and

$$f(z) = w + \frac{1}{a_m(z - z_0)^m + a_{m+1}(z - z_0)^{m+1} + \cdots}$$

which clearly has only a pole at z_0, again giving a contradiction. This completes the proof.

It would be surprising if $f(z)$ could come arbitrarily close to every possible value without actually attaining all, or almost all, possible values; it can in fact be proved (Picard's theorem) that, in any neighborhood of an isolated essential singularity, a single-valued function attains every possible complex value with at most one exception.[1]

Point at Infinity

The stereographic projection of Sec. 1-5 provided a topological interpretation of the "point at infinity" in the complex z plane. Given a function $f(z)$ which is analytic for sufficiently large $|z|$—that is, for points z sufficiently close to the point at ∞—it will be convenient to be able to describe the behavior of $f(z)$ at ∞ in terms of concepts already introduced, such as pole, branch point, etc. To do this, we first carry out the substitution $w = 1/z$, which maps the point at ∞ into the origin of the w plane; next we define $g(w) = f(1/w)$. Then, by definition, the "behavior of $f(z)$ at ∞" is taken to be that of $g(w)$ at the origin. Thus $1/z$ is analytic at ∞, z has a simple pole at ∞, e^z has an isolated essential singularity at ∞, and $z^{1/2}$ has a branch point at ∞.

Let $f(z)$ be single-valued and analytic for sufficiently large (finite) values of z; then it may be expanded in a Laurent series valid outside some circle,

[1] For a proof, see E. C. Titchmarsh, "The Theory of Functions," 2d ed., p. 283, Oxford University Press, New York, 1939.

and the following statements are now seen to follow from results already proved:

1. If $f(z)$ is bounded near ∞ (equivalently, if there are no positive powers $\geqq 1$ in its Laurent series), then it is analytic at ∞—on the assumption that we assign the obvious value to $f(z)$ at the point ∞ itself.

2. If the positive-powers portion of its Laurent series terminates with z^m, then $f(z)$ has a pole of order m at ∞. Given any positive number M, a sufficiently large circle around the origin can be found so that outside this circle $|f(z)| > M$.

3. If there are an infinite number of positive powers in its Laurent series, $f(z)$ has an essential singularity at ∞. Outside every circle, $f(z)$ comes arbitrarily close to every complex value.

Either the Taylor series for an entire function may terminate, in which case the function is a polynomial, or it may contain an infinite number of terms, in which case the function is said to be *transcendental*. These two kinds of entire functions provide special cases of (2) and (3) above, respectively.

Liouville's theorem shows that a constant is the only function which is analytic everywhere, including the point at ∞. Consider next a function $f(z)$ whose only singularities (including that at $z = \infty$) are poles. Then there must be only a finite number of such poles, for otherwise a limit point of poles would exist (perhaps at ∞), but by Exercise 4 of this section this limit point cannot be a pole. Let the finite poles occur at $z_1, \ldots, z_n$; let their multiplicities be $\alpha_1, \ldots, \alpha_n$, and consider

$$g(z) = f(z)(z - z_1)^{\alpha_1} \cdots (z - z_n)^{\alpha_n}$$

Then $g(z)$ is analytic in the finite plane, and since it has at worst a pole at ∞, it must be a polynomial. Consequently $f(z)$ is the quotient of two polynomials, that is, $f(z)$ is a *rational* function. Thus any function whose only singularities (including that at ∞) are poles is a rational function.

As a general rule in discussing the behavior of analytic functions, statements as to analyticity, singularity, etc., will not be taken to include the point at ∞ unless specific reference to this point is made.

Zeros and Poles

Let $f(z)$ be analytic everywhere inside and on a simple closed curve C, except that (1) there may be a finite number of poles inside C and (2) $f(z)$ does not vanish anywhere on C. Then the integral

$$\frac{1}{2\pi i} \int_C \frac{f'(z)}{f(z)}\, dz$$

can be evaluated by replacing C by a sum of contours surrounding each zero or pole of $f(z)$, since these are the only singularities of the integrand. By use of local Laurent-series expansions, these individual integrals are easily evaluated, and we find that

$$\frac{1}{2\pi i}\int_C \frac{f'(z)}{f(z)}\,dz = N - P \tag{2-34}$$

where N is the number of zeros inside C and P is the number of poles inside C, each zero or pole being counted a number of times equal to its multiplicity.

Since the integrand can also be written as $d/dz\,[\ln f(z)]$, and since

$$\ln f(z) = \ln|f(z)| + i \arg f(z)$$

we see that *N-P is also equal to* $(1/2\pi)$ *times the change in the argument of* $f(z)$ *resulting from a complete traversal of* C. For the special case $P = 0$, this fact provides a useful tool for determining the number of zeros of $f(z)$ within a contour.

Equation (2-34) may also be used to prove a useful theorem due to Rouché. Let $f(z)$ and $g(z)$ be analytic inside and on C, with $|g(z)| < |f(z)|$ on C. Then it follows that $f(z)$ *and* $f(z) + g(z)$ *have the same number of zeros inside* C. For neither $f(z)$ nor $f(z) + g(z)$ has a zero on C, so that Eq. (2-34) applies; its "argument" interpretation, visualized in terms of the complex values $f(z)$ or $f(z) + g(z)$ circling the origin, then gives the result at once.

Inverse of an Analytic Function

Rouché's theorem can be used to give an easy proof for the existence of a local inverse function for any analytic function whose derivative is nonzero. Let $w = f(z)$ be analytic at z_0, with $w_0 = f(z_0)$ and with $f'(z_0) \neq 0$. Then there exists some contour C circling z_0 such that on and within that contour $f(z)$ does not, except at z_0 itself, take on the value w_0. Thus, on C, $|f(z) - w_0|$ has some nonzero lower bound m. Consider now those values of w satisfying $|w - w_0| < m$; it is easily seen that for any such value, say w_1, there is one and only one point z_1 inside C such that $f(z_1) = w_1$. For on C, the functions $f(z) - w_0$ and (the constant) $w_0 - w_1$ satisfy the conditions of Rouché's theorem, so that, inside C, $f(z) - w_1$ has the same number of zeros as $f(z) - w_0$ itself—i.e., exactly one zero, since $f(z) - w_0$ vanishes only at z_0 [and there to the first order since $f'(z_0) \neq 0$].

Associating such a unique z with each value of w in $|w - w_0| < m$ defines a single-valued function $z = g(w)$ for this range of values of w. Since the limit $\Delta w/\Delta z$ exists at z_0 as $\Delta z \to 0$, and is not zero, so does the reciprocal of this limit, so that $g(w)$ is an analytic function of w. Thus, *if* $f'(z_0) \neq 0$, *a local inverse function exists and is single-valued and analytic.* [The same result could be obtained by considering the real and imaginary parts of $f(z)$, u and v, to be a pair of new variables replacing x and y, and using the fact from real-variable theory that an inverse transformation exists if the jacobian determinant does not vanish.]

The above result guarantees the existence (under the stated conditions) of an inverse function $z = g(w)$, analytic for w in some neighborhood of w_0. It is often necessary to determine the coefficients b_n of the power-series expansion,

$$z = g(w) = z_0 + b_1(w - w_0) + b_2(w - w_0)^2 + \cdots$$

One way of doing this is of course to treat the b_n as unknowns and to substitute this series for $z - z_0$ into the (presumed known) power series for $w = f(z)$,

$$w - w_0 = a_1(z - z_0) + a_2(z - z_0)^2 + \cdots$$

Collecting the coefficients of the various powers of $w - w_0$, we obtain a sequence of equations from which the b_n may be determined. It is clear, however, that this method can be tedious; consequently it is useful to observe that an analytical expression for $g(w)$ can be obtained by a modification of Eq. (2-34).

By an argument similar to that used to prove Eq. (2-34), we find easily that, for any given value of w sufficiently close to w_0,

$$g(w) = \frac{1}{2\pi i}\int_C \frac{\zeta f'(\zeta)\,d\zeta}{f(\zeta) - w}$$

[The contour C and the permitted range of values for w are the same as those used in proving the existence of an inverse function; hence $f(\zeta) - w$ vanishes at only one point inside C, and there to the first order.] Differentiation gives

$$g'(w) = \frac{1}{2\pi i}\int_C \frac{\zeta f'(\zeta)\,d\zeta}{[f(\zeta) - w]^2}$$

which may be integrated by parts to give

$$\begin{aligned} g'(w) &= \frac{1}{2\pi i}\int_C \frac{d\zeta}{f(\zeta) - w} \\ &= \frac{1}{2\pi i}\int_C \frac{d\zeta}{f(\zeta) - w_0}\left[1 - \frac{w - w_0}{f(\zeta) - w_0}\right]^{-1} \end{aligned}$$

Expanding and integrating term by term then give the desired coefficients as

$$\begin{aligned} nb_n &= \frac{1}{2\pi i}\int_C \frac{d\zeta}{[f(\zeta) - w_0]^n} \\ &= \frac{1}{(n-1)!}\left(\frac{d^{n-1}}{d\zeta^{n-1}}\left\{\frac{(\zeta - z_0)^n}{[f(\zeta) - w_0]^n}\right\}\right)_{\zeta = z_0} \end{aligned} \tag{2-35}$$

Equation (2-35) is called *Lagrange's formula;* Exercise 12 below provides one example of its utility.

EXERCISES

5. Expand the function $\exp [t(z + z^{-1})]$ in a Laurent series around the origin of the z plane; express the coefficients as simple trigonometric integrals.

6. Obtain a Laurent series for $z^{1/2}\{1 + \sin z\}^{-1}$ in some annular region centered on the point $-\pi/2$. Describe the character of each singularity of this function, including that at the point at ∞.

7. Show that $[\sin (z^2)]^{-1}$ has only one double pole. Discuss the character of the singularities of $[(z^2 + 2)^2(z - i)]^{-1}$, $\operatorname{cotan}^2 z$, and $[(z^2 - 1)^{1/2} + (z + i)]^{-1}$. In the function

$$\arcsin z = -i \ln [iz + (1 - z^2)^{1/2}]$$

there is no value of z for which the argument of the ln function vanishes. Does this imply that the only singularities are the two branch points ± 1, so that a Laurent expansion around the origin is possible for $|z| > 1$?

8. Show that no limit point (including the point at ∞) of poles can be a pole. Discuss the behavior of $\csc (1/z)$ as $z \to 0$.

9. Use an easy method to determine the quadrants in which the roots of $2z^4 + z^3 + 2z^2 + 1 = 0$ lie.

10. Modify Eq. (2-34) so as to hold for

$$\frac{1}{2\pi i} \int_C g(z) \frac{f'(z)}{f(z)}\, dz$$

where $g(z)$ is analytic on and inside C, and where $f(z)$ has a finite number of zeros and poles inside C.

11. Let $f(z)$ be analytic and nonconstant inside a closed contour C, on which $|f(z)| = \text{const.}$ Show that $f(z)$ has at least one zero inside C. Is it true that, if $f(z)$ has n zeros inside this contour, then $f'(z)$ has $n - 1$ zeros inside C?

12. Show that one root of the algebraic equation $z^3 + 3z - w = 0$ is given by $z = \frac{1}{3}wF(\frac{1}{3}, \frac{2}{3}; \frac{3}{2}; -\frac{1}{4}w^2)$ for $|w| < 2$. Here F is the hypergeometric series of Sec. 2-6 [Hille].

2-8 *Analytic Continuation*

In practice, an analytic function is usually defined by means of some mathematical expression—such as a polynomial, an infinite series, or an integral. Ordinarily, there is some region within which the expression is meaningful and does yield an analytic function. Outside this region, the expression may cease to be meaningful, and the question then arises as to whether or not there is any way of extending the definition of the function so that this "extended" function is analytic over a larger region. A simple example is

$$f(z) = \sum_{n=0}^{\infty} z^n$$

which describes an analytic function for $|z| < 1$ but which diverges for $|z| > 1$. However, the function

$$g(z) = \frac{1}{1 - z}$$

is analytic over the whole plane (except at $z = 1$), and it coincides with $f(z)$ inside the unit circle. Such a function $g(z)$, which coincides with a given analytic $f(z)$ over that region for which $f(z)$ is defined but which also is analytic over some extension of that region, is said to be an *analytic continuation* of $f(z)$. It is useful to think of $f(z)$ and $g(z)$ as being one and the same function and to consider that the formula defining $f(z)$ failed to provide the values of $f(z)$ in the extended region because of some defect in the mode of description rather than because of some limitation inherent in $f(z)$ itself.

The method of analytic continuation used in the above example is usually not available, because of the difficulty of replacing a given mathematical expression by an equivalent one possessing a wider range of validity. A more general method is to use a chain of circles within which power-series expansions are computed. Let $f(z)$, as defined, be analytic within a region R (Fig. 2-3). Choose a point z_0 near the boundary of R, and expand $f(z)$

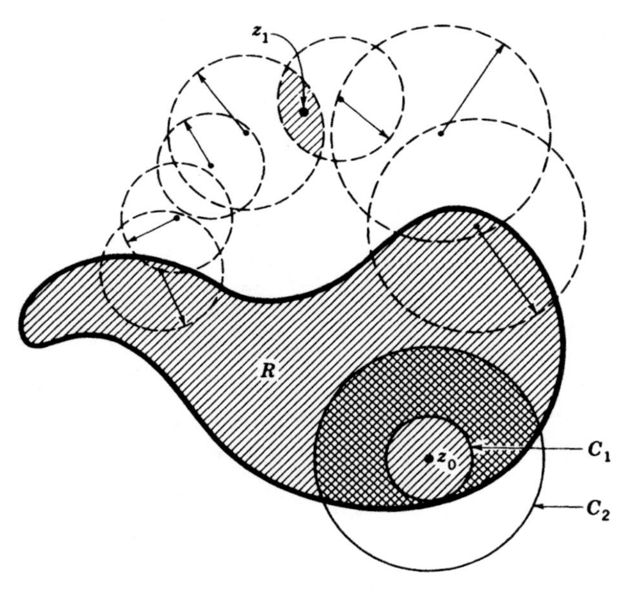

Fig. 2-3 Analytic continuation.

in a Taylor series about z_0. The results of Sec. 2-6 ensure that this series will converge, and will represent $f(z)$, within the largest circle C_1 with center z_0 that lies entirely in R. However, it may well happen that this series, so constructed, actually converges within some larger circle C_2, still with center z_0. The series then necessarily represents an analytic function within C_2, which coincides with $f(z)$ within C_1, and so this method provides a possible mechanism for *continuing* $f(z)$ to a larger region. The method could now be repeated indefinitely, choosing each time appropriate values of z_0 near the boundary of the current region; this is called the *circle-chain method.*[1] However, some immediate questions arise. The procedure gives a new representation for $f(z)$ inside that part of C_2 which lies outside C_1 but inside R (shown crosshatched in Fig. 2-3); does this new representation there agree with the original values of $f(z)$? Also, if z_1 is some point outside R, then there may be two or more possible chains of circles (shown dotted in Fig. 2-3) which start in R and which eventually continue $f(z)$ onto a region containing z_1; will $f(z_1)$ turn out to be the same in each case?

Identity Theorem for Single-valued Analytic Functions

Let each of two functions $f_1(z)$ and $f_2(z)$ be analytic in a common region R. Let $f_1(z)$ and $f_2(z)$ coincide in some portion R' of R (R' may be a subregion, a segment of a curve, or even only an infinite set of points having a limit point in R). *Then* $f_1(z)$ *and* $f_2(z)$ *coincide throughout* R.

For with any point z_0 in R' as center (if R' is an infinite set of points, z_0 must be a limit point of this set), draw the largest circle C contained entirely in R. Each of $f_1(z)$ and $f_2(z)$ can be represented by a Taylor series about z_0, convergent inside this circle, and since the two series give the same values for all points in R', the identity theorem for power series requires the two series to be identical. But then $f_1(z)$ and $f_2(z)$ are identical inside C. Now choose a new point z_0, inside C but close to its boundary, and repeat the procedure; it is clear that all of R may eventually be filled up in this way, so that $f_1(z) \equiv f_2(z)$ within R.

The most important consequence of this identity theorem is that *analytic continuation is locally unique.* More precisely, let $f_1(z)$ be analytic in a region R_1, and let R_2 be another region which has a region R in common with R_1. Then, if a function $f_2(z)$ can be found which is analytic in R_2, and which coincides with $f_1(z)$ in R, it follows at once from the identity theorem that $f_2(z)$ *is unique.*

Singular Point

Let $f(z)$ be analytic at some point z_0, so that it can be expanded in a Taylor series about z_0. Let this series have radius of convergence r. Starting with this circle of radius r, within which $f(z)$ is analytic, it may be possible

[1] A method which, incidentally, is fine in theory but rarely useful in any practical case!

to analytically continue $f(z)$ outside this circle, by choosing new points within the circle to serve as new centers for series expansion. On the other hand, the process may fail for some, or for all, paths of continuation. In particular, let ζ be a point on the circumference of the circle which we cannot include in a new region of analyticity; for example, suppose that we find that no Taylor series centered on a point lying on the line joining z_0 to ζ has a radius of convergence sufficiently large to include ζ. Then we say that ζ is a singular point of $f(z)$.

It is now clear that, if a power series has a finite radius of convergence, then the function $f(z)$ represented by that power series must have a singularity on the circle of convergence—for otherwise $f(z)$ could be analytically continued a finite distance outside the circle everywhere around the circumference; but a function analytic in this larger region would have its Taylor series converge inside a circle enclosing the previous circle. Thus *at least one singularity of a function represented by a power series lies on the boundary of the circle of convergence of that series.*

To return now to the two possible paths of Fig. 2-3, it is also clear that, *if there are no singular points in between the two paths of analytic continuation, then the result of analytic continuation is the same for each path.* For, otherwise, we could cover the in-between region with small overlapping subregions and could use the fact that analytic continuation is locally unique. This result is sometimes called the *monodromy theorem.*

Thus, if we started with one branch of the function $f(z) = z^{1/2}$, defined for only a small part of the plane on one side of the origin, and then continued $f(z)$ analytically by two different paths lying on opposite sides of the origin—here a singular point—we should not expect the two values of $f(z)$ so obtained necessarily to coincide; nevertheless, the process of analytic continuation is still valid, and it gives us in fact all branches of this multiple-valued function. It is just as true for multiple-valued functions as it is for single-valued functions that, starting with a knowledge of $f(z)$ over a small region, analytic continuation can be used to construct $f(z)$ over the entire plane or Riemann surface. This fact, that the total structure of an analytic function is deducible from a small "sample" of the function, is a remarkable illustration of the strength of the concept of analyticity.

Exercises

1. Suppose that the real part u, and its normal derivative u_y, of an analytic function $f(z)$ are known on a part of the real axis. Discuss the following proposed procedure for determining $f(z)$: (*a*) Form the expression $u_x - iu_y$; (*b*) integrate this expression along the real axis, and denote the result by $p(x) + iq(x)$; (*c*) replace x by z to obtain $f(z) = p(z) + iq(z)$.

2. Let two regions R_1 and R_2 be adjacent to one another, with a portion Γ of their boundaries in common. Let $f_1(z)$ be analytic in R_1,

$f_2(z)$ in R_2; let each function be continuous onto Γ, and let $f_1(z) = f_2(z)$ on Γ. Show that the combined function is analytic over the combined region. [*Hint:* Use Morera's theorem.]

3. Let $f(z)$ be analytic in a region of the upper half plane adjacent to a part Γ of the real axis, with $f(z)$ continuous onto Γ and purely real on Γ. Then show that $f(z)$ can be analytically continued onto an image region in the lower half plane, by setting $f(z) = f^*(z^*)$, for points z in the lower half plane. (This is *Schwarz's reflection principle.*) Devise a nontrivial example. Extend this reflection principle so as to apply to an arbitrary curve.

4. Which portions of Sec. 2-8 can be made to apply, in some equivalent form, to harmonic functions? Is schwarzian reflection possible?

2-9 *Entire and Meromorphic Functions*

An *entire* function is one which is analytic everywhere[1] in the plane; a *meromorphic* function is one whose only singularities are poles. These two kinds of function are closely related, since the reciprocal of an entire function is meromorphic; the reciprocal function has a pole wherever the original function has a zero. It is natural to ask whether (1) an entire function can be constructed which has zeros of specified orders at certain assigned points, and only at those points, or whether (2) a meromorphic function can be constructed which has poles of specified characters at certain assigned points, and only at those points. If the number of points is finite, then it is almost obvious that this can indeed be done; if the number of points is infinite, it becomes a matter of some delicacy to construct such functions. For the entire and meromorphic cases, appropriate methods have been devised by Weierstrass and Mittag-Leffler, respectively; a discussion of these methods, and their consequences, will be given in this section.

Let us pause, however, to note that the most general entire function which has no zeros has the form $e^{g(z)}$, where $g(z)$ is an arbitrary entire function. This is easily shown by observing that, if $f(z)$ is entire and if $f(z)$ does not vanish anywhere, then any branch of its logarithm is also entire [one way of making this evident is to expand any branch of $\ln f(z)$ in a power series near the origin, near which it is certainly analytic and single-valued; this series must converge out to the first singularity of the ln function, which is here at ∞ since $f(z)$ is nowhere zero]. An immediate consequence of this result is the fact that, if $g(z)$ is an arbitrary entire function, then $h(z)e^{g(z)}$ is the most general entire function whose zeros correspond in location and order with those of a given entire function $h(z)$.

[1] In accordance with a previously adopted convention, the point at infinity is not included in either of these definitions.

Meromorphic Function with Prescribed Poles

Consider first a finite number of points $z_1, z_2, \ldots, z_n$ in the plane. Let $p_j(z)$, $j = 1, 2, \ldots, n$, be a polynomial in z, with $p_j(0) = 0$. To construct a meromorphic function having poles only at the z_j, and such that at each z_j its principal part is

$$p_j\left(\frac{1}{z - z_j}\right)$$

we merely form the sum

$$\sum_{j=1}^{n} p_j\left(\frac{1}{z - z_j}\right) \tag{2-36}$$

which obviously has the desired property. If the number of points is infinite, the sum (2-36) may still be appropriate. If, for example, the points are $1, -1, 2, -2, 3, -3, \ldots$, in order, and if $p_j(z) = z$ for all j, then the sum becomes

$$\sum_{j=1}^{\infty} \frac{1}{z - z_j} = 2z\left(\frac{1}{z^2 - 1^2} + \frac{1}{z^2 - 2^2} + \frac{1}{z^2 - 3^2} + \cdots\right)$$

which converges uniformly for any bounded region of the plane from which small circles around the poles have been deleted and so represents an analytic function—except at each z_j, where it has the principal part $1/(z - z_j)$. Often, however, the sum (2-36) will not converge as $n \to \infty$, so that some sort of modification of this sum is necessary. *A theorem due to Mittag-Leffler says in effect that such a modification is always possible* and moreover provides a specific recipe for that modification.

First, let the points z_j be numbered in order of nondecreasing distance from the origin, i.e., so that $|z_r| \leqq |z_s|$ if $r < s$. We require that the points have no limit point in the finite plane (for then the desired result is of course impossible) so that $|z_j| \to \infty$ as $j \to \infty$. If the origin is one of the points, label it z_0. For any $j \neq 0$, the function $p_j[1/(z - z_j)]$ is analytic in any circle centered at the origin and of radius $< |z_j|$; in particular, this function may be expanded in a power series in z which converges uniformly in the circle $|z| < \frac{1}{2}|z_j|$. Thus enough terms of this series may be taken so that the sum of these terms, denoted by $g_j(z)$, is within $(\frac{1}{2})^j$ of $p_j[1/(z - z_j)]$ for all z in the circle $|z| < \frac{1}{2}|z_j|$; that is,

$$\left| g_j(z) - p_j\left(\frac{1}{z - z_j}\right) \right| < \frac{1}{2^j}$$

for $|z| < \frac{1}{2}|z_j|$. Consider now the series

$$f(z) = p_0\left(\frac{1}{z}\right) + \sum_{j=1}^{\infty}\left[p_j\left(\frac{1}{z - z_j}\right) - g_j(z)\right] \tag{2-37}$$

For any chosen value of z different from one of the z_j, this series has the sum of the absolute values of its terms bounded and so is absolutely convergent. For z in any bounded set (with small circles around the z_j deleted) it is also uniformly convergent. Thus $f(z)$ is analytic for all z, except at each z_j, where it has the desired principal part; it is therefore the solution of the problem.

If $h(z)$ is any other function satisfying the conditions of the problem, then the difference between $f(z)$ and $h(z)$ is entire, so that the most general meromorphic function having the desired properties is obtained by adding an arbitrary entire function to the function $f(z)$ given by Eq. (2-37).

In any particular case, it may not be necessary to construct the $g_j(z)$ in so elaborate a manner as that used in the general proof of Mittag-Leffler's theorem. For example, the reader may easily show that, if each pole is simple, so that

$$p_j(z) = a_j z$$

and if there exists some integer m such that the series

$$\sum_{j=1}^{\infty} \frac{|a_j|}{|z_j|^m}$$

converges, then all the $g_j(z)$ may be taken to be of degree $m - 2$.

Rather than be given a set of $\{z_j\}$ and be asked to construct a meromorphic function having poles at these z_j, we may be asked to expand a given meromorphic function (for example, $\csc z$) in the form (2-37). Mittag-Leffler's theorem assures us that such an expansion does exist; moreover, the location of the z_j, the polynomials $p_j(z)$, and the convergence functions $g_j(z)$ are usually easy to find. However, this procedure gives a representation for the desired function in which some additive entire function must still be determined, and it may not be so easy to find this entire function. A powerful method is that of contour integration; this method is appropriately postponed to Chap. 3, where a number of such partial-fraction expansions will be obtained.

There is, however, one particular case of general utility. Let C be a circle surrounding all the (assumed *simple*) zeros z_j of an nth-order polynomial $Q(z)$. If $P(z)$ is any other polynomial, and if z is any point inside C not coincident with any of the z_j, then by surrounding each of the z_j with a small circle and using Cauchy's integral theorem, we see easily that

$$\frac{1}{2\pi i}\int_C \frac{P(\zeta)\,d\zeta}{Q(\zeta)(\zeta - z)} = \frac{P(z)}{Q(z)} + \sum_{j=1}^{n} \frac{P(z_j)}{Q'(z_j)(z_j - z)}$$

If now the degree of $P(z)$ is less than n, the left-hand integral vanishes as $C \to \infty$ and we have Lagrange's formula for the partial-fraction decomposition of a rational function,

$$\frac{P(z)}{Q(z)} = \sum_{j=1}^{n} \frac{P(z_j)/Q'(z_j)}{z - z_j}$$

Exercises

1. Obtain a meromorphic function, in the form (2-37), whose poles coincide in location and principal part with those of $\tan z$. Repeat for $\tan^2 z$.

2. To what extent can the Mittag-Leffler theorem be generalized to hold for an infinite number of powers in the prescribed principal parts at the z_j (that is, essential singularities, rather than poles)?

3. In Lagrange's partial-fraction formula, the numerator of each term is seen to equal $\lim_{z \to z_j} [(z - z_j)P(z)/Q(z)]$. Explain why this should be so. Generalize the formula so as to include the possibility of double zeros of $Q(z)$.

Entire Function with Prescribed Zeros

As in Mittag-Leffler's theorem, let $\{z_j\}$ be a sequence of points tending to infinity. With each point, associate a positive integer α_j (reserve $j = 0$ for the origin, if it is included); it is desired to construct an entire function $f(z)$ having a zero of order α_j at each point z_j and vanishing nowhere else. Although the reciprocal of $f(z)$ would be meromorphic, having a pole of order α_j at each z_j, and such a meromorphic function $g(z)$ could be found by use of Mittag-Leffler's theorem, the reciprocal of $g(z)$ would in general not provide the desired $f(z)$ because of the possibility that $g(z)$ so constructed might have zeros. However, we note that $f'(z)/f(z)$ would be meromorphic, with a simple pole at each z_j, and with residue α_j there, and this provides an opportunity to use the Mittag-Leffler theorem.

Let us first construct, then, a function $h(z)$ which is everywhere analytic, except that it has a simple pole with residue α_j at each point z_j. In using the Mittag-Leffler theorem, polynomials $g_j(z)$ for $j \neq 0$ must be chosen from the first part of the Taylor-series expansion of $\alpha_j/(z - z_j)$, that is, from

$$-\frac{\alpha_j}{z_j}\left[1 + \frac{z}{z_j} + \left(\frac{z}{z_j}\right)^2 + \cdots\right]$$

As before, choose the order n_j of each polynomial $g_j(z)$ so that $|g_j(z) - \alpha_j/(z - z_j)| < 2^{-j}$ for $|z| < \frac{1}{2}|z_j|$. Then

$$h(z) = \frac{\alpha_0}{z} + \sum_{j=1}^{\infty} \left\{\frac{\alpha_j}{z - z_j} + \frac{\alpha_j}{z_j}\left[1 + \frac{z}{z_j} + \left(\frac{z}{z_j}\right)^2 + \cdots + \left(\frac{z}{z_j}\right)^{n_j}\right]\right\} \tag{2-38}$$

Choose arbitrarily any real positive number r, and consider the region R formed from $|z| \leqq r$ by excluding small circles centered on any of the points z_j that otherwise would have fallen inside this circle. Let the last of these points inside the circle be z_m. We shall now obtain a form for our desired function $f(z)$ which is valid for the region R; it will turn out that the form is independent of r and so has universal validity.

Break the series (2-38) into two parts, the first containing all values of $j \leqq m$ and the second all remaining terms. Denote this second part by $h_1(z)$. It is uniformly convergent and may be integrated term by term to yield a function $p(z)$ which is analytic within R; within an unimportant arbitrary constant, $p(z)$ is given by the uniformly convergent series

$$p(z) = \sum_{j=m+1}^{\infty} \left\{ \alpha_j \ln\left(1 - \frac{z}{z_j}\right) + \alpha_j \left[\frac{z}{z_j} + \frac{1}{2}\left(\frac{z}{z_j}\right)^2 + \cdots + \frac{1}{n_j + 1}\left(\frac{z}{z_j}\right)^{n_j+1} \right] \right\}$$

where we define the branch of the ln function to be that satisfying $\ln 1 = 0$. The function $e^{p(z)}$ is also analytic and has the form of a uniformly convergent infinite product

$$\prod_{j=m+1}^{\infty} \left\{ \left(1 - \frac{z}{z_j}\right) \exp\left[\frac{z}{z_j} + \cdots + \frac{1}{n_j + 1}\left(\frac{z}{z_j}\right)^{n_j+1} \right] \right\}^{\alpha_j} \tag{2-39}$$

For z in R, we now define a function $f(z)$ by

$$f(z) = z^{\alpha_0} \prod_{j=1}^{\infty} \left\{ \left(1 - \frac{z}{z_j}\right) \exp\left[\frac{z}{z_j} + \cdots + \frac{1}{n_j + 1}\left(\frac{z}{z_j}\right)^{n_j+1} \right] \right\}^{\alpha_j} \tag{2-40}$$

The terms for $j \leqq m$ form a finite product, and we have seen from (2-39) that the remainder of the product converges to an analytic function. Moreover the final form is independent of r and so holds for all z; it also vanishes at each z_j to the order α_j and nowhere else since the meromorphic function $h(z)$ had a pole nowhere else.

The most general solution is now obtained by multiplying Eq. (2-40) by an arbitrary entire function without zeros—i.e., by $e^{g(z)}$, where $g(z)$ is an arbitrary entire function. Thus we have shown that *an entire function which has zeros of specified orders at a set of points z_j (having no limit point other than ∞), and which vanishes nowhere else, does exist;* its most general form is given by multiplying Eq. (2-40) by $e^{g(z)}$. This result is called *Weierstrass's factor theorem.*

In any particular case, it may be possible to omit, or at least simplify, the exponential "convergence-producing" terms in Eq. (2-40). For example, if an entire function is to have first-order zeros at $z = 0, \pm 1, \pm 2, \ldots$

and is to vanish nowhere else, then the infinite product

$$z\left(1 - \frac{z^2}{1^2}\right)\left(1 - \frac{z^2}{2^2}\right) \cdots = z \prod_{j=1}^{\infty} \left(1 - \frac{z^2}{j^2}\right) \tag{2-41}$$

is convergent to an analytic function and so suffices.

It has been mentioned that meromorphic functions can sometimes be obtained in partial-fraction form by contour integration. The logarithmic derivative of an entire function having prescribed zeros is meromorphic, and so contour integration may here also provide a method of expanding the entire function in an infinite product. For example, the function $\sin \pi z$ has first-order zeros at $z = 0, \pm 1, \pm 2, \ldots$ and nowhere else and so must have the form of (2-41), multiplied by $e^{g(z)}$, where $g(z)$ is some entire function. The easiest way of finding this entire function is to realize that

$$\frac{d}{dz}[\ln (\sin \pi z)] = \pi \cot \pi z$$

and to use the fact (shown by contour integration in Chap. 3) that

$$\pi \cot \pi z = \frac{1}{z} + \sum_{j=1}^{\infty} \frac{2z}{z^2 - j^2}$$

Then $\quad \ln (\sin \pi z) = \ln z + \sum_{j=1}^{\infty} \left[\ln\left(1 - \frac{z}{j}\right) + \ln\left(1 + \frac{z}{j}\right)\right] + \text{const}$

so that

$$\sin \pi z = Kz \prod_{j=1}^{\infty} \left(1 - \frac{z^2}{j^2}\right) \tag{2-42}$$

where K is a constant, easily seen to equal π, from the behavior of each side of the equation for small values of z. Equation (2-42) was first obtained by Euler, by very different means; if $z = \frac{1}{2}$, we obtain the historically famous result of Wallis that

$$\frac{\pi}{2} = \frac{2 \cdot 2}{1 \cdot 3} \cdot \frac{4 \cdot 4}{3 \cdot 5} \cdot \frac{6 \cdot 6}{5 \cdot 7} \cdots$$

EXERCISES

4. By considering the product of an appropriate entire function and a meromorphic function, show that an entire function can be constructed which attains prescribed values on any set of points $\{z_j\}$ tending to ∞.

5. Show that the most general function having first-order zeros at $z = 0, -1, -2, \ldots$ and vanishing nowhere else is

$$ze^{q(z)} \prod_{j=1}^{\infty} \left[\left(1 + \frac{z}{j}\right) e^{-z/j}\right] \tag{2-43}$$

where $q(z)$ is an arbitrary entire function.

6. Let w and w' be any two nonzero complex numbers, whose ratio is not real. Show that, within the usual function $e^{q(z)}$, an entire function having first-order zeros at all lattice points $\alpha w + \beta w'$, where (α,β) is any pair of positive or negative real integers, is given by

$$z \prod_{\alpha,\beta} \left\{\left(1 - \frac{z}{\alpha w + \beta w'}\right) \exp\left[\frac{z}{\alpha w + \beta w'} + \frac{1}{2}\left(\frac{z}{\alpha w + \beta w'}\right)^2\right]\right\}$$

where the infinite product is taken over all possible nonzero integral values of α and β. This function is called *Weierstrass's sigma function;* it plays a role in the theory of elliptic functions.

2-10 Results Concerning the Modulus of $f(z)$

There are a number of formulas relating the modulus of $f(z)$ on a contour—such as a circle, half-plane, or sector and the locations of its zeros and poles. We shall here give only some representative results and refer to the literature[1] for a more complete treatment.

We have remarked previously that $\ln |f(z)|$ is harmonic in any region in which $f(z)$ does not vanish. In particular, let $f(z)$ be analytic and nonzero for $|z| \leqq R$; then the mean-value theorem requires that

$$\ln |f(0)| = \frac{1}{2\pi} \int_0^{2\pi} \ln |f(Re^{i\theta})| \, d\theta \tag{2-44}$$

Jensen's formula provides a generalization of Eq. (2-44) to the case in which $f(z)$ has zeros inside $|z| \leqq R$.

Let $f(z)$ be analytic for $|z| \leqq R$, with $f(0) \neq 0$, and with its zeros inside the circle $|z| = R$ located at points $z_1, z_2, \ldots, z_n$, where the notation is chosen so that the moduli of these points, $a_1, a_2, \ldots, a_n$, are in increasing order. (A double zero will require two identical z_j's in the sequence, and similarly for higher-order zeros.) Then

$$g(z) = \frac{f(z)}{(z - z_1)(z - z_2) \cdots (z - z_n)}$$

is analytic and nonzero for $|z| \leqq R$, so that the mean-value theorem for $\ln |g(z)|$ gives

$$\begin{aligned}\ln |f(0)| &- (\ln a_1 + \ln a_2 + \cdots + \ln a_n) \\ &= \frac{1}{2\pi} \int_0^{2\pi} [\ln |f(Re^{i\theta})| - (\ln |Re^{i\theta} - z_1| + \cdots + \ln |Re^{i\theta} - z_n|)] \, d\theta\end{aligned}$$

[1] See, for example, M. L. Cartwright, "Integral Functions," Cambridge University Press, New York, 1956, or R. P. Boas, "Entire Functions," Academic Press Inc., 1954.

The reader may show that[1]

$$\frac{1}{2\pi}\int_0^{2\pi} \ln |Re^{i\theta} - z_j|\, d\theta$$

is independent of z_j and hence (set $z_j = 0$) has the value $\ln R$. It follows that

$$\ln |f(0)| + \ln \frac{R^n}{a_1a_2 \cdots a_n} = \frac{1}{2\pi}\int_0^{2\pi} \ln |f(Re^{i\theta})|\, d\theta \tag{2-45}$$

which is *Jensen's formula.* The second term on the left-hand side is necessarily positive, so that the theorem tells us that the average value of $\ln |f|$ on the circumference of the circle $|z| = R$ is increased by the presence of zeros of $f(z)$ within the circle; the closer these zeros are to the origin, the more this average value is increased. Also, the larger the number of zeros within the circle $|z| = R$, the more rapidly the average value of $\ln |f|$ on the circumference of the circle must grow as R increases.

The second term on the left-hand side can also be written

$$\ln \frac{R^n}{a_1a_2 \cdots a_n} = \int_0^R \frac{n(t)}{t}\, dt \tag{2-46}$$

where $n(t)$ denotes the number of zeros of $f(z)$ for $|z| \leqq t$; again $f(0) \neq 0$, so that $n(0) = 0$ and the integral in Eq. (2-46) converges.

An analogous argument may be used to extend the theorem to the case in which $f(z)$ has poles at points inside $|z| = R$, as well as the previous zeros. Let the moduli of the locations of these poles, arranged in increasing order, be $b_1, b_2, \ldots, b_m$. Then the reader will have no difficulty in showing that

$$\ln |f(0)| + \ln \left(\frac{b_1b_2 \cdots b_m}{a_1a_2 \cdots a_n} R^{n-m}\right) = \frac{1}{2\pi}\int_0^{2\pi} \ln |f(Re^{i\theta})|\, d\theta \tag{2-47}$$

(we require the point $z = 0$ to be neither a zero nor a pole). Again, a multiple pole will occur in the b_j sequence a number of times equal to its order.

Exercises

1. If $f(z)$ is analytic and nonzero for $|z| \leqq R$, then the value of $\ln |f(z)|$ at any point $z = re^{i\alpha}$, $r < R$, may be expressed in terms of the boundary values of $\ln |f(z)|$ by means of Poisson's formula. Show that if, however, $f(z)$ has zeros at $a_1, a_2, \ldots, a_n$ and poles at $b_1, b_2, \ldots, b_m$, then the appropriate generalization of Poisson's formula

[1] One method is to differentiate with respect to r_j and to use Poisson's formula [cf. Exercise 3 of Sec. 2-5 and Eq. (2-24); set $u \equiv 1$] to show that the result is zero.

is given by the *Poisson-Jensen formula,*

$$\ln |f(re^{i\alpha})| = \frac{1}{2\pi} \int_0^{2\pi} \frac{(R^2 - r^2) \ln |f(Re^{i\theta})| \, d\theta}{R^2 - 2Rr \cos (\theta - \alpha) + r^2} - \sum_{j=1}^{n} \ln \left| \frac{R^2 - a_j^* re^{i\alpha}}{R(re^{i\alpha} - a_j)} \right| + \sum_{j=1}^{m} \ln \left| \frac{R^2 - b_j^* re^{i\alpha}}{R(re^{i\alpha} - b_j)} \right| \qquad (2\text{-}48)$$

Show also that Jensen's theorem is a special case of Eq. (2-48). [*Hint:* In deriving Eq. (2-48), observe that $|z - a_j| = |a_j^* z/R - R|$ for $z = Re^{i\theta}$; the logarithm of the right-hand term is harmonic inside the circle, so that Poisson's formula may be applied to it.]

2. If $f(z)$ is analytic for $|z| \leqq R$, with $f(0) = 1$, and if $|f(z)| \leqq M(R)$ for $|z| = R$, show that

$$n(r) < \frac{\ln M(r)}{\ln (R/r)}$$

for $0 < r < R$.

Carleman's Formula

Let $f(z)$ be analytic for $\operatorname{Re} z \geqq 0$, $\rho \leqq |z| \leqq R$. Denote the boundary of this region by Γ. Let $f(z)$ vanish at the points $z_1, z_2, \ldots, z_n$ inside Γ, with $z_j = r_j \exp (i\theta_j)$; $f(z)$ is, however, to have no zeros on Γ. Then Carleman's formula provides a relationship between $|f|$ on Γ and the values of the r_j and the θ_j; the fact that the arguments of the z_j are involved is of particular interest.

To derive the formula, consider

$$I = \operatorname{Re} \left[\frac{1}{2\pi i} \int_\Gamma \ln f(z) \left(\frac{1}{z^2} + \frac{1}{R^2} \right) dz \right]$$

where we start the integration at $z = iR$, with some choice for the branch of $\ln f(z)$. We can evaluate I by either (1) considering the contribution from each part of the contour separately or (2) integrating by parts to give

$$I = \operatorname{Re} \left\{ \frac{1}{2\pi i} \left[\ln f(z) \left(-\frac{1}{z} + \frac{z}{R^2} \right) \right]_\Gamma + \frac{1}{2\pi i} \int_\Gamma \frac{f'(z)}{f(z)} \left(\frac{1}{z} - \frac{z}{R^2} \right) dz \right\}$$

[the real part of the first term vanishes, since $\ln f$ alters by a multiple of $2\pi i$ during a traversal of Γ and the second term can be calculated by summing residues; cf. Eq. (2-34)]. Equating the two results, we obtain *Carleman's formula*

$$\sum_{j=1}^{n} \left(\frac{1}{r_j} - \frac{r_j}{R^2} \right) \cos \theta_j = \frac{1}{\pi R} \int_{-\pi/2}^{\pi/2} \ln |f(Re^{i\theta})| \cos \theta \, d\theta + \frac{1}{2\pi} \int_\rho^R \left(\frac{1}{y^2} - \frac{1}{R^2} \right) \ln |f(iy) f(-iy)| \, dy + A \qquad (2\text{-}49)$$

where A is the contribution to I from the semicircle contour, $|z| = \rho$. Observe that A is bounded as R increases.

EXERCISES

3. Let $f(z)$ be analytic and bounded for $\text{Re } z \geqq 0$, with zeros in the right half-plane at points $r_j e^{i\theta_j}$, $j = 1, 2, \ldots$. Then the series $\sum_{j=1}^{\infty} (\cos \theta_j)/r_j$ converges.

4. Let $f(z)$ be analytic for $\text{Re } z \geqq 0$, and let $f(0) = 1$. Obtain a version of Carleman's formula valid for the case $\rho = 0$; show that A becomes $-\frac{1}{2} \text{Re } f'(0)$.

Phragmén-Lindelöf Theorem

Let $f(z)$ be analytic for $\text{Re } z \geqq 0$, and let $|f(iy)| < M$, where M is a given constant. Write $z = re^{i\theta}$, and let $|f(z)| < K \exp(r^\beta)$, $\beta < 1$, uniformly in θ, for some constant K and for a sequence $r = r_j \to \infty$. Then the Phragmén-Lindelöf theorem states that $|f(z)| < M$ for all z in $\text{Re } z \geqq 0$.

To prove the theorem, we need merely define $F(z) = f(z) \exp(-\epsilon z^\gamma)$, where $\epsilon > 0$ is arbitrarily small and where $\beta < \gamma < 1$, and apply the maximum principle to deduce that $|F| < M$ throughout the sector. Letting $\epsilon \to 0$, we obtain the desired result.

By use of the transformation $z = \zeta^\rho$, we can obtain a version of the theorem valid for any sector. In particular, let $f(z)$ be analytic for $|\arg z| \leqq \frac{1}{2}\pi/\alpha$, and with $|f(z)| < M$ on the boundary. Let $|f(z)| < K \exp(r^\beta)$, uniformly in θ, for a sequence $r = r_j \to \infty$. Then either $|f(z)| < M$ throughout the sector, or else $\beta \geqq \alpha$. Thus, the more narrow we make the sector on whose sides $f(z)$ is bounded, the more rapidly we must permit $|f(z)|$ to grow as $z \to \infty$ [if we discard the case in which $f(z)$ is bounded everywhere].

EXERCISE

5. (*a*) If $f(z)$ is analytic and bounded in a sector, and if $f(z) \to a$ as $z \to \infty$ along either of the two bounding rays, then show that $f(z)$ must $\to a$ uniformly in the whole angle.

(*b*) Show also (Montel's theorem) that the hypothesis of (*a*) can be replaced by the requirement that $f(z) \to a$ as $z \to \infty$ along one ray in the interior of the sector.

(*c*) Discuss the case in which $f(z) \to a$ along one bounding ray and $f(z) \to b$ along the other.

(*d*) Show (Carlson) that if $f(z)$ is analytic for $\text{Re } z \geqq 0$, with $|f(z)| < K_1 e^{\gamma r}$, uniformly in θ as $y \to \infty$, and with $|f(iy)| < K_2 e^{-\alpha|y|}$, $\alpha > 0$, as $y \to \infty$, then $f(z) \equiv 0$.

3

contour integration

3-1 Illustrative Examples

Throughout the sciences it is repeatedly necessary to evaluate integrals. Frequently, the use of Cauchy's integral formula (Sec. 2-3) is the most direct and uncomplicated approach to such a task. Skill in the exploitation of this tool is related to a familiarity with a few standard devices, several of which are illustrated concisely in this chapter. In fact, this chapter is largely a collection of problems and solutions chosen to illustrate techniques of continuing value.

Consider first the real integral

$$I = \int_0^\infty \frac{dx}{1 + x^2} \equiv \lim_{R \to \infty} I_R \tag{3-1}$$

where $$I_R = \int_0^R \frac{dx}{1 + x^2}$$

I can be written as

$$I = \frac{1}{2} \int_{-\infty}^{\infty} \frac{dz}{1 + z^2} \tag{3-2}$$

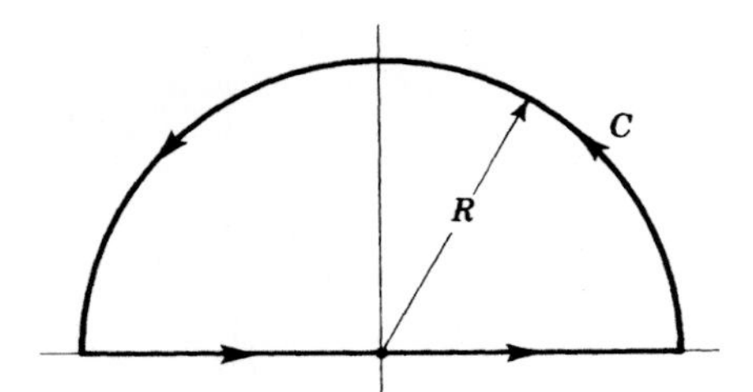

Fig. 3-1 Contour for integral of Eq. (3-3).

where the path is the line $\text{Im } z \equiv y = 0$. In order to use Cauchy's formula, we must consider an integral along a closed contour; we choose to experiment with

$$I_1 = \frac{1}{2} \oint_C \frac{dz}{1 + z^2} \tag{3-3}$$

where C is depicted in Fig. 3-1. We know that

$$I_1 = I_R + I_\theta$$

where I_θ is given by

$$I_\theta = \frac{1}{2} \int_R^{Re^{i\pi}} \frac{dz}{1 + z^2} = \frac{1}{2} \int_0^\pi \frac{iRe^{i\theta}\, d\theta}{1 + R^2 e^{2i\theta}}$$

Clearly, the hope is that I_1 and I_θ are easy to evaluate, especially in the limit $R \to \infty$; otherwise the experiment fails. In fact, of course, Cauchy's integral formula gives $I_1 = \pi/2$ (since the only pole within Γ is at $z = i$ and has residue $-i/2$) and

$$I_\theta \leq \int_0^\pi \frac{R\, d\theta}{2(R^2 - 1)}$$

which tends to zero as $R \to \infty$. Thus

$$I = \frac{\pi}{2}$$

Consider, next, the integral

$$A = \int_0^\infty \frac{dx}{1 + x^3} \tag{3-4}$$

The odd integrand precludes the use of the foregoing contour, but a simple modification salvages the technique. Define

$$A' = \int_C \frac{dz}{1 + z^3} \tag{3-5}$$

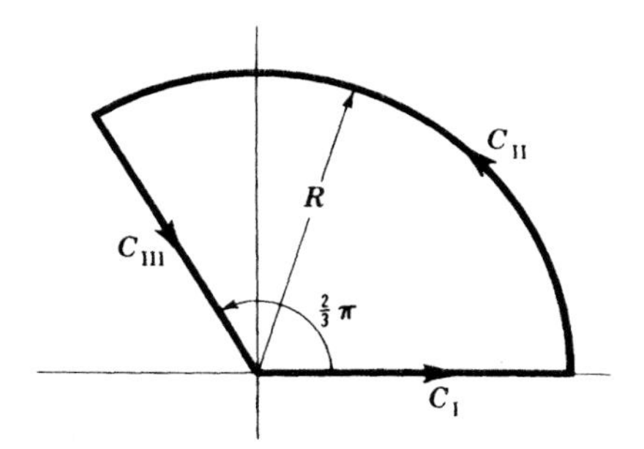

Fig. 3-2 Contour for integral of Eq. (3-5).

where C is depicted in Fig. 3-2. Its choice is dictated by the fact that

$$[1 + (r\, e^{2\pi i/3})^3]^{-1} \equiv (1 + r^3)^{-1}$$

Denote by A_I, A_{II}, A_{III} the contributions to A' associated with each line segment C_I, C_{II}, C_{III} of Fig. 3-2; observe that

$$\lim_{R \to \infty} (A_I + A_{III}) = (1 - e^{2\pi i/3})A \tag{3-6}$$

and

$$A_{II} \leq \int_0^{2\pi/3} \frac{R}{R^3 - 1}\, d\theta$$

The residue of the only pole enclosed by C is $e^{-2\pi i/3}/3$ so that

$$A = \frac{2\pi i e^{-2\pi i/3}}{3(1 - e^{2\pi i/3})} = \frac{2\pi}{3\sqrt{3}} \tag{3-7}$$

A more subtle extension arises when we consider the integral

$$B = \int_0^\infty \frac{dx}{x^2 + 3x + 2} \tag{3-8}$$

In this instance there is no radial line other than $\theta = 0, 2\pi, \ldots$ for which $z^2 + 3z + 2$ is a constant multiple of $r^2 + 3r + 2$, and there is no contour C for which the integral

$$B_1 = \oint_C \frac{dz}{z^2 + 3z + 2}$$

will lead to the desired result. However, the integral

$$B_2 = \int_C \frac{\ln z\, dz}{z^2 + 3z + 2} \tag{3-9}$$

with C as given in Fig. 3-3, will serve our purpose. In fact, we see that, as

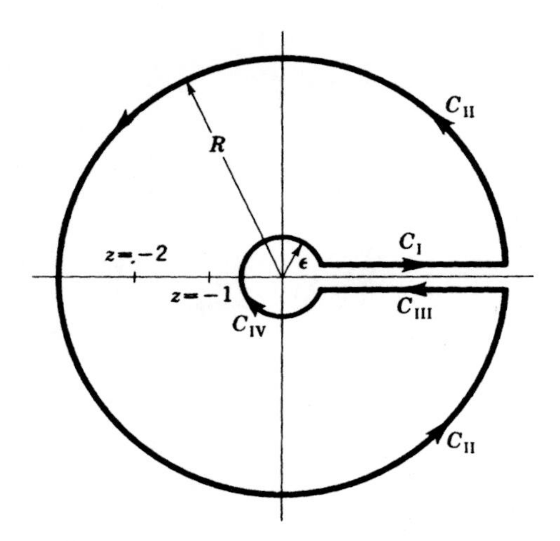

Fig. 3-3 Contour for integral of Eq. (3-9).

$R \to \infty$ and $\epsilon \to 0$, the contributions from C_{II} and C_{IV} will shrink to zero and

$$B_2 = \int_{C_I + C_{III}} \frac{\ln z \, dz}{z^2 + 3z + 2}$$
$$= -2\pi i B \tag{3-10}$$

Alternatively, Cauchy's integral formula leads to

$$B_2 = 2\pi i \left(\frac{\ln 2e^{i\pi}}{-1} + \frac{\ln e^{i\pi}}{1} \right) = -2\pi i \ln 2$$

Thus, $B = \ln 2$.

The foregoing results could have been obtained without the use of complex-variable methods, but one can anticipate the utility of the device in more intricate problems. The reader should note that the method used to evaluate B could have been used to obtain Eq. (3-7), but he should also note that, when it works at all, the method we did use requires less labor. We also note that we might have used $z^\alpha/(z^2 + 3z + 2)$ as the integrand of (3-9). The reader should verify that (3-10) would have emerged after a comparable effort (including a limit process in which $\alpha \to 0$) had been expended.

An integral which requires a similar approach is

$$D(a) = \int_0^\infty \frac{\cos ax}{1 + x^2} \, dx \tag{3-11}$$

where a is real and positive. It is clear that the device of the first problem (wherein x is replaced by z and the contour becomes the real axis plus a large semicircle) fails because the integral along the semicircle of radius R does not vanish as $R \to \infty$. However, when we recall that $\cos ax = \operatorname{Re} e^{iaz}$ on $z = x + i0$, we ask whether the integral

$$D^{(1)} = \frac{1}{2} \oint_C \frac{e^{iaz}}{1 + z^2}\, dz \tag{3-12}$$

with the contour C of Fig. 3-1 may serve our purpose. Expressed as the sum of the contributions from C_1 and C_2, $D^{(1)}$ becomes

$$D^{(1)} = D_R(a) + D_\theta$$

where $$D_\theta = \frac{1}{2} \int_0^\pi e^{-aR \sin \theta + iaR \cos \theta}(1 + R^2 e^{2i\theta})^{-1} Rie^{i\theta}\, d\theta$$

Thus, D_θ vanishes as $R \to \infty$, and

$$D(a) = \lim_{R \to \infty} D_R(a) = D^{(1)} = \frac{\pi}{2} e^{-a} \tag{3-13}$$

The reader should now complete the description of $D(a)$ for *all* real a. He should also calculate $D_{nm}(a)$, for all real a, where

$$D_{nm}(a) = \int_0^\infty \frac{\cos ax}{(1 + x^{2m})^n}\, dx \tag{3-14}$$

Jordan's Lemma

A frequently encountered group of integrals, of which (3-12) and (3-14) are examples, is typified by

$$E = \int_\Gamma e^{iaz} g(z)\, dz \tag{3-15}$$

where $g(z)$ is single-valued, a is real, and the path Γ coincides with the real axis except for indentations at the singularities of $g(z)$. We take $a > 0$; the case $a < 0$ is similar. In the treatment of such integrals, as illustrated in the foregoing, one frequently is concerned with

$$I_\theta = \int_0^\pi ie^{iaR \cos \theta - aR \sin \theta} g(Re^{i\theta}) Re^{i\theta}\, d\theta$$

and in many such integrals $Rg(Re^{i\theta})$ does not vanish as $R \to \infty$. We note, however, that for $|g(Re^{i\theta})| \le G(R)$,

$$\begin{aligned} |I_\theta| &\le \int_0^\pi e^{-aR \sin \theta} RG(R)\, d\theta \\ &\le 2 \int_0^{\pi/2} RG(R) e^{-2aR\theta/\pi}\, d\theta \\ &= \frac{\pi G(R)}{a} (1 - e^{-aR}) \end{aligned}$$

and $I_\theta \to 0$ as $R \to \infty$ provided only that $G(R) \to 0$ as $R \to \infty$.

This result is *Jordan's lemma;* it is broadly useful, not only in the context given above, but also in connection with integrals for which $g(z)$ is multivalued and for which isolated pieces of the semicircle are under consideration.

EXERCISES

1. Using the contour of Fig. 3-1, show that, if a, b, c are real with $b^2 < 4ac$, then

$$\int_{-\infty}^{\infty} \frac{dx}{ax^2 + bx + c} = \frac{2\pi}{(4ac - b^2)^{1/2}}$$

2. Using the contour of Fig. 3-1, evaluate (using Jordan's lemma where necessary)

$$I = \int_0^{\infty} \frac{x \sin x}{a^2 + x^2} \, dx$$

3. For real $a > 0$, show that

$$\int_{-i\infty}^{i\infty} \frac{a^{z+1}}{z+1} \, dz = \begin{cases} 2\pi i & \text{if } a > 1 \\ 0 & \text{if } 0 < a < 1 \end{cases}$$

Why do the integrals along the semicircular path vanish?

4. Evaluate

$$\int_0^{\infty} \frac{dx}{1 + x^{300}}$$

5. If p and s are positive integers with $s \leq 2p - 2$, show that

$$\int_{-\infty}^{\infty} \frac{x^s \, dx}{1 + x^{2p}} = [1 + (-1)^s] \frac{\pi}{2p \sin [(\pi/2p)(s + 1)]}$$

6. Show that

$$\int_0^{\infty} \frac{dx}{2 + x + x^3} = \frac{1}{8} \ln 2 + \frac{3(\pi - \arctan \sqrt{7})}{4 \sqrt{7}}$$

Obtain a similar result for any *finite* upper limit of integration.

7. Let $f(x)$ be a polynomial, none of whose zeros are real and positive. Find

$$I = \int_0^{\infty} \frac{dx}{f(x)}$$

Consider now the integrals

$$F_1(a) = \int_0^{\infty} \cos (ax^2) \, dx \tag{3-16}$$

$$F_2(a) = \int_0^{\infty} \sin (ax^2) \, dx \tag{3-17}$$

We note that

$$F_1(a) + iF_2(a) \equiv F(a) = \int_0^{\infty} \exp (iax^2) \, dx$$

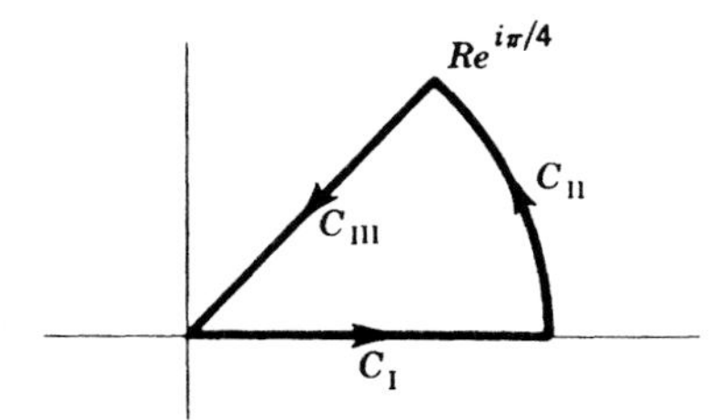

Fig. 3-4 Contour for integral of Eq. (3-18).

and that, by Cauchy's integral theorem,

$$F^{(1)}(a) = \oint_C \exp(iaz^2)\, dz = 0 \tag{3-18}$$

where C is the contour of Fig. 3-4. Using $C = C_{\mathrm{I}} + C_{\mathrm{II}} + C_{\mathrm{III}}$, we obtain

$$F^{(1)}(a) = F_R(a) + I_2 - e^{i\pi/4} \int_0^R \exp(-at^2)\, dt$$

where $\lim_{R\to\infty} F_R(a) = F(a)$ and where I_2 is the contribution from the integration along C_{II}, which, as the reader may verify, vanishes in the limit $R \to \infty$. Since[1]

$$\int_0^\infty \exp(-at^2)\, dt = \frac{1}{2}\sqrt{\frac{\pi}{a}}$$

we have

$$F_1(a) = F_2(a) = \frac{1}{2}\sqrt{\frac{\pi}{2a}} \tag{3-19}$$

The integral (for real b)

$$G(b) = \int_0^\infty \frac{\sin bx}{x}\, dx = \frac{1}{2}\int_{-\infty}^{\infty} \frac{\sin bx}{x}\, dx \tag{3-20}$$

requires a variation on the above procedures. The temptation is to proceed as we did with Eq. (3-11), but we note that the integral

$$G^{(1)}(b) = \frac{1}{2}\int_{-\infty}^{\infty} \frac{e^{ibz}}{z}\, dz$$

has an integrand which is singular on the potentially useful path of integration. Although, in this rudimentary example, it is not necessary to do

[1] An easy way in which to evaluate $I = \int_0^\infty \exp(-ax^2)\, dx$ is to set

$$I^2 = \iint_0^\infty \exp[-a(x^2 + y^2)]\, dx\, dy$$

and then change to polar coordinates.

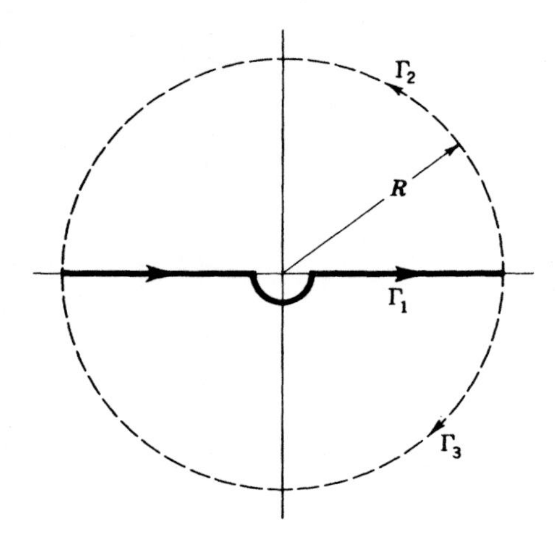

Fig. 3-5 Contour for integral of Eq. (3-21).

so, we can avoid this difficulty by studying the integral

$$G^{(2)}(b) \equiv \frac{1}{2} \int_{-\infty}^{\infty} \frac{e^{ibz} - e^{-ibz}}{2iz} \, dz \tag{3-21}$$

Using Cauchy's integral theorem, we have

$$G^{(2)}(b) = \frac{1}{2} \lim_{R \to \infty} \int_{\Gamma_1} \frac{e^{ibz} - e^{-ibz}}{2iz} \, dz$$

where Γ_1 deviates from the real axis only along the indentation of Fig. 3-5. We also see, using Jordan's lemma, that for $b > 0$,

$$\lim_{R \to \infty} \int_{\Gamma_1} \frac{e^{ibz}}{z} \, dz = \lim_{R \to \infty} \int_{\Gamma_1 + \Gamma_2} \frac{e^{ibz}}{z} \, dz$$

and

$$\lim_{R \to \infty} \int_{\Gamma_1} \frac{e^{-ibz}}{z} \, dz = \lim_{R \to \infty} \int_{\Gamma_1 + \Gamma_3} \frac{e^{-ibz}}{z} \, dz$$

The first of these has value $2\pi i$ by Cauchy's integral formula, and the second vanishes by Cauchy's integral theorem. Thus, for $b > 0$,

$$G(b) = G^{(2)}(b) = \frac{\pi}{2} \tag{3-22}$$

Since $\sin bx$ is odd, we obtain directly

$$G(b) = \begin{cases} \dfrac{\pi}{2} & b > 0 \\ 0 & b = 0 \\ -\dfrac{\pi}{2} & b < 0 \end{cases} \tag{3-23}$$

The foregoing problem is typical of many integrals in which the representation of the full integral as the sum of several pieces must be preceded by a contour modification. Another, somewhat different, example of this kind of integral is

$$H(a) = \int_{-\infty}^{\infty} \frac{e^{ia\xi}\,d\xi}{\sqrt{\xi + i} + \sqrt{\xi + 3i}} \tag{3-24}$$

where a is real. The arguments of $\sqrt{\xi + i}$ and $\sqrt{\xi + 3i}$ lie in the range $(0,\pi)$. Clearly, when $a > 0$, $H(a) = 0$. For $a < 0$, it is desirable to "rationalize the denominator" and write

$$H(a) = \int_{-\infty}^{\infty} \frac{\sqrt{\xi + 3i}\, e^{ia\xi}}{2i}\,d\xi - \int_{-\infty}^{\infty} \frac{\sqrt{\xi + i}\, e^{ia\xi}}{2i}\,d\xi \tag{3-25}$$

but, unfortunately, these individual integrals do not exist. A useful procedure, then, is to consider (still for $a < 0$)

$$H^{(1)}(a) = \oint_{\Gamma} \frac{e^{ia\xi}\,d\xi}{\sqrt{\xi + 3i} + \sqrt{\xi + i}} = 0$$

where Γ is depicted in Fig. 3-6. By Jordan's lemma, the contributions along $|z| = R$ tend to zero as $R \to \infty$ and

$$H(a) + \lim_{R\to\infty} \int_{\Gamma_1(R)} \frac{e^{ia\xi}\,d\xi}{\sqrt{\xi + i} + \sqrt{\xi + 3i}} = 0 \tag{3-26}$$

where $\Gamma_1(R)$ is the keyhole-shaped portion of the path. Thus

$$H(a) = \int_{\Gamma_1(\infty)} \frac{e^{ia\xi}\sqrt{\xi + i}}{2i}\,d\xi - \int_{\Gamma_1(\infty)} \frac{e^{ia\xi}\sqrt{\xi + 3i}}{2i}\,d\xi$$

since each of these integrals does exist.

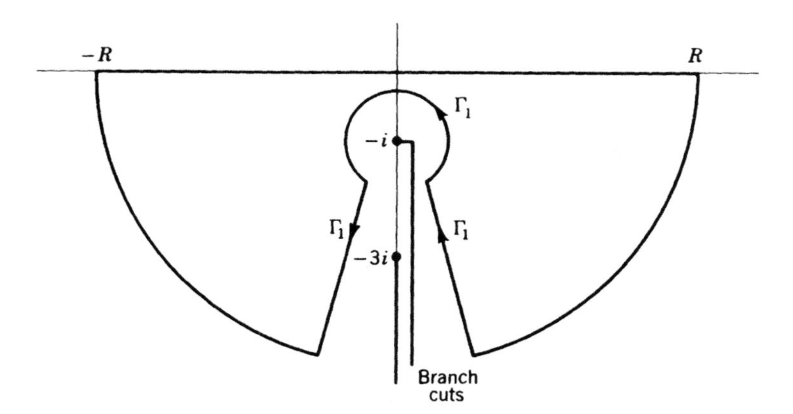

Fig. 3-6 Contour for integral of Eq. (3-24).

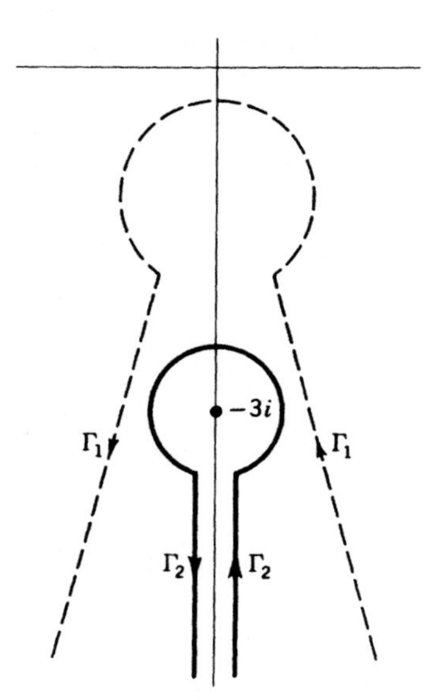

Fig. 3-7 **Modification of Γ_1, in Fig. 3-6.**

The second of these integrals can be simplified by another application of Cauchy's theorem so that

$$H_1(a) = \int_{\Gamma_1(\infty)} \frac{\sqrt{\xi + 3i}\, e^{ia\xi}}{2i}\, d\xi = \int_{\Gamma_2} \frac{\sqrt{\xi + 3i}\, e^{ia\xi}}{2i}\, d\xi \tag{3-27}$$

where Γ_2 is depicted in Fig. 3-7 and where the radius of the little circle is any real positive number ϵ. When we write $\xi = -i(3 + \alpha)$ and let $\epsilon \to 0$,

$$\begin{aligned} H_1(a) &= e^{-i\pi/4} \int_0^\infty \sqrt{\alpha}\, e^{a(\alpha+3)}\, d\alpha \\ &= 2e^{3a-i\pi/4} \frac{\partial}{\partial a} \int_0^\infty \exp\,(as^2)\, ds \\ &= \frac{\sqrt{\pi}}{2} (-a)^{-3/2} e^{3a-i\pi/4} \end{aligned}$$

It follows that

$$H(a) = \frac{\sqrt{\pi}}{2} (-a)^{-3/2} e^{-i\pi/4} (e^a - e^{3a}) \tag{3-28}$$

Integrals whose integrands contain trigonometric functions can often be evaluated by writing $e^{i\theta} = z$ and using the unit circle in the z plane as the path of integration. To illustrate this technique and make it more precise, let

$$I(a,b) = \int_0^{2\pi} \frac{d\theta}{a + b\cos\theta} \tag{3-29}$$

with a, b real and $a > b > 0$. We note that, on the unit circle,

$$\frac{1}{2}\left(z + \frac{1}{z}\right) = \cos\theta$$

and $dz/iz = d\theta$. Thus

$$I(a,b) = \int_\Gamma \frac{dz}{[a + (b/2)(z + z^{-1})]iz} = \frac{2}{ib}\oint_\Gamma \frac{dz}{z^2 + (2a/b)z + 1} \tag{3-30}$$

where Γ is the unit circle. The integrand has poles at

$$z_1 = -\frac{a}{b} + \left[\left(\frac{a}{b}\right)^2 - 1\right]^{1/2} \qquad z_2 = -\frac{a}{b} - \left[\left(\frac{a}{b}\right)^2 - 1\right]^{1/2}$$

and, of these, only z_1 is inside the unit circle; computing the residue there, we obtain

$$I(a,b) = \frac{2\pi}{\sqrt{a^2 - b^2}} \tag{3-31}$$

This procedure requires only minor modification when we evaluate

$$J = \int_0^\alpha \frac{d\theta}{a + b\cos\theta} \tag{3-32}$$

where the real number α lies in $(0,2\pi)$. We note that

$$J = \frac{2}{ib}\int_C \frac{dz}{z^2 + 2(a/b)z + 1}$$

where C is that arc of the unit circle from $z = 1$ to $z = e^{i\alpha}$. Defining z_1 and z_2 as before, we have

$$\begin{aligned} J &= \frac{1}{i\sqrt{a^2 - b^2}}\int_C \left(\frac{1}{z - z_1} - \frac{1}{z - z_2}\right) dz \\ &= \frac{1}{i\sqrt{a^2 - b^2}}[\ln(z - z_1) - \ln(z - z_2)]_1^{e^{i\alpha}} \end{aligned}$$

Since J must be real, we need evaluate only the imaginary parts of the logarithms (the real parts must cancel); thus, we obtain

$$J = \frac{1}{\sqrt{a^2 - b^2}}\arctan\frac{\sqrt{a^2 - b^2}\sin\alpha}{b + a\cos\alpha} \tag{3-33}$$

For $\alpha = 2\pi$, Eq. (3-31) can be recovered.

An integral requiring a different technique is

$$K(\lambda,a) = \int_0^\infty e^{i\lambda t}J_0(\alpha t)\,dt \tag{3-34}$$

with α and λ real. Here $J_0(z)$ is the Bessel function of order zero for which a familiar integral representation (cf. Chap. 5) is

$$J_0(z) = \frac{1}{2\pi}\int_0^{2\pi} e^{iz\sin\theta}\,d\theta \tag{3-35}$$

We can write

$$K(\lambda,\alpha) = \frac{1}{2\pi}\int_0^{\infty} e^{i\lambda t}\,dt \int_0^{2\pi} e^{i\alpha t\sin\theta}\,d\theta \tag{3-36}$$

Temporarily, we replace λ by $\lambda + i\epsilon$, $\epsilon > 0$; we can then invert the order of integration[1] to obtain

$$K(\lambda + i\epsilon, \alpha) = \frac{-1}{2\pi}\int_0^{2\pi} \frac{d\theta}{i(\lambda + i\epsilon + \alpha\sin\theta)} \tag{3-37}$$

The problem is now essentially that of Eq. (3-29). Using residues, and making $\epsilon \to 0$, we obtain

$$K(\lambda,\alpha) = i(\lambda^2 - \alpha^2)^{-\frac{1}{2}} \tag{3-38}$$

where, as a function of λ for given α, the branch lines are drawn downward from α, $-\alpha$, and where, for real λ, α and $\lambda > \alpha > 0$, K is a "positive imaginary" number.

Another variation of the use of Cauchy's integral formula is exhibited in the treatment of the integral

$$L(a) = \int_{-\infty}^{\infty} \frac{\theta\,d\theta}{\sinh\theta - a} \tag{3-39}$$

where a is a complex number with $\operatorname{Im} a \neq 0$. The singularities of the integrand lie along two vertical lines in the θ plane at the points

$$\theta_j = p_0 + iq_0 + 2n\pi i$$

and

$$\theta_j' = -p_0 + i(\pi - q_0) + 2n\pi i$$

where $\theta_0 = p_0 + iq_0$ is either root of $\sinh\theta - a = 0$, for which $0 < q_0 < 2\pi$.

We consider first the contour integral

$$L_1(a) = \int_\Gamma \frac{z\,dz}{\sinh z - a} \tag{3-40}$$

[1] The legitimacy of the interchange of order of integration, for $\epsilon > 0$, is assured by the uniform convergence with respect to θ of the infinite-range integral (see E. C. Titchmarsh, "The Theory of Functions," 2d ed., chap. I, Oxford University Press, New York, 1939, or E. T. Whittaker and G. N. Watson, "A Course of Modern Analysis," 4th ed., chap. 4, Cambridge University Press, New York, 1927). In general, throughout the rest of the work, we shall not pause to establish the validity of interchange of order of integration in any particular case; this may be very difficult, and, in cases where the process appears dubious, it is usually more efficient to proceed formally and then to verify the final result by some direct process.

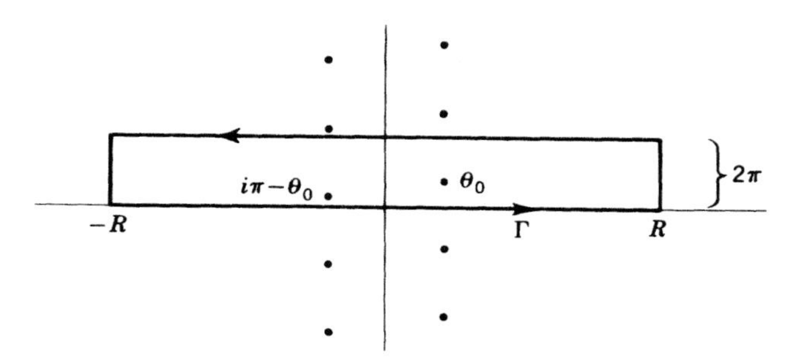

Fig. 3-8 Contour for integral of Eq. (3-40).

where Γ is the contour of Fig. 3-8. As $R \to \infty$, the contributions from the vertical segments vanish and the contributions from the horizontal segments combine to give

$$\int_{-\infty}^{+\infty} \frac{d\theta}{\sinh \theta - a} = -\frac{1}{2\pi i} L_1(a) \tag{3-41}$$

The integral in Eq. (3-41) can thus be evaluated in terms of the residues in Eq. (3-40). Comparing (3-40) and (3-41) (and using some hindsight the authors gained when they had completed the solution), we can anticipate that the integral (3-39) is related to the contour integral

$$L_2(a) = \int_\Gamma \frac{z(z - 2\pi i)}{\sinh z - a}\, dz \tag{3-42}$$

with the same path Γ as in Eq. (3-40). It then follows that

$$L_2(a) = -4\pi i L(a)$$

and so, evaluating the residues at the two relevant poles, we find

$$L(a) = \frac{\pi}{2} \frac{\pi + 2i\theta_0}{\cosh \theta_0} \tag{3-43}$$

EXERCISES

8. Evaluate $\int_0^{2\pi} \ln (a + b \cos \theta)\, d\theta$ for $a > b > 0$.

9. Evaluate

$$\int_0^\infty \frac{d\theta}{\sinh^{1/4} \theta \cosh \theta}$$

10. If α and β are real and positive, show that

$$\int_0^\infty \frac{\cos \alpha x\, dx}{x + \beta} = \int_0^\infty \frac{x e^{-\alpha\beta x}\, dx}{1 + x^2}$$

Which of these two would be easier to compute numerically?

11. Show that

$$\frac{1}{2\pi i}\int_{c-i\infty}^{c+i\infty}\frac{\exp\,[-b(z^2+a^2)^{1/2}+zt]}{(z^2+a^2)^{1/2}}\,dz = J_0[a(t^2-b^2)^{1/2}]$$

Here a and b are real and > 0, and t is real with $t > b$. The path of integration is a vertical line to the right of all singularities. [One possible method is to replace the path of integration with a closed path around $(-ia,ia)$, to evaluate this new integral by first allowing b to be purely imaginary, and then to use analytic continuation.] Use the fact that $J_0(z) = \dfrac{1}{2\pi}\displaystyle\int_0^{2\pi} e^{iz\sin\theta}\,d\theta$.

12. Show that

$$(a)\quad \int_0^{2\pi} e^{\cos\theta}\cos\,(n\theta-\sin\,\theta)\,d\theta = \frac{2\pi}{n!}$$

$$(b)\quad \int_0^{2\pi} e^{\cos\theta}\sin\,(n\theta-\sin\,\theta)\,d\theta = 0$$

13. By use of contour integrals (rather than series expansions), show that, if $-\pi < \alpha < \pi$,

$$(a)\quad \int_0^{\infty}\frac{\cosh\alpha x}{\cosh\pi x}\,dx = \frac{1}{2}\sec\frac{\alpha}{2}$$

$$(b)\quad \int_0^{\infty}\frac{\sin\alpha x}{\sinh\pi x}\,dx = \frac{1}{2}\tanh\frac{\alpha}{2}$$

14. Evaluate

$$\int_0^{\pi}\frac{x\sin x}{1-2\alpha\cos x+\alpha^2}\,dx$$

for α real and for each of the two cases $|\alpha| < 1$, $|\alpha| > 1$.

15. Show that, if α is real and $0 < \alpha < 3$, and if β is real and > 0, then

$$\int_0^{\infty}\frac{x^{\alpha-1}\sin\,(\pi\alpha/2-x)}{x^2+\beta^2}\,dx = \frac{\pi}{2}e^{-\beta}\beta^{\alpha-2}$$

16. Evaluate

$$(a)\quad \int_0^{\infty}\frac{\ln\,(1+x^2)}{1+x^2}\,dx$$

$$(b)\quad \int_0^{\infty}\frac{(\ln x)^2}{1+x^2}\,dx$$

17. Show that

$$\int_0^{\pi}\ln\,(\sin x)\,dx = -\pi\ln 2$$

18. If α and β are arbitrary complex numbers, and if the path of integration is the vertical line $\operatorname{Re} z = c > 0$, express

$$\frac{1}{2\pi i}\int_{c-i\infty}^{c+i\infty}\frac{(z+\alpha)^{1/2}e^{zt}\,dz}{z+\beta}$$

in terms of error functions. The path of integration is to the right of all singularities, and t is a real number.

Note: The error function erf z and the complementary error function erfc z are defined by

$$\operatorname{erf} z = \frac{2}{\sqrt{\pi}}\int_0^z \exp(-y^2)\,dy \qquad \operatorname{erfc} z = \frac{2}{\sqrt{\pi}}\int_z^\infty \exp(-y^2)\,dy$$

19. If the path of integration is the vertical line $\operatorname{Re} z = c > 0$, which is to the right of all singularities of the integrand, show that

$$\int_{c-i\infty}^{c+i\infty}\frac{ze^{zt}\ln[(z^2+a^2)^{1/2}]\,dz}{z^2+a^2}$$

can be expressed in terms of a selection from the following functions:

(*a*) *The exponential integral*

$$Ei(z) = \int_z^\infty \frac{e^{-t}}{t}\,dt$$

(*b*) *The cosine integral*

$$Ci(z) = \int_z^\infty \frac{\cos t}{t}\,dt$$

(*c*) *The sine integral*

$$Si(z) = \int_0^z \frac{\sin t}{t}\,dt$$

In these definitions, z is a complex number, and ∞ is the point at infinity on the positive real axis. Functions (*a*) and (*b*) are multiple-valued, each incrementing by $2\pi i$ for each circuit of the origin. The constant

$$\gamma = \int_0^1 \frac{1-e^{-t}}{t}\,dt - \int_1^\infty \frac{e^{-t}}{t}\,dt$$

is called *Euler's constant,* or the *Euler-Mascheroni constant,* and has the value 0.577215665 . . . . The reader is cautioned that the definitions of the above functions are not consistent throughout the literature.

20. Evaluate

$$\int_0^\infty e^{i\lambda t}J_0(\alpha t)J_0(\beta t)\,dt$$

where α, β, λ are real and positive. (The reader will be encouraged by the statement made in a well-known text that the evaluation of such an integral would be exceedingly complicated.)

We turn now to the integral

$$M(\beta,t,\nu) = \frac{1}{2\pi i}\int_{-\infty}^{\infty} \frac{e^{it\alpha+\beta/(\alpha-i)}}{(\alpha - i)^{\nu}}\,d\alpha \tag{3-44}$$

where t and ν are real and positive, and where β is complex. First, let ν be an integer. Then the integrand has no branch points, and the path of integration may be replaced by a circle of radius ρ surrounding the point $\alpha = i$. We write

$$\alpha = i + \rho e^{i(\theta-\sigma)}$$

where σ is a real constant. The numerator of the integrand is simplified by the choices

$$\rho^2 = \frac{|\beta|}{t} \qquad \sigma = \frac{\pi}{4} - \frac{1}{2}\arg\beta$$

Equation (3-44) becomes

$$M(\beta,t,\nu) = \frac{e^{-t}}{2\pi}\left(\frac{\beta}{it}\right)^{(1-\nu)/2}\int_0^{2\pi}\exp\left[2i\sqrt{\frac{\beta t}{i}}\cos\theta + i(1-\nu)\theta\right]d\theta \tag{3-45}$$

The integral is now in simpler form; as we shall see in Chap. 5, it may in fact be expressed in terms of the Bessel function $J_{\nu-1}$ of order $\nu - 1$,

$$M(\beta,t,\nu) = e^{-t}\left(\frac{\beta}{it}\right)^{(1-\nu)/2}(-i)^{\nu-1}J_{\nu-1}\left(-2\sqrt{\frac{\beta t}{i}}\right) \tag{3-46}$$

Second, let ν be nonintegral. The transformation $\alpha = i + \rho e^{-i\sigma}z$, where ρ and σ are the same numbers as before, leads to

$$M(\beta,t,\nu) = \frac{e^{-t}}{2\pi i}\left(\frac{\beta}{it}\right)^{(1-\nu)/2}\int_\Gamma \exp\left[\sqrt{t\beta i}\left(z + \frac{1}{z}\right)\right]z^{-\nu}\,dz$$

where Γ is the z-plane image of the real axis in the α plane. An examination of the behavior of the exponent of this integrand shows that Γ may be deformed into the path Γ_1 of Fig. 3-9, with a branch cut along the negative half of the real axis. If the circular portion of Γ_1 is taken as the unit circle, and if the straight-line portions of Γ_1 are made to coincide, we obtain

$$\begin{aligned} M(\beta,t,\nu) = \frac{e^{-t}}{2\pi}\left(\frac{\beta}{it}\right)^{(1-\nu)/2}&\left\{\int_0^{2\pi}\exp\left[2i\sqrt{\frac{\beta t}{i}}\cos\theta + i(1-\nu)\theta\right]d\theta\right. \\ &\left. + 2\sin\pi\nu\int_0^{\infty}\exp\left[-2i\sqrt{\frac{\beta t}{i}}\cosh\omega + (1-\nu)\omega\right]d\omega\right\} \end{aligned} \tag{3-47}$$

[which is again expressible in terms of $J_{\nu-1}(-2\sqrt{\beta t/i})$ via Eq. (3-46)].

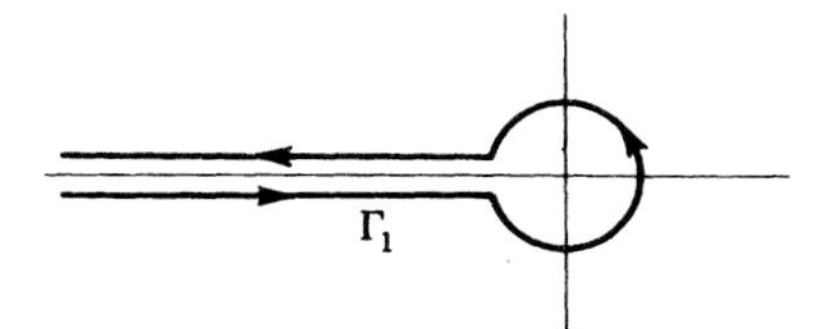

Fig. 3-9 Contour for integral of Eq. (3-47).

Another interesting integral in which changes of variable (conformal mapping) play a role is

$$N(x,y,a,k) = \int_{-\infty}^{\infty} \frac{|y|}{y} \frac{e^{i\xi x - |y|[(\xi+ia)(\xi-ik)]^{1/2}}}{(\xi - ik)^{1/2}} \, d\xi \tag{3-48}$$

where a, k, x, y are real numbers. In particular, a and k are positive, and the branch of the exponents is that for which convergence is assured.

The substitution

$$\zeta = \xi + \frac{i(a-k)}{2} \qquad \text{with } \sigma = \frac{a+k}{2}$$

yields

$$N(x,y,a,k) = e^{x(a-k)/2} \int_{-\infty+i(a-k)/2}^{\infty+i(a-k)/2} \frac{|y|}{y} \frac{\exp\,(i\zeta x - |y|\sqrt{\zeta^2+\sigma^2})}{\sqrt{\zeta - i\sigma}} \, d\zeta \tag{3-49}$$

and the path for this integral can be replaced by one along which $\operatorname{Im}\zeta = 0$. The substitutions $x = r\cos\theta$, $|y| = r\sin\theta$, $\zeta = \sigma\sinh\beta$ transform Eq. (3-49) into

$$N = \frac{\sigma^{1/2}|y|}{y} e^{x(a-k)/2} \int_{-\infty}^{\infty} e^{i\sigma r(\sinh\beta\cos\theta + i\cosh\beta\sin\theta)} \frac{\cosh\beta}{\sqrt{\sinh\beta - i}} \, d\beta \tag{3-50}$$

If we now write $z = \sinh(\beta + i\theta)$ and use Cauchy's integral theorem, we can again recover an integral along the line $\operatorname{Re} z = 0$. Upon noting that $\cosh\beta/\sqrt{\sinh\beta - i} = \sqrt{\sinh\beta + i}$, this integral is

$$N = \sigma^{1/2} \frac{|y|}{y} e^{x(a-k)/2} \int_{-\infty}^{\infty} \frac{e^{i\sigma r z}\sqrt{\sinh\beta + i}}{\cosh(\beta + i\theta)} \, dz \tag{3-51}$$

and since

$$\sqrt{\sinh\beta + i} = \sqrt{-2i}\,\sinh\frac{\beta + i\theta - i\theta + i\pi/2}{2}$$

Eq. (3-51) can be rewritten

$$N = (i\sigma)^{1/2} \frac{|y|}{y} e^{x(a-k)/2} \int_{-\infty}^{\infty} e^{i r\sigma z} \left[\frac{\cos(\theta/2)}{(1+iz)^{1/2}} + \frac{\sin(\theta/2)}{(1-iz)^{1/2}} \right] dz \tag{3-52}$$

The branch cuts run from i to $i\infty$ and from $-i$ to $-i\infty$. The second term in the bracket cannot contribute to the integral (for the contour may be closed in the upper half-plane) and the single remaining integration yields

$$N = 2\frac{|y|}{y}\sqrt{\frac{\pi i}{r}}\,e^{-r(a+k)/2+x(a-k)/2}\cos\frac{\theta}{2} \tag{3-53}$$

A more compact formulation of this result is

$$N = 2\sqrt{\pi i}\,e^{-r(a+k)/2+x(a-k)/2}\,\mathrm{Re}\,(x+iy)^{-1/2} \tag{3-54}$$

where the argument of the complex number $x + iy$ lies in the range $(0,2\pi)$.

EXERCISES

21. Let it be desired to simplify simultaneously two square roots $\sqrt{\xi+\alpha}$ and $\sqrt{\xi+\beta}$ by a trigonometric substitution. Define a new variable ζ by $\xi + \alpha = A(1 + \cos\zeta)$, and show that, if $A = (\alpha - \beta)/2$, then

$$\sqrt{\xi+\alpha} = \sqrt{2A}\cos\frac{\zeta}{2} \qquad \sqrt{\xi+\beta} = \sqrt{-2A}\sin\frac{\zeta}{2}$$

22. Show directly that

$$[z + (z^2 + a^2)^{1/2}]^{1/2} = 2^{-1/2}[(z + ia)^{1/2} + (z - ia)^{1/2}]$$

and interpret this result trigonometrically.

23. Use a change of variable to evaluate (in terms of the erfc function)

$$I = \frac{1}{2\pi i}\int_{c-i\infty}^{c+i\infty}\frac{\exp(zt - az^{1/2})}{z^{3/2}(z^{1/2}+b)}\,dz$$

where the path of integration is a vertical line c units to the right of the imaginary axis, with $c > 0$, and where a and b are real.

24. Evaluate

$$\frac{1}{2\pi i}\int_{c-i\infty}^{c+i\infty}\frac{e^{zt}\ln[(z+\alpha)^{1/2} + (z+\beta)^{1/2}]}{(z+\beta)^{1/2}}\,dz$$

in terms of the exponential integral of Exercise 19. The path of integration is a vertical line to the right of all singularities.

25. Evaluate

$$\frac{1}{2\pi}\int_{-\infty}^{\infty}\frac{e^{ixt}e^{\alpha/(c+ix)}}{(c+ix)^{5/2}}\,dx$$

where c, t are positive real numbers. The path of integration is the real axis.

26. Given below are several functions $F_n(\lambda)$ and paths of integration γ_n. Evaluate, for all real values of x, each of the integrals

$$f_n(x) = \frac{1}{\sqrt{2\pi}} \int_{\gamma_n} e^{-i\lambda x} F_n(\lambda)\, d\lambda$$

[In each case, choose branch lines for $F_n(\lambda)$ which do not cross γ_n.]

For each $f_n(x)$ so obtained, evaluate also

$$G_n(\lambda) = \frac{1}{\sqrt{2\pi}} \int_{-\infty}^{\infty} e^{i\lambda x} f_n(x)\, dx$$

for each value of λ for which this integral exists; compare $G_n(\lambda)$ with $F_n(\lambda)$. In what domain is each $G_n(\lambda)$ analytic?

(*a*) $F_1(\lambda) = \lambda^\alpha$, α constant with $\operatorname{Re} \alpha < 0$. γ_1 is the curve

$$\operatorname{Im} \lambda = -\frac{1}{1 + |\lambda|^2}$$

(*b*) $F_2(\lambda) = \lambda^\alpha$, α constant with $\operatorname{Re} \alpha < 0$. γ_2 is the curve

$$\operatorname{Im} \lambda = \frac{1}{1 + |\lambda|^2}$$

(*c*) $F_3(\lambda) = \dfrac{1}{\sqrt{\lambda}\,(\lambda - \beta)}$, β constant. γ_3 is parallel to the real axis and lies below the points $\lambda = 0$ and $\lambda = \beta$.

(*d*) Same as (*c*), except that γ_3 lies above the points $\lambda = 0$, $\lambda = \beta$.

(*e*) Same as (*c*), except that γ_3 is deformed to pass above $\lambda = 0$ and below $\lambda = \beta$.

(*f*) $F_4(\lambda) = \lambda^{-3/2} \ln \lambda$. γ_4 is parallel to the real axis and lies below $\lambda = 0$.

(*g*) $F_5(\lambda) = \lambda^{-1/2} \exp(-\sqrt{i\lambda})$. γ_5 is parallel to the real axis and lies below $\lambda = 0$.

(*h*) $F_6(\lambda) = \dfrac{\lambda \exp(-\sqrt{i\lambda})}{1 + \beta\sqrt{\lambda}}$, β constant. Choose any γ_6 proceeding from $-\infty$ to ∞.

3-2 *Series and Product Expansions*

It was shown in Sec. 2-9 that, as a generalization of the fact that a rational function can be written in partial-fraction form, any meromorphic function can be expressed in a form which exhibits the principal part at each pole. However, the practical problem of actually obtaining such an expansion for a given meromorphic function is often best attacked by use of contour integration. The method is one which simultaneously demonstrates convergence.

To illustrate the method, we study the integral

$$I_n = \frac{1}{2\pi i} \int_{C_n} \frac{d\xi}{\xi(\xi - z) \sin \xi} \tag{3-55}$$

where the contour of integration is the square with corners at $\xi = (n + \frac{1}{2})(\pm 1 \pm i)\pi$, n being any positive integer. If $\xi = x + iy$, then

$$|\sin \xi|^2 = \sin^2 x + \sinh^2 y$$

so that $|1/\sin \xi| \leq 1$ on C_n; it follows that $I_n \to 0$ as $n \to \infty$. On the other hand, I_n is equal to the sum of the residues of the integrand at all enclosed poles. Letting $n \to \infty$, we obtain

$$0 = \sum_{n=1}^{\infty} \left[\frac{(-1)^n}{n\pi(n\pi - z)} + \frac{(-1)^n}{n\pi(n\pi + z)} \right] + \frac{1}{z \sin z} - \frac{1}{z^2}$$

from which

$$\frac{1}{\sin z} = \frac{1}{z} - 2z \sum_{n=1}^{\infty} \frac{(-1)^n}{n^2\pi^2 - z^2} \tag{3-56}$$

This is the desired decomposition in partial fractions of $1/\sin z$. Note that the method consists in finding contours C_n on which the function of interest, $f(\xi)$ (here, $1/\sin \xi$), is bounded for all n, and then inserting a function of ξ and z into the denominator of the integrand so that we obtain in the integral one term in $1/\sin z$ and other terms in $1/(z - n\pi)$. If $f(\xi)$ were not bounded on contours such as C_n, then we might try $f(\xi)/[\xi^p(\xi - z)]$ in I_n, where p is an appropriate positive integer.

Expansions similar to that of Eq. (3-56) can be obtained for other trigonometric and hyperbolic functions; some of these are listed in Exercise 2 below. For particular values of z, such decompositions can provide useful numerical identities. Thus, setting $z = \pi/2$ in Eq. (3-56), we obtain

$$\frac{\pi}{4} = 1 - \frac{1}{3} + \frac{1}{5} - \frac{1}{7} + \frac{1}{9} \cdots$$

(usually attributed to Leibniz), and setting $z = \alpha\pi$ (with α nonintegral), we have

$$\frac{1}{2\alpha}\left(\frac{\pi}{\sin \alpha\pi} - \frac{1}{\alpha}\right) = \frac{1}{1^2 - \alpha^2} - \frac{1}{2^2 - \alpha^2} + \frac{1}{3^2 - \alpha^2} \cdots$$

For $\alpha \to 0$, we find

$$\frac{\pi^2}{12} = 1 - \frac{1}{4} + \frac{1}{9} - \frac{1}{16} \cdots$$

Again, similar results can be obtained from the formulas of Exercise 2 below.

Contour integrals can also be used directly to sum certain numerical series. In this regard, we observe that $\pi \cot \pi z$ and $\pi \csc \pi z$ have simple poles at the integral points $z = n$, with residues 1 and $(-1)^n$, respectively. For example, to sum the series $\sum_{n=1}^{\infty} (n^4 + \alpha^4)^{-1}$, we study the integral

$$P_q = \frac{1}{2\pi i} \int_{C_q} \frac{\pi \cot \pi z}{z^4 + \alpha^4} dz \tag{3-57}$$

where C_q is again the square contour with corners $(q + \frac{1}{2})(\pm 1 \pm i)$, with $m = 1, \ldots, 4$. The presence of $\cot \pi z$ in the integrand provides a pole at each integer value of z, and the residues are such that

$$\sum_{n=1}^{\infty} \frac{1}{n^4 + \alpha^4} = -\frac{1}{2\alpha^4} + \frac{\pi\sqrt{2}}{4\alpha^3} \frac{\sinh \pi\alpha\sqrt{2} + \sin \pi\alpha\sqrt{2}}{\cosh \pi\alpha\sqrt{2} - \cos \pi\alpha\sqrt{2}} \tag{3-58}$$

For suitable entire functions $f(z)$, the method used to obtain Eq. (3-56) can be applied to the function $f'(z)/f(z)$ to obtain an infinite-product representation of $f(z)$. For example, if $f(z) = \sin z$, then $f'(z)/f(z)$ is $\cot z$, whose partial-fraction representation is given in Eq. (3-62) below. That equation may be rewritten in the form

$$\frac{d}{dz}\left(\ln \frac{\sin z}{z}\right) = \sum_{n=1}^{\infty} \frac{d}{dz} [\ln (z^2 - n^2\pi^2)]$$

and integration from 0 to z, along any path which does not pass through any of the poles, gives

$$\ln \frac{\sin z}{z} - \ln 1 = \sum_{n=1}^{\infty} [\ln (z^2 - n^2\pi^2) - \ln (-n^2\pi^2)]$$

Although the values of the logarithms depend on the path chosen, the multiplicity of possible values disappears when we take exponentials of each side to give

$$\sin z = z \prod_{n=1}^{\infty} \left(1 - \frac{z^2}{n^2\pi^2}\right) \tag{3-59}$$

Other infinite products are given in Exercise 4 below. Again, the choice of particular numerical values for z can lead to interesting results; e.g., setting $z = \pi/2$ in Eq. (3-59) gives Wallis's product

$$\frac{\pi}{2} = \frac{2 \cdot 2 \cdot 4 \cdot 4 \cdot 6 \cdot 6 \cdot 8 \cdot 8 \cdots}{1 \cdot 3 \cdot 3 \cdot 5 \cdot 5 \cdot 7 \cdot 7 \cdot 9 \cdots}$$

Exercises

1. Replace the denominator of the integrand in Eq. (3-55) by $(\xi - z)^2 \sin \xi$ so as to obtain an expansion for $(\cot z)\,(\csc z)$. This method can also be used for some of the expansions in the next exercise.

2. Show that

$$\frac{1}{\cos z} = 2\pi \sum_{n=0}^{\infty} \frac{(-1)^n(n + \frac{1}{2})}{(n + \frac{1}{2})^2\pi^2 - z^2} \tag{3-60}$$

$$\tan z = 2z \sum_{n=0}^{\infty} \frac{1}{(n + \frac{1}{2})^2\pi^2 - z^2} \tag{3-61}$$

$$\cot z = \frac{1}{z} + 2z \sum_{n=1}^{\infty} \frac{1}{z^2 - n^2\pi^2} \tag{3-62}$$

$$\frac{1}{\sin^2 z} = \sum_{n=-\infty}^{\infty} \frac{1}{(z - n\pi)^2} \tag{3-63}$$

$$\frac{1}{\cos^2 z} = \sum_{n=-\infty}^{\infty} \frac{1}{[z - (n + \frac{1}{2})\pi]^2} \tag{3-64}$$

3. Use the results of Exercise 2 to obtain expansions for the corresponding hyperbolic functions.

4. Show that

$$\cos z = \prod_{n=0}^{\infty} \left[1 - \frac{z^2}{(n + \frac{1}{2})^2\pi^2}\right] \tag{3-65}$$

$$\cos z - \sin z = \prod_{n=0}^{\infty} \left[1 - (-1)^n \frac{2z}{(n + \frac{1}{2})\pi}\right]$$

$$e^{\alpha z} - e^{\beta z} = (\alpha - \beta) z e^{\frac{1}{2}(\alpha+\beta)z} \prod_{n=1}^{\infty} \left[1 + \frac{(\alpha - \beta)^2 z^2}{4\pi^2 n^2}\right]$$

5. Let $f(z)$ be an entire function, with the property that

$$\frac{f'(\xi)}{f(\xi)\xi^p(\xi - z)}$$

is suitable for use in the integrand of Eq. (3-55), where p is a positive integer. Assuming that $f(0) \neq 0$, obtain the general form of the infinite-product expansion, and comment on any similarity to that of Weiertrass's factor theorem, as given by Eq. (2-40).

6. In terms of the Bernoulli numbers B_n of Eq. (1-7), show that, for any positive integer p,

$$\sum_{n=1}^{\infty} \frac{1}{n^{2p}} = (-1)^{p+1} \frac{(2\pi)^{2p} B_{2p}}{2(2p)!} \tag{3-66}$$

and obtain from this the value of the sum

$$\sum_{n=0}^{\infty} \frac{1}{(2n+1)^{2p}}$$

Note that Eq. (3-66) shows that the Bernoulli numbers of even index alternate in sign. As $p \to \infty$, we deduce that

$$B_{2p} \sim (-1)^{p+1} \frac{2(2p)!}{(2\pi)^{2p}}$$

7. Show that, if p is a positive integer and if the Euler numbers E_n are defined by Eq. (1-7), then

$$1 - \frac{1}{3^{2p+1}} + \frac{1}{5^{2p+1}} - \frac{1}{7^{2p+1}} + \cdots = \frac{(-1)^p \pi^{2p+1} E_{2p}}{2^{2p+2}(2p)!} \tag{3-67}$$

Again, this shows that the Euler numbers of even index must alternate in sign, and as $p \to \infty$, we have

$$E_{2p} \sim (-1)^p \frac{2^{2p+2}(2p)!}{\pi^{2p+1}}$$

8. Prove that

$$\frac{\pi}{8} = \frac{1}{1 \cdot 3} + \frac{1}{5 \cdot 7} + \frac{1}{9 \cdot 11} + \cdots$$

$$\frac{\pi}{4} - \frac{1}{2} = \frac{1}{1 \cdot 3} - \frac{1}{3 \cdot 5} + \frac{1}{5 \cdot 7} - \frac{1}{7 \cdot 9} + \cdots$$

$$\frac{7\pi^4}{720} = 1 - \frac{1}{2^4} + \frac{1}{3^4} - \frac{1}{4^4} + \cdots$$

$$\frac{\pi(\pi - 2)}{32} = \frac{1}{(2^2 - 1)^2} + \frac{1}{(6^2 - 1)^2} + \frac{1}{(10^2 - 1)^2} + \frac{1}{(14^2 - 1)^2} + \cdots$$

$$\frac{19\pi^7}{56{,}700} = \frac{\coth \pi}{1^7} + \frac{\coth 2\pi}{2^7} + \frac{\coth 3\pi}{3^7} + \cdots$$

9. By integrating $(\pi \cot \pi z)/z^3$ around an appropriate contour in the right half-plane, show that

$$\sum_{n=1}^{\infty} \frac{1}{n^3} = \int_0^{\infty} \frac{(\pi/4) \operatorname{sech}^2 \pi y - y \tanh \pi y}{(\frac{1}{4} + y^2)^2} \, dy$$

10. Show that an integral representation for the Bernoulli numbers is given by

$$B_{2n} = (-1)^{n+1} 4n \int_0^{\infty} \frac{y^{2n-1} \, dy}{e^{2\pi y} - 1} \tag{3-68}$$

11. The *Bernoulli polynomials*[1] $\varphi_n(x)$ are defined by

$$\frac{ze^{xz}}{e^z - 1} = \sum_{n=0}^{\infty} \frac{\varphi_n(x)}{n!} z^n \tag{3-69}$$

from which it follows easily that

$$\varphi_n(x) = \sum_{r=0}^{n} \binom{n}{r} B_{n-r} x^r \tag{3-70}$$

Here the B_r are the previous Bernoulli numbers, and the binomial coefficient $\binom{n}{r}$ is equal to $n![r!(n-r)!]^{-1}$. The first few are

$$\varphi_0(x) = 1 \qquad \varphi_1(x) = x - \tfrac{1}{2} \qquad \varphi_2(x) = x^2 - x + \tfrac{1}{6}$$
$$\varphi_3(x) = x^3 - \tfrac{3}{2}x^2 + \tfrac{1}{2}x \qquad \varphi_4 = x^4 - 2x^3 + x^2 - \tfrac{1}{30}$$
$$\varphi_5(x) = x^5 - \tfrac{5}{2}x^4 + \tfrac{5}{3}x^3 - \tfrac{1}{6}x$$
$$\varphi_6(x) = x^6 - 3x^5 + \tfrac{5}{2}x^4 - \tfrac{1}{2}x^2 + \tfrac{1}{42}$$

Prove the following results:

(*a*) $\varphi_n'(x) = n\varphi_{n-1}(x)$

(*b*) $\varphi_n(x+1) - \varphi_n(x) = nx^{n-1}$

(*c*) $\int_x^{x+1} \varphi_n(t)\, dt = x^n$

(*d*) $\sum_{p=1}^{m} p^n = \dfrac{\varphi_{n+1}(m+1) - B_{n+1}}{n+1} \quad (n \geq 2)$

(*e*) $\varphi_n(1-x) = (-1)^n \varphi_n(x)$

12. Show that, for $0 \leq x \leq 1$,

$$\sum_{p=1}^{\infty} \frac{\cos 2\pi px}{p^{2n+1}} = \frac{(-1)^{n+1}(2\pi)^{2n}\varphi_{2n}(x)}{2(2n)!} \tag{3-71}$$

$$\sum_{p=1}^{\infty} \frac{\sin 2\pi px}{p^{2n+1}} = \frac{(-1)^{n+1}(2\pi)^{2n+1}\varphi_{2n-1}(x)}{2(2n+1)!} \tag{3-72}$$

Express $\varphi_{2n+1}(\tfrac{1}{4})$ in terms of an Euler number.

[1] The *Euler polynomials* $\psi_n(x)$ are defined analogously, by

$$e^{xz} \operatorname{sech} z = \sum_{n=0}^{\infty} \frac{\psi_n(x)}{n!} z^n$$

They satisfy relations very similar to those given for $\varphi_n(x)$ in this and the next exercises; a listing of properties for the B_n, E_n, φ_n, and ψ_n will be found on pages 35 to 44 of Erdélyi, Magnus, Oberhettinger, and Tricomi, "Higher Transcendental Functions," vol. 1, McGraw-Hill Book Company, New York, 1953.

Use Eqs. (3-71) and (3-72) to express

$$\int_0^\infty \frac{t^{2n}\sin 2\pi x\, dt}{\cosh 2\pi t - \cos 2\pi x}$$

(where $0 < x < 1$) in terms of a Bernoulli polynomial. Note the possibility of expressing $(\sin 2\pi x)/(\cosh 2\pi t - \cos 2\pi x)$ as a difference of two tangents.

13. Let

$$I = \int_0^\infty \frac{\sin 2\lambda j}{\lambda} \cos\lambda \left[\frac{1 - 2\alpha \sin^2(\lambda/2)}{1 + 2\alpha\sin^2(\lambda/2)}\right]^n d\lambda$$

where j, n are integers and $\alpha > 0$. Using

$$\int_{-\infty}^{\infty} = \cdots + \int_{-3\pi}^{-\pi} + \int_{-\pi}^{\pi} + \int_{\pi}^{3\pi} + \cdots$$

and the periodicity of the integrand (apart from the factor $1/\lambda$), as well as the expansion formula for the cotangent, obtain a contour integral around the origin for I. Evaluate I for low values of j, n.

3-3 Integral Representations of Functions

Contour-integration methods can often be used to obtain an integral representation of a given function. Such a representation can help not only in deriving some of the properties of the function but also in providing one or more equivalent formal expressions for the function, among which may be selected the one which is advantageous for a particular application. In some cases, such as those in which the desired function is defined only by means of a differential or difference equation, contour integration may in fact provide the only practical method for finding a description of the function in closed form.

Legendre Polynomials

We shall define the Legendre polynomial $P_n(z)$ by *Rodrigues'* formula

$$P_n(z) = \frac{1}{2^n n!}\frac{d^n}{dz^n}[(z^2 - 1)^n] \tag{3-73}$$

It is clear from this formula that $P_n(z)$ is a polynomial of degree n. The first few such polynomials are

$P_0(z) = 1 \qquad P_1(z) = z \qquad P_2(z) = \tfrac{1}{2}(3z^2 - 1) \qquad P_3(z) = \tfrac{1}{2}(5z^3 - 3z)$
$P_4(z) = \tfrac{1}{8}(35z^4 - 30z^2 + 3) \qquad P_5(z) = \tfrac{1}{8}(63z^5 - 70z^3 + 15z)$

Since
$$(z^2 - 1)^n \equiv \frac{1}{2\pi i}\oint \frac{(\xi^2 - 1)^n}{\xi - z}\, d\xi$$

we can construct the contour-integral representation

$$P_n(z) = \frac{1}{2\pi i n!} \frac{d^n}{dz^n} \oint \frac{(t^2 - 1)^n}{2^n(t - z)} \, dt = \frac{1}{2\pi i} \int_C \frac{(t^2 - 1)^n \, dt}{2^n(t - z)^{n+1}} \tag{3-74}$$

where the contour C in the complex t plane circles the point z once in the counterclockwise direction. This result, known as *Schläfli's integral*, can conveniently be used to derive a number of the properties of $P_n(z)$. For example, the value of

$$\int_C \frac{d}{dt} \left[\frac{(t^2 - 1)^{n+1}}{(t - z)^{n+2}} \right] dt$$

is zero, since $(t^2 - 1)^{n+1}/(t - z)^{n+2}$ returns to its original value when t circles z once. Carrying out the differentiation and using Eq. (3-74), we see that $y = P_n(z)$ satisfies *Legendre's equation*[1]

$$(1 - z^2)y'' - 2zy' + n(n + 1)y = 0 \tag{3-75}$$

This proof shows that, even if n is not integral, the right-hand side of Eq. (3-74) still gives a solution of Eq. (3-75), provided that C is a closed contour for which the expression $(t^2 - 1)^{n+1}/(t - z)^{n+2}$ returns to its original value after a complete circuit. One such contour is that which includes each of the points $t = 1$, $t = z$, but not $t = -1$; this approach leads to *Legendre functions of the first kind.*[2] It is interesting that the Schläfli integral expression turns out to have a wider range of validity, insofar as permissible values of n for Legendre's equation are concerned, than has Rodrigues' formula, from which it was derived.

The reader may also use Schläfli's integral to derive a number of recursion relations for the $P_n(z)$, typified by

$$\begin{aligned}
(2n + 1)P_n(z) &= P'_{n+1}(z) - P'_{n-1}(z) \\
(z^2 - 1)P'_n(z) &= nzP_n(z) - nP_{n-1}(z) \\
(n + 1)P_{n+1}(z) - (2n + 1)zP_n(z) &+ nP_{n-1}(z) = 0 \\
nP_n(z) &= zP'_n(z) - P'_{n-1}(z)
\end{aligned}$$

Let us turn now to the construction of a *generating function* for the $P_n(z)$. By this we mean a function $\varphi(\zeta,z)$, such that

$$\varphi(\zeta,z) = P_0(z) + \zeta P_1(z) + \zeta^2 P_2(z) + \zeta^3 P_3(z) + \cdots$$

[1] The reader may be reminded that this equation arises when the method of separation of variables is applied to a rotationally symmetric problem involving the laplacian in spherical polar coordinates. See Sec. 5-5.

[2] See Sec. 5-5.

for small enough $|\zeta|$. Using Eq. (3-74),

$$\begin{aligned}\varphi(\zeta,z) &= \frac{1}{2\pi i}\int_C dt \sum_{n=0}^{\infty} \frac{\zeta^n(t^2-1)^n}{2^n(t-z)^{n+1}} \\ &= \frac{1}{2\pi i}\int_C \frac{dt}{t-z}\left[1 - \frac{\zeta(t^2-1)}{2(t-z)}\right]^{-1} \\ &= \frac{1}{2\pi i}\int_C \frac{dt}{t-z-(\zeta/2)(t^2-1)}\end{aligned}$$

The denominator vanishes for two values of t, but only one of these points is inside C (since otherwise the series $[1 - \zeta(t^2-1)/2(t-z)]^{-1}$ would not converge), and calculation of the residue gives

$$\varphi(\zeta,z) = (1 - 2z\zeta + \zeta^2)^{-\frac{1}{2}}$$

where the principal value is to be taken. We have then, for sufficiently small $|\zeta|$,

$$\frac{1}{\sqrt{1-2z\zeta+\zeta^2}} = P_0(z) + \zeta P_1(z) + \zeta^2 P_2(z) + \cdots \qquad (3\text{-}76)$$

From Eq. (3-76), we see that $P_{2n+1}(0) = 0$, $P_n(1) = 1$, $P_n(-1) = (-1)^n$, and $P_n(-z) = (-1)^n P_n(z)$.

Laplace and Mehler-Dirichlet Integrals

In Eq. (3-74), let the contour C be a circle with center z and radius $|z^2-1|^{\frac{1}{2}}$. Set $t = z + (z^2-1)^{\frac{1}{2}}e^{i\theta}$, where the range of θ is $-\pi$ to π. Then

$$\begin{aligned}P_n(z) &= \frac{1}{2^{n+1}\pi}\int_{-\pi}^{\pi}\left[\frac{z^2-1+2z(z^2-1)^{\frac{1}{2}}e^{i\theta}+(z^2-1)e^{2i\theta}}{(z^2-1)^{\frac{1}{2}}e^{i\theta}}\right]^n d\theta \\ &= \frac{1}{2\pi}\int_{-\pi}^{\pi}[z+(z^2-1)^{\frac{1}{2}}\cos\theta]^n\,d\theta \\ &= \frac{1}{\pi}\int_0^{\pi}[z+(z^2-1)^{\frac{1}{2}}\cos\theta]^n\,d\theta \qquad (3\text{-}77)\end{aligned}$$

which is called *Laplace's first integral* for $P_n(z)$. The choice of branch for $(z^2-1)^{\frac{1}{2}}$ is immaterial. Notice that Eq. (3-77) allows us to write, for arbitrary complex α and β,

$$\int_0^{\pi}(\alpha+\beta\cos\theta)^n\,d\theta = \pi(\alpha^2-\beta^2)^{n/2}P_n[\alpha(\alpha^2-\beta^2)^{-\frac{1}{2}}] \qquad (3\text{-}78)$$

A quite different result is obtained if we start with Eq. (3-76) rather than with Eq. (3-74). If C is a small contour around the origin, not enclosing either of the singularities

$$\zeta_1 = z + (z^2 - 1)^{1/2} \qquad \zeta_2 = z - (z^2 - 1)^{1/2}$$

then
$$P_n(z) = \frac{1}{2\pi i} \int_C \frac{d\zeta}{\zeta^{n+1} \sqrt{1 - 2z\zeta + \zeta^2}} \tag{3-79}$$

where the square root has the value unity for $\zeta = 0$. Since $\zeta_1\zeta_2 = 1$, these two points lie on the same side of the imaginary axis and on opposite sides of the real axis. The point $\zeta = z$ bisects the line joining them. Choose the line joining ζ_1 to ζ_2 as a branch cut for the integrand, and assume that this line does not pass through the origin (it does so only if z lies on the imaginary axis). Since the integral around a very large circle centered on the origin must vanish, the contour C may be replaced by a dumbbell-shaped contour enclosing the branch cut and coinciding with it in the limit. This dumbbell contour must be traversed in the clockwise direction; setting

$$\zeta = z + (z^2 - 1)^{1/2} \cos \theta$$

on the branch cut, the reader may show that

$$P_n(z) = \pm \frac{1}{\pi} \int_0^\pi \frac{d\theta}{[z + (z^2 - 1)^{1/2} \cos \theta]^{n+1}} \tag{3-80}$$

where the $+$ or $-$ sign is to be taken when z is to the right or left, respectively, of the imaginary axis. This result is called *Laplace's second integral.* Again the sign of $(z^2 - 1)^{1/2}$ is immaterial. For z on the imaginary axis, the integral is singular; hence the entire imaginary axis is a *line of singularities* for the integral.

Next, let z be real with $|z| \leq 1$. Set $z = \cos \theta$; then ζ_1 and ζ_2 become $e^{i\varphi}$ and $e^{-i\varphi}$. Let the branch cut now be the arc of the unit circle joining these points. Set $\zeta = e^{i\theta}$; then by a computation similar to that leading to Eq. (3-80) we find

$$P_n (\cos \varphi) = \frac{2}{\pi} \int_0^\varphi \frac{\cos (n + \frac{1}{2})\theta \, d\theta}{\sqrt{2 \cos \theta - 2 \cos \varphi}} \tag{3-81}$$

where $0 < \varphi \leq \pi$. Equivalently, Eq. (3-81) may be written

$$P_n (\cos \varphi) = \frac{2}{\pi} \int_\varphi^\pi \frac{\sin (n + \frac{1}{2})\theta \, d\theta}{\sqrt{2 \cos \varphi - 2 \cos \theta}} \tag{3-82}$$

Equations (3-81) and (3-82) are known as the *Mehler-Dirichlet integrals.*

We shall return to Legendre functions in Chap. 5.

EXERCISES

1. Show that:

(*a*) All zeros of $P_n(z)$ are real and lie in $(-1,1)$.

(*b*) $\int_{-1}^{1} P_m(x)P_n(x)\,dx = \begin{cases} 0 & \text{if } m \neq n \\ \dfrac{2}{2n+1} & \text{if } m = n \end{cases}$

(*c*) $|P_n(\cos\theta)| \le 1$ for real θ

2. Obtain Laplace's second integral, confining your attention to real z with $|z| > 1$, and then using analytic continuation. Obtain Laplace's first integral by starting with Eq. (3-79) and replacing ζ by $1/\zeta$.

3. The *Hermite polynomials* $H_n(z)$ may be defined by

$$H_n(z) = (-1)^n \exp\left(\frac{z^2}{2}\right) \frac{d^n}{dz^n}\left[\exp\left(\frac{-z^2}{2}\right)\right]$$

Find a simple generating function for these polynomials. Find also one for the Tchebycheff polynomials

$$T_n(z) = \frac{(-1)^n 2^n n!}{(2n)!}(1-z^2)^{1/2}\frac{d^n}{dz^n}[(1-z^2)^{n-1/2}]$$

4. Starting with Eq. (3-79), for z real and $|z| < 1$, take branch cuts parallel to the imaginary axis to ∞, and prove that

$$\begin{aligned} P_n(\cos\theta) &= \frac{2}{\pi}\int_0^\infty \operatorname{Im}\left[(\cos\theta - i\sin\theta\cosh\varphi)^{-n-1}\right] d\varphi \\ &= \frac{2}{\pi}\int_0^\infty \frac{\sin[(n+1)\arctan(\tan\theta\cosh\varphi)]\,d\varphi}{(1+\sin^2\theta\sinh^2\varphi)^{(n+1)/2}} \end{aligned} \tag{3-83}$$

Note the special case

$$P_{2n}(0) = (-1)^n \frac{2}{\pi}\int_0^\infty \operatorname{sech}^{2n+1}\varphi\,d\varphi \tag{3-84}$$

Solution of a Difference Equation

In later sections, we shall use contour integration to solve certain differential equations; here we shall take the opportunity of discussing a *difference* equation. Such equations, although not as common as differential equations, are frequently encountered in probability theory; they also arise when numerical methods are used for solving differential equations. We shall consider a difference-equation analog of Bessel's equation,

$$x^2y'' + xy' + (x^2 - \mu^2)y = 0 \tag{3-85}$$

where x, y, μ are real, and where a prime denotes differentiation with respect to x. To solve this equation numerically in the interval $(0,L)$, the interval may be divided into n subintervals of length h, and the mesh-point values

y_j of the dependent variable treated as unknowns. The most natural approximations are

$$y'' \sim \frac{y_{j+1} - 2y_j + y_{j-1}}{h^2} \qquad y' \sim \frac{y_{j+1} - y_{j-1}}{2h}$$

Upon setting $x_j = jh$, Eq. (3-85) may then be replaced by

$$(j^2 + \tfrac{1}{2}j)y_{j+1} + (h^2j^2 - 2j^2 - \mu^2)y_j + (j^2 - \tfrac{1}{2}j)y_{j-1} = 0 \qquad (3\text{-}86)$$

which is to hold for $j = 1, 2, 3, \ldots (n - 1)$. The origin and the point $x = L$ are indexed by 0, n, respectively. Equation (3-86) involves three adjacent unknowns, namely, y_{j+1}, y_j, and y_{j-1}; consequently it is said to be a *second-order difference equation.* As might be expected, such an equation has properties which are closely analogous to the corresponding second-order differential equation; for example, there are exactly two linearly independent solutions.

We shall now assume that we can find a solution for Eq. (3-86) in the form

$$y_j = \int e^{ijt}f(t)\,dt \qquad (3\text{-}87)$$

where the path of integration is to be some contour in the complex t plane; this path of integration and the complex function $f(t)$ are as yet unspecified. The first term of Eq. (3-86) becomes

$$(j^2 + \tfrac{1}{2}j)y_{j+1} = \int[(j + 1)^2 - \tfrac{3}{2}(j + 1) + \tfrac{1}{2}]e^{i(j+1)t}f(t)\,dt$$

and if we denote the end points of the integration contour by A, B, the right-hand side may be integrated by parts to give

$$\begin{aligned}(j^2 + \tfrac{1}{2}j)y_{j+1} = {} & [e^{i(j+1)t}[(-ij + \tfrac{1}{2}i)f(t) + f'(t)]]_A^B \\ & + \int e^{i(j+1)t}[-f''(t) - \tfrac{3}{2}if'(t) + \tfrac{1}{2}f(t)]\,dt \qquad (3\text{-}88)\end{aligned}$$

We shall next anticipate that we can find a function $f(t)$ and a path of integration such that all end contributions arising from this and subsequent integrations by parts vanish. One way of doing this, for example, might be to use a closed contour for which the various functions involved return to their original values after a complete traversal of the contour. Then only the integral remains in Eq. (3-88); upon computing the other terms in Eq. (3-86) similarly, Eq. (3-86) becomes

$$\begin{aligned}\int e^{ijt}[e^{it}(-f'' - \tfrac{3}{2}if' + \tfrac{1}{2}f) + (-h^2f'' + 2f'' - \mu^2f) \\ + e^{-it}(-f'' + \tfrac{3}{2}if' + \tfrac{1}{2}f)]\,dt = 0\end{aligned}$$

This equation will be satisfied for all values of j if $f(t)$ is chosen to satisfy

$$(2 - h^2 - 2\cos t)f'' + 3(\sin t)f' + (\cos t - \mu^2)f = 0 \qquad (3\text{-}89)$$

Consequently we shall try to find a solution of Eq. (3-89) and use that solution as $f(t)$. Equation (3-89) may be simplified by the substitution

$$f(t) = g(t)[(\sin \tfrac{1}{2}t)^2 - (\tfrac{1}{2}h)^2]^{-\frac{1}{2}}$$

which gives

$$4g''[(\sin \tfrac{1}{2}t)^2 - (\tfrac{1}{2}h)^2] + g' \sin t - \mu^2 g = 0 \tag{3-90}$$

(For the time being, we need not concern ourselves with the branch of the complex-valued function $[(\sin \frac{1}{2}t)^2 - (\frac{1}{2}h)^2]^{-\frac{1}{2}}$; any branch will do insofar as finding a solution of Eq. (3-89) is concerned. Later on, when the contour in the t plane is discussed, we shall indeed have to consider the branches of any multiple-valued functions that appear.) Define now a new independent variable $\xi(t)$ by means of

$$\frac{d\xi}{dt} = \tfrac{1}{2}[(\tfrac{1}{2}h)^2 - (\sin \tfrac{1}{2}t)^2]^{-\frac{1}{2}} \tag{3-91}$$

Equation (3-90) then becomes

$$\frac{d^2 g}{d\xi^2} + \mu^2 g = 0$$

so that, finally, one solution of Eq. (3-89) is

$$f(t) = [(\sin \tfrac{1}{2}t)^2 - (\tfrac{1}{2}h)^2]^{-\frac{1}{2}} e^{-i\mu\xi(t)}$$

where $\xi(t)$ is given by Eq. (3-91). We can now substitute back into Eq (3-87) to obtain

$$y_j = \int \exp \{i[x_j\tau - \mu\xi(\tau)]\} \left(\frac{d\xi}{d\tau}\right) d\tau \tag{3-92}$$

where we have written $t = h\tau$, $x_j = jh$. Note that

$$\frac{d\xi}{d\tau} = \left\{1 - \left[\frac{\sin (\tau h/2)}{h/2}\right]^2\right\}^{-\frac{1}{2}} \tag{3-93}$$

It is now necessary to choose an appropriate path of integration in the τ plane.

For $|h| < 2$, the function $d\xi/d\tau$ has branch points only on the real axis, the two closest to the origin being $\pm\alpha$, where

$$\left|\sin \frac{\alpha h}{2}\right| = \tfrac{1}{2}h$$

If the branch cut is taken as the segment of the real axis lying between $-\alpha$ and α, then $d\xi/d\tau$ is single-valued outside this line for any contour which does not circle one of the remaining branch points. The function ξ is multiple-valued; in fact, each time this branch cut is circled counterclockwise, ξ increases by the amount

$$A = 2\int_{-\alpha}^{\alpha} \frac{d\tau}{\left\{1 - \left[\frac{\sin(\tau h/2)}{h/2}\right]^2\right\}^{1/2}}$$

as may be seen by shrinking the contour onto the interval $(-\alpha,\alpha)$.† We have taken the sign of $d\xi/d\tau$ to be positive on the lower part of the contour. With the change in variable

$$\frac{\sin(\tau h/2)}{h/2} = \sin\theta$$

we obtain

$$A = 4\int_0^{\pi/2} \frac{d\theta}{\sqrt{1-(h/2)^2\sin^2\theta}} = 4K\left(\frac{h}{2}\right)$$

where K is the complete elliptic integral of the first kind.

Consider now the τ-plane contour shown in Fig. 3-10. At the two ends of the contour, $\tau = i\infty$, so that the exponential decay factor $\exp(ix_j\tau)$ justifies the assumption concerning vanishing of end contributions in the previous integrations by parts (note that $d\xi/d\tau \to 0$ as $t \to i\infty$). The contour is supposed shrunken onto the interval $(-\alpha,\alpha)$ and the imaginary axis, the circles surrounding $\pm\alpha$ yielding no contribution in the limit.

If the value of ξ on the left-hand side of the imaginary axis is denoted by $\xi_1(\tau)$, then its value on the right-hand side is $\xi_1(\tau) + 4K(h/2)$. Upon setting $\tau = i\varphi$ and $\xi_1 = -i\zeta$, the contribution of the imaginary-axis portion of the contour to Eq. (3-92) may be verified by the reader to be

$$-2\sin 2\mu K\int_0^{\infty} \exp\left[-x_j\varphi - \mu\zeta(\varphi) - 2\mu iK\right]\frac{d\zeta}{d\varphi}\,d\varphi$$

where
$$\frac{d\zeta}{d\varphi} = \left\{1 + \left[\frac{\sinh(\varphi h/2)}{h/2}\right]^2\right\}^{-1/2}$$

† The reader will find it a useful exercise to verify this result.

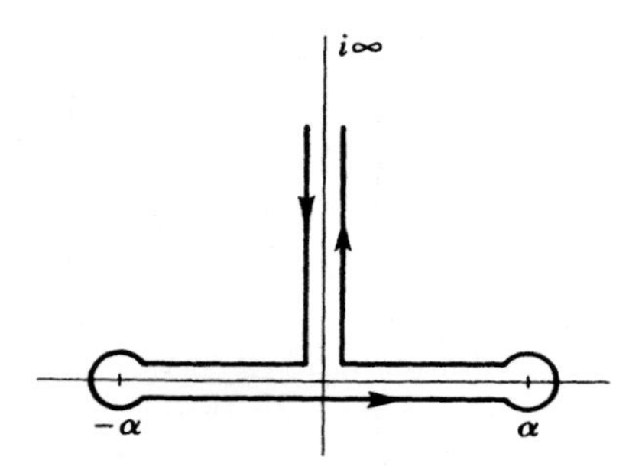

Fig. 3-10 Contour in τ plane.

using the positive square root. Upon writing

$$\sinh \frac{\varphi h}{2} = \tfrac{1}{2}h \sinh \omega \tag{3-94}$$

the contribution to y_j becomes

$$-2 \sin 2\mu K \int_0^\infty \exp\left[-x_j\varphi - \mu\zeta(\omega) - 2\mu i K\right] \frac{d\zeta}{d\omega}\, d\omega \tag{3-95}$$

where

$$\frac{d\zeta}{d\omega} = \left[1 + \left(\frac{h}{2}\right)^2 \sinh^2 \omega\right]^{-\frac{1}{2}} \tag{3-96}$$

The function ζ still contains an arbitrary additive constant; we shall arbitrarily set $\zeta(0) = i\xi_1(0) = -2iK$.

Consider next the contribution of the horizontal-line segments. Let $\xi_2(\tau)$ denote the value of ξ on the lower horizontal segment; then the reader may check that, on the top left segment, $\xi(\tau) = \xi_2(-\tau) - 2K$, and on the top right segment, $\xi(\tau) = \xi_2(-\tau) + 2K$. With these results, the contribution of the horizontal segments is

$$2 \int_0^\pi \cos\left[\mu\rho(\theta) + x_j\tau(\theta)\right] \frac{d\rho}{d\theta}\, d\theta \tag{3-97}$$

where

$$\sin \frac{h\tau}{2} = \frac{h}{2} \sin \theta \tag{3-98}$$

$$\frac{d\rho}{d\theta} = \left[1 - \left(\frac{h}{2}\right)^2 \sin^2 \theta\right]^{-\frac{1}{2}} \qquad \text{with } \rho(0) = 0 \tag{3-99}$$

Adding both contributions and multiplying by $\pi/2$ for future convenience, we write the final solution as

$$\begin{aligned} y(x_j) = {} & \frac{1}{\pi} \int_0^\pi \cos\left[\mu\rho(\theta) - x_j\tau(\theta)\right] \frac{d\rho}{d\theta}\, d\theta \\ & - \frac{1}{\pi} \sin 2\mu K \left(\frac{h}{2}\right) \int_0^\infty \exp\left[-x_j\varphi(\omega) - \mu\zeta(\omega)\right] \frac{d\zeta}{d\omega}\, d\omega \end{aligned} \tag{3-100}$$

where

$$\begin{aligned} \sin \tfrac{1}{2}h\tau &= \tfrac{1}{2}h \sin \theta \qquad \text{with } \tau(0) = 0 \\ \rho &= \int_0^\theta \left[1 - (\tfrac{1}{2}h)^2 \sin^2 \theta\right]^{-\frac{1}{2}} d\theta \\ \sinh \tfrac{1}{2}\varphi h &= \tfrac{1}{2}h \sinh \omega \qquad \text{with } \varphi(0) = 0 \\ \zeta &= \int_0^\omega \left[1 + (\tfrac{1}{2}h)^2 \sinh^2 \omega\right]^{-\frac{1}{2}} d\omega \end{aligned}$$

A second solution is obtained by reversing the sign of μ; unless $2\mu K(h/2)$ is an integral multiple of π, those two solutions are independent.[1] As $h \to 0$, we have $K \to \pi/2$, $\rho \to \theta$, $t \to \sin \theta$, $\varphi \to \sinh \omega$, $\zeta \to \omega$, and both $d\rho/d\theta$ and

[1] See C. Pearson, *J. Math. Phys.*, *39:* 287 (1960).

$d\zeta/d\omega \to 1$; consequently the solution becomes

$$J_\mu = \frac{1}{\pi}\int_0^\pi \cos(\mu\theta - x\sin\theta)\,d\theta - \frac{1}{\pi}\sin\mu\pi\int_0^\infty \exp(-\mu t - x\sinh t)\,dt \tag{3-101}$$

Equation (3-101), due to Schläfli, gives an integral representation for the two solutions J_μ and $J_{-\mu}$ of Bessel's differential equation (3-85); we shall refer to it in Chap. 5.

4
conformal mapping

4-1 Two-dimensional Potential Problems

In the physical sciences, one often encounters boundary-value problems for Laplace's equation in the form

$$\Delta\varphi \equiv \varphi_{xx} + \varphi_{yy} = 0 \qquad \text{in } D \tag{4-1}$$

$$A(t)\varphi + B(t)\frac{\partial\varphi}{\partial n} = C(t) \qquad \text{on } \Gamma \tag{4-2}$$

where D is a simply or multiply connected domain in the xy plane and Γ is the boundary of D; A, B, and C are prescribed functions of a real parameter t that specifies position on Γ, and $\partial\varphi/\partial n$ is the outward normal derivative of $\varphi(x,y)$ on Γ.

If the domain D is not finite, the boundary condition (4-2) may have to be supplemented or replaced by conditions at infinity which are usually of the form

$$\varphi_x \to U \qquad \varphi_y \to V \qquad \text{as } x^2 + y^2 \to \infty \tag{4-3}$$

where U and V are real constants (more often than not, $U = V = 0$).

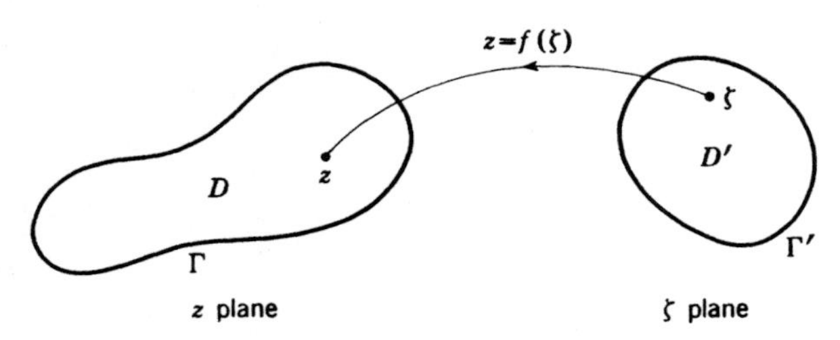

Fig. 4-1 Mapping of D' into D.

Our purpose in this chapter is to exhibit, largely by means of examples, the power of conformal-mapping techniques in treating the above boundary-value problem and some of its extensions. The basic ideas underlying the method are the following (see Chap. 2):

1. If a function $\Omega(z)$ is analytic in a domain D of the z plane ($z = x + iy$), then $\varphi(x,y) \equiv \operatorname{Re}\Omega$ and $\psi(x,y) \equiv \operatorname{Im}\Omega$ are conjugate harmonic functions in D.

2. Conversely, to any function $\varphi(x,y)$ harmonic in a domain D, there corresponds a function $\Omega(z)$ analytic in D such that $\varphi = \operatorname{Re}\Omega$. The conjugate harmonic function $\psi = \operatorname{Im}\Omega$ and hence Ω itself are determined uniquely to within an arbitrary additive constant by the Cauchy-Riemann relations

$$\varphi_x = \psi_y \qquad \varphi_y = -\psi_x \tag{4-4}$$

3. Let $f(\zeta) \equiv f(\xi + i\eta)$ be analytic in a domain D' of the ζ plane, and let the transformation $z = f(\zeta)$ map D' onto the domain D in the z plane and therefore map the boundary Γ' of D' into the boundary Γ of D. If $\Omega(z)$ is analytic in the domain D, then

$$\chi(\zeta) \equiv \Omega(f(\zeta)) = \Phi(\xi,\eta) + i\Psi(\xi,\eta)$$

is analytic in D'. Hence $\Phi \equiv \varphi(x(\xi,\eta),y(\xi,\eta))$, and $\Psi \equiv \psi(x(\xi,\eta),y(\xi,\eta))$ are conjugate harmonic functions in D'.

4. At any point ζ in the region[1] $D' + \Gamma'$, the values of Φ, Ψ, and χ are respectively equal to the values of φ, ψ, and Ω at the image point $z = f(\zeta)$ in the region $D + \Gamma$ (see Fig. 4-1). Moreover, as will be seen presently, the outward normal direction at a point ζ_0 on the boundary Γ' ordinarily corresponds to the outward normal direction at the image point z_0 on the boundary Γ, so that $\partial\Phi/\partial n'$ at the point ζ_0 is directly related to $\partial\varphi/\partial n$ at the point z_0.

Thus, if $f(\zeta)$ is analytic in a domain D' and the transformation $z = f(\zeta)$ maps D' onto the domain D, the change of variable $(x,y) \rightarrow (\xi,\eta)$ generated by the relation $x + iy = f(\xi + i\eta)$ transforms the boundary-value problem

[1] $D' + \Gamma'$ denotes the set of points formed by the union of the sets D' and Γ'.

defined by Eqs. (4-1) and (4-2) into a new problem defined by the formally similar equations

$$\Phi_{\xi\xi} + \Phi_{\eta\eta} = 0 \qquad \text{in } D' \tag{4-5}$$

$$A'(t')\Phi + B'(t')\frac{\partial \Phi}{\partial n'} = C'(t') \qquad \text{on } \Gamma' \tag{4-6}$$

where A', B', and C' are known functions of a real parameter t' that specifies position on the boundary Γ' of D'. [Note that, under a change of variable $(x,y) \to (\xi,\eta)$ which does not satisfy the above conditions, the transformed function Φ would *not* in general satisfy Laplace's equation in terms of the new variables.] If the original domain D is unbounded and if $f(\zeta_0) = \infty$, the condition at infinity [given by Eq. (4-3)] transforms into a condition at the point ζ_0 in a manner that will be brought out in subsequent examples.

Given a boundary-value problem for a domain D in the z plane, we shall exploit the above properties by seeking a transformation $z = f(\zeta)$ which leads to a new problem whose solution is already known or can be found by elementary means. Since the boundary-value problem can be solved easily for domains of certain standard shapes (e.g., a half-plane, exterior of unit circle, etc.), our procedure will generally consist in selecting a suitable domain D' and then looking for a transformation $z = f(\zeta)$ that maps D' onto D or, for the inverse transformation, say $\zeta = F(z)$, that maps D onto D'. To recover the solution $\varphi(x,y)$ of the original problem, we merely apply the inverse change of variable $(\xi,\eta) \to (x,y)$ generated by the relation $\xi + i\eta = F(x + iy)$, so that

$$\varphi(x,y) = \Phi(\xi(x,y),\eta(x,y))$$

In order that this inverse transformation be unambiguous, it is necessary that $F(z)$ be single-valued in the domain D. Ordinarily the function $f(\zeta)$ will also be single-valued in D' so that the transformation $z = f(\zeta)$ of D' onto D is one-one.

The Complex Potential

Because of its most prevalent physical interpretation, the solution φ of Eqs. (4-1) and (4-2) is termed a *real potential*, and its level curves $\varphi(x,y) = \text{const}$ are called *equipotentials*. For the sake of conciseness, we shall refer to the conjugate harmonic function ψ as the *stream function* and to its level curves $\psi(x,y) = \text{const}$ as *streamlines* (this nomenclature corresponds to the hydrodynamic interpretation of φ as a velocity potential).

The corresponding analytic function $\Omega(z) = \varphi + i\psi$ is called the *complex potential*. The potential Ω (or its real part φ) is said to describe a *field* in the domain D; direct physical significance is ascribed to the two-dimensional vector field which has φ_x and φ_y as x and y components, respectively. The derivative $\Omega'(z)$ will be termed the *(complex) intensity*. By virtue of the

Cauchy-Riemann relations (4-4) the intensity can be expressed in the form:

$$\Omega'(z) = \varphi_x - i\varphi_y \tag{4-7}$$

We have already noted that knowledge of the real potential φ is equivalent to knowledge of the complex potential Ω, except for an arbitrary additive constant in Ω. In physical contexts one is normally interested only in the complex intensity, or in differences of potential or of stream function. The specific value of the arbitrary constant is then unimportant, and in the sequel, some convenient choice of this constant will usually be made without further comment.

In the solution of Laplace's equation by mapping techniques, it is convenient to work with the complex potentials $\Omega(z)$ and $\chi(\zeta)$ rather than with the real potentials φ and Φ. This makes for a more symmetric and concise treatment of the problem and, at the same time, leads to more compact formulas for the change of variable and for the real potential and stream function. Moreover, one can then apply directly the powerful apparatus provided by complex-variable theory, e.g., the method of residues.

In physical interpretations of potential theory, significant roles are played by line integrals of the forms

$$\mathfrak{C} \equiv \int_C (\varphi_x\,dx + \varphi_y\,dy) = \int_C d\varphi \tag{4-8}$$

$$\mathfrak{F} \equiv \int_C (\varphi_x\,dy - \varphi_y\,dx) = \int_C d\psi \tag{4-9}$$

where C is any simple curve in the region $D + \Gamma$. The quantity $\mathfrak{F}$ is called the *flux* across the curve C. When C is a simple closed contour, $\mathfrak{C}$ is called the *circulation* around C. Equations (4-8) and (4-9) are equivalent to the single equation

$$\mathfrak{C} + i\mathfrak{F} = \int_C \Omega'(z)\,dz \tag{4-10}$$

If the coefficients $A(t)$, $B(t)$, and $C(t)$ in Eq. (4-2) are continuous on the boundary Γ of D, the solution of Eqs. (4-1) and (4-2) is essentially rendered unique by requiring φ to be continuous in the whole region $D + \Gamma$ (cf. Sec. 2-5). On the other hand, if the boundary data are not continuous, φ cannot be continuous in $D + \Gamma$, and therefore other kinds of auxiliary conditions will have to be imposed in order to guarantee uniqueness. These features are illustrated in the following simple examples.

Example 1

Γ_1 and Γ_2 are the half-lines $\arg z = 0$ and $\arg z = 3\pi/2$, respectively, and D is the domain $0 < \arg z < 3\pi/2$ (Fig. 4-2*a*). Find the bounded solution of Laplace's equation $\Delta\varphi = 0$ in D, subject to the conditions $\varphi = a$ on Γ_1 and $\varphi = a + k$ on Γ_2, where a and k are real constants.

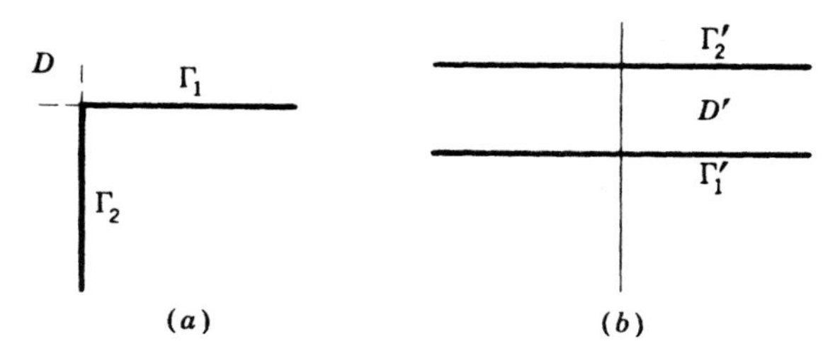

Fig. 4-2 (a) z plane; (b) ζ plane.

Recalling the mapping properties of the logarithmic function (Chap. 1), we apply the transformation

$$\zeta = \frac{2}{3\pi} \ln z \qquad 0 < \arg z < \frac{3\pi}{2} \tag{4-11}$$

to map the half-line Γ_1 into the line Γ_1' ($\eta = 0$) and the half-line Γ_2 into the line Γ_2' ($\eta = 1$); the sectorial domain D maps onto the infinite strip $D': 0 < \eta < 1$ (Fig. 4-2b). By inspection, Laplace's equation in the strip D' with boundary conditions $\Phi = a$ on Γ_1', $\Phi = a + k$ on Γ_2' has the bounded solution

$$\Phi = k\eta + a \equiv \text{Re}\,(-ik\zeta + a)$$

The corresponding complex potential is

$$\chi(\zeta) = -ik\zeta + a \tag{4-12}$$

From Eqs. (4-11) and (4-12), the complex potential for the original problem in the z plane is

$$\Omega(z) = -\frac{2ik}{3\pi} \ln z + a \qquad 0 < \arg z < \frac{3\pi}{2} \tag{4-13}$$

thus implying the real potential $\varphi = (2k/3\pi) \arg z + a$ and the stream function $\psi = (-2k/3\pi) \ln |z|$. The equipotentials are half-lines $\arg z =$ const, and the streamlines are arcs of the family of concentric circles $|z| =$ const. Note that the intensity $\Omega'(z) = -2ik/2\pi z \to \infty$ as $z \to 0$ in D.

That the real part of $\Omega(z)$ as given by Eq. (4-13) is the unique bounded solution of the problem can be shown most easily by considering the equivalent problem in the strip D' of the ζ plane. The difference of any two permissible solutions vanishes on the two horizontal boundaries; if the infinite strip is replaced by a sequence of rectangles formed by the horizontal boundaries and two vertical lines at $\xi = \pm R$, then, since the difference function is bounded on these vertical lines, it is clear that its value at any interior point must approach zero as $R \to \infty$.

The functions $\Lambda_k(z) = iz^{2k/3}$ ($k = \pm 1, \pm 2, \ldots$) are all analytic in D and have the property that $\operatorname{Re} \Lambda_k = 0$ on Γ_1 and on Γ_2. If we had not imposed the condition that φ be bounded in D, any of the functions

$$\Omega_k(z) \equiv \Omega(z) + \Lambda_k(z) \qquad k = \pm 1, \pm 2, \ldots$$

would be an acceptable complex potential for the problem. However, $\operatorname{Re} \Omega_k$ is unbounded at $z = \infty$ when $k > 0$ and at $z = 0$ when $k < 0$, and such solutions are therefore not admissible.

Example 2

D is the domain defined in Example 1. Solve Laplace's equation $\Delta\varphi = 0$ in D subject to the conditions (1) $\partial\varphi/\partial n = 0$ on Γ_1 and on Γ_2 and (2) φ continuous in $D + \Gamma_1 + \Gamma_2$ except at the points $z = 0$ and $z = \infty$, where φ has at most logarithmic singularities.

The problem, as stated, still has no unique solution; we shall shortly see that an arbitrary constant k appears. Condition (1) and the Cauchy-Riemann equations (4-4) imply that $\partial\psi/\partial s = 0$ on Γ_1 and on Γ_2, where $\partial/\partial s$ denotes differentiation with respect to length of arc s along the boundary. Hence ψ is constant on Γ_1 and on Γ_2. We identify Γ_1 arbitrarily as the streamline $\psi = 0$; Γ_2 is then the streamline $\psi = k$, where the value of the constant k must be specified by means of auxiliary data (such as a given flux condition; we note that the total flux across any curve joining Γ_1 to Γ_2 is k).

If φ has at most logarithmic singularities, ψ must be bounded in D. Once k has been chosen, ψ can then be determined. In fact ψ is just equal to the real potential of Example 1 for the special case $a = 0$ that is,

$$\psi = \frac{2k}{3\pi} \arg z \equiv \operatorname{Im}\left(\frac{2k}{3\pi} \ln z\right)$$

The corresponding complex potential is

$$\Omega(z) = \frac{2k}{3\pi} \ln z \tag{4-14}$$

and the real potential is $\varphi = (2k/3\pi) \ln |z|$, which has the desired property of being only logarithmically singular at $z = 0$ and $z = \infty$.

We have ensured uniqueness of the solution by requiring that the stream function ψ be bounded in D, thereby ruling out potential functions φ which are algebraically infinite at $z = 0$ or $z = \infty$ (cf. Example 1). The solution φ then has the mildest possible singularities on the boundary. In applications of potential theory, this condition usually appears as a consequence of some physical constraint, e.g., finiteness of the total energy.

A Generalization of the Boundary-value Problem

For formal analytic purposes and also for physical applications, it is desirable to generalize the boundary-value problem defined in Eqs. (4-1) and (4-2) by permitting the potential to be singular at one or more points within D. For example, the solution of a whole class of boundary-value problems for Laplace's equation in a given domain D can be expressed simply in terms of an appropriate Green's function $g(z;z_0) \equiv g(x,y;x_0,y_0)$ with the following properties (see Chap. 7):

1. $g(z;z_0)$ is harmonic everywhere in D except at the arbitrary point $z = x_0 + iy_0$.
2. g satisfies certain homogeneous conditions on the boundary Γ of D.
3. g is logarithmically singular at the point $z = z_0$,

$$\frac{g(z;z_0)}{\ln |z - z_0|} \rightarrow \text{const} \qquad \text{as } z \rightarrow z_0 \tag{4-15}$$

For the extended boundary-value problem, Laplace's equation (4-1) is valid everywhere in the domain D except at certain singular points, say $z_1, z_2, \ldots, z_n$. The data must now include a specification of the asymptotic form of φ, or equivalently of Ω, as $z \rightarrow z_\lambda$ ($\lambda = 1, 2, \ldots, n$). Ordinarily, the singularities of Ω are logarithmic singularities or poles. For reasons of physical interpretation, the statement "There is a point source of strength q (real) at the point $z = z_0''$ will be used to mean that

$$\frac{\Omega(z)}{q \ln (z - z_0)} \rightarrow 1 \qquad \text{as } z \rightarrow z_0 \tag{4-16}$$

Similarly the statement "There is a point vortex of strength p (real) at $z = z_0''$ means that

$$\frac{\Omega(z)}{ip \ln (z - z_0)} \rightarrow 1 \qquad \text{as } z \rightarrow z_0 \tag{4-17}$$

It is worth remarking that analytic functions enter into our treatment of two-dimensional potential problems in two essentially distinct ways, (1) as complex potentials $\Omega(z)$, $\chi(\zeta)$, etc., and (2) as transformation functions $z = f(\zeta)$, $\zeta = F(z)$, etc.

EXERCISES

1. Sketch the equipotentials and streamlines of the field in the whole z plane described by each of the following complex potentials (α, β are real constants, and c is a complex constant):

(*a*) $\Omega = (\alpha - i\beta)z$, which represents a *uniform field* of constant intensity $\Omega' = \alpha - i\beta$.

(*b*) $\Omega = \alpha \log(z - c)$, which represents the field due to a *point source* of strength α at $z = c$.

(*c*) $\Omega = i\beta \log(z - c)$, which represents the field due to a *point vortex* of strength β at $z = c$.

(*d*) $\Omega = (\alpha + i\beta) \log(z - c)$.

2. In a field described by a complex potential $\Omega(z)$, any segment of an equipotential $\operatorname{Re} \Omega = k$ may be identified as a new boundary arc on which the real potential φ is required to assume the constant value k. An analogous remark applies to streamlines. Use the results of Exercise 1 to find the complex potential for each of the following domains D and stated conditions:

(*a*) D: $-1 < x + y < +1$; $\varphi = \lambda$ on $x + y = 1$, $\varphi = \mu$ on $x + y = -1$, and φ bounded in D.

(*b*) D as in (*a*): $\psi = \lambda$ on $x + y = 1$, $\psi = \mu$ on $x + y = -1$, and ψ bounded in D.

(*c*) D: $R_1 < |z| < R_2$; $\varphi = \lambda$ on $|z| = R_1$, $\varphi = \mu$ on $|z| = R_2$, and φ continuous in D.

3. Find (and sketch) the equipotentials and streamlines of the field in the whole z plane due to two point sources of strength q and $-q$ at the points $z = a$ and $z = -a$, respectively ($q > 0$, $a > 0$).

4. Repeat Exercise 3, with both point sources now of strength q.

5. D is the half-plane $\operatorname{Im} z > 0$, and there is a point source of strength $q > 0$ at the point $z = ia$, with $a > 0$. Use Exercise 3 to find the solution to Laplace's equation $\Delta\varphi = 0$ in D, subject to the conditions that (*a*) $\varphi = 0$ on the line $\operatorname{Im} z = 0$ and (*b*) φ is continuous in the region $\operatorname{Im} z \geq 0$, except at the point $z = ia$.

6. Repeat Exercise 5, with the stream function ψ replacing the real potential φ in condition (*a*).

Note: Exercises 5 and 6 provide two of the simplest examples of the so-called *method of images*.

7. A dipole of (complex) strength μ at the point z_0 is defined as the limit, as a (real) $\to 0$ and $q \to \infty$, of two sources of strength q and $-q$ at the points $z_0 + ae^{i\theta}$ and $z_0 - ae^{i\theta}$, respectively, the limit being taken in such a way that the product $2qae^{i\theta} = \mu$ remains constant. Show that the field in the whole z plane due to the dipole is $\Omega = -\mu/(z - z_0)$. Sketch the equipotentials and streamlines of the field.

8. A quadrupole (2^2 pole) of strength k at $z = z_0$ is defined in a manner analogous to the dipole (2^1 pole), as the limit of two equal and opposite dipoles. Find the complex potential of the field in the whole z plane due to the quadrupole, and sketch the equipotentials and streamlines.

9. Find the complex potential in the whole z plane due to a 2^n pole at $z = z_0$ ($n = 3, 4, \ldots$), the 2^n pole being defined as the limit of two 2^{n-1} poles.

10. $\Omega_1(z)$ is the complex potential in the whole z plane due to some distribution of sources, dipoles, etc., all of which are exterior to the circle $|z| = a$. Show that $\Omega(z) \equiv \Omega_1(z) + \Omega_1^*(a^2/z^*)$ is a complex potential for the domain $|z| > a$, with the same distribution of singularities and subject to the condition $\partial\varphi/\partial n = 0$ on $|z| = a$.†

In the following example, we derive a complex potential for the domain exterior to a circle. The solution of a wide class of problems (of a kind that occurs primarily in fluid mechanics) can be reduced to the solution of this problem.

Example 3

Γ is the circle $|z| = a$, and D is the domain $|z| > a$. Find the most general form for the complex potential $\Omega(z)$ in D subject to the conditions (1) $\partial\varphi/\partial n = 0$ on Γ, (2) $\Omega'(z)$ is single-valued and analytic in D, and (3) $\Omega'(z) \to Q \equiv U - iV$ (const) as $z \to \infty$.

As in Example 2, we interpret condition (1) as requiring that the stream function $\psi = 0$ on Γ. Conditions (2) and (3) imply that $\Omega'(z)$ can be represented by a Laurent series that converges for $|z| > a$,

$$\Omega'(z) = Q + \sum_{k=1}^{\infty} \frac{\alpha_k}{z^k}$$

and so
$$\Omega(z) = Qz + \alpha_0 + \alpha_1 \ln z - \sum_{k=1}^{\infty} \frac{\alpha_{k+1}}{kz^k} \tag{4-18}$$

where $\alpha_k \equiv \lambda_k + i\mu_k$ $(k = 0, 1, 2, \ldots)$ are constants to be determined so as to satisfy the condition $\psi = 0$ on Γ.

Writing $z = re^{i\theta}$ in Eq. (4-18), separating out $\psi = \operatorname{Im} \Omega$, and then setting $r = a$ and correspondingly $\psi = 0$, we obtain the equation

$$a(U \sin\theta - V\cos\theta) + \mu_0 + \lambda_1\theta + \mu_1 \ln a + \sum_{k=1}^{\infty} \frac{1}{ka^k}(\lambda_{k+1} \sin k\theta - \mu_{k+1}\cos k\theta) = 0$$

valid for all values of θ. Since ψ is single-valued on Γ, it follows that $\lambda_1 = 0$ and so also that $\mu_0 = -\mu_1 \ln a$, $\lambda_2 = -a^2U$, $\mu_2 = -a^2V$, $\lambda_k = \mu_k = 0$ for $k > 2$. The complex potential in D then has the general form

$$\Omega(z) = Qz + \frac{Q^*a^2}{z} - \frac{i\gamma}{2\pi} \ln \frac{z}{a} \tag{4-19}$$

where γ $(= -2\pi\mu_1)$ is an arbitrary real constant.

† This result is very useful in hydrodynamics. See L. M. Milne-Thomson, "Theoretical Hydrodynamics," p. 154, The Macmillan Company, New York, 1960.

If C is any simple closed contour in the region $|z| \geq a$, it follows directly from Eqs. (4-19) and (4-10) that the flux across C vanishes and that the circulation around C is γ or 0 according to whether C does or does not enclose the circle Γ. We then say that the field has the (constant) circulation γ.

We note that the points in D at which the intensity vanishes are given by the formula

$$z = \frac{i\gamma}{4\pi Q} \pm \sqrt{\frac{Q^* a^2}{Q} - \left(\frac{\gamma}{4\pi Q}\right)^2}$$

Exercises

11. If $a = 1$, $V = 0$, and $Q = U > 0$, sketch the streamlines of the field in Example 3. Distinguish the three cases $|\gamma/4\pi U| \gtreqless 1$.

12. Show that, in any transformation $z = f(\zeta)$ of a potential problem from the z plane into the ζ plane, the circulation around a simple closed contour C in the z plane equals the circulation around the image contour C' in the ζ plane and that the flux across C equals the flux across C'.

13. Deduce from Example 3 the general form of the complex potential in the domain $|z| > a$ if (*a*) $\varphi \equiv \operatorname{Re} \Omega = 0$ on the circle $|z| = a$, (*b*) $\Omega'(z)$ is single-valued in $|z| > a$, and (*c*) $\Omega'(z) \to E \equiv E_1 - iE_2$ (const) as $z \to \infty$. Interpret the value of the arbitrary constant in terms of line integrals.

14. In applications of potential theory to fluid mechanics (boundary condition $\partial\varphi/\partial n = 0$), a real-valued function $p(z)$, the pressure at the point z, is defined in terms of the intensity $\Omega'(z)$ by an equation of the form

$$p(z) + \tfrac{1}{2}\rho|\Omega'(z)|^2 = K$$

where K and ρ are specified positive constants. If D is the domain exterior to a simple closed curve Γ on which $\psi = $ const, three quantities X, Y, and M are defined as the line integrals,

$$X = -\oint_\Gamma p\,dy \qquad Y = \oint_\Gamma p\,dx \qquad M = \oint_\Gamma p(x\,dx + y\,dy)$$

(*a*) Show that (Blasius)

$$R \equiv X - iY = \tfrac{1}{2}\rho i \oint_\Gamma [\Omega'(z)]^2\,dz$$

$$M = \operatorname{Re}\left\{-\tfrac{1}{2}\rho \oint_\Gamma z[\Omega'(z)]^2\,dz\right\}$$

(*b*) Find R and M in terms of Q and γ for the field in the domain $|z| > a$ described by Eq. (4-19) of Example 3.

4-2 Conformal Transformation

We begin by studying transformations without specific reference to the solution of boundary-value problems. As was exemplified in Chap. 1, any functional relation $z = f(\zeta)$ can be represented geometrically as a transformation or mapping of the ζ plane onto the z plane. When the function $f(\zeta)$ is multivalued, we confine attention to some single-valued branch of $f(\zeta)$ and the associated mapping of the cut ζ plane onto the z plane. If $f(\zeta)$ is analytic at all points of a continuous curve C' in the ζ plane, the transformation maps C' into a continuous curve C in the z plane; moreover, if C' is smooth (i.e., has a continuously turning tangent), then C is also smooth.

The transformation $z = f(\zeta)$ has the important property that it is generally *conformal* except at certain *critical points*. A transformation is said to be conformal at a point ζ_0 if the angle of intersection of any two smooth curves through ζ_0 is equal, in both magnitude and sense of rotation, to the angle of intersection of the image curves through the point $z_0 = f(\zeta_0)$.

To establish the conformal property and to discover the nature of the critical points, we consider a small displacement $\delta\zeta = \zeta - \zeta_0$ from the point ζ_0 and the corresponding displacement $\delta z = z - z_0$ from the point $z_0 = f(\zeta_0)$. Since $f(\zeta)$ is analytic at ζ_0, δz can be represented by the Taylor series,

$$\delta z = \sum_{n=1}^{\infty} \frac{f^{(n)}(\zeta_0)}{n!} (\delta\zeta)^n \tag{4-20}$$

We distinguish the alternatives $f'(\zeta_0) \neq 0$ and $f'(\zeta_0) = 0$.

If $f'(\zeta_0) \neq 0$, then, as $\delta\zeta \to 0$,

$$\delta z \to f'(\zeta_0)\, \delta\zeta \tag{4-21}$$

and so

$$\arg \delta z \to \arg \delta\zeta + \arg f'(\zeta_0)$$

Hence the transformation rotates all infinitesimal line elements $\delta\zeta$ at the point ζ_0 through exactly the same angle $\arg f'(\zeta_0)$. The conformality of the transformation at $\zeta = \zeta_0$ is a direct consequence of this property.

Equation (4-21) also implies that, when $f'(\zeta_0) \neq 0$ and $\delta\zeta \to 0$,

$$|\delta z| \to |f'(\zeta_0)|\, |\delta\zeta|$$

The transformation therefore stretches all infinitesimal line elements at the point ζ_0 by exactly the same factor $|f'(\zeta_0)|$. This property, together with the conformality, implies that any infinitesimal figure at ζ_0 transforms into a similar infinitesimal figure at $z_0 = f(\zeta_0)$ and that the area of the figure is increased by the factor

$$|f'(\zeta_0)|^2 \equiv \left| \frac{\partial(x,y)}{\partial(\xi,\eta)} \right|_{\zeta=\zeta_0}$$

On the other hand, if $f'(\zeta_0) = 0$, it follows from Eq. (4-20) that, as $\delta\zeta \to 0$,

$$\delta z \to \frac{1}{m!} f^{(m)}(\zeta_0)(\delta\zeta)^m$$

where $f^{(m)}(\zeta_0)$ is the first nonvanishing derivative at $\zeta = \zeta_0$. Thus, as $\delta\zeta \to 0$,

$$\arg \delta z \to m \arg \delta\zeta + \arg f^{(m)}(\zeta_0) \tag{4-22}$$

The angle between any two infinitesimal line elements at the point ζ_0 is increased by the factor m, and therefore the transformation is not conformal at the point. Solving Eq. (4-22) for $\arg \delta\zeta$, we find

$$\arg \delta\zeta \to \frac{1}{m} [\arg \delta z - \arg f^{(m)}(\zeta_0) + 2k\pi] \qquad k = 0, 1, \ldots, m - 1$$

Thus there are m distinct infinitesimal line elements $\delta\zeta$ at ζ_0 that map into each infinitesimal line element δz at $z_0 = f(\zeta_0)$; the inverse transformation $\zeta = F(z)$ is then multivalued, and $F(z)$ has a branch point of order m at the point z_0.

A transformation $z = f(\zeta)$ is now seen to be conformal at all points ζ where f is analytic and $f'(\zeta) \neq 0$. The critical points are those points at which $f'(\zeta) = 0$ or ∞; critical points with $f'(\zeta) = 0$ correspond to branch points of the inverse function $F(z)$.

If the transformation $z = f(\zeta)$ is conformal at a point ζ_0, it is also one-one in a neighborhood of that point. On the other hand, if the transformation is conformal in a domain D' in the ζ plane, the mapping of D' onto its image domain D in the z plane need not be one-one. (For example, the transformation $z = \sin \zeta$ is conformal in the infinite strip $1 < \text{Im } \zeta < 2$; however, if ζ_0 is any point in the strip, the points $\zeta_0 + 2k\pi$, $k = \pm 1, \pm 2, \ldots$, also lie in the strip and all map into the same point $z_0 = \sin \zeta_0$.) In fact, the transformation $z = f(\zeta)$ can generally be interpreted as providing a one-one conformal mapping of the domain D' onto a domain in the Riemann surface for the inverse function $F(z)$ (see Chap. 1). However, when the image domain lies on a single sheet of that Riemann surface, the transformation can be interpreted as providing a conformal one-one mapping of D' onto a domain D in a simple z plane; the function $f(\zeta)$ is then said to be *simple*, or *schlicht*, in the domain D'. In the solution of Laplace's equation by mapping techniques, we shall ordinarily introduce transformations that are conformal and one-one in the relevant domains (however, see Sec. 4-7).

A knowledge of the mapping properties of the elementary functions is very useful for applications of conformal mapping. The reader should acquire familiarity with the ways in which particular families of curves (e.g., the coordinate lines) in one complex plane map into corresponding families of curves in the other complex plane under such transformations. Some of these mapping properties have already been explored in Chap. 1; additional examples are included among the following exercises.

EXERCISES

1. If the function $f(\zeta)$ is analytic in a domain D' and if the transformation $z = f(\zeta)$ maps D' biuniformly (i.e., one-one) onto a domain D in the z plane, show that the transformation is conformal in D'. (Use this result in Exercise 2.)

2. If $F(z)$ is analytic within and on a simple closed contour C and assumes any value at most once on C, show that the transformation $\zeta = F(z)$ maps the domain interior to C conformally and biuniformly onto the domain interior to C'.

3. Study the mapping properties of the following functions (compare Chap. 1): (*a*) e^z, (*a'*) $\log z$, (*b*) z^{10}, (*b'*) $z^{1/10}$, (*c*) z^{-10}, (*c'*) $z^{-1/10}$, (*d*) $\cos z$, (*d'*) $\arccos z$, (*e*) $\tan z$, (*e'*) $\arctan z$, (*f*) $\sinh z$, (*f'*) $\operatorname{arcsinh} z$, (*g*) z^{1+i}.

4. (*a*) If the domain D in the z plane can be mapped conformally onto a domain D' in the t plane, and if D' can be mapped conformally onto a domain D'' in the ζ plane, show that D can be mapped conformally onto D''.

(*b*) If the domain D in the z plane and the domain D'' in the ζ plane can both be mapped conformally onto the domain D' in the t plane, show that D can be mapped conformally onto D''.

5. Find the critical points of all the transformations defined by the equation $\cos z = \sinh \zeta$.

6. If $\zeta^2 f'(\zeta)$ tends to a finite nonzero limit as $\zeta \to \infty$, show that the transformation $z = f(\zeta)$ is conformal at $\zeta = \infty$.

7. (*a*) If the transformation $z = f(\zeta)$ maps $|\zeta| < R$ conformally onto a simply connected domain D of area A in the z plane, show that

$$A = \pi \sum_{n=1}^{\infty} \frac{|f^{(n)}(0)|^2}{(n-1)!n!} R^{2n}$$

(*b*) If $f(\zeta)$ is also analytic on the boundary $|\zeta| = R$, show that the boundary Γ of the domain D has length

$$L = R \int_0^{2\pi} |f'(Re^{i\varphi})| \, d\varphi$$

and that $L \geq 2\pi R |f'(0)|$.

8. Find the area of the domain in the ζ plane which is the image of the rectangular domain $0 < x < \pi/4$, $0 < y < 3$ under the transformation $\zeta = \sin z$.

9. (*a*) Show that the transformation $\zeta = 2z^{-1/2} - 1$ maps the domain exterior to the parabola $y^2 = 4(1 - x)$ conformally onto the domain $|\zeta| < 1$. Explain carefully why this transformation does not, at the same time, map the domain interior to the parabola onto the domain $|\zeta| > 1$.

(*b*) Show that the transformation $\zeta = \tan^2 \frac{1}{4}\pi \sqrt{z}$ maps the domain exterior to the parabola $y^2 = 4(1 - x)$ onto the domain $|\zeta| < 1$.

10. (*a*) Let $g(\zeta)$ be analytic and arg $g(\zeta)$ be constant on a linear segment L' that contains the point ζ_0 (not an end point). If $f'(\zeta) = (\zeta - \zeta_0)^k g(\zeta)$, $-3 < k < +1$, show that the transformation $z = f(\zeta)$ maps L' into a continuous curve composed of two linear segments that subtend an angle $\pi(k + 1)$ at $z_0 = f(\zeta_0)$.

(*b*) Study the mapping properties of the transformation in the special case that $k = -1$, $g(\zeta) \equiv 1$, and ζ_0 is real.

11. If $f'(\zeta) = a(\zeta - p)^\lambda(\zeta - q)^\mu$, where a is a complex constant and p, q, λ, μ are real constants with $0 < (|\lambda|, |\mu|) < 1$, show that the transformation $z = f(\zeta)$ maps the ξ axis into a continuous curve composed of three linear segments in the z plane. Distinguish between the cases $p < q$ and $p > q$.

12. If the transformation $\zeta = F(z)$ is conformal at $z = z_0$, show that (*a*) a point source of strength q at the point z_0 transforms into a point source of the same strength q at the point $\zeta_0 = F(z_0)$; (*b*) a dipole of strength μ at the point z_0 transforms into a dipole of strength $\mu/F'(z_0)$ at the point ζ_0. Find the transformation properties of a 2^n pole at the point z_0.

Mapping of Domains. Riemann's Theorem

A central question in conformal-mapping theory concerns the existence of a transformation $\zeta = F(z)$ which maps a given domain D in the z plane conformally onto a given domain D' in the ζ plane. It is clear that the transformation will generally not exist without some restrictions on the character of one domain relative to the other. Thus, by examining the mapping of simple closed contours in D onto corresponding contours in D', the reader may convince himself that equal connectivity of the domains D and D' is a necessary condition. However, equal connectivity is by no means a sufficient condition, except for the special case of simple connectivity. A detailed discussion of necessary and sufficient conditions for the existence of a conformal transformation would involve us in a considerable digression. We shall therefore content ourselves with a brief statement of the main results.[1] We note that, in order to establish that two domains D and D' can be mapped conformally one on the other, it is sufficient to show that D and D' can each be mapped conformally onto some "standard domain" D'' (cf. Exercise 4).

[1] The reader interested in pursuing this topic will find a detailed account in C. Carathéodory, "Conformal Representation," Cambridge University Press, New York, 1932. See also Exercise 16 below.

For simply connected domains,[1] the main theorem is Riemann's mapping theorem, which asserts that any two simply connected domains, each having more than one boundary point, can be mapped conformally one on the other. The interior of the unit circle provides a convenient standard domain for this case. The Riemann theorem may then be stated in the following form:

If a simply connected domain D in the z plane has more than one boundary point, there exists a three-parameter family of transformations that map D conformally onto the domain $|\zeta| < 1$.

A unique transformation $\zeta = F(z)$ is specified by any three independent "real" conditions, e.g.:

1. Any given point z_0 in D is to map into the point $\zeta = 0$, and an arbitrary direction at $z = z_0$ is to map into the direction of the positive ζ axis at $\zeta = 0$.

2. A given point z_0 in D is to map into $\zeta = 0$, and a point z_1 on the boundary Γ of D is to map into a given point ζ_1 on the circle $|\zeta| = 1$.

3. Three arbitrary points z_1, z_2, z_3 on the boundary Γ of D are to map into any three points ζ_1, ζ_2, ζ_3 on the circle $|\zeta| = 1$, provided that the two sets of points appear in the same order when the two boundaries are described in the positive sense.

Although Riemann's theorem guarantees the existence of the desired transformation, it provides no direct assistance in discovering that transformation.

The annular domain between two concentric circles, $R_1 < |\zeta| < R_2$, provides a convenient standard, doubly connected domain. Any doubly connected domain can be mapped conformally onto such an annular domain with a particular, characteristic ratio of the radii (see Exercise 15).

For domains of still higher connectivity, the situation is more complex. A convenient standard domain for the *n-tuply* connected case consists of the whole z plane exterior to n slits, which may be parallel line segments, concentric circular arcs, etc. The relative lengths and locations of the slits are then characteristic features of the standard domain.

We note the important fact that the potential problem for a simply connected domain D may be regarded as essentially solved when we have found a transformation that maps D onto the interior of the unit circle. The solution of the transformed problem can then be expressed simply in terms of an integral of Poisson type [cf. Eq. (2-24)].

EXERCISES

13. If $\zeta = G(z)$ is a particular transformation that maps a simply connected domain D in the z plane onto $|\zeta| < 1$, obtain an explicit

[1] If the point at ∞ is included in the domain D exterior to a simple closed curve, then D is simply connected, in the sense that any simple closed curve enclosing D can be shrunk to zero size at ∞. Thus Riemann's theorem covers the case in which D is mapped into the interior of the unit circle.

representation of the full three-parameter family of transformations that map D onto $|\zeta| < 1$.

14. (*a*) A transformation $\zeta = F(z;z_0)$ maps the simply connected domain D in the z plane onto the circular domain $|\zeta| < 1$, the point $\zeta = 0$ being the image of the arbitrary point z_0 in D. If $g(z;z_0)$ is the Green's function for Laplace's equation in D, with the condition $g = 0$ on the boundary Γ of D, show that

$$g(z;z_0) = \ln |F(z;z_0)|$$

(*b*) Let $\zeta = F(z)$ be any transformation that maps the simply connected domain D onto the domain $|\zeta| < 1$. Show that the Green's function is also given by

$$g(z;z_0) = -\ln \left| \frac{1 - F^*(z_0)F(z)}{F(z) - F(z_0)} \right|$$

15. A transformation $\zeta = F(z)$ gives a one-one conformal map of the domain $1 < |z| < R_1$ onto the domain $1 < |\zeta| < R_2$. If $p(z) = \ln R_1 \ln F(z) - \ln R_2 \ln z$, show that $\operatorname{Im} p$ is constant in $1 < |z| < R_1$ and hence that $R_1 = R_2$.

16. The equivalence between the Green's function (which corresponds to a concentrated source at z_0, with zero potential on Γ) and the mapping function has been pointed out in Exercise 14. Assuming —perhaps on physical grounds—that such a Green's function exists for a particular region, show that the degree of choice given by Riemann's mapping theorem is indeed appropriate.

17. Find the form of the most general transformation $\zeta = f(z)$ that maps the unit circle onto itself, with $f(z_0) = 0$. [*Hint:* Note that $w = (z - z_0)/(z_0^* z - 1)$ is one such transformation, and use Schwarz's lemma given in Exercise 3 of Sec. 2-6.]

18. Let Γ be a simple closed curve in the z plane and D its exterior. Show that there exists a real number $a > 0$ such that D may be mapped conformally into the exterior of a circle of radius a in the ζ plane by a function $z = f(\zeta)$ satisfying $f'(\infty) = 1$. [*Hint:* Preliminary mappings $w = 1/z$ and $t = 1/\zeta$ may be useful.]

4-3 *Bilinear Transformations*

The main properties of the bilinear transformation

$$\zeta = \frac{az + b}{cz + d} \qquad z = \frac{-d\zeta + b}{c\zeta - a} \qquad ad - bc \neq 0$$

were discussed in Chap. 1. Bearing in mind that straight lines are circles through the point at infinity, we note that a bilinear transformation (1)

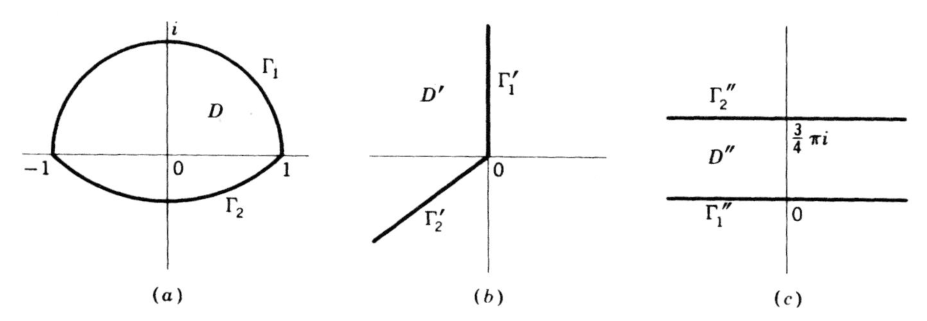

Fig. 4-3 (*a*) z plane; (*b*) w plane; (*c*) ζ plane.

maps the whole z plane conformally and biuniformly onto the whole ζ plane, (2) maps circles into circles, and (3) maps inverse points with respect to any circle into inverse points with respect to the transformed circle. We now study some examples in which bilinear transformations play a central role in reducing the problem to elementary form.

Example 4

D is the domain (Fig. 4-3*a*) bounded by the circular arcs

$$\Gamma_1: |z| = 1 \qquad \operatorname{Im} z \geq 0$$
$$\Gamma_2: |z - i| = \sqrt{2} \qquad \operatorname{Im} z \leq 0$$

Find the bounded solution of Laplace's equation $\Delta\varphi = 0$ in D subject to the conditions that $\varphi = a$ on Γ_1 and $\varphi = a + k$ on Γ_2 where a, k are real constants.

We reduce the problem to elementary form by mapping the domain D onto an infinite strip. As a first step, the bilinear transformation

$$w = \frac{z-1}{z+1} \tag{4-23}$$

maps D onto the sectorial domain D': $\pi/2 < \arg w < 5\pi/4$ in the w plane (Fig. 4-3*b*). The subsequent transformation

$$\zeta = \ln w - i\frac{\pi}{2} \tag{4-24}$$

then maps D' onto the infinite strip D'': $0 < \operatorname{Im} \zeta < 3\pi/4$ in the ζ plane (Fig. 4-3*c*). Combining Eqs. (4-23) and (4-24), we find that D is mapped onto D'' by the transformation

$$\zeta = \log\frac{z-1}{z+1} - i\frac{\pi}{2} \qquad \frac{\pi}{2} < \arg\frac{z-1}{z+1} < \frac{5\pi}{4} \tag{4-25}$$

By inspection, the bounded solution of Laplace's equation in the domain D'' is

$$\Phi = a + \frac{4k}{3\pi}\eta \equiv \operatorname{Re}\left(a - i\frac{4k}{3\pi}\zeta\right)$$

The complex potential in the ζ plane is

$$\chi(\zeta) = a - i\frac{4k}{3\pi}\zeta \tag{4-26}$$

From Eqs. (4-26) and (4-25), the complex potential for the original problem in the z plane is

$$\begin{gathered}\Omega(z) = a - \frac{2k}{3} - i\frac{4k}{3\pi}\ln\frac{z-1}{z+1} \\ \frac{\pi}{2} < \arg\frac{z-1}{z+1} < \frac{5\pi}{4}\end{gathered} \tag{4-27}$$

Note that the complex intensity $\Omega'(z) \to \infty$ as $z \to \pm 1$.

The one essential feature of the initial transformation (4-23) is the choice of denominator on the right-hand side to make $w = \infty$ the image of one of the points common to the circular arcs Γ_1 and Γ_2. This ensures that Γ_1, Γ_2 map into the straight segments Γ_1', Γ_2', respectively. The numerator in Eq. (4-23) was chosen for convenience so as to locate the sectorial domain D' with its vertex at the origin $w = 0$. Note that, since the transformation (4-23) is conformal at the point $z = 1$, the angle $3\pi/4$ between Γ_1' and Γ_2' at $w = 0$ conforms in magnitude and sense to the angle between Γ_1 and Γ_2 at $z = 1$. The term $-i(\pi/2)$ in the transformation (4-24) was included merely to place the strip D'' in a convenient location.

The equipotentials of the field described by Eq. (4-27) are arcs of the family of intersecting coaxial circles with $z = +1$ and $z = -1$ as common points,

$$\arg\frac{z-1}{z+1} = k \qquad \frac{\pi}{2} < k < \frac{5\pi}{4}$$

The streamlines are arcs of the orthogonal family of nonintersecting coaxial circles with $z = +1$ and $z = -1$ as limit points,

$$\left|\frac{z-1}{z+1}\right| = \text{const}$$

Example 5

D is the domain defined in Example 4. Find the complex potential $\Omega_1(z) = \varphi + i\psi$ in D subject to the conditions that (1) ψ is bounded in

D, (2) $\partial\varphi/\partial n = 0$ on Γ_1 and on Γ_2, and (3) the flux across any curve joining a point on Γ_1 to a point on Γ_2 is k (constant).

As in Example 2, we identify Γ_1 as the streamline $\psi = 0$; then Γ_2 is the streamline $\psi = k$. The solution is obtained directly from the solution of Example 4 by setting $a = 0$ in Eq. (4-27) and then writing $\Omega_1(z) = i\Omega(z)$. Thus

$$\begin{gathered}\Omega_1(z) = \frac{4k}{3\pi}\ln\frac{z-1}{z+1} - i\frac{2k}{3} \\ \frac{\pi}{2} < \arg\frac{z-1}{z+1} < \frac{5\pi}{4}\end{gathered} \tag{4-28}$$

In Examples 1, 2, 4, and 5 the complex potentials contain logarithmic terms which are singular at points of discontinuity of the boundary data. Formally, these terms correspond to point sources or vortices at those boundary points. In Eq. (4-28), for example, the sources at the points $z = \pm 1$ have just the strengths $\pm 2k/3\pi$, respectively, necessary to provide a flux k across an infinitesimal circular arc at $z = 1$ from Γ_1 to Γ_2.

Example 6

D is the annular domain (Fig. 4-4a) between the two circles

$$\Gamma_1\colon |z| = 1 \qquad \Gamma_2\colon |z-1| = 5/2$$

Solve $\Delta\varphi = 0$ in D, subject to the conditions $\varphi = a$ on Γ_1 and $\varphi = b$ on Γ_2, where a and b are constants.

Let α, β, k be constants, with k real and >0. Then the equation $|z - \alpha| = k|z - \beta|$ describes a circle, with respect to which the points

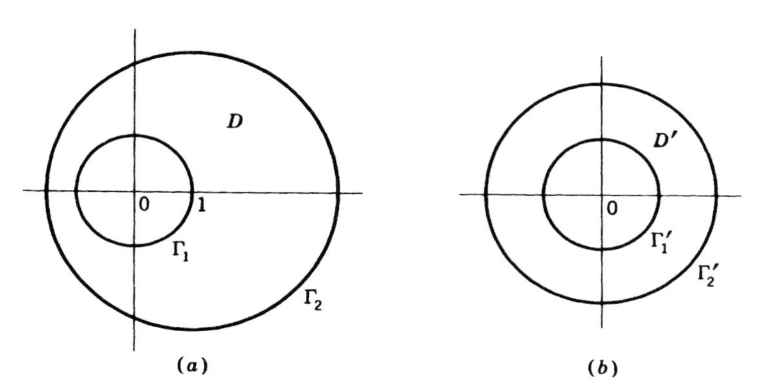

Fig. 4-4 (a) z plane; (b) ζ plane.

α and β are inverse to one another (i.e., they lie on the same radial line from the center of the circle, and the product of their distances from the center is equal to the square of the radius). If α and β are held fixed and k allowed to vary, we obtain a family of noninteresting circles. The given circles Γ_1 and Γ_2 can be regarded as members of such a family if we choose α and β (real, by symmetry) so as to be inverse points for each of Γ_1 and Γ_2; thus

$$(0-\alpha)(0-\beta) = 1$$
$$(1-\alpha)(1-\beta) = 25/4$$

and so $\alpha = -1/4$, $\beta = -4$. The family of circles is then defined by

$$\left|\frac{z+1/4}{z+4}\right| = k \tag{4-29}$$

and Γ_1, Γ_2 correspond to $k = 1/4, 1/2$, respectively.

We now apply the bilinear transformation

$$\zeta = \frac{z+1/4}{z+4} \tag{4-30}$$

which clearly maps all circles constituting the family (4-29) into a family of concentric circles. The domain D is mapped into D': $1/4 < |\zeta| < 1/2$ in the ζ plane (Fig. 4-4*b*). The solution of the potential problem in the ζ plane (cf. Exercise 2*c* of Sec. 4-1) is

$$\Phi = (2b-a) + \frac{b-a}{\ln 2}\ln|\zeta| = \text{Re}\left[(2b-a) + \frac{b-a}{\ln 2}\ln\zeta\right] \tag{4-31}$$

From Eqs. (4-30) and (4-31), the complex potential in the z plane is

$$\Omega(z) = (2b-a) + \frac{b-a}{\ln 2}\ln\frac{z+1/4}{z+4} \tag{4-32}$$

thus implying the real potential

$$\varphi = (2b-a) + \frac{b-a}{\ln 2}\ln\left|\frac{z+1/4}{z+4}\right|$$

and the stream function

$$\psi = \frac{b-a}{\ln 2}\arg\frac{z+1/4}{z+4}$$

The equipotentials are members of the family of nonintersecting circles defined by Eq. (4-29). The streamlines are arcs of the orthogonal family of intersecting circles with $z = -1/4$ and $z = -4$ as common points.

In physical applications we are often interested in evaluating certain overall properties of the solution to a potential problem. In the present case, for example, φ may be interpreted as an electrostatic potential. The configuration of Fig. 4-4a then represents a capacitor composed of two infinite circular cylinders with parallel axes (perpendicular to the z plane). The capacitance per unit length of the capacitor is then defined by the equation

$$C = \frac{\left| (1/4\pi)\oint_{\Gamma_1} (\partial\varphi/\partial n)\, ds \right|}{|b - a|}$$

From Eq. (4-9) or (4-10) for the flux, we find that

$$\oint_{\Gamma_1} \frac{\partial\varphi}{\partial n}\, ds = -\oint_{\Gamma_2} \frac{\partial\varphi}{\partial n}\, ds = \frac{2\pi(b - a)}{\ln 2}$$

and hence that $$C = \frac{1}{2 \ln 2}$$

Note that the capacitance is an intrinsic property of the capacitor and is independent of the values of φ on Γ_1 and on Γ_2.

***Example* 7**

D is the circular domain $|z| < 1$, and a source of strength q (real) is located at the origin $z = 0$ (Fig. 4-5a). Solve Laplace's equation $\Delta\varphi = 0$ in D subject to the conditions (1) $\varphi = 0$ on the lower semicircle Γ_1: $|z| = 1$, $\operatorname{Im} z < 0$; (2) $\partial\varphi/\partial n = 0$ on the upper semicircle Γ_2: $|z| = 1$, $\operatorname{Im} z > 0$; and (3) φ is bounded everywhere in D except at the point $z = 0$.

As usual, we interpret condition (2) as requiring that Γ_2 be the streamline $\psi = 0$. The transformation

$$z = \frac{t - i}{t + i} \qquad t = -i\frac{z + 1}{z - 1} \tag{4-33}$$

maps the domain D onto the upper half-plane $\operatorname{Im} t > 0$ (Fig. 4-5b). The complex potential $\Lambda(t)$ in the t plane corresponds to a source of

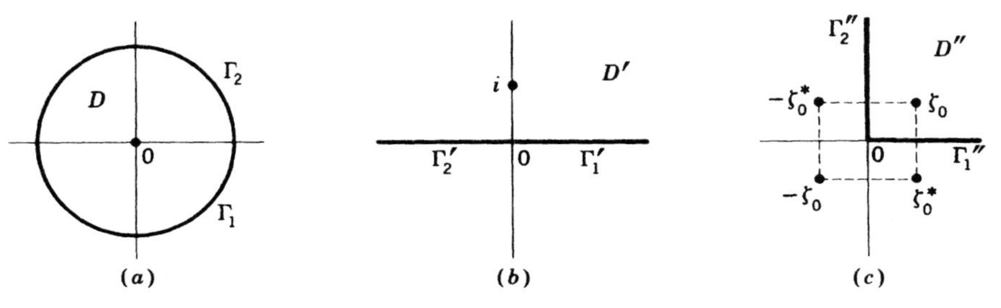

***Fig.* 4-5** (a) z plane; (b) t plane; (c) ζ plane.

strength q at the point $t = i$ and satisfies the conditions $\operatorname{Re} \Lambda = 0$ on the half-line Γ_1' and $\operatorname{Im} \Lambda = 0$ on the half-line Γ_2'.

To handle these mixed boundary conditions, we apply the further transformation

$$\zeta = t^{1/2} \tag{4-34}$$

to map the half-plane $\operatorname{Im} t > 0$ onto the quarter-plane D'': $\operatorname{Re} \zeta > 0$, $\operatorname{Im} \zeta > 0$ (Fig. 4-5*c*). The complex potential $\chi(\zeta)$ in D'' corresponds to a source of strength q at the point $\zeta_0 = e^{i\pi/4}$ and satisfies the conditions $\Phi \equiv \operatorname{Re} \chi = 0$ on the positive ξ axis Γ_1'' and $\Psi \equiv \operatorname{Im} \chi = 0$ on the positive η axis Γ_2''.

The solution of the boundary-value problem in the ζ plane is easily obtained by the method of images. If we introduce an image source of strength $-q$ at the point ζ_0^*, the ξ axis is the equipotential $\Phi = 0$ (but the η axis is not a streamline). By introducing two further image sources of strength q and $-q$ at the points $-\zeta_0^*$ and $-\zeta_0$, respectively (Fig. 4-5*c*), we ensure that the η axis becomes the streamline $\Psi = 0$, without disturbing the equipotential character of the ξ axis. The complex potential for the domain D'' in the ζ plane is then

$$\chi(\zeta) = q \ln \frac{(\zeta - \zeta_0)(\zeta + \zeta_0^*)}{(\zeta - \zeta_0^*)(\zeta + \zeta_0)}$$

Hence, noting that $\zeta_0 = e^{i\pi/4}$, and using Eqs. (4-33) and (4-34), we obtain after a little reduction the complex potential for the original problem in the z plane,

$$\Omega(z) = \tfrac{1}{2} q \ln \frac{\{[(z+1)/(z-1)]^{1/2} + 1\}\{[(z+1)/(z-1)]^{1/2} - i\}}{\{[(z+1)/(z-1)]^{1/2} - 1\}\{[(z+1)/(z-1)]^{1/2} + i\}} \tag{4-35}$$

Example 8

Γ_1 is the circle $|z + i| = 1$, Γ_2 is the circle $|z - i| = 1$, and D is the domain exterior to both Γ_1 and Γ_2 (Fig. 4-6*a*). Find the complex potential $\Omega(z)$ in D subject to the conditions (1) $\operatorname{Im} \Omega = 0$ on Γ_1 and Γ_2, (2) $\Omega'(z) \to Q$ (const) as $z \to \infty$, and (3) the circulation in the field $= \gamma$.

We shall map D onto the domain exterior to the unit circle; this will permit us to use the result of Example 3. As a first step, we apply the bilinear transformation

$$w = \frac{iz - 2}{z} \tag{4-36}$$

to map D onto the infinite strip D': $0 < \operatorname{Im} w < 2$ in the w plane (Fig. 4-6*b*). [The transformation (4-36) was chosen to make the

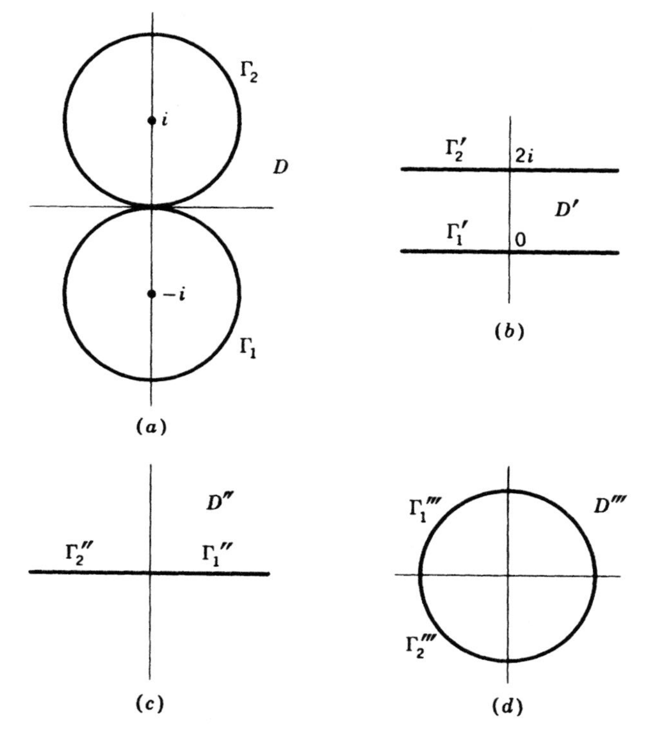

Fig. 4-6 (*a*) z plane; (*b*) w plane; (*c*) t plane; (*d*) ζ plane.

points $w = -i, \infty, i$ correspond to the points $z = -i, 0, \infty$, respectively. This ensures that both Γ_1, Γ_2 map into straight lines and in particular that the domain exterior to Γ_1 maps into the upper half-plane.] The subsequent transformation

$$t = e^{\pi w/2} \tag{4-37}$$

maps the infinite strip D' onto the half-plane $\operatorname{Im} t > 0$ (Fig. 4-6*c*). The point $z = \infty$ in D corresponds to $w = i$ in D' and to $t = i$ in $\operatorname{Im} t > 0$.

Finally we apply the transformation

$$\zeta = \frac{t + i}{t - i} \tag{4-38}$$

to map the half-plane $\operatorname{Im} t > 0$ onto the domain $|\zeta| > 1$ in such a way that the point $z = \infty$ in D corresponds to $t = i$ and therefore to $\zeta = \infty$ (Fig. 4-6*d*). Combining Eqs. (4-36), (4-37), and (4-38), we obtain the transformation that maps the domain D in the z plane onto the domain $|\zeta| > 1$,

$$\zeta = -\coth\frac{\pi}{2z} \tag{4-39}$$

The complex potential $\chi(\zeta)$ in the ζ plane satisfies the condition $\text{Im}\,\chi = 0$ on the circle $|\zeta| = 1$ and corresponds to circulation γ in the field. To find the condition at infinity in the ζ plane, we note that

$$\chi'(\zeta) = \frac{d\Omega(z)/dz}{d\zeta/dz} \to \frac{Q}{\lim\limits_{z\to\infty}(d\zeta/dz)} \qquad \text{as } z \to \infty$$

and so $\chi'(\zeta) \to -\pi Q/2$ as $\zeta \to \infty$. Hence, from Eq. (4-19) of Example 3, we find

$$\chi(\zeta) = -\frac{2}{\pi}\left(Q\zeta + \frac{Q^*}{\zeta}\right) - \frac{i\gamma}{2\pi}\ln\zeta$$

It then follows from Eq. (4-39) that the complex potential for the original problem in the z plane is

$$\Omega(z) = \frac{2}{\pi}\left(Q\coth\frac{\pi}{2z} + Q^*\tanh\frac{\pi}{2z}\right) - \frac{i\gamma}{2\pi}\ln\coth\frac{\pi}{2z}$$

EXERCISES

1. Find the bounded solution of Laplace's equation $\Delta\varphi = 0$ for each of the following domains D and stated boundary conditions:

(*a*) D is defined by the inequalities $\text{Re}\,z > 0$ and $|z - ia| > R$, $0 < R < a$; $\varphi = \lambda$ on $\text{Re}\,z = 0$, $\varphi = \mu$ on $|z - ia| = R$.

(*b*) D is the domain exterior to both the circles $|z - a| = R_1$ and $|z + a| = R_2$, with $0 < R_1 < a$, $0 < R_2 < a$; $\varphi = \lambda$ on $|z - a| = R_1$, $\varphi = \mu$ on $|z - a| = R_2$. In each case sketch the equipotentials and streamlines, and, interpreting φ as an electrostatic potential, find the capacitance per unit height of the corresponding capacitor.

2. The domain D is defined by the inequalities $\text{Re}\,z > 0$ and $|z - 1| > 1$. Find the complex potential $\Omega(z)$ in D subject to the conditions that (1) the boundary forms a streamline and (2) $\Omega'(z) \to iV$ as $z \to \infty$ in D. Find the points at which $\Omega'(z) = 0$ and the points at which $|\Omega'(z)|$ is a maximum. Sketch the streamlines of the field.

3. D is the circular domain $|z| < 1$, and there is a source of (real) strength q at the center $z = 0$. Solve Laplace's equation in D with the conditions that (1) $\varphi = 0$ on the arc $|z| = 1$, $0 < \arg z < \alpha$; (2) $\psi = 0$ on the arc $|z| = 1$, $\alpha < \arg z < 2\pi$; and (3) φ is bounded everywhere in D except at $z = 0$.

4. Find the Green's function for Laplace's equation in the domain $|z| < 1$ subject to the conditions (1) $\varphi = 0$ on the lower semicircle

$|z| = 1$, $\operatorname{Im} z < 0$; and (2) $\psi = 0$ on the upper semicircle $|z| = 1$, $\operatorname{Im} z > 0$. Solve the problem by two methods, (*a*) directly and (*b*) by using the result of Exercise 3.

5. Γ_1 and Γ_2 are the circular arcs defined in Example 4, and D is the domain exterior to the contour $\Gamma_1 + \Gamma_2$. Find the complex potential $\Omega = \varphi + i\psi$ in D subject to the conditions (1) $\psi = 0$ on Γ_1 and Γ_2, (2) $\Omega'(z) \to Q$ (const) as $z \to \infty$; and (3) the circulation $= \gamma$.

6. Equation (2-26) can be used to express a complex potential $\Omega(z)$ for the domain $|z| < 1$ in terms of the boundary values $\varphi(e^{i\theta}) \equiv \operatorname{Re} \Omega(e^{i\theta})$,

$$\Omega(z) = \frac{1}{2\pi} \int_0^{2\pi} \varphi(e^{i\theta}) \frac{e^{i\theta} + z}{e^{i\theta} - z} \, d\theta$$

Deduce that a complex potential $\chi(\zeta) = \Phi + i\Psi$ for the half-plane $\operatorname{Im} \zeta > 0$ is given in terms of the boundary values $\Phi(\xi,0)$ by the formula

$$\chi(\zeta) = \frac{-1}{\pi i} \int_{-\infty}^{+\infty} \Phi(\tau,0) \frac{1 + \tau\zeta}{(1 + \tau^2)(\zeta - \tau)} \, d\tau$$

7. Let $\xi_1 < \xi_2 < \cdots < \xi_n$ and $c_0, c_1, \ldots, c_n$ be real constants. If the complex potential $\chi(\zeta) = \Phi + i\Psi$ in the half-plane $\operatorname{Im} \zeta > 0$ satisfies the conditions $\chi'(\zeta) \to 0$ as $\zeta \to \infty$ and

$$\Phi(\xi,0) = \begin{cases} c_0 & \xi < \xi_1 \\ c_\lambda & \xi_\lambda < \xi < \xi_{\lambda+1} \qquad \lambda = 1, 2, \ldots, n-1 \\ c_n & \xi > \xi_n \end{cases}$$

show that

$$\chi(\zeta) = c_n - \frac{i}{\pi} \sum_{\lambda=0}^{n-1} (c_\lambda - c_{\lambda+1}) \ln (\zeta - \xi_{\lambda+1}) \tag{4-40}$$

with $\qquad 0 < \arg (\zeta - \xi_\lambda) < \pi \qquad \lambda = 1, 2, \ldots, n$

8. If, in the specification of the boundary data of Exercise 7, $\Phi(\xi,0)$ is replaced by $\Psi(\xi,0)$, show that the complex potential is given by the formula:

$$\chi(\zeta) = ic_n - \frac{1}{\pi} \sum_{\lambda=0}^{n-1} (c_\lambda - c_{\lambda+1}) \ln (\zeta - \xi_{\lambda+1}) \tag{4-41}$$

The results of Exercises 7 and 8 will be used repeatedly in subsequent sections of this chapter.

4-4 The Schwarz-Christoffel Transformation

We now turn to a class of boundary-value problems in which the domains have rectilinear boundaries. We begin by looking for a transformation $\zeta = F(z)$ that maps the simply connected domain D interior to an n-sided polygon in the z plane onto the half-plane $\text{Im}\, \zeta > 0$; the polygonal boundary must of course map into the ξ axis. In the procedure to be followed here we actually construct the inverse transformation $z = f(\zeta)$ that maps the half-plane $\text{Im}\, \zeta > 0$ onto the domain D. For the present, we shall confine our attention to finite closed polygons, although most of the subsequent applications will involve polygons that have one or more vertices at infinity.

Let the n vertices of the polygon, enumerated in some counterclockwise order, be the points $z_1, z_2, \ldots, z_n$, and let their image points on the ξ axis be $\xi_1, \xi_2, \ldots, \xi_n$, respectively. Without loss of generality, we may take $\xi_1 < \xi_2 < \cdots < \xi_n$. According to Riemann's mapping theorem, just three of the points ξ_j may be chosen arbitrarily. The n sides of the polygon are denoted by $\Gamma_1, \Gamma_2, \ldots, \Gamma_n$ (Fig. 4-7a), and their respective image segments on the ξ axis are denoted by $\Gamma'_1, \Gamma'_2, \ldots, \Gamma'_n$ (Fig. 4-7b). (If $\xi_1 \neq -\infty$ and $\xi_n \neq +\infty$, Γ'_n consists of the two segments $\xi < \xi_1$ and $\xi > \xi_n$.) The interior angle of the polygon at the vertex z_j is denoted by α_j ($j = 1, 2, \ldots, n$).

We define $\gamma_j = \pi - \alpha_j$ ($j = 1, \ldots, n$) and note that a counterclockwise rotation through the angle γ_j is required to bring the direction of Γ_{j-1} into coincidence with the direction of Γ_j. We also adduce the simple but

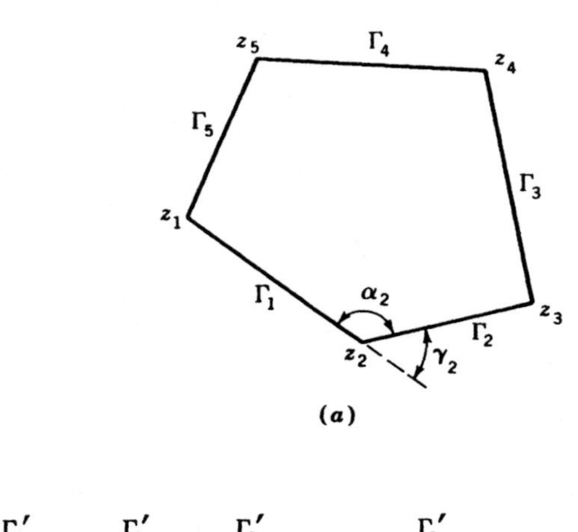

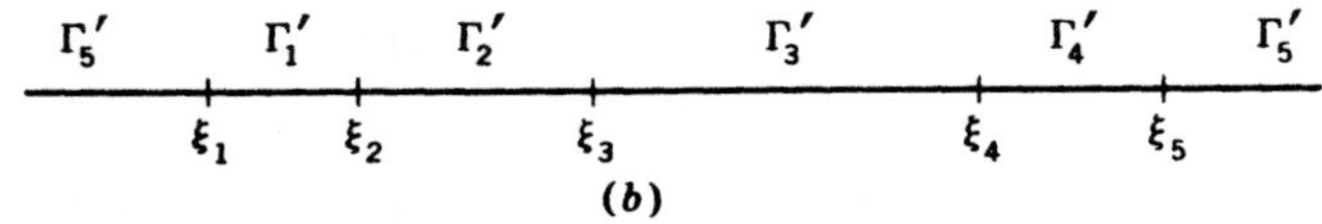

Fig. 4-7 (a) z plane; (b) ζ plane.

important relation

$$\sum_{j=1}^{n} \alpha_j = (n-2)\pi \qquad \text{or} \qquad \sum_{j=1}^{n} \left(1 - \frac{\alpha_j}{\pi}\right) \equiv \sum_{j=1}^{n} \frac{\gamma_j}{\pi} = 2 \tag{4-42}$$

The function $z = g(\zeta)$, with $g'(\zeta) = \zeta^{-k}$ (k real), maps the real ζ axis into a pair of straight-line segments in the z plane, intersecting each other at an angle of $k\pi$. This observation suggests that, for the case of finite ξ_1 and ξ_n, an appropriate transformation function (more exactly, the derivative of such a function) for our problem is given by

$$f'(\zeta) = a \prod_{j=1}^{n} (\zeta - \xi_j)^{\alpha_j/\pi - 1} \tag{4-43}$$

where a is a complex constant and where $0 < \arg(\zeta - \xi_j) < \pi$, for all j.

To verify that $f(\zeta)$ obtained by integrating Eq. (4-43) has the desired properties, we need merely note that, when the variable point ζ is on the segment Γ'_j of the ξ axis,

$$\arg(\zeta - \xi_k) = \begin{cases} 0 & \text{for } k \leq j \\ \pi & \text{for } k > j \end{cases}$$

and so

$$\arg f'(\zeta) = \arg a - \sum_{k=j+1}^{n} (\pi - \alpha_k) \tag{4-44}$$

Hence $\arg f'(\zeta)$ remains constant as the point ζ traverses the segment Γ'_j (so that the image curve Γ_j is also a straight-line segment); as ζ traverses Γ_{j+1}, $\arg f'(\zeta)$ has a new constant value, which is larger than before by the amount $\pi - \alpha_{j+1}$ added discontinuously at the point $\zeta = \xi_{j+1}$. To satisfy the condition $0 < \arg(\zeta - \xi_{j+1}) < \pi$, we calculate the change in $\arg(\zeta - \xi_{j+1})$ by considering a small upper half-plane semicircle centered on ξ_{j+1}. Thus the interior angle at z_{j+1} is indeed α_{j+1}.

Integration of Eq. (4-43) leads to the *Schwarz-Christoffel formula* for the desired transformation that maps the half-plane $\operatorname{Im} \zeta > 0$ onto the interior D of the polygon,

$$z = f(\zeta) = a \int_{\zeta_0}^{\zeta} \prod_{j=1}^{n} (\zeta - \xi_j)^{\alpha_j/\pi - 1} \, d\zeta + b \tag{4-45}$$

where b and ζ_0 are constants. In Exercise 2 the reader is asked to show that the transformation defined by Eq. (4-45) maps the half-plane $\operatorname{Im} \zeta > 0$ into a *closed* polygon; this completes the check on the validity of Eq. (4-45). An alternative and more formal derivation of Eq. (4-45), based on Schwarz's reflection principle, is outlined in Exercise 6.

The function $f(\zeta)$ is analytic in the whole region $\operatorname{Im} \zeta \geq 0$ except at the n critical points $\xi_1, \xi_2, \ldots, \xi_n$. The initial point ζ_0 in Eq. (4-45) can be chosen arbitrarily in this region and may for example be identified as the origin or as one of the critical points.

If the n interior angles $\alpha_1, \alpha_2, \ldots, \alpha_n$ are given, the shape of the polygon in the z plane (i.e., the ratio of its sides) is determined uniquely by the choice of the n points $\xi_1, \xi_2, \ldots, \xi_n$ in Eq. (4-45) and is independent of the particular values of the constants a, b and of ζ_0. Once the initial point ζ_0 has been chosen, the constant b in Eq. (4-45) controls the location of the polygon in the z plane. The constant a controls the scale of the polygon through the value of $|a|$, and its orientation through the value of $\arg a$.

For bounded polygonal domains, there are very few cases in which the mapping function $f(\zeta)$ and its inverse $F(z)$ can be evaluated in terms of standard functions. One of these cases is treated in the following example.

Example 9

Find a transformation that maps the interior of the rectangle with vertices at the points $z = \pm 1, \pm 1 + ic$ (Fig. 4-8*a*) onto the half-plane $\text{Im}\, \zeta > 0$; c is a positive constant.

In the formula (4-45) with $n = 4$, just three of the points $\xi_1, \ldots, \xi_4$ may be chosen arbitrarily. However, because of the symmetry of the rectangular shape, we anticipate that a symmetric choice of the ξ_j is

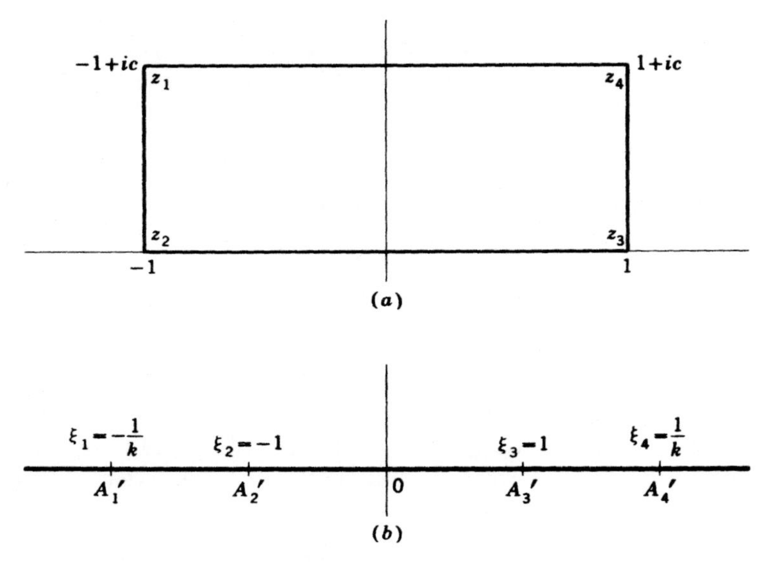

Fig. 4-8 (*a*) z plane; (*b*) ζ plane.

admissible. Hence, writing

$$z_1 = -1 + ic \qquad z_2 = -1 \qquad z_3 = +1 \qquad z_4 = 1 + ic$$

we choose $$\xi_1 = -\frac{1}{k} \qquad \xi_2 = -1 \qquad \xi_3 = +1 \qquad \xi_4 = +\frac{1}{k}$$

where k is a constant to be determined ($k < 1$). The symmetric location of the rectangle in the z plane suggests the choice $b = \zeta_0 = 0$. Since the interior angles of the rectangle are $\alpha_1 = \alpha_2 = \alpha_3 = \alpha_4 = \pi/2$, Eq. (4-45) yields the transformation formula

$$z = f(\zeta) = a \int_0^\zeta \frac{d\zeta}{\sqrt{(\zeta^2 - 1)(\zeta^2 - 1/k^2)}} \tag{4-46}$$

where the appropriate branch of the many-valued integrand is to be chosen so that, for Im $\zeta > 0$,

$$0 < \arg (\zeta - \xi_j) < \pi \qquad j = 1, \ldots, 4 \tag{4-47}$$

Writing $b = ka$, we can express the transformation in terms of the standard form for an elliptic integral of the first kind,

$$z = f(\zeta) \equiv b \int_0^\zeta \frac{d\zeta}{\sqrt{(1 - \zeta^2)(1 - k^2\zeta^2)}} \equiv bE(k,\zeta) \tag{4-48}$$

where the branch of the integrand must be chosen to conform with Eqs. (4-46) and (4-47).

The constants b and k in Eq. (4-48) are evaluated by requiring that the transformation yield the correct position for the vertices of the polygon in the z plane. Thus, with $O'A_3'$ as path of integration, the requirement $f(1) = 1$ yields the equation

$$bK(k) = 1 \tag{4-49}$$

where $K(k)$ is the complete elliptic integral of the first kind,

$$K(k) = E(k,1) \equiv \int_0^1 \frac{d\xi}{\sqrt{(1 - \xi^2)(1 - k^2\xi^2)}}$$

Similarly, with $O'A_3'A_4'$ as path of integration, the requirement $f(1/k) = 1 + ic$ yields the condition

$$1 + ic = f(1) + b \int_1^{1/k} \frac{d\zeta}{\sqrt{(1 - \zeta^2)(1 - k^2\zeta^2)}}$$

On the segment $A_3'A_4'$, $\sqrt{1 - k^2\zeta^2}$ is real, and $\sqrt{1 - \zeta^2}$ is pure imaginary. Hence, noting the choice of sign dictated by Eq. (4-47) and using $f(1) = 1$, we find that

$$bK_1(k) = c \tag{4-50}$$

where $$K_1(k) = \int_1^{1/k} \frac{d\xi}{\sqrt{(\xi^2 - 1)(1 - k^2\xi^2)}}$$

Equations (4-49) and (4-50) suffice to determine the unknown real constants k and b in Eq. (4-48) and thus the transformation $z = f(\zeta)$ that maps the half-plane $\operatorname{Im} \zeta > 0$ onto the interior of the rectangle. The required inverse transformation is obtained by inverting the functional relation (4-48),

$$\zeta = sn\left(\frac{z}{b}\,k\right) \tag{4-51}$$

where $sn(t,k)$, the jacobian elliptic function, is defined by the inversion process.[1]

EXERCISES

1. In mapping the half-plane $\operatorname{Im} \zeta > 0$ onto the interior of an n-sided polygon, it is often convenient to take $\xi_1 = -\infty$ (or $\xi_n = +\infty$). Show that, if $\xi_1 = -\infty$, the Schwarz-Christoffel formula has the form

$$z = b + a \int_{\zeta_0}^{\zeta} \prod_{j=2}^{n} (\zeta - \xi_j)^{\alpha_j/\pi - 1}\, d\zeta \tag{4-52}$$

2. If $-\infty < \xi_1 < \xi_2 < \cdots < \xi_n < +\infty$, what conditions must the real numbers $k_1, k_2, \ldots, k_n$ satisfy in order that the transformation

$$z = \int_{\zeta_0}^{\zeta} \prod_{j=1}^{n} (\zeta - \xi_j)^{-k_j}\, d\zeta$$

should map the half-plane $\operatorname{Im} \zeta > 0$ onto the interior of a finite closed polygon in the z plane? Consider polygons with n and with $n + 1$ sides. [*Hint:* Use contour integration to show that the integral representing any "closure gap" has a zero value for certain $\{k_j\}$.]

3. For some purposes it is desirable to map the interior of a polygon in the z plane onto the interior of the unit circle in the ζ plane. Show that $|\zeta| < 1$ is mapped conformally onto the interior of an n-sided polygon with interior angles $\alpha_1, \alpha_2, \ldots, \alpha_n$ by a transformation of the form

$$z = g(\zeta) \equiv B + A \int_{\zeta_0}^{\zeta} \prod_{j=1}^{n} (\zeta - \zeta_j)^{\alpha_j/\pi - 1}\, d\zeta \tag{4-53}$$

where $\zeta_1, \zeta_2, \ldots, \zeta_n$ are n points in counterclockwise order on the circle $|\zeta| = 1$ and correspond respectively to the vertices $z_1, z_2, \ldots, z_n$ of the polygon.

[1] A general discussion of properties of elliptic functions is given in E. T. Whittaker and G. N. Watson, "Modern Analysis," chap. 22, Cambridge University Press, New York, 1952.

4. (*a*) Taking advantage of the symmetry, find a transformation that maps $|\zeta| < 1$ conformally onto the interior of a regular n-sided polygon in the z plane.

(*b*) Deduce that the transformation

$$z = \int_0^\zeta (1 - \zeta^4)^{-\frac{1}{2}}\, d\zeta$$

maps $|\zeta| < 1$ onto the interior of a square, and find the length of the sides of that square.

(*c*) Show that the transformation

$$z = \int_0^\zeta \zeta^{-\frac{1}{2}}(1 - \zeta^2)^{-\frac{1}{2}}\, d\zeta$$

maps $\operatorname{Im} \zeta > 0$ onto the interior of a square. Find the length of the sides of the square in terms of the beta function defined by Eq. (5-28).

5. If $k > 1$, sketch the domain in the z plane which is the image of the half-plane $\operatorname{Im} \zeta > 0$ under the transformation

$$z = \int_0^\zeta \frac{d\zeta}{\zeta^{\frac{1}{2}}(1 - \zeta^2)^{\frac{1}{2}}(1 - k^2\zeta^2)^{\frac{1}{2}}}$$

6. (Alternative derivation of the Schwarz-Christoffel formula)

(*a*) Let $z = f(\zeta)$ be any transformation which maps the half-plane $\operatorname{Im} \zeta > 0$ conformally onto the domain D interior to a finite closed polygon with interior angles $\alpha_1, \alpha_2, \ldots, \alpha_n$ in the z plane. If $f(\zeta)$ is continued analytically into the half-plane $\operatorname{Im} \zeta < 0$ across the segment Γ'_j (Fig. 4-7*b*), show that $z = f(\zeta)$ then maps the half-plane $\operatorname{Im} \zeta < 0$ conformally onto the interior D_1 of a polygon obtained by reflecting D in its side Γ_j (Fig. 4-7*a*).

(*b*) If $g(\zeta)$ is the analytic continuation of $f(\zeta)$ from $\operatorname{Im} \zeta < 0$ into the half-plane $\operatorname{Im} \zeta > 0$ across the segment Γ'_k ($k \neq j$), how is $g(\zeta)$ related to the original function $f(\zeta)$ in the half-plane $\operatorname{Im} \zeta > 0$?

(*c*) Show that $f''(\zeta)/f'(\zeta)$ is single-valued in the whole ζ plane and that its only singularities are simple poles at the points $\xi_1, \xi_2, \ldots, \xi_n$. Hence (using Liouville's theorem) show that

$$\frac{f''(\zeta)}{f'(\zeta)} = \sum_{j=1}^{n} \frac{\alpha_j/\pi - 1}{\zeta - \xi_j}$$

and deduce Eq. (4-45).

7. The domain D is defined by the inequalities

$$|z - 2\sqrt{2}| < 4 \qquad |z + 2\sqrt{2}| < 4 \qquad |z - i\sqrt{2}| > \sqrt{2}$$

Find the bounded solution of Laplace's equation $\Delta\varphi = 0$ in D, subject to the conditions that $\varphi = -1$ on the arc $|z - 2\sqrt{2}| = 4$, $\varphi = +1$ on the arc $|z + 2\sqrt{2}| = 4$, and $\varphi = 0$ on the arc $|z - i\sqrt{2}| = \sqrt{2}$.

Degenerate Polygons

The arguments used to establish the Schwarz-Christoffel formulas (4-45) and (4-52) are not valid when the polygon is degenerate, i.e., has one or more vertices at infinity. Nevertheless we shall apply these formulas for such degenerate polygons (thus admitting angles 0 and 2π and even negative angles), but with the provision that any transformation constructed in this way should subsequently be checked to ensure that it does have the desired mapping properties.

Example 10

The semi-infinite strip D in the z plane (Fig. 4-9*a*) is defined by the inequalities

$$D: \ |\text{Re } z| < k \qquad \text{Im } z > 0$$

where k is a positive constant. Find the (least singular) complex potential $\Omega = \varphi + i\psi$ in D subject to the conditions (1) $\psi(x,0) = 0$, $-k < x < +k$; (2) $\psi(-k,y) = -1$, $\psi(+k,y) = +1$ for $y > 0$; and (3) $\Omega'(z) \to 0$ as $z \to \infty$.

As indicated in Fig. 4-9*a*, the domain D may be regarded as a degenerate triangle with interior angles

$$\alpha_1 = 0 \qquad \alpha_2 = \alpha_3 = \frac{\pi}{2}$$

To map the half-plane Im $\zeta > 0$ onto D, we use Eq. (4-52), with $n = 3$, and choose

$$\zeta_0 = 0 \qquad \xi_1 = -\infty \qquad \xi_2 = -1 \qquad \xi_3 = +1$$

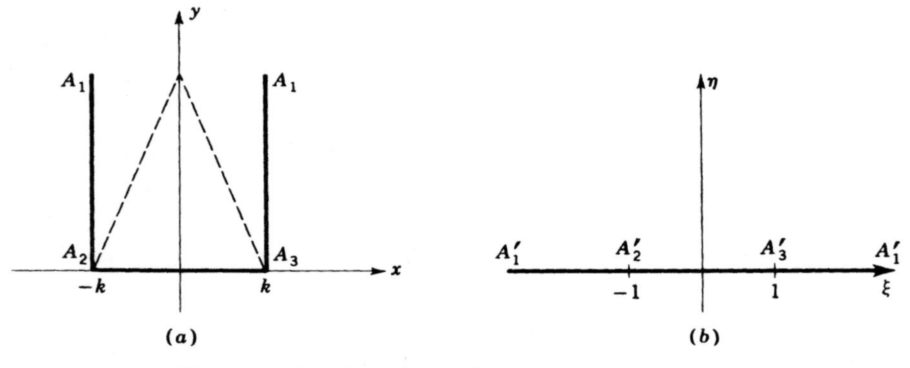

Fig. 4-9 (*a*) z plane; (*b*) ζ plane.

The required transformation then has the form:

$$z = f(\zeta) \equiv a \int_0^\zeta \frac{d\zeta}{(1 - \zeta^2)^{1/2}} + b = a \arcsin \zeta + b \tag{4-54}$$

In Eq. (4-54), we have written the integrand in the form $(1 - \zeta^2)^{-1/2}$ instead of $(\zeta^2 - 1)^{-1/2}$, a factor $\pm i$ being thereby absorbed into the constant a.

The constants a and b are determined by requiring that the points A_2', A_3' correspond to the points A_2, A_3, respectively, that is, $f(-1) = -k$ and $f(+1) = +k$. Identifying the branch of $\arcsin \zeta$ by the condition

$$-\frac{\pi}{2} < \arcsin \xi < +\frac{\pi}{2} \qquad \text{on } A_2'A_3'$$

we find that $a = 2k/\pi$ and $b = 0$. The transformation that maps the half-plane $\operatorname{Im} \zeta > 0$ onto the domain D is then

$$z = \frac{2k}{\pi} \arcsin \zeta$$

and the inverse transformation is

$$\zeta = \sin \frac{\pi z}{2k} \tag{4-55}$$

From Eq. (4-41), the complex potential in the ζ plane is

$$\chi(\zeta) = i - \frac{1}{\pi} \ln (\zeta^2 - 1) \tag{4-56}$$

Substituting for ζ from Eq. (4-55), we obtain the complex potential for the original problem in the z plane,

$$\Omega(z) = i - \frac{1}{\pi} \ln [\sin^2 (\pi z/2k) - 1] \tag{4-57}$$

It is often necessary to perform a sequence of transformations in order to map a given domain D conformally onto a desired standard domain. More often than not, it then turns out that one or more of the intermediate transformations cannot be inverted explicitly in terms of known functions. In such cases we have to content ourselves with a solution of the boundary-value problem in terms of a complex parameter. These features are illustrated by the following example.

Example 11

The figure Γ consists of the circle $|z| = a$ with a tangential line segment $y = a$, $0 \leq x \leq 2a$ (Fig. 4-10a), and D is the domain exterior to Γ.

Find the complex potential $\Omega \equiv \varphi + i\psi$ in D subject to the conditions (1) $\varphi = 0$ on Γ; (2) $\Omega'(z) \to E \equiv E_1 - iE_2$ (const) as $z \to \infty$; and (3) the net flux across any closed curve in D vanishes.

We first apply the bilinear transformation

$$t = (1 + i)\,\frac{z - a}{z - ia} \tag{4-58}$$

to map D onto the domain D', which consists of the half-plane $\operatorname{Im} t > 0$ with a semi-infinite horizontal slit along the line $\operatorname{Im} t = 1$, $\operatorname{Re} t \leq 0$ (Fig. 4-10*b*). The inverse transformation is

$$z = ia\,\frac{t - (1 - i)}{t - (1 + i)} \equiv ia - \frac{2a}{t - (1 + i)} \tag{4-59}$$

[The factor $z - a$ in Eq. (4-58) was chosen arbitrarily to make the image of the circle $|z| = a$ pass through the origin $t = 0$. The factor $1 + i$ has been included to yield a convenient orientation of the domain D'.]

As indicated by the labeling $A_1'A_2'A_3'A_1'$ in Fig. 4-10*b*, the domain D' may be regarded as the interior of a degenerate triangle. The interior angles of the triangle at the vertices A_2' and A_3' are $\alpha_2 = 2\pi$ and $\alpha_3 = 0$, respectively. To map the domain D' onto the half-plane $\operatorname{Im} \zeta > 0$, we use the Schwarz-Christoffel formula (4-52) with $n = 3$ and with

$$\xi_1 = -\infty \qquad \xi_2 = -1 \qquad \xi_3 = 0$$

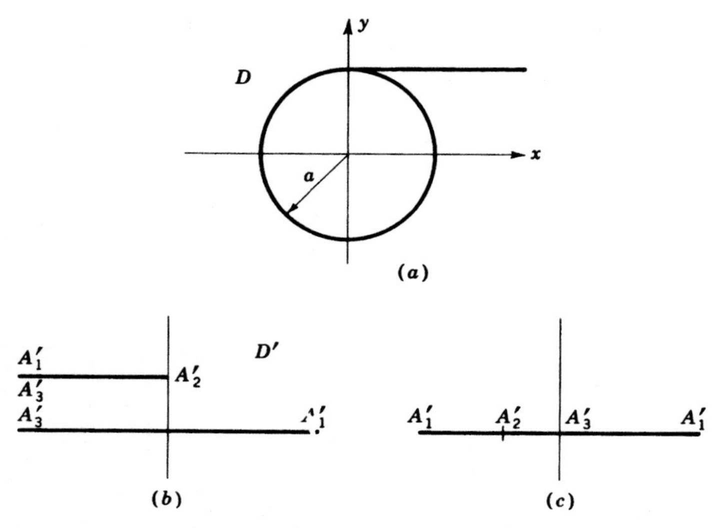

Fig. 4-10 (*a*) z plane; (*b*) t plane; (*c*) ζ plane.

(The choice $\xi_1 = -\infty$ enables us to sidestep the question of assigning a value to the strange angle α_1 at the vertex A_1'.) The required transformation then has the form

$$t = f(\zeta) \equiv b \int \frac{\zeta + 1}{\zeta} d\zeta + c = b(\zeta + \ln \zeta) + c \tag{4-60}$$

If the branch of the logarithm in Eq. (4-60) is specified by the condition that, for $\operatorname{Im} \zeta > 0$,

$$\ln \zeta = \ln |\zeta| + i \arg \zeta \qquad 0 < \arg \zeta < \pi$$

the constants b and c are real. As the point ζ traverses an infinitesimal upper semicircle about $\zeta = 0$ in the clockwise direction, $f(\zeta)$ changes by $-ib\pi$ and, at the same time, the point t goes from A_3' on the slit to A_3' on the real t axis (Fig. 4-10*b*). That is, t changes by $-i$; hence $b = 1/\pi$. The requirement $f(-1) = i$ then implies that $c = 1/\pi$. The half-plane $\operatorname{Im} \zeta > 0$ is thus mapped onto the domain D' by the transformation

$$t = \frac{1}{\pi} (\zeta + \ln \zeta + 1) \tag{4-61}$$

From Eqs. (4-59) and (4-61), the half-plane $\operatorname{Im} \zeta > 0$ is mapped onto the original domain D in the z plane by the transformation

$$z = ia - \frac{2\pi a}{\zeta + \ln \zeta + 1 - \pi(1 + i)} \tag{4-62}$$

Let the point $\zeta = \zeta_0$ correspond to $z = \infty$. The (unique) value of ζ_0 can be found by numerical solution of the transcendental equation

$$\zeta_0 + \ln \zeta_0 + 1 - \pi(1 + i) = 0 \tag{4-63}$$

The transformation (4-62) can then be written in the form

$$z = ia - \frac{2\pi a}{\zeta - \zeta_0 + \ln \zeta - \ln \zeta_0} \tag{4-64}$$

The condition at infinity in the z plane transforms into a condition at the point ζ_0 in the ζ plane. The derivative of the complex potential $\chi(\zeta)$ in the ζ plane can be expressed in the form:

$$\frac{d\chi}{d\zeta} = \frac{d\Omega}{dz}\frac{dz}{d\zeta} = \frac{2\pi a(1 + 1/\zeta)}{(\zeta - \zeta_0 + \ln \zeta - \ln \zeta_0)^2} \frac{d\Omega}{dz}$$

It follows that

$$\lim_{\zeta \to \zeta_0} (\zeta - \zeta_0)^2 \frac{d\chi}{d\zeta} = 2\pi E a \left(1 + \frac{1}{\zeta_0}\right)^{-1}$$

which implies a dipole of strength $2\pi Ea(1 + 1/\zeta_0)^{-1}$ at the point ζ_0. Condition (3) of the data implies that there is not also a point source at the point ζ_0.

The problem has now been reduced to that of finding the field in the half-plane Im $\zeta > 0$ due to a dipole of strength $2\pi Ea(1 + 1/\zeta_0)^{-1}$ at the point ζ_0 and with the condition that $\Phi \equiv \operatorname{Re} \chi = 0$ on the ξ axis. The complex potential $\chi(\zeta)$ is easily found by the method of images,

$$\chi(\zeta) = -2\pi a \left[\frac{E\zeta_0}{(1 + \zeta_0)(\zeta - \zeta_0)} - \frac{E^*\zeta_0^*}{(1 + \zeta_0^*)(\zeta - \zeta_0^*)} \right] \tag{4-65}$$

The transformation (4-64) cannot be inverted explicitly in terms of standard functions. We therefore accept Eqs. (4-64) and (4-65) as providing a parametric representation of the solution of the original problem in terms of the complex parameter ζ, Im $\zeta > 0$.

In arriving at the transformation (4-61) from the Schwarz-Christoffel formula, we were able to avoid assigning a value to the strange angle α_1 by choosing $\xi_1 = -\infty$. For the sake of completeness, however, we note that, when such angles at infinity are not immediately obvious (e.g., the angle α_3), they can often be found from the relation (4-42) for the interior angles. In the present case this formula implies $\alpha_1 = \pi - 2\pi - 0 = -\pi$.

For polygonal domains that involve several strange angles it is preferable to evaluate the exterior angles $\gamma_j = \pi - \alpha_j$. We recall that γ_j represents the angle of rotation (in counterclockwise direction) required to bring the direction of Γ_{j-1} into coincidence with the direc-

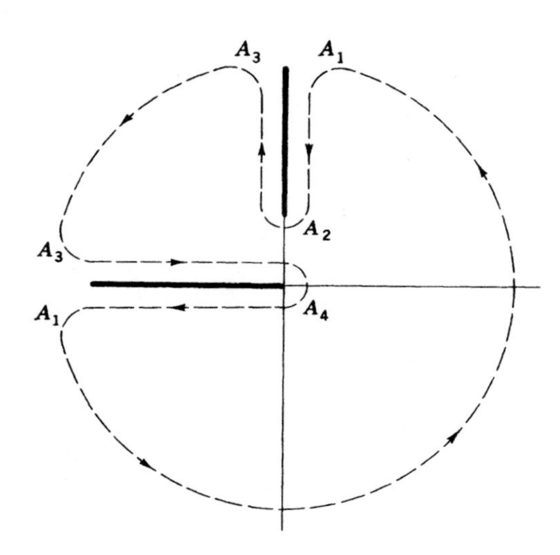

Fig. 4-11

tion of Γ_j. For example, if D is the domain exterior to the two half-lines $y = 0$, $x \leq 0$ and $x = 0$, $y \geq a > 0$ (Fig. 4-11), we find that

$$\gamma_1 = \frac{5\pi}{2} \qquad \gamma_2 = -\pi \qquad \gamma_3 = +\frac{3\pi}{2} \qquad \gamma_4 = -\pi$$

Note that $\gamma_1 + \gamma_2 + \gamma_3 + \gamma_4 = 2\pi$, in conformity with Eq. (4-42).

The task of mapping a simply connected domain onto a half-plane can often be simplified by taking advantage of any symmetry that may exist in the shape of the domain and by using the Schwarz reflection principle (Chap. 2). Similarly, if a multiply connected domain D is sufficiently symmetric, the solution of a potential problem for D can sometimes be reduced to the solution of a problem for a related simply connected domain. This kind of reduction is illustrated by the following example.

Example 12

D is the domain exterior to the two horizontal slits $y = \pm 1$, $-a \leq x \leq +a$ (Fig. 4-12a). Find the complex potential $\Omega \equiv \varphi + i\psi$ in D subject to the conditions (1) $\varphi = +1$ on the upper slit, $\varphi = -1$

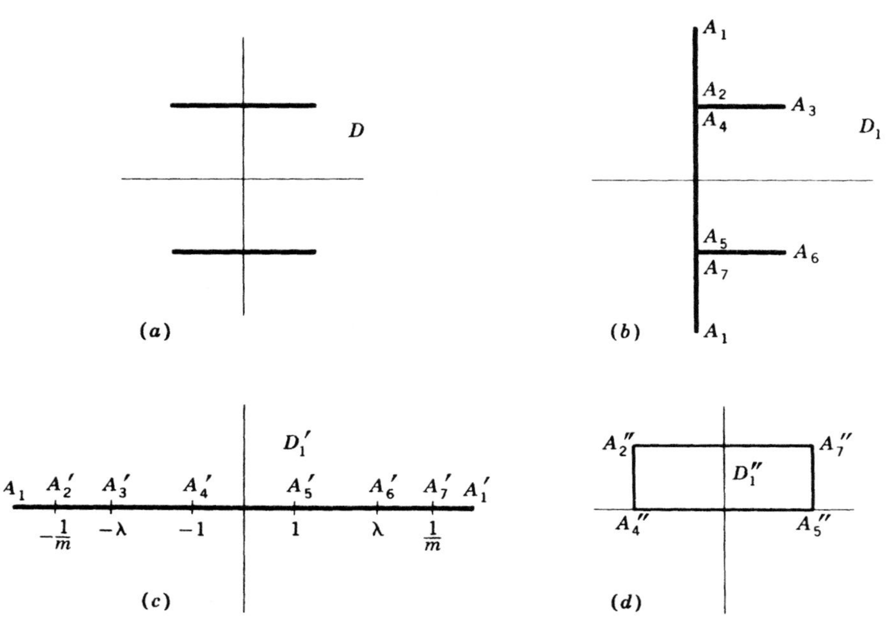

Fig. 4-12 (a) z plane; (b) z plane; (c) ζ plane; (d) t plane.

on the lower slit; (2) $\Omega'(z) \to 0$ as $z \to \infty$; and (3) $\Omega(z)$ is at most logarithmically singular on the boundary.

By symmetry, the segment $x = 0$, $-1 < y < +1$ is a streamline which we shall label $\psi = 0$. Similarly, the segments $x = 0$, $y < -1$ and $x = 0$, $y > +1$ together constitute a streamline $\psi = \beta$, where the appropriate value of β will have to be determined (symmetry requires β to be the same on each segment). Since the equipotentials and streamlines are clearly symmetric about the line $x = 0$, it is sufficient to solve a potential problem for the simply connected domain D_1 (Fig. 4-12*b*) consisting of the half-plane Re $z > 0$ exterior to the two projecting slits $y = \pm 1$, $0 \leq x \leq a$. The complex potential $\Omega_1(z)$ in D_1 satisfies the mixed conditions

$$\begin{aligned} \psi &= \beta \text{ (unknown)} && \text{on } A_1A_2 \text{ and } A_7A_1 \\ \varphi &= +1 && \text{on } A_2A_3A_4 \\ \psi &= 0 && \text{on } A_4A_5 \\ \varphi &= -1 && \text{on } A_5A_6A_7 \\ \Omega_1'(z) &\to 0 && \text{as } z \to \infty \text{ in } D_1 \end{aligned}$$

As a first step, we map the half-plane Im $\zeta > 0$ (Fig. 4-12*c*) onto the domain D_1. The interior angles at the vertices $A_1, \ldots, A_7$ (Fig. 4-12*b*) are

$$\alpha_1 = -\pi \qquad \alpha_2 = \alpha_7 = \frac{\pi}{2} \qquad \alpha_3 = \alpha_6 = 2\pi \qquad \alpha_4 = \alpha_5 = \frac{\pi}{2}$$

where α_1 has been found by using Eq. (4-42) with $n = 7$. Since the domain D_1 is symmetric about the x axis, we use the Schwarz-Christoffel formula (4-45) with the symmetric choice,

$$\xi_1 = -\infty \qquad \xi_2 = -\frac{1}{m} \qquad \xi_3 = -\lambda \qquad \xi_4 = -1 \qquad \xi_5 = +1$$

$$\xi_6 = +\lambda \qquad \xi_7 = +\frac{1}{m}$$

where the constants m, λ are to be determined ($1/m > \lambda > 1$). The required transformation is

$$z = f(\zeta) \equiv p \int_0^\zeta \frac{\zeta^2 - \lambda^2}{\sqrt{(\zeta^2 - 1)(m^2\zeta^2 - 1)}}\, d\zeta \tag{4-66}$$

where p is a constant. Equation (4-66) can be expressed in terms of the standard forms for the elliptic integrals of the first and second kinds,

$$\begin{aligned} z &= f(\zeta) \\ &\equiv \frac{p}{m^2}\left[\int_0^\zeta \sqrt{\frac{m^2\zeta^2 - 1}{\zeta^2 - 1}}\, d\zeta + (1 - m^2\lambda^2)\int_0^\zeta \frac{d\zeta}{\sqrt{(\zeta^2 - 1)(m^2\zeta^2 - 1)}}\right] \end{aligned}$$

and so, using standard notation,

$$z = f(\zeta) \equiv \frac{p}{m^2} [E(m,\zeta) + (1 - m^2\lambda^2)F(m,\zeta)] \tag{4-67}$$

The quantities m, λ, and p can be found explicitly from the condition

$$f(1) = -i \qquad f(\lambda) = a - i \qquad f\left(\frac{1}{m}\right) = -i$$

The complex potential $\chi(\zeta)$ in the half-plane $\operatorname{Im} \zeta > 0$ (Fig. 4-12c) must satisfy the conditions

$$\begin{array}{ll} \Psi = \beta \text{ (unknown)} & \text{on } A_1'A_2' \text{ and } A_7'A_1' \\ \Phi = +1 & \text{on } A_2'A_4' \\ \Psi = 0 & \text{on } A_4'A_5' \\ \Phi = -1 & \text{on } A_5'A_7' \\ \chi'(\zeta) \to 0 & \text{as } \zeta \to \infty \end{array}$$

solve this mixed problem, we map the half-plane $\operatorname{Im} \zeta > 0$ onto the rior D_1'' of a rectangle in the t plane (Fig. 4-12d), the vertices A_2'', A_4'', A_5'', A_7'' corresponding, respectively, to the symmetrically placed points A_2', A_4', A_5', A_7' in the ζ plane. From Example 9, it follows that the transformation

$$t = F(m,\zeta) \equiv \int_0^\zeta \frac{d\zeta}{(\zeta^2 - 1)^{1/2}(m^2\zeta^2 - 1)^{1/2}} \tag{4-68}$$

maps the half-plane $\operatorname{Im} \zeta > 0$ onto the interior of the rectangle $A_2''A_4''A_5''A_7''$ with vertices at the points $t = \pm b$, $\pm b + ic$, where b and c are positive numbers defined by the relations

$$b = K(m) \equiv E(m,1)$$

$$ic = F\left(m, \frac{1}{m}\right) - b \equiv \int_1^{1/m} \frac{d\zeta}{(\zeta^2 - 1)^{1/2}(m^2\zeta^2 - 1)^{1/2}}$$

The complex potential $\Lambda(t)$ in D_1'' satisfies the conditions:

$$\begin{array}{ll} \operatorname{Im} \Lambda = \beta \text{ (unknown)} & \text{on } A_2''A_7'' \\ \operatorname{Re} \Lambda = +1 & \text{on } A_2''A_4'' \\ \operatorname{Im} \Lambda = 0 & \text{on } A_4''A_5'' \\ \operatorname{Re} \Lambda = -1 & \text{on } A_5''A_7'' \end{array}$$

By inspection, the potential in the t plane is

$$\Lambda(t) = -\frac{t}{b} \tag{4-69}$$

and the hitherto unknown value of β is found to be $\beta = -c/b$.

From Eqs. (4-67), (4-68), and (4-69), the complex potential for the problem in the domain D_1 of the z plane can be expressed in the parametric form

$$
\begin{aligned}
z &= \frac{p}{m^2}\,[E(m,\zeta) + (1 - m^2\lambda^2)F(m,\zeta)] \\
\Omega &= -\frac{1}{b}\,E(m,\zeta)
\end{aligned}
\tag{4-70}
$$

where ζ plays the role of a complex parameter, $\operatorname{Im}\zeta > 0$.

Exercises

In the following exercises a, b, c are given positive constants, and λ, μ, ν are real constants:

8. Γ_1 is the half-line $y = -a$, $x \leq 0$, Γ_2 is the half-line $y = +a$, $x \leq 0$, and D is the domain exterior to Γ_1 and Γ_2. Find the bounded solution of Laplace's equation $\Delta\varphi = 0$ in D subject to the conditions $\varphi = \lambda$ on Γ_1 and $\varphi = \mu$ on Γ_2. Solve the problem in two ways, (*a*) directly, and (*b*) by using the symmetry of the domain and the reflection principle. Sketch the equipotentials and streamlines of the field.

9. Γ_1 is the half-line $y = 0$, $x \leq -a$, Γ_2 is the half-line $y = 0$, $x \geq +a$, and D is the domain exterior to Γ_1 and Γ_2. Find the complex potential $\Omega = \varphi + i\psi$ in D subject to the conditions (1) ψ is bounded in D and (2) $\psi = \lambda$ on Γ_1, $\psi = \mu$ on Γ_2. Sketch the equipotentials and streamlines.

10. Γ_1 is the line $y = 0$, $-\infty < x < +\infty$, Γ_2 is the line $y = a$, $-\infty < x < +\infty$, and Γ_3 is the half-line $y = b$, $x \leq 0$ $(0 < b < a)$. D is the domain between Γ_1 and Γ_2 and exterior to Γ_3. Find the bounded solution of the equation $\Delta\varphi = 0$ in D subject to the conditions (1) $\varphi = 0$ on Γ_1 and Γ_2 and (2) $\varphi = \lambda$ on Γ_3.

11. Γ_1 consists of the line $y = 0$, $-\infty < x < \infty$, with a projecting segment $x = 0$, $0 \leq y \leq b$, and Γ_2 is the line $y = a$, $-\infty < x < +\infty$ $(0 < b < a)$. D is the domain between Γ_1 and Γ_2. Find the complex potential $\Omega = \varphi + i\psi$ in D if $\psi = 0$ on Γ_1, $\psi = \lambda$ on Γ_2, and ψ is bounded in D. At which points is $|\Omega'(z)| = \infty$?

12. Γ_1 is defined as in Exercise 11, and Γ_3 is the mirror image of Γ_1 obtained by reflection in the line $y = a$. D is the domain between Γ_1 and Γ_3. Find the complex potential $\Omega = \varphi + i\psi$ in D if $\psi = 0$ on Γ_1, $\psi = 2\lambda$ on Γ_2, and ψ is bounded in D.

13. A curve Γ_1 is composed of the three linear segments

$$y = -a,\ x \leq 0 \qquad x = 0,\ -a \leq y \leq -b \qquad y = -b,\ x \geq 0\ (a > b > 0)$$

and D is the domain bounded below by Γ_1. Find the complex potential $\Omega = \varphi + i\psi$ in D if (1) $\psi = 0$ on Γ_1 and (2) $\Omega'(z) \to U > 0$ as $z \to \infty$ in D.

14. (*a*) Γ_1 is defined as in Exercise 13, Γ_2 is the x axis, and D is the domain between Γ_1 and Γ_2. Find the complex potential $\Omega(z) = \varphi + i\psi$ in D subject to the conditions (1) ψ is bounded in D and (2) $\psi = 0$ on Γ_1, $\psi = \lambda$ on Γ_2.

(*b*) Γ_3 is the mirror image of Γ_1 obtained by reflection in the x axis, and D_1 is the domain between Γ_1 and Γ_3. Find $\Omega = \varphi + i\psi$ in D (ψ bounded) if $\psi = 0$ on Γ_1 and $\psi = \mu$ on Γ_3.

15. Γ_1 is the line $y = -a$, $-\infty < x < +\infty$, Γ_2 is the line $y = +a$, $-\infty < x < +\infty$, and Γ_3 is a continuous curve composed of the three straight segments

$$y = -b,\ x \leq 0 \qquad x = 0,\ -b \leq y \leq +b \qquad y = +b,\ x \leq 0 \quad (a > b > 0)$$

D is the domain bounded by Γ_1, Γ_2, and Γ_3. Find the complex potential $\Omega = \varphi + i\psi$ in D subject to the conditions that ψ is bounded in D, the net flux between Γ_1 and Γ_3 is λ, and the net flux between Γ_3 and Γ_2 is 2λ. Sketch the streamlines of the field, and find the points at which $|\Omega'(z)| = 0$ and $|\Omega'(z)| = \infty$.

16. Γ_1 is the half-line $y = 0$, $x \leq 0$, Γ_2 is the half-line $y = a$, $x \leq 1$, and Γ_3 is the half-line $y = b$, $x \leq 2$ $(0 < a < b)$. D is the domain exterior to Γ_1, Γ_2, and Γ_3. Find the bounded solution of Laplace's equation in D subject to the conditions that $\varphi = \lambda$ on Γ_1, $\varphi = \mu$ on Γ_2, and $\varphi = \nu$ on Γ_3.

17. The curve Γ_1 consists of the two half-lines $y = 0$, $x \geq 0$ and $x = 0$, $y \geq 0$, and Γ_2 is the half-line $y = a$, $x \geq 1$. D is the domain bounded by Γ_1 and Γ_2. Find the bounded solution of $\Delta\varphi = 0$ in D if $\varphi = \lambda$ on Γ_1 and $\varphi = \mu$ on Γ_2.

18. The curve Γ_1 consists of the two half-lines $\arg z = 3\pi/4$ and $\arg z = -3\pi/4$, and the curve Γ_2 consists of the two half-lines $\arg z = \pi/3$ and $\arg z = -\pi/3$. D is the domain between Γ_1 and Γ_2. Find the bounded solution of the equation $\Delta\varphi = 0$ in D if $\varphi = \lambda$ on Γ_1 and $\varphi = \mu$ on Γ_2.

19. Γ_1 consists of three segments

$$y = -a,\ x \leq -1 \qquad x = -1,\ -a \leq y \leq +a \qquad y = +a,\ x \leq -1$$

and Γ_2 is the mirror image of Γ_1 in the y axis. D is the domain exterior to Γ_1 and Γ_2. Find the bounded solution of Laplace's equation in D if $\varphi = \lambda$ on Γ_1 and $\varphi = \mu$ on Γ_2. Is the capacity finite, and if so, what is its value?

20. The curve Γ_1 consists of the two half-lines $y = +1$, $x \leq 0$ and $\arg z = \pi/4$, and Γ_2 is the mirror image of Γ_1 in the x axis.

D is the domain between Γ_1 and Γ_2. Find the complex potential $\Omega = \varphi + i\psi$ in D (ψ bounded) if $\psi = -\lambda$ on Γ_2 and $\psi = +\lambda$ on Γ_1 ($\lambda > 0$).

21. Γ_1 is the line segment $y = 0$, $-a \le x \le -1$, Γ_2 is the line segment $y = 0$, $1 \le x \le a(a > 1)$, and D is the doubly connected domain exterior to Γ_1 and Γ_2. Find the complex potential $\Omega = \varphi + i\psi$ in D subject to the conditions (1) $\partial\varphi/\partial n = 0$ on Γ_1 and on Γ_2, (2) $\Omega'(z) \to Q \equiv U - iV$ (const) as $z \to \infty$, and (3) the circulation vanishes around every closed contour that encloses Γ_1 or Γ_2. [*Hint:* Superpose solutions corresponding to $\Omega'(z) \to U$ and $\Omega'(z) \to -iV$ as $z \to \infty$.]

22. Repeat Exercise 21 with Γ_1 replaced by the segment $x = -1$, $-a \le y \le +a$ and Γ_2 replaced by the segment $x = +1$, $-a \le y \le +a$.

Mapping of the Exterior of a Polygon

We now look for a transformation $z = f(\zeta)$ that maps the half-plane $\operatorname{Im} \zeta > 0$ onto the domain D_1 exterior to a finite, closed, n-sided polygon in the z plane. As the point ζ describes the ξ axis from $-\infty$ to $+\infty$, the image point z describes the polygonal boundary once in the clockwise direction (Fig. 4-13a). We retain the notation used in mapping the interior of the polygon (Fig. 4-7a and b) with the single modification that the vertices

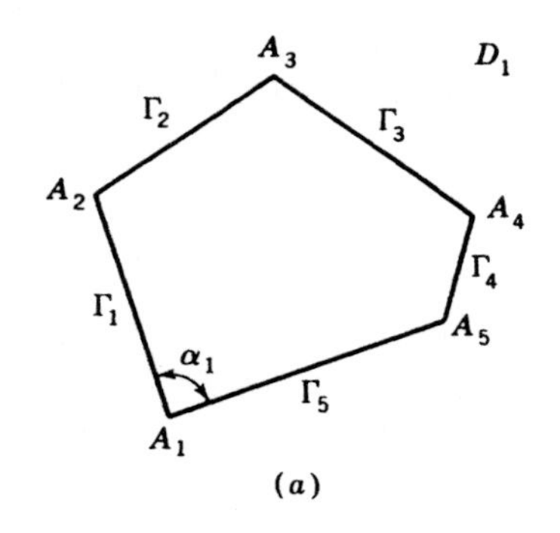

(a)

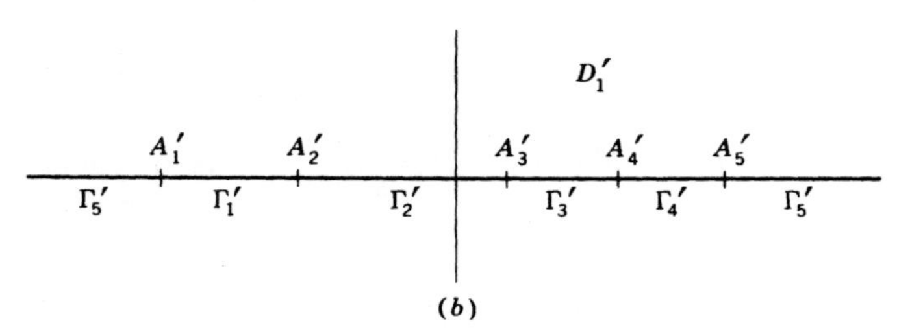

(b)

Fig. 4-13 (a) z plane; (b) ζ plane.

$z_1, z_2, \ldots, z_n$ and sides $\Gamma_1, \Gamma_2, \ldots, \Gamma_n$ of the polygon are now enumerated in clockwise order (Fig. 4-13*a* and *b*).

The angle interior to the domain D_1 at the vertex z_j is now $\beta_j = 2\pi - \alpha_j$, so that $\beta_j/\pi - 1 = 1 - \alpha_j/\pi$. For a reason which will become clear presently, it is convenient to represent the required transformation in the form

$$z = f(\zeta) \equiv \int_{\zeta_0}^{\zeta} g(\zeta) \prod_{j=1}^{n} (\zeta - \xi_j)^{1-\alpha_j/\pi}\, d\zeta + b \tag{4-71}$$

where ζ_0 and b are constants and where $g(\zeta)$ is a function yet to be determined. We then have

$$f'(\zeta) = g(\zeta) \prod_{j=1}^{n} (\zeta - \xi_j)^{1-\alpha_j/\pi} \equiv g(\zeta)h(\zeta) \tag{4-72}$$

The factor $h(\zeta)$ in Eq. (4-72) already possesses two of the properties demanded of the function $f'(\zeta)$. Its argument is constant on each segment Γ_j' of the ξ axis, and increases by an amount $-(\pi - \alpha_j)$ as the point ζ goes from Γ_j' to Γ_{j+1}'. We therefore require that

$$\arg g(\zeta) = \text{const} \qquad \text{on } \operatorname{Im} \zeta = 0 \tag{4-73}$$

To characterize $g(\zeta)$ further, we note that $z = \infty$ lies in the domain D_1. Let the point $\zeta = c$ (c finite and $\operatorname{Im} c > 0$) correspond to $z = \infty$. Then

$$f(c) = \int_{\zeta_0}^{c} g(\zeta)h(\zeta)\, d\zeta + b = \infty$$

where the path of integration is any curve from ζ_0 to c in the half-plane $\operatorname{Im} \zeta > 0$. Since $h(\zeta)$ is finite on the whole path of integration, it follows that

$$\lim_{\zeta \to c} g(\zeta) = \infty \tag{4-74}$$

Finally, since Eq. (4-42) implies that $h(\zeta) \sim \zeta^2$ as $\zeta \to \infty$, and since the transformation is conformal at $\zeta = \infty$ (in order that there be no change in angle at $\zeta = \pm\infty$ on the real axis), it follows that

$$\lim_{\zeta \to \infty} \zeta^4 g(\zeta) = \text{finite and} \neq 0 \tag{4-75}$$

Equations (4-73), (4-74), and (4-75) imply that $g(\zeta)$ has the form

$$g(\zeta) = \frac{a}{(\zeta - c)^2(\zeta - c^*)^2}$$

where a is a complex constant. From Eq. (4-71), the required transformation formula is

$$z = f(\zeta) \equiv a \int_{\zeta_0}^{\zeta} \frac{\prod_{j=1}^{n} (\zeta - \xi_j)^{1-\alpha_j/\pi}}{(\zeta - c)^2(\zeta - c^*)^2}\, d\zeta + b \tag{4-76}$$

Since $f'(\zeta)$ has a double pole at the point $\zeta = c$ ($\text{Im } c > 0$), the transformation is not conformal at that point.

EXERCISES

23. Find a transformation that maps the circular domain $|\zeta| < 1$ onto the exterior of the rectangle with vertices at the points $z = \pm 1$, $z = \pm 1 + ia$ ($a > 0$).

24. Examine the limiting forms of the transformation in Exercise 23 (1) as $a \to 0$ and (2) as $a \to \infty$.

25. Show that the exterior of the unit circle, $|\zeta| > 1$, is mapped onto the domain D exterior to a finite, closed, n-sided polygon in the z plane by a transformation of the form

$$z = f(\zeta) = a \int_0^\zeta \prod_{j=1}^{n} (\zeta - \zeta_j)^{1-\alpha_j/\pi} \frac{d\zeta}{\zeta^2} + b \tag{4-77}$$

where $\zeta_1, \ldots, \zeta_n$ are points on the unit circle $|\zeta| = 1$, enumerated in a clockwise sense.

The Rounding Off of Corners

When the boundary Γ of a domain D has a sharp reentrant corner (for example, A in Fig. 4-14*a*), the intensity $\Omega'(z)$ may become infinite at that corner even if the real potential φ (or the streamfunction ψ) is continuous at that point. Strictly speaking, such reentrant corners should not occur in realistic models of physical situations. We could avoid the difficulty by replacing the sharp corner by a small circular arc (Fig. 4-14*b*) in such a way that the boundary has a continuously turning tangent. Generally, however, this procedure results in a considerable complication of the mathematical problem. For example, in order to map the domain of Fig. 4-14*b* onto the half-plane $\text{Im } \zeta > 0$, we would have to solve a complicated differential- or integral-equation problem.

A simpler alternative method for rounding off corners consists in first solving the problem with the reentrant corner and subsequently replacing

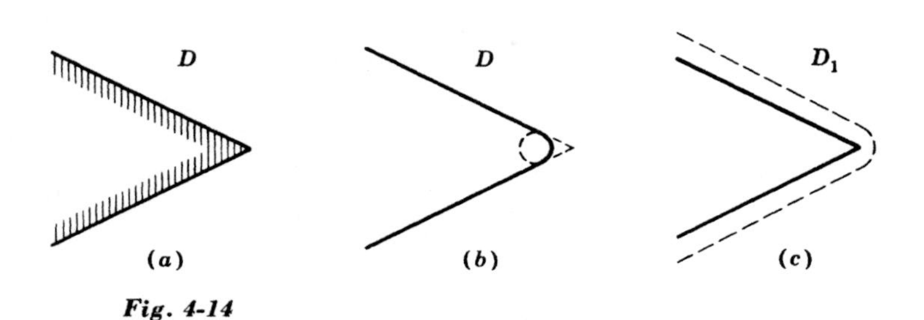

Fig. 4-14

the original domain D with a modified domain D_1 in the following way: If part of the boundary that contains the corner is an equipotential, we replace that whole segment of the boundary by a neighboring equipotential (Fig. 4-14c). However, although this device produces only a slight modification of the boundary in the neighborhood of the corner, it frequently results in considerable modifications at large distances from the corner.

A more general procedure consists in replacing the factor $(\zeta - \xi_j)^{\alpha_j/\pi - 1}$ associated with a reentrant corner in the Schwarz-Christoffel formula (4-45) by a factor of the form

$$(\zeta - \xi_j')^{\alpha_j/\pi - 1} + k(\zeta - \xi_j'')^{\alpha_j/\pi - 1} \qquad (4\text{-}78)$$

where k, ξ_j', ξ_j'' are real constants and $\xi_{j-1} < \xi_j' < \xi_j < \xi_j'' < \xi_{j+1}$. The modified boundary segment is of course only approximately circular.

Example 13

The domain D in the z plane (Fig. 4-15a) is bounded by the line $x = 0$ and by the half-lines $x = 1$, $y \leq 0$ and $y = 0$, $x \geq 1$. Find a transformation that maps the half-plane $\operatorname{Im} \zeta > 0$ onto some domain D_1 which differs from D by having the corner at the point $z = 1$ rounded off.

The domain D is the interior of a degenerate triangle $A_1A_2A_3A_1$ with interior angles $\alpha_1 = -\pi/2$, $\alpha_2 = 0$, $\alpha_3 = 3\pi/2$. In the modified form of the Schwarz-Christoffel formula with $n = 3$, we choose

$$\xi_1 = -\infty \qquad \xi_2 = 0 \qquad \xi_3 = 1 \qquad \xi_3' = \lambda \qquad \xi_3'' = \mu$$
$$0 < \lambda < 1 < \mu$$

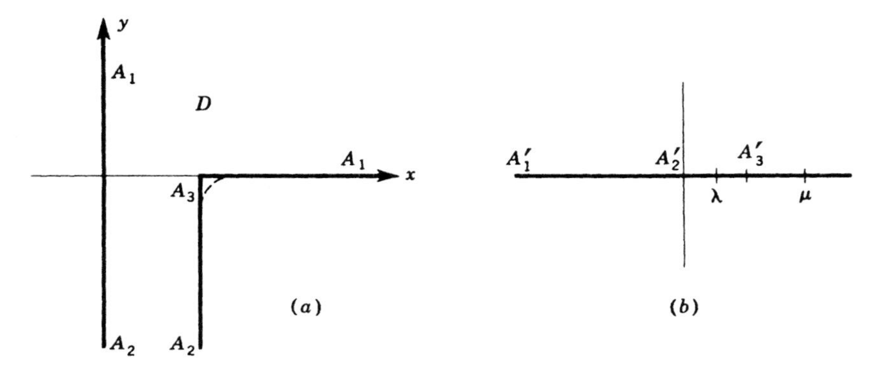

Fig. 4-15 (a) z plane; (b) ζ plane.

where λ and μ are for the moment unspecified. The transformation is then

$$z = f(\zeta) \equiv b + a \int [(\zeta - \lambda)^{1/2} + k(\zeta - \mu)^{1/2}] \frac{d\zeta}{\zeta}$$

$$= b + 2a\left\{(\zeta - \lambda)^{1/2} - \lambda^{1/2} \arctan\left(\frac{\zeta - \lambda}{\lambda}\right)^{1/2} + k\left[(\zeta - \mu)^{1/2} - \mu^{1/2} \arctan\left(\frac{\zeta - \mu}{\mu}\right)^{1/2}\right]\right\} \tag{4-79}$$

Exercising due care in the definition of branches for the multivalued functions, we evaluate the constant a from the property that, as ζ traverses an infinitesimal upper semicircle at $\zeta = 0$, $f(\zeta)$ changes from $0 - i\infty$ to $1 - i\infty$,

$$a = \frac{1}{\pi(\sqrt{\lambda} + k\sqrt{\mu})}$$

The values of the remaining constants λ, μ, k, b depend in part on the location of the end points of the modified segment of the boundary in the z plane. For example, let us require that the end points of the modified segment be at the points $z = 1 - i\epsilon$ and $z = 1 + \epsilon$, that is, $f(\lambda) = 1 - i\epsilon$ and $f(\mu) = 1 + \epsilon$. Then λ, μ, k, and b satisfy the relations

$$1 - i\epsilon = \frac{2ik}{(\pi\sqrt{\lambda} + k\sqrt{\mu})}\left[(\mu - \lambda)^{1/2} - \sqrt{\mu}\,\text{arctanh}\sqrt{\frac{\mu - \lambda}{\mu}}\right] + b \tag{4-80}$$

$$1 + \epsilon = \frac{2}{\pi(\sqrt{\lambda} + k\sqrt{\mu})}\left[(\mu - \lambda)^{1/2} - \sqrt{\lambda}\,\arctan\sqrt{\frac{\mu - \lambda}{\lambda}}\right] + b \tag{4-81}$$

The degree of arbitrariness in the choice of the disposable constants may be further reduced by requiring the point $\zeta = -1$, say, to correspond to a definite point on the line $x = 0$ (say the point $z = 0$). The possible transformations and corresponding domains are thereby reduced to a two-parameter family.

EXERCISES

26. Complete Example 13 by choosing $f(-1) = 0$, $\epsilon = \frac{1}{2}$, $\mu - \lambda = \frac{1}{2}$ and then determining corresponding values of k, λ, and μ. Estimate the maximum curvature on the modified boundary segment.

27. Γ_1 is the half-line $\arg z = \pi/4$, Γ_2 is the half-line $\arg z = 7\pi/4$, and D is the domain $\pi/4 < \arg z < 7\pi/4$. On the half-line Γ_2 and on the part of the half-line Γ_1 from $z = 0$ to $z = 1 + i$, the real potential

$\varphi = 0$; on the remaining part of the half-line Γ_1, $\varphi = 1$. Round off the corner at $z = 0$ in some way, and calculate the maximum value of $|\Omega'(z)|$ on the rounded-off segment.

4-5 The Joukowsky Transformation

Example 14

D is the domain in the z plane exterior to a slit A_1A_2 along the real axis from $x = -a$ to $x = +a$ (Fig. 4-16). Find the complex potential $\Omega = \varphi + i\psi$ in D subject to the conditions (1) $\psi = 0$ on the slit, (2) $\Omega'(z) \to Q \equiv U - iV$ (const) as $z \to \infty$, and (3) the circulation in the field is γ (real).

We look for a transformation $z = f(\zeta)$ that maps the domain $|\zeta| > 1$ onto D. To this end, we regard D formally as the domain exterior to a two-sided polygon A_1A_2 with interior angles $\alpha_1 = \alpha_2 = 0$. Equation (4-77), with the choice $\zeta_1 = -1$ and $\zeta_2 = +1$, implies that

$$z = f(\zeta) \equiv b \int (\zeta^2 - 1)\frac{d\zeta}{\zeta^2} + c = b\left(\zeta + \frac{1}{\zeta}\right) + c$$

where the constants b and c are to be evaluated by using the relations $f(-1) = -a$ and $f(+1) = +a$. The resulting transformation is the so-called *Joukowsky transformation*,

$$z = \tfrac{1}{2}a\left(\zeta + \frac{1}{\zeta}\right) \tag{4-82}$$

It has the property that $z = \infty$ corresponds to $\zeta = \infty$.

To find the inverse transformation $\zeta = F(z)$, we solve Eq. (4-82) for ζ,

$$\zeta = \frac{1}{a}\left(z \pm \sqrt{z^2 - a^2}\right) \tag{4-83}$$

and interpret $\sqrt{z^2 - a^2}$ as that branch of the function obtained by cutting the z plane along the segment A_1A_2 and requiring that

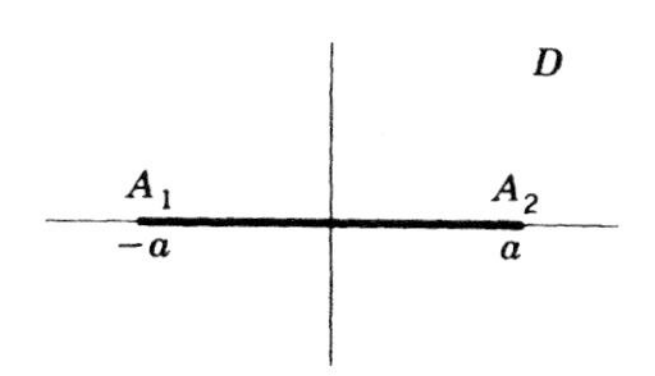

Fig. 4-16 z plane.

$\sqrt{z^2 - a^2}$ be positive for z real and greater than a. When $z \to \infty$ in Eq. (4-83), $\zeta \to \infty$ if the upper sign is chosen and $\zeta \to 0$ if the lower sign is chosen. Since $\zeta = \infty$ corresponds to $z = \infty$, the appropriate inverse transformation is

$$\zeta = F(z) \equiv \frac{1}{a}(z + \sqrt{z^2 - a^2}) \tag{4-84}$$

[We note that the discarded inverse transformation

$$\zeta = \frac{1}{a}(z - \sqrt{z^2 - a^2})$$

maps the domain D onto the *interior* of the unit circle $|\zeta| = 1$.]

The transformed potential $\chi(\zeta)$ for the domain $|\zeta| > 1$ satisfies the conditions:

1. $\Psi \equiv \text{Im}\,\chi = 0$ on the circle $|\zeta| = 1$.
2. $\dfrac{d\chi}{d\zeta} \equiv \dfrac{d\Omega}{dz}\dfrac{dz}{d\zeta} \to \frac{1}{2}aQ$ as $\zeta \to \infty$.
3. The circulation $= \gamma$.

It follows immediately from Eq. (4-19) that

$$\chi(\zeta) = \tfrac{1}{2}a\left(Q\zeta + \frac{Q^*}{\zeta}\right) - \frac{i\gamma}{2\pi}\ln\zeta$$

Applying the inverse transformation (4-84) and writing $Q = U - iV$, we obtain the complex potential for the original problem in the z plane,

$$\Omega(z) = Uz - iV\sqrt{z^2 - a^2} - \frac{i\gamma}{2\pi}\ln(z + \sqrt{z^2 - a^2}) \tag{4-85}$$

The complex intensity in the z plane is

$$\Omega'(z) = U - \frac{i(Vz + \gamma/2\pi)}{\sqrt{z^2 - a^2}} \tag{4-86}$$

If the circulation $\gamma \neq \pm 2\pi Va$, $|\Omega'|$ is infinite at both end points of the segment A_1A_2. On the other hand, if $\gamma = -2\pi Va$, Eq. (4-86) reduces to the form

$$\Omega'(z) = U - iV\sqrt{\frac{z - a}{z + a}} \tag{4-87}$$

Hence $\Omega'(a) = U$, and $|\Omega'|$ is finite at $z = a$ (but is of course still infinite at $z = -a$).

The choice $\gamma = -2\pi Va$ constitutes a special case of the *Kutta condition* by which the circulation γ in a field is selected in such a way as to make the intensity Ω' finite at a sharp corner, where it would otherwise be infinite. This kind of condition is often imposed in airfoil theory, as a means of determining the circulation around a wing.

The Joukowsky transformation

$$z = \tfrac{1}{2}a\left(\zeta + \frac{1}{\zeta}\right) \tag{4-88}$$

plays an important role in applications of conformal mapping, especially in fluid mechanics. Some useful properties of the transformation are exhibited in the following exercises.

EXERCISES

1. (*a*) Show that the Joukowsky transformation (4-88) can be expressed in the useful alternative form:

$$\frac{z-a}{z+a} = \left(\frac{\zeta - 1}{\zeta + 1}\right)^2 \tag{4-89}$$

(*b*) Show that the transformation doubles angles at the points $\zeta = \pm 1$.

2. Show that the transformation (4-88) maps the family of circles $|\zeta| = \rho$ into the family of confocal ellipses

$$\frac{x^2}{\tfrac{1}{4}a^2(\rho + 1/\rho)^2} + \frac{y^2}{\tfrac{1}{4}a^2(\rho - 1/\rho)^2} = 1$$

with $z = \pm a$ as common foci. Note that the circles $|\zeta| = \rho$ and $|\zeta| = 1/\rho$ map into the same ellipse and that the circle $|\zeta| = 1$ maps into the degenerate ellipse $y = 0$, $-a \leq x \leq +a$.

3. Show that the transformation (4-88) maps the family of half-lines $\arg \zeta = \varphi$ into the family of confocal hyperbolas (orthogonal to the family of ellipses in Exercise 2),

$$\frac{x^2}{a^2 \cos^2 \varphi} - \frac{y^2}{a^2 \sin^2 \varphi} = 1$$

Consider, in particular, the curves corresponding to $\varphi = 0, \pi/2, \pi$, and $3\pi/2$. Note that the half-lines $\arg \zeta = \varphi$ and $\arg \zeta = \varphi + \pi$ map into the same hyperbola.

4. If λ is real, $|\zeta - i\lambda| = \sqrt{\lambda^2 + 1}$ represents a circle Γ' through the points $\zeta = \pm 1$. Show that the transformation (4-88) or (4-89) maps the domain in the ζ plane exterior to Γ' onto the domain exterior to a circular arc in the z plane.

5. D' is the domain in the ζ plane exterior to the circle Γ': $|\zeta + k| = 1 + k$ $(k > 0)$. Show that the Joukowsky transformation (4-88) maps D' onto the domain in the z plane exterior to a profile Γ of

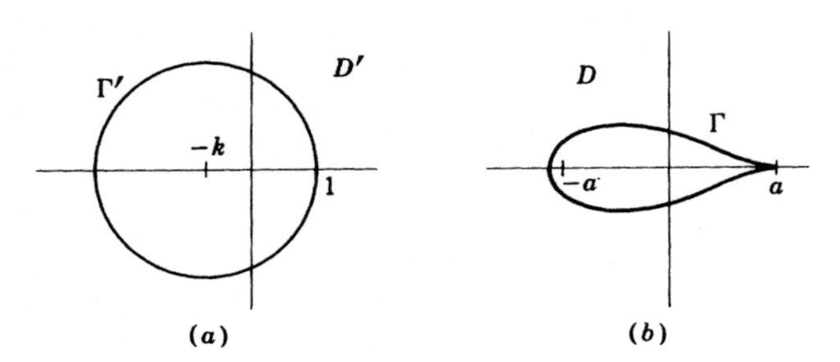

Fig. 4-17 (a) ζ plane; (b) z plane.

the form shown in Fig. 4-17; Γ has a cusp with horizontal tangent at $z = +a$.

6. D_1' is the domain in the ζ plane exterior to the circle Γ_1',

$$|\zeta + i\lambda + k(i\lambda + 1)| = (1 + k)\sqrt{\lambda^2 + 1}$$

where k and λ are real and $k > 0$. Show that the transformation (4-88) maps D_1' onto the domain exterior to a profile Γ_1 of the form shown in Fig. 4-18; Γ_1 has a cusp at the point $z = a$.

7. Examine the mapping properties of the *Kármán-Trefftz* mapping

$$\frac{z - ka}{z + ka} = \left(\frac{\zeta - a}{\zeta + a}\right)^k$$

where $1 < k < 2$. (Note that $k = 2$ gives the Joukowsky case.) In particular, compare the cusp angle (cf. Exercises 5 and 6) with that of the Joukowsky mapping.

Example 15

The figure Γ consists of the circle $|z| = a$ with two radial slits $y = 0$, $a \le |x| \le 2a$ (Fig. 4-19a), and D is the domain exterior to Γ. Find the complex potential $\Omega = \varphi + i\psi$ in D subject to the conditions

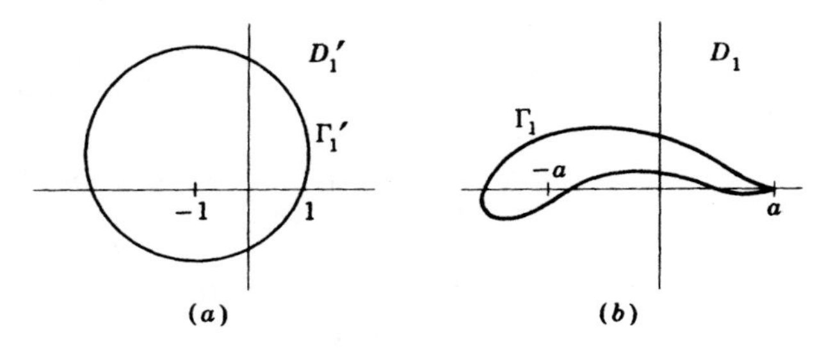

Fig. 4-18 (a) ζ plane; (b) z plane.

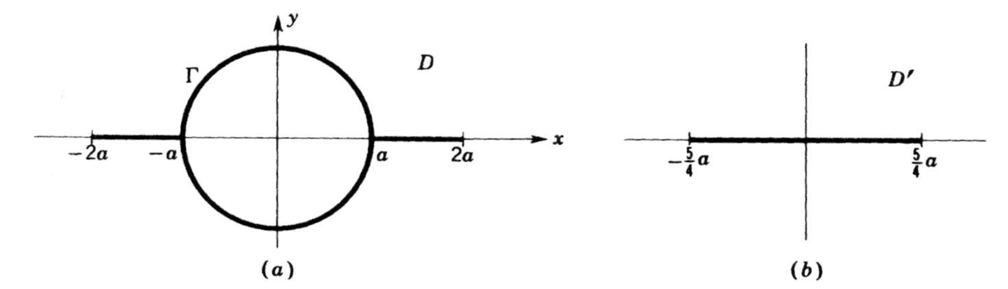

Fig. 4-19 (*a*) z plane; (*b*) t plane.

(1) $\psi = 0$ on Γ, (2) $\Omega'(z) \to Q \equiv U - iV$ (const) as $z \to \infty$, and (3) the circulation $\gamma = 0$.

We look for a transformation that maps the domain D onto the domain $|\zeta| > 1$. As a first step, the transformation

$$t = \frac{1}{2}\left(z + \frac{a^2}{z}\right) \tag{4-90}$$

maps D onto the domain D' in the t plane exterior to the slit $\operatorname{Im} t = 0$, $-\frac{5}{4}a \leq \operatorname{Re} t \leq +\frac{5}{4}a$ (Fig. 4-19*b*). A subsequent transformation

$$t = \tfrac{5}{8}a\left(\zeta + \frac{1}{\zeta}\right) \tag{4-91}$$

then maps the domain D' onto the domain $|\zeta| > 1$. The appropriate inverse of the transformation (4-91) is

$$\zeta = \frac{1}{5a}\,(4t + \sqrt{16t^2 - 25a^2}) \tag{4-92}$$

Combining Eqs. (4-90) and (4-92), we find the transformation that maps the domain D in the z plane onto $|\zeta| > 1$,

$$\zeta = \frac{1}{5az}\,[2(z^2 + a^2) + \sqrt{(4z^2 - a^2)(z^2 - 4a^2)}] \tag{4-93}$$

By a process which is by now familiar, we obtain as the solution to our problem

$$\Omega(z) = \frac{Q}{4z}\,[2(z^2 + a^2) + \sqrt{(4z^2 - a^2)(z^2 - 4a^2)}] + \frac{25a^2Q^*z}{8(z^2 + a^2) + 4\sqrt{(4z^2 - a^2)(z^2 - 4a^2)}} \tag{4-94}$$

EXERCISES

8. Γ is the circular arc $|z + ia| = \sqrt{2}a$, $\operatorname{Im} z \leq 0$, and D is the domain exterior to Γ. Find the complex potential $\Omega = \varphi + i\psi$ in D subject to the conditions (1) $\psi = 0$ on Γ, (2) $\Omega'(z) \to Q = U - iV$ (const) as $z \to \infty$, and (3) the circulation is adjusted to ensure that $\Omega'(z) = 0$ at $z = a$. Sketch the streamlines of the field for each of the cases (*a*) $U > 0$, $V = 0$, (*b*) $U = 0$, $V > 0$, and (*c*) $U = V > 0$.

9. A figure Γ is composed of the circle $|z| = a$ and a radial line segment $y = 0$, $a \leq x \leq b$. The complex potential $\Omega(z)$ for the domain D exterior to Γ satisfies the condition $\operatorname{Im} \Omega = 0$ on Γ and $\Omega'(z) \to Q = U - iV$ (const) as $z \to \infty$. Find the circulation required to make the intensity finite at the point $z = b$. Sketch the streamlines (*a*) when $U > 0$, $V = 0$, and (*b*) when $U = V > 0$.

10. Γ_1 is the figure of Exercise 9, with an additional radial projection $x = 0$, $a \leq y \leq c$. Repeat Exercise 9, with Γ and D replaced by Γ_1 and D_1.

11. Γ_3 is the circle $|z| = 1$, with two radial projections of length $a > 0$ along the directions of the positive and negative x axes and with two radial projections of length $b > 0$ along the directions of the positive and negative y axes. Find the potential Ω for the domain exterior to Γ_3 subject to the conditions (1) $\operatorname{Im} \Omega = 0$ on Γ_1, (2) $\Omega'(z) \to Q = U - iV$ as $z \to \infty$, and (3) the circulation $= \gamma$.

12. The figure Γ consists of the semicircle $|z| = a$, $\operatorname{Im} z \leq 0$, and two straight segments $x = 0$, $-b \leq y \leq -a$ $(b > a)$ and $x = 0$, $-a \leq y \leq c$ $(c > -a)$, and D is the domain exterior to Γ. Find the complex potential $\Omega = \varphi + i\psi$ in D subject to the conditions (1) $\varphi = 0$ on Γ, (2) $\Omega'(z) \to E = E_1 - iE_2$ (const) as $z \to \infty$, and (3) the net flux across any closed curve that surrounds Γ has the constant value λ (real).

4-6 *The Hodograph*

In certain physical problems involving Laplace's equation $\Delta\varphi = 0$, the basic domain is not specified completely in the original data but must in fact be discovered in the process of solving the problem. Problems of this kind occur primarily in fluid mechanics when surfaces of discontinuity or free surfaces are present in two-dimensional flow fields. The boundary of the domain consists of one or more streamlines whose shapes may not be known in advance.

Example 16

A domain D in the z plane is bounded by a curve $E_2C_2C_1E_1$ (Fig. 4-20*a*), where C_1C_2 is the straight segment $x = 0$, $-a \leq y \leq +a$, and C_1E_1,

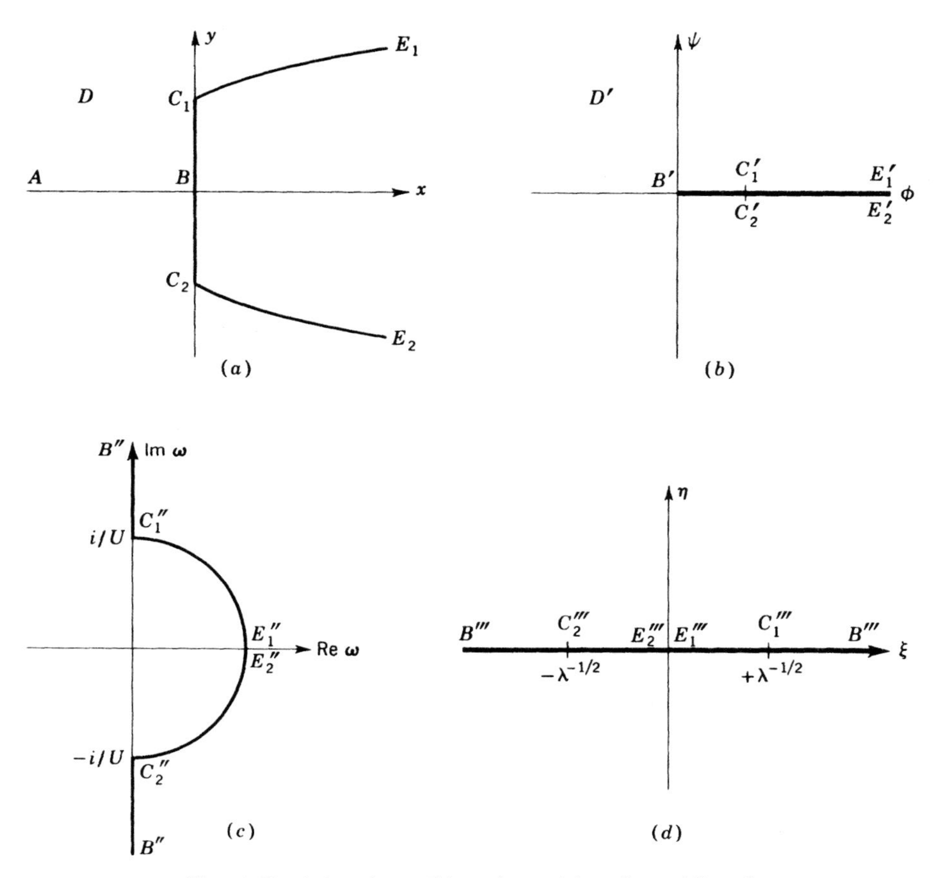

Fig. 4-20 (*a*) z plane; (*b*) Ω plane; (*c*) ω plane; (*d*) ζ plane.

C_2E_2 are as yet unspecified curves with E_1, E_2 at infinity. (When the boundary is described from E_2 to E_1, the domain D lies on the left.) Find the curves C_1E_1, C_2E_2 and the corresponding complex potential $\Omega = \varphi + i\psi$ in D so as to satisfy the following conditions:

1. ψ = const on the boundary $E_1C_1C_2E_2$.

2. $\Omega'(z) \to U > 0$ as $z \to \infty$ in D.

3. $|\Omega'(z)| = U$ on CE_1 and on CE_2.

4. The solution is to be symmetrical with respect to the real axis, in the sense that $\Omega(z) = [\Omega(z^*)]^*$.

(The corresponding physical problem is that of an ideal fluid moving against a flat plate extending from C_1 to C_2; the pressure in the stationary fluid region between the existing streamlines C_1E_1 and C_2E_2 is constant, and by Bernoulli's theorem this requires the velocity along C_1E_1 and C_2E_2 to be U.)

From the symmetry of the problem, the negative x axis AB is part of the streamline $\psi = 0$ that divides at B into two symmetric parts BC_1E_1 and BC_2E_2. Since the streamline through B has two different directions, we must have $\Omega'(B) = 0$. We seek a solution having the following (physically motivated) properties:

1. As the point z traverses the whole streamline $\psi = 0$ from left to right, the real potential φ increases monotonically from $-\infty$ at A to $+\infty$ at E_1 and E_2. By the suitable choice of an arbitrary constant, we may take $\varphi = 0$ at the point $z = 0$.

2. As the point z describes any equipotential from a point on ABC_1E_1 into the upper half-plane, ψ increases monotonically from 0 to ∞.

We now treat the functions $\Omega(z)$ and $\Omega'(z)$ as mapping functions and ask for the shapes of the regions into which these two functions map D. Considering first the relation $\Omega = \Omega(z)$, we see that this must provide a mapping of D onto the Ω plane cut along the positive φ axis, as shown in Fig. 4-20*b*. On the other hand, the relation

$$\omega = \frac{dz}{d\Omega} = \frac{1}{\Omega'(z)} \tag{4-95}$$

maps D onto some region D'' in the ω plane; a little thought will show that the boundary of D'' is as shown in Fig. 4-20*c*.

The essential feature of the present method, due to Helmholtz and Kirchhoff, is to recognize at this point that the mapping between the Ω and ω planes can now be determined, in the form

$$\omega = h(\Omega) \tag{4-96}$$

say. Once $h(\Omega)$ has been found, we use

$$z = \int_0^{\Omega} h(\Omega)\, d\Omega \tag{4-97}$$

to obtain the desired relation between z and Ω. The location of the curves C_1E_1 and C_2E_2 can be derived subsequently from Eq. (4-97).

To find $h(\Omega)$, we first apply the transformation $\zeta = \Omega^{-1/2}$ to map the cut Ω plane onto $\operatorname{Im} \zeta < 0$ (Fig. 4-20*d*); here λ denotes the unknown value of φ at C_1, C_2. The half-plane $\operatorname{Im} \zeta < 0$ can then be mapped onto D'' in the ω plane by a Joukowsky transformation in which $\zeta = \lambda^{-1/2}, -\lambda^{-1/2}$ are the images of $\omega = iU^{-1}, -iU^{-1}$, respectively,

$$\omega = \frac{1}{U}\,(i\lambda^{1/2}\zeta + \sqrt{1 - \lambda\zeta^2})$$

The substitution $\zeta = \Omega^{-1/2}$ then leads to

$$\omega = h(\Omega) \equiv \frac{\sqrt{\Omega - \lambda} + i\sqrt{\lambda}}{U\sqrt{\Omega}}$$

and from Eq. (4-97) it follows that

$$z = \frac{i}{U}\left[\tfrac{1}{2}\lambda \arccos\left(1 - \frac{2\Omega}{\lambda}\right) + \sqrt{\Omega(\lambda - \Omega)} + 2\sqrt{\lambda\Omega}\right] \quad (4\text{-}98)$$

Since $\Omega = \lambda$ corresponds to $z = \pm ia$, the value of λ can be determined (due care being exercised in the interpretation of multivalued functions),

$$\lambda = \frac{2Ua}{\pi + 4} \quad (4\text{-}99)$$

The problem is now solved, at least in implicit form. In parametric form, the curve CE has the equation

$$\begin{aligned} x &= \frac{1}{U}\left[\sqrt{\varphi(\varphi - \lambda)} - \lambda \ln\left(\sqrt{\frac{\varphi}{\lambda}} + \sqrt{\frac{\varphi}{\lambda} - 1}\right)\right] \\ y &= a + \frac{2}{U}(\sqrt{\lambda\varphi} - \lambda) \end{aligned} \quad (4\text{-}100)$$

where $\lambda < \varphi < \infty$.

Exercises

1. Γ_1 is the half-line $y = 0$, $x \leq -a$, and Γ_2 is the half-line $y = 0$, $x \geq a$ $(a > 0)$. The symmetric domain D (Fig. 4-21) is bounded by Γ_1, Γ_2 and by two unspecified curves Γ_3 and Γ_4 extending from $z = -a$ and $z = +a$, respectively, to infinity in the lower half-plane. $\Gamma_1 + \Gamma_3$ and $\Gamma_2 + \Gamma_4$ are streamlines, and the net flux across any curve between these streamlines is $Q > 0$. Find the curves Γ_3 and Γ_4 and the limiting value of the distance between these curves at ∞.

2. Repeat Exercise 1 when Γ_1 is the half-line $y = -(x + a)$, $x \leq -a$ and Γ_2 is the half-line $y = x - a$, $x \geq a$.

3. Repeat Exercise 1 when Γ_1 is the half-line $x = -a$, $y \geq 0$ and Γ_2 is the half-line $x = +a$, $y \geq 0$.

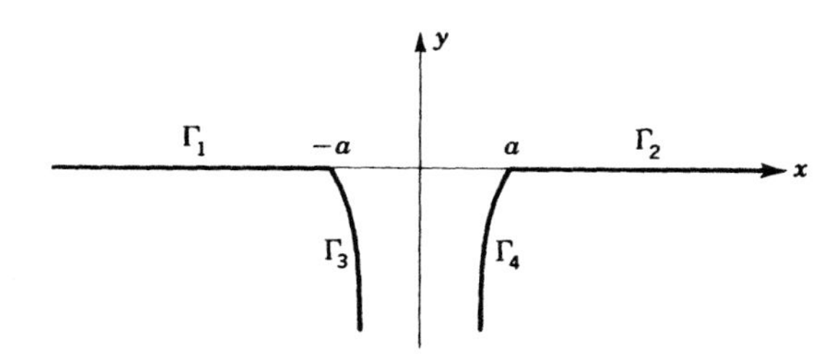

Fig. 4-21

4. Solve the problem of Example 16 when the vertical linear segment C_1C_2 is replaced by the segment $x = y$, $-1 \leq x \leq +1$. (Note that the problem is now not symmetrical.)

4-7 *Periodic Domains and Fields*

The following three examples illustrate the way in which it is sometimes possible to simplify a conformal-mapping problem by taking advantage of domain symmetries.

Example 17

Let D be the domain $|z| < R$ (Fig. 4-22*a*). A source of strength Q (real) is located at the point $z = 0$, and n sources each of strength q are located at the n points $z_k = ae^{i(2k+1)\pi/n}$, with $k = 0, 1, \ldots, n-1$ and $0 < a < R$. Find the complex potential $\Omega = \varphi + i\psi$ in D subject to the condition $\varphi = 0$ on the boundary $|z| = R$. This problem can be solved by a direct application of the method of images, but for illustrative purposes we shall first simplify the domain by an appropriate conformal mapping.

We apply the transformation

$$t = z^n \qquad z = t^{1/n} \tag{4-101}$$

to map the domain D onto the circular domain $|t| < R^n$ (Fig. 4-22*b*). The n points z_k all map into the single point $t = -a^n$.

We note that the transformation (4-101) actually maps D onto the domain $|t| < R^n$ on a Riemann t surface of n sheets; the n points z_k map into n distinct (but corresponding) points on the Riemann surface, one on each sheet. Since the transformation is conformal at the points

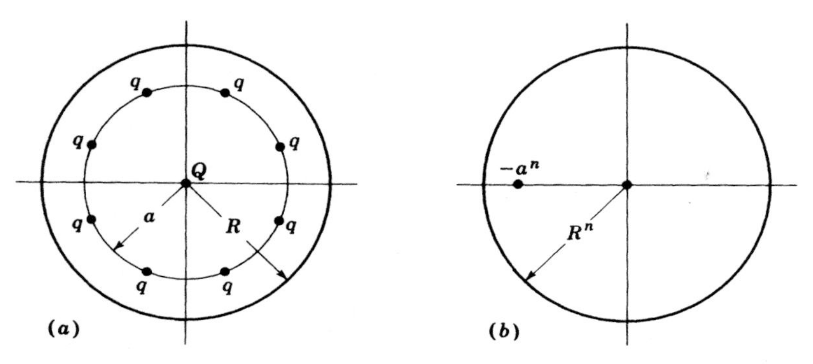

Fig. 4-22 (*a*) z plane; (*b*) t plane.

z_k, each source maps into a source of the same strength q on the corresponding sheet of the Riemann surface. Since $t = 0$ is a branch point of order n, the source of strength Q at $z = 0$ contributes a source of strength Q/n at $t = 0$ in each sheet of the Riemann t surface. The configurations are identical in all n sheets, and the real potential φ has the same values on both sides of any branch line. It is therefore sufficient to restrict our considerations to a single t sheet.

In the reduced problem for the domain $|t| < R^n$ in the simple t plane, there is a source of strength q at $t = -a^n$ and a source of strength Q/n at $t = 0$; the real potential vanishes on $|t| = R^n$. This problem can be solved at once by the method of images (which is here equivalent to the fact that, if ζ_0 is any point inside a circle of radius ρ centered on the origin, then $\ln [(\zeta - \zeta_0)/(\zeta - \rho^2/\zeta_0^*)]$ has a constant real part for ζ on the circle) to give, in terms of the original variables,

$$\Omega(z) = Q \ln \frac{z}{R} + q \ln \frac{R^n(z^n + a^n)}{R^{2n} + z^n a^n} \tag{4-102}$$

In the immediate neighborhoods of the points $z = 0$ and $z = z_k$, the equipotentials are approximately circles centered on these points. The complex potential $\Omega(z)$ of Eq. (4-102) thus also provides an approximate solution to a potential problem for the domain D_1 defined by

$$|z| < R \qquad |z| > \delta \qquad |z - z_k| > \epsilon \qquad k = 0, 1, \ldots, n - 1$$

where $\delta \ll a$ and $\epsilon \ll a$. The constants Q, q in Eq. (4-102) can then be chosen to make the real potential φ satisfy (approximately) the conditions $\varphi = 0$ on $|z| = R$, $\varphi = \lambda$ on $|z| = \delta$, and $\varphi = \mu$ on $|z - z_k| = \epsilon$ $(k = 0, 1, \ldots, n - 1)$, where λ and μ are prescribed real constants.

Example 18

A figure Γ in the z plane (Fig. 4-23a) consists of the circle $|z| = 1$ with N radial projections defined by

$$\arg z = \frac{(2k + 1)\pi}{N} \qquad 1 \leqq |z| \leqq a; \qquad k = 0, 1, \ldots, N - 1$$

with $a > 1$. Find the complex potential $\Omega = \varphi + i\psi$ in the domain D exterior to Γ subject to the conditions (1) $\varphi = 0$ on Γ, (2) the circulation in the field vanishes, and (3) $\Omega'(z) \to Q = U - iV$ (const) as $z \to \infty$.

To solve this problem, we shall map D onto the domain $|\zeta| > 1$, taking advantage of the symmetries of D (we note that the boundary-value problem itself does not possess these symmetries; hence the z^n

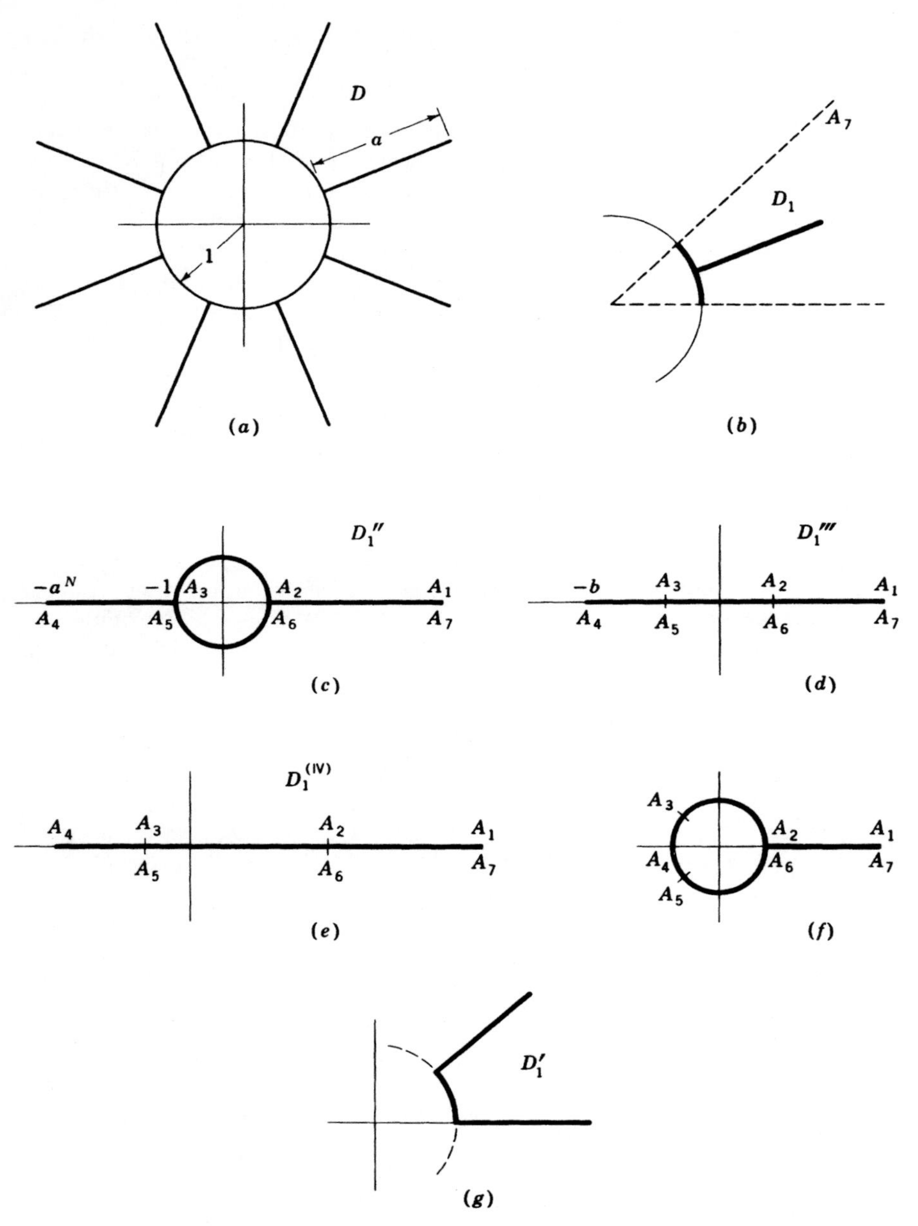

Fig. 4-23 (*a*) z plane; (*b*) z plane; (*c*) t plane; (*d*) w plane; (*e*) λ plane; (*f*) ρ plane; (*g*) ζ plane.

mapping of Example 17 would not be directly useful). We shall base the derivation of the mapping function on an analytic-continuation technique, using the Schwarz reflection principle.

To this end we consider the subdomain D_1 in the z plane (Fig. 4-20b) obtained from D by introducing two half-lines

$$\arg z = 0, \ |z| \geqq 1 \qquad \text{and} \qquad \arg z = \frac{2\pi}{N}, \ |z| \geqq 1$$

as auxiliary boundaries. We then look for a transformation $\zeta = f(z)$ that maps the domain D_1 onto the domain D_1' defined by

$$|\zeta| > 1 \qquad 0 < \arg \zeta < \frac{2\pi}{N}$$

Since $g(z) = f(z)/z$ is real on the boundary $\arg z = 2\pi/N$, it may be continued by schwarzian reflection across this boundary and it therefore follows that the analytic continuation of $f(z)$ across the half-line $\arg z = 2\pi/N$, $|z| > 1$, provides a transformation that maps the mirror image of the domain D_1 in the line $\arg z = 2\pi/N$ onto the mirror image of the domain D_1' in the line $\arg \zeta = 2\pi/N$. The $(N - 1)$fold repetition of this reflection process then yields a transformation that maps the original domain D_1 conformally onto the domain $|\zeta| > 1$. (It may be verified a posteriori that there is no discontinuity on the half-line $\arg z = 0$, $|z| > 1$.)

To begin with, the transformation

$$t = z^N \qquad 0 < \arg z < \frac{2\pi}{N} \tag{4-103}$$

maps D_1 onto a domain D_1'' in the t plane cut from $t = 1$ to $t = \infty$ along the positive real t axis (Fig. 4-23c). D_1'' is the domain in the cut plane exterior to the figure consisting of the circle $|t| = 1$ with a single projecting line segment $\operatorname{Im} t = 0$, $-a^N \leqq \operatorname{Re} t \leqq -1$; the bounding half-lines of D_1 map into the two sides of the cut. The transformation

$$w = \frac{1}{2}\left(t + \frac{1}{t}\right) \tag{4-104}$$

maps D_1'' onto a domain D_1''' in the w plane cut from $w = 1$ to $w = \infty$ along the positive real w axis (Fig. 4-23d). D_1''' is the domain exterior to the line segment $\operatorname{Im} w = 0$, $-b \leqq \operatorname{Re} w \leqq \infty$, where we have written

$$b = \tfrac{1}{2}(a^N + a^{-N}) \tag{4-105}$$

To center the line segment A_2A_4 at the origin, we apply the transformation

$$\lambda = w + \tfrac{1}{2}(b - 1) \tag{4-106}$$

which maps D_1''' onto the domain $D_1^{(iv)}$ in the λ plane (shown in Fig. 4-23*e*) exterior to the line segment

$$\text{Im}\,\lambda = 0 \qquad -\tfrac{1}{2}(b+1) \leqq \text{Re}\,\lambda \leqq \infty$$

We now apply the transformation

$$\lambda = \frac{1}{4}(b+1)\left(\rho + \frac{1}{\rho}\right) \tag{4-107}$$

to map $D_1^{(iv)}$ onto the domain $D_1^{(v)}$ exterior to the circle $|\rho| = 1$ in the ρ plane cut from $\rho = 1$ to $\rho = \infty$ along the positive real ρ axis (Fig. 4-23*f*). Finally the transformation

$$\zeta = \rho^{1/N} \qquad 0 < \arg \rho < 2\pi \tag{4-108}$$

maps $D_1^{(v)}$ onto the domain D_1' in the ζ plane (Fig. 4-23*g*).

Combining Eqs. (4-103) to (4-108), we find that the domain D_1 in the z plane is mapped onto the domain D_1' in the ζ plane by the transformation

$$\zeta = (b+1)^{-N}\{z^N + z^{-N} + b - 1 \\ + [(z^N + z^{-N} + 2b)(z^N + z^{-N} - 2)]^{1/2}\}^{1/N} \tag{4-109}$$

with $0 < \arg z < 2\pi/N$. The same transformation (4-109) with $0 \leqq \arg z \leqq 2\pi$ then maps the original domain D in the z plane onto the domain $|\zeta| > 1$.

The solution of the potential problem is now straightforward and is left as an exercise to the reader.

Example 19

Let D be the domain in the z plane (Fig. 4-24*a*) exterior to an infinite array of equally spaced line segments (slits),

$$y = kb \qquad 0 \leqq x \leqq a \qquad k = 0, \pm 1, \pm 2, \ldots$$

The complex potential $\Omega = \varphi + i\psi$ in D satisfies the conditions (1) each slit is (part of) a streamline, (2) $\Omega'(z) \to Q_1 \equiv U - iV_1$ ($U > 0$, $V_1 > 0$) as $z \to \infty$ in the sector $\pi/2 < \arg z < 3\pi/2$, and (3) $\Omega'(z)$ is finite at all the points $z_k = a + ikb$ ($k = 0, \pm 1, \pm 2, \ldots$). Find the limit of the intensity $\Omega'(z)$ as $z \to \infty$ in the right half-plane.

Condition (3) implies the existence of nonzero circulation, say γ, around each individual slit. By integrating the intensity $\Omega'(z)$ around a rectangular contour with vertices at the points $z = \pm R \pm (i/2)b$, noting the periodicity of the field, and letting $R \to \infty$, we find that, as $z \to \infty$ in the right half-plane, $\Omega'(z) \to Q_2 = Q_1 - i\gamma/b$. If we write $Q_2 = U - iV_2$, then $V_2 = V_1 + \gamma/b$. The value of γ will subsequently be chosen so that condition (3) is satisfied.

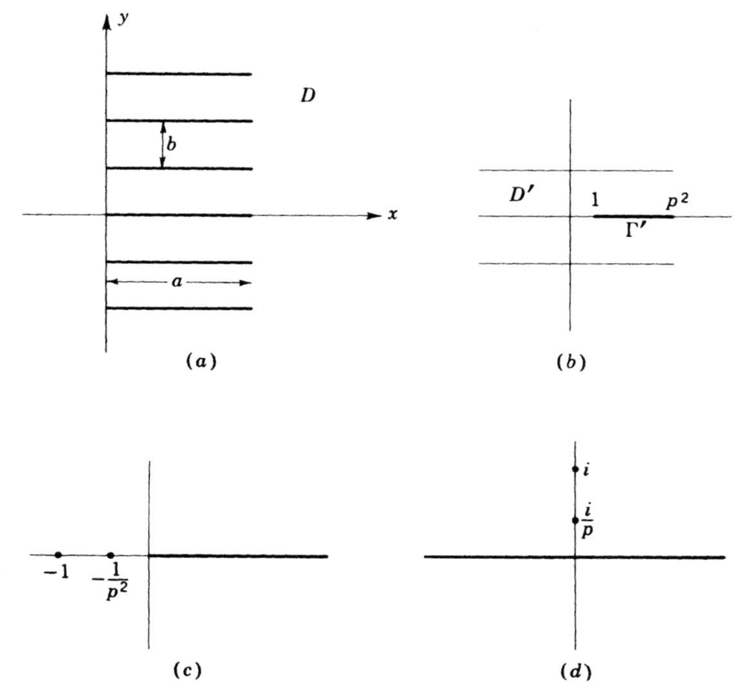

Fig. 4-24 (*a*) z plane; (*b*) t plane; (*c*) w plane; (*d*) ζ plane.

We apply the exponential transformation

$$t = e^{2\pi z/b} \tag{4-110}$$

which maps each infinite strip congruent to $-b/2 < y < +b/2$ onto the whole t plane. All the slits in the z plane map into a single slit Γ' in the t plane (Fig. 4-24*b*),

$$\Gamma': \quad \operatorname{Im} t = 0 \qquad 1 \leqq \operatorname{Re} t \leqq p^2 \qquad p = e^{\pi a/b}$$

The solution of the original problem is now reduced to a potential problem for the domain D' in the t plane exterior to the slit Γ' and with known singularities at $t = 0$ and $t = \infty$. Since we are eventually going to solve for a periodic field, we do not cut the t plane along the half-line $\operatorname{Im} t = 0$, $t < 0$.

We next apply the successive transformations

$$w = \frac{1 - t}{t - p^2} \qquad t = \frac{p^2 w + 1}{w + 1} \tag{4-111}$$

and

$$\zeta = w^{1/2} \qquad w = \zeta^2 \qquad 0 < \arg \zeta < \pi \tag{4-112}$$

(An alternative method of completing this calculation is contained in Exercise 1.) The transformation (4-111) maps the domain D' in the t plane onto the w plane cut along the positive real axis (Fig. 4-24c) and maps the points $t = 0$ and $t = \infty$ into $w = -1/p^2$ and $w = -1$, respectively. The subsequent transformation (4-112) then maps the cut w plane onto the half-plane $\operatorname{Im} \zeta > 0$ (Fig. 4-24d) and maps the points $w = -1/p^2$ and $w = -1$ into the points $\zeta = i/p$ and $\zeta = i$, respectively.

It is easily verified that the complex potential $\chi(\zeta)$ for the problem in the half-plane $\operatorname{Im} \zeta > 0$ has singularities at $\zeta = i/p$ and $\zeta = i$ defined by

$$\chi'(\zeta) \to \frac{bQ_1}{2\pi} \frac{1}{\zeta - i/p} \qquad \text{as } \zeta \to \frac{i}{p}$$

$$\chi'(\zeta) \to -\frac{bQ_2}{2\pi} \frac{1}{\zeta - i} \qquad \text{as } \zeta \to i$$

which correspond to sources of (complex) strength $bQ_1/2\pi$ and $-bQ_2/2\pi$ at the points $\zeta = i/p$ and $\zeta = i$, respectively. Since the ξ axis is a streamline, say $\operatorname{Im} \chi = 0$, it follows from the method of images that

$$\chi(\zeta) = \frac{b}{2\pi} \left[Q_1 \ln \left(\zeta - \frac{i}{p} \right) + Q_1^* \ln \left(\zeta + \frac{i}{p} \right) - Q_2 \ln (\zeta - i) - Q_2^* \ln (\zeta + i) \right] \tag{4-113}$$

Upon noting that $z = a$ corresponds to $\zeta = \infty$, that $Q_1 = U - iV_1$, $Q_2 = U - iV_2$, and $V_2 = V_1 + \gamma/b$, it follows from Eq. (4-113) that

$$\frac{d\chi}{d\zeta} = \frac{b}{\pi\zeta^2} \left[\left(\frac{1}{p} - 1 \right) V_1 - \frac{\gamma}{b} \right] + 0 \left(\frac{1}{\zeta^3} \right) \qquad \text{as } \zeta \to \infty$$

Also $\dfrac{d\zeta}{dz} = 0(\zeta^3)$ as $\zeta \to \infty$. Therefore, in order to make $\Omega'(z) = \dfrac{d\chi}{d\zeta}\dfrac{d\zeta}{dz}$ finite at $z = a$, we must choose

$$\gamma = bV_1 \left(\frac{1}{p} - 1 \right)$$

and so

$$V_2 = V_1 e^{-\pi a/b} \tag{4-114}$$

EXERCISES

1. Do parts (a) and (b) by two methods—conformal mapping, and directly.

(a) Find the complex potential in the whole z plane due to an infinite array of sources each of strength $q > 0$ at the points $z = ka$ ($k = 0$,

$\pm 1, \pm 2, \ldots$), where $a > 0$. Sketch the equipotentials and streamlines of the field.

(*b*) Find the complex potential in the whole z plane due to an infinite array of point sources each of strength $q > 0$ at the points $z = 2ka$ and an infinite array of point sources each of strength $-q$ at the points $(2k + 1)a$, where $a > 0$ and $k = 0, \pm 1, \pm 2, \ldots$. Sketch the field.

(*c*) Sketch the field in each of (*a*) and (*b*) resulting from the addition of the complex potential Vz to the previous potential.

2. A figure Γ in the z plane consists of the x axis with projecting line segments $x = ka$, $0 \leqq y \leqq b$ $(k = 0, \pm 1, \pm 2, \ldots)$, and D is the domain bounded below by Γ. Find the complex potential $\Omega = \varphi + i\psi$ in D subject to the conditions (1) $\psi = 0$ on Γ, (2) $\Omega'(x + iy) \to U > 0$ as $y \to +\infty$.

3. D is the domain of Exercise 2. Find the bounded solution of the Laplace equation $\Delta\varphi = 0$ in D subject to the conditions (1) $\varphi = 0$ on the horizontal segments of Γ and (2) $\varphi = 1$ on the vertical line segments.

4. D is the domain exterior to the infinite array of line segments in the z plane defined by the equation $z = ikb + re^{i\beta}$ with $0 < \beta < \pi/2$, $0 \leqq r \leqq a$ $(k = 0, \pm 1, \ldots)$. The complex potential $\Omega = \varphi + i\psi$ in D satisfies the conditions (1) each given line segment is part of a streamline, (2) $\Omega'(z) \to Q_1 \equiv U - iV_1$ $(U > 0, V_1 > 0)$ as $z \to \infty$ in the left half-plane, and (3) $\Omega'(z)$ is finite at all the points $z_k = ikb + ae^{i\beta}$ $(k = 0, \pm 1, \pm 2, \ldots)$. If $\Omega'(z) \to Q_2$ as $z \to \infty$ in the right half-plane, find Q_2 and sketch the streamlines of the field.

5. The figure Γ consists of the circle $|z| = 1$ with 10 inner radial projections $\arg z = k\pi/5$, $\frac{2}{3} \leqq |z| \leqq 1$ $(k = 0, 1, \ldots, 9)$, and D is the domain interior to Γ. Find the complex potential $\Omega = \varphi + i\psi$ in D when there is a vortex of strength q at the origin, with the boundary Γ a streamline. Repeat with the vortex of strength q at the point $z = \frac{1}{3}$ instead of at the origin.

6. D is the domain in the z plane exterior to the circle $|z| = \epsilon$ and to N radial half-lines $\arg z = 2k\pi/N$, $a \leqq |z| \leqq \infty$ $(k = 0, 1, \ldots, N - 1)$, where $a > 0$ and $\epsilon \ll a$. Find (approximately) the bounded solution of Laplace's equation $\Delta\varphi = 0$ in D if $\varphi = 0$ on the circle $|z| = \epsilon$ and $\varphi = 1$ on each of the N half-lines.

7. D is the domain between the lines $\operatorname{Im} z = a > 0$ and $\operatorname{Im} z = -b$ $(b > 0)$ and exterior to the infinite array of circles $|z - k| = \epsilon$ $(k = 0, \pm 1, \pm 2, \ldots)$, where $\epsilon \ll 1, a, b$. Find approximately the bounded solution of Laplace's equation $\Delta\varphi = 0$ in D if $\varphi = 0$ on each of the circles, $\varphi = \lambda$ on the line $\operatorname{Im} z = a$ and $\varphi = \mu$ on the line $\operatorname{Im} z = -b$; λ and μ are real constants.

4-8 *Integral Equations and Approximation Techniques*

One way in which to find an unknown mapping function is to solve an integral equation in which that mapping function occurs. A useful tool in obtaining such an integral equation is Green's third identity, which we proceed to derive.

From Exercise 7 of Sec. 2-5, we know that, if u and v are real and harmonic in a two-dimensional region D_1 with boundary Γ_1, then

$$\int_{\Gamma_1} \left(u \frac{\partial v}{\partial n} - v \frac{\partial u}{\partial n} \right) ds = 0 \tag{4-115}$$

where s is arc length and where $\partial/\partial n$, here and in future, indicates the outward normal derivative. Consider now a region D with boundary Γ. Let z_0 be any point inside D, and let D_1 be the region obtained by deleting from D a circular disk centered at z_0 and of (small) radius ρ_0; let Γ_1 consist of the original contour Γ, plus the disk contour C. If we now set $v(z) = \ln \rho$ in Eq. (4-115), where z is in D_1 and where $\rho = |z - z_0|$, we obtain

$$\int_\Gamma \left[u \frac{\partial}{\partial n} (\ln \rho) - (\ln \rho) \frac{\partial u}{\partial n} \right] ds - \int_C \left(u \frac{1}{\rho} ds - \ln \rho \frac{\partial u}{\partial \rho} \right) ds = 0$$

where we have used $\partial/\partial n = -\partial/\partial \rho$ on C. If we now make $\rho_0 \to 0$, this equation leads at once to *Green's third identity*,

$$2\pi u(z_0) = \int_\Gamma \left[u \frac{\partial}{\partial n} (\ln \rho) - (\ln \rho) \frac{\partial u}{\partial n} \right] ds \tag{4-116}$$

Equation (4-116) expresses the value of a harmonic function u at any interior point in terms of the boundary values of u and $\partial u/\partial n$. If the contour Γ has a continuously turning tangent (no corners), then an analogous calculation in which a half disk rather than a full disk is deleted from D shows that, for z_0 now a point on Γ, Eq. (4-116) must be replaced by

$$\pi u(z_0) = (P) \int_\Gamma \left[u \frac{\partial}{\partial n} (\ln \rho) - (\ln \rho) \frac{\partial u}{\partial n} \right] ds \tag{4-117}$$

The symbol (P) indicates that the integral is to be interpreted in its principal-value sense—i.e., as the limit as $\rho_0 \to 0$ of the integral over the contour Γ deleted by the removal of that part of Γ lying within a circle of radius ρ_0 centered on z_0.

We shall now use Eq. (4-117) to obtain an integral equation for the function $f(z)$ that maps the interior of a single closed curve Γ (again, no corners) onto the interior of the unit circle. Without loss of generality, let $f(0) = 0$, where the point $z = 0$ is inside Γ. We know that $|f| = 1$ on Γ, so that the harmonic function $\ln |f|$ vanishes on Γ. However, $\ln f$ is singular at $z = 0$;

this suggests that we consider

$$g(z) = \frac{f(z)}{z} \tag{4-118}$$

which is nonsingular within Γ. Let $z = re^{i\theta}$, and let $\ln f = p + iq$, where p and q are real. Then $p = 0$ on Γ, and $\ln r$ is known on Γ, so that $\text{Re}\,[\ln g(z)] = p - \ln r$ is known on Γ. In Eq. (4-117), let

$$u = \text{Im}\,[\ln g(z)] = q - \theta$$

then the Cauchy-Riemann conditions require

$$\frac{\partial}{\partial n}(q - \theta) = -\frac{\partial}{\partial s}(p - \ln r)$$

so that Eq. (4-117) becomes

$$\pi(q - \theta) = (P)\int_\Gamma (q - \theta)\frac{\partial}{\partial n}\ln \rho\, ds - (P)\int_\Gamma \ln \rho \frac{\partial}{\partial s}\ln r\, ds \tag{4-119}$$

which is an integral equation for the function $q - \theta$ on Γ. Once Eq. (4-119) has been solved for $q - \theta$ on Γ, we can use

$$\begin{aligned} \ln g(z) &= \frac{1}{2\pi i}\int_\Gamma \frac{\ln g(\zeta)}{\zeta - z}\,d\zeta \\ &= \frac{1}{2\pi i}\int_\Gamma \frac{-\ln r + i(q - \theta)}{\zeta - z}\,d\zeta \end{aligned} \tag{4-120}$$

to determine $g(z)$ and so $f(z)$ at any point z interior to Γ.

In Eq. (4-119), ρ is the distance from that point z_1 on Γ at which the left-hand side of the equation is evaluated to a point z_2 lying in ds. If we write $z_2 - z_1 = \rho e^{i\alpha}$ (see Fig. 4-25a), then (since $\ln \rho$ and α are conjugate functions) the kernel function $(\partial/\partial n)(\ln \rho)$ may be written as $\partial\alpha/\partial s$. Since

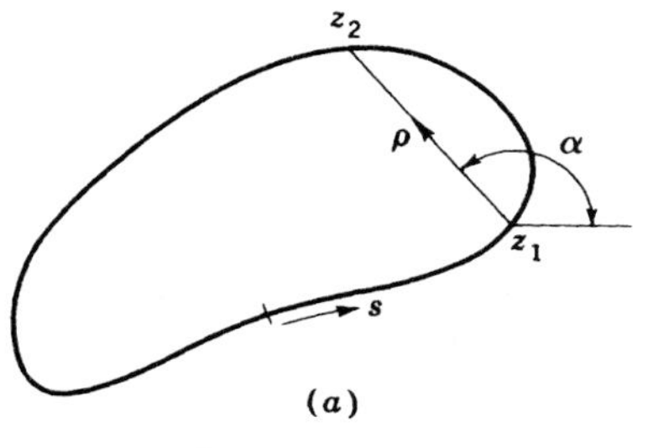

(a)

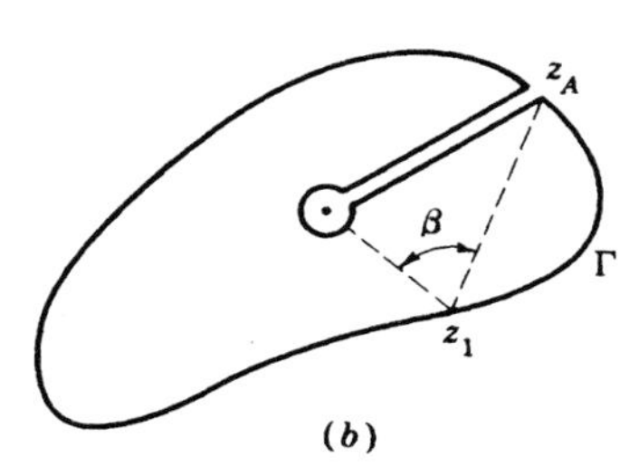

(b)

Fig. 4-25

Γ has no corners, $\partial\alpha/\partial s = 0$ at z_1, so that the first integral in Eq. (4-119) is not singular, and the symbol (P) may be dropped.

A version of Eq. (4-119) is associated with the names of Lichtenstein[1] and Gershgorin[2]; a similar equation was studied earlier by Neumann and Poincaré[3] in connection with an existence proof for the solution of the Dirichlet problem. The solution of Eq. (4-119) is unique only within an additive constant, since $\int_\Gamma d\alpha = \pi$. The arbitrary constant corresponds to the indeterminacy in the mapping; we have specified only that $f(0) = 0$, so that the image of a boundary point can be chosen freely. Equation (4-119) may be solved by iteration,[4] by collocation (with an appropriate mechanical quadrature formula used for the integral), or by any of a number of other methods. In practice, one might try first to find a simple mapping which would map Γ into an almost circular region Γ' and then use the integral-equation technique to map Γ' into a circle. (Since the iteration process requires only one step if Γ' is a circle, it can be expected to be very rapid if Γ' is close to a circle.) In searching for such a preliminary $\Gamma \to \Gamma'$ mapping, a table of conformal mappings[5] can be useful.

We can obtain an integral equation basically different from Eq. (4-119) by borrowing two results from Chap. 8. Again, we consider the function $f(z)$ which maps the interior of Γ into the interior of the unit circle, with $f(0) = 0$. The first result we need (Exercise 22 of Sec. 8-6) is that, given a function $h(z)$ analytic in D, then a *real* function $w(t)$, defined for t on Γ, can always be found such that

$$h(z) = \frac{1}{2\pi i} \int_\Gamma \frac{w(t)\,dt}{t - z} + ik \tag{4-121}$$

for all z in D, where k is some real constant. The second result we need [Eq. (8-130)] is that the limiting form of Eq. (4-121) as $z \to z_1$, where z_1 is a point on Γ, is given by

$$\lim_{z \to z_1} h(z) = \tfrac{1}{2} w(z_1) + \frac{1}{2\pi i} (P) \int_\Gamma \frac{w(t)\,dt}{t - z_1} + ik \tag{4-122}$$

We now choose

$$\begin{aligned} h(z) &= \ln f(z) - \ln z \\ &= p + iq - \ln r - i\theta \end{aligned}$$

[1] L. Lichtenstein, *Arch. Math. Phys.*, **25**: 179 (1917).

[2] S. Gershgorin, *Mat. Sbornik N.S.*, **40**: 48 (1933).

[3] See, for example, H. Poincaré, *Acta Math.*, **20**: 59 (1897).

[4] Cf. Sec. 8-6. When Eq. (4-119) is written in the form of Eq. (8-112), the coefficient λ becomes $1/\pi$, which is an eigenvalue; the corresponding eigenfunction is a constant. The iteration process will nevertheless converge for any Γ (Poincaré, *loc. cit.*), provided that an appropriate constant is subtracted off each time. The simple case in which Γ is a circle may clarify the situation; here, convergence in this sense is immediate.

[5] See, for example, H. Kober, "Dictionary of Conformal Representations," Dover Publications, Inc., New York, 1952.

and take the real part of Eq. (4-122) to obtain an integral equation for $w(z_1)$,

$$w(z_1) = -\frac{1}{\pi}(P)\int_\Gamma w(z_2)\,d\alpha - \ln r(z_1) \tag{4-123}$$

where $\alpha(z_1,z_2)$ is shown in Fig. 4-25*a*. Once Eq. (4-123) has been solved, we can find $f(z)$ for z in D by setting $h(z) = \ln f(z) - \ln z$ in Eq. (4-121). Although $d\alpha = (\partial/\partial n)(\ln \rho)\,ds$, so that the kernels of Eqs. (4-119) and (4-123) are the same, the signs of the coefficients $1/\pi$ are different. It may be shown by use of the Plemelj formulas of Chap. 8 that the solution of Eq. (4-119) is unique; the eigenvalue situation of Eq. (4-119) is thus avoided.

Another integral equation, developed for use in airfoil theory by Theodorsen and Garrick,[1] can be easily derived from Eq. (2-26). We consider again the interior problem, but in inverse form; i.e., we want the function $z = h(\zeta)$ that maps the interior of the unit circle in the ζ plane into the interior of a simple closed curve Γ. We take $h(0) = 0$; for z on Γ, write $z = re^{i\theta}$, and let the curve Γ be given by $r = g(\theta)$. Thus $g(\theta)$ is a known function. Let

$$w(\zeta) = \ln\frac{h(\zeta)}{\zeta} = u + iv \tag{4-124}$$

Then, from Eq. (2-26), $$w(\zeta) = iv(0) + \frac{1}{2\pi}\int_0^{2\pi} u(\varphi)\,\frac{t+\zeta}{t-\zeta}\,d\varphi \tag{4-125}$$

where $t = e^{i\varphi}$. We shall further constrain $h(\zeta)$ by requiring $h'(0)$ to be real, so that $v(0) = 0$ in Eq. (4-125). Upon using now Eq. (4-122) to calculate the limiting value as $\zeta \to t_0$, where $t_0 = \exp(i\varphi_0)$, Eq. (4-125) leads to

$$v(\varphi_0) = \frac{1}{2\pi}(P)\int_0^{2\pi} u(\varphi)\cot\frac{1}{2}(\varphi_0 - \varphi)\,d\varphi \tag{4-126}$$

Now $u(\varphi) = \ln g(\theta)$ and $v(\varphi) = \theta(\varphi) - \varphi$, so that Eq. (4-126) becomes

$$\theta(\varphi_0) - \varphi_0 = \frac{1}{2\pi}(P)\int_0^{2\pi} \ln g(\theta)\cot\frac{1}{2}(\varphi_0 - \varphi)\,d\varphi \tag{4-127}$$

which is a nonlinear integral equation for the function $\theta(\varphi)$. Equation (4-127) may be solved by iteration; once $\theta(\varphi)$ and so also $u(\varphi) = \ln g[\theta(\varphi)]$ are known, Eq. (4-125) gives the desired mapping function. Other nonlinear integral equations can be obtained by making different choices for $w(\zeta)$ from that given by Eq. (4-124); for example, Friberg[2] makes the choice

$$w(\zeta) = \ln\frac{\zeta f'(\zeta)}{f(\zeta)}$$

to obtain an integral-differential equation for $\theta(\varphi)$.

[1] T. Theodorsen and I. E. Garrick, *NACA Rept.* 452, 1933.

[2] M. S. Friberg, thesis, University of Minnesota, 1951. See S. Warschawski, *Proc. Symposium Appl. Math., Am. Math. Soc.*, **6:** 228 (1956).

EXERCISES

1. Let $\zeta = f(z)$ map the exterior of a simple closed curve Γ onto the exterior of the unit circle. [Then $\arg f(z)$ must increase by 2π as z traverses Γ.] Again, let the point $z = 0$ be inside Γ. For large z, let $f(z) \sim z$. Obtain the analog of Eq. (4-119) for the function $q - \theta$ appropriate to this exterior mapping problem. This new integral equation has a unique solution [cf. the discussion following Eq. (4-123)]; is this fact compatible with the flexibility prescribed in Riemann's mapping theorem? Obtain also the analog of Eq. (4-120).

2. (*a*) The integral equation actually obtained by Gershgorin differs slightly from Eq. (4-119); it may be derived as follows: Let z_A be a chosen fixed point on Γ (Fig. 4-25*b*), and modify the contour Γ so as to exclude the origin, as shown. Apply Eq. (4-117) to the modified contour, with $u = \text{Im}(\ln f) = q$, to obtain the integral equation

$$\pi q(z_1) = \int_\Gamma q(z_2) \frac{\partial}{\partial n} \ln \rho \, ds + 2\pi\beta$$

where β is the angle intercepted at z_1 by the line from 0 to z_A. Give a criterion for determining the sign of β. Note that q is not periodic, whereas the quantity $q - \theta$ of Eq. (4-119) is periodic.

(*b*) Derive Eq. (4-119), and also the result of Exercise 2*a*, by starting with Eq. (2-12).

(*c*) Choose a fixed point α within Γ, and let $f(0) = 0$ as before. Using a contour consisting of Γ plus a dumbbell-shaped contour enclosing the straight line from $z = 0$ to $z = \alpha$, use Eq. (2-12) for the function $\ln [f(z)/(z - \alpha)]$ (which is analytic and single-valued in the region between the two contours) to obtain a new integral equation involving q. Relate it to Eq. (4-119).

3. Obtain an integral equation for the mapping function $f(z)$ between a simple closed z-plane curve Γ and the unit ζ-plane circle by considering Γ to be the limiting form of a polygon with the number of sides tending to ∞ and by using the Schwarz-Christoffel formula.

4. Use any integral-equation method to obtain the mapping between the unit circle and an ellipse of small eccentricity ϵ; solve the integral equation, at least approximately.

Approximation Techniques

The most common approximation procedure in determining a mapping function is probably that of obtaining a numerical solution to one of the preceding integral equations. This topic is part of numerical analysis rather than complex-variable theory, and so we shall content ourselves with some references[1] to the extensive literature on the subject.

In previous sections, we have occasionally pointed out other approximation possibilities—such as the rounding off of a sharp corner by the use of an adjacent equipotential contour or by a modification in the Schwarz-Christoffel formula. A related example, in which a short parabolic arc is approximated by a circular arc, will be found in Exercise 6 below.

We consider next, however, a somewhat different kind of approximation procedure, based on a variational principle. Let D be the region interior to a simple closed curve Γ, and let $w = f(z)$ be analytic in $D + \Gamma$. Then the area of D', the w-plane image of D, is given by

$$\begin{aligned} I &= \int_D |f'(z)|^2 \, dA \\ &= \int_D (u_x^2 + u_y^2) \, dA \end{aligned} \tag{4-128}$$

where dA is the element of area in D and where $f = u + iv$. Let the point $z = 0$ be inside D. Then the variational principle states that, among all those functions $f(z)$ for which $f'(0) = 1$, that one which minimizes I maps Γ into a circle in the w plane. This variational principle differs from the usual kind of variational principle (cf. Exercise 8 of Sec. 2-5, where the same integral occurs) in that we are given information about $f(z)$ at only a single point rather than on a contour and are asked to find the boundary values of $f(z)$ rather than the differential equation it satisfies. (In fact, with reference to the expression for I in terms of u, we already know that u is harmonic.)

To derive this principle, we start with the observation that if $w = g(\zeta)$ is analytic in $|\zeta| \leqq R$, with $g'(0) = 1$, then

$$\int |g'(\zeta)|^2 \, dA \geqq \pi R^2 \tag{4-129}$$

where the integration is over the area of the circle $|\zeta| \leqq R$. To prove Eq. (4-129), we need merely write $g'(\zeta) = 1 + a_1\zeta + a_2\zeta^2 + \cdots$, set $\zeta = \rho e^{i\varphi}$, and integrate first with respect to φ and then with respect to ρ. Equality in Eq. (4-129) can occur only if $g' \equiv 1$.

Next, let $w = f(z)$ be any function satisfying $f'(0) = 1$; denote the w-plane image of D by D'. From the mapping theorem, we know that there exists a function $\zeta = h(z)$ mapping D into the interior of a circle in the ζ plane; simply by multiplying by a constant if necessary, we can adjust $h(z)$ so that $h'(0) = 1$. Let the radius of the ensuing circle be R. We require, of course, that $h'(z)$ not vanish anywhere in D. Any point z in D

[1] G. Birkhoff, D. Young, and E. Zarantello, *Proc. Symposium Appl. Math., Am. Math. Soc.*, *4:* 117 (1953).

Warschawski, *op. cit.*, p. 219.

B. Noble, "Mathematical Research Center Symposium on Non-linear Integral Equations," University of Wisconsin Press, Madison, Wis., p. 215, 1964.

thus has two images—a point w in D' and a point ζ in the disk. Let $w = g(\zeta)$ represent the relationship between this pair of image points; $g(\zeta)$ is clearly analytic, with $g'(0) = 1$. Thus Eq. (4-129) applies, and the area of D' must be greater than or equal to the area of the circular disk, with equality holding only if $g \equiv 1$. The variational principle follows.

Armed with this minimum principle, we can determine approximately the desired function $f(z)$ that maps D into the interior of a circle by expressing $f(z)$ in terms of a number of undetermined parameters and then choosing the parameters so as to minimize I. Such a *Rayleigh-Ritz* procedure can give a good approximation[1] to $f(z)$ if the chosen expression for $f(z)$ is adequately flexible.

EXERCISES

5. (*a*) Devise an alternative variational principle based on the integral of $|f'|$ around the contour Γ.

(*b*) Discuss the variational approach to the exterior-mapping problem.

6. A figure Γ in the z plane is composed of three linear segments: $x = 0, |y| \leqq 2; y = +1, |x| \leqq \epsilon; y = -1, |x| \leqq \epsilon$, and D is the domain exterior to Γ. Let $\epsilon \ll 1$. It is required to find the complex potential $\Omega = \varphi + i\psi$ subject to the conditions (1) $\psi = 0$ on Γ, (2) $\Omega'(z)$ is single-valued in D and $\Omega'(z) \to U > 0$ as $z \to \infty$, and (3) the circulation is zero. A Schwarz-Christoffel transformation is feasible but complicated; obtain therefore an approximate solution by proceeding as follows. From symmetry, only the top half-plane need be considered. Use $w = z^2 + (1 - \epsilon^2)$ to map Γ into a straight line plus a parabolic arc, and then approximate the parabolic arc by a circular arc in order to be able to use the Joukowsky transformation $w = t - \epsilon^2/t$, etc.

7. Discuss integral-equation methods and variational principles for multiply connected regions.

4-9 *The Biharmonic Equation*

A real function $u(x,y)$ is said to be *biharmonic* in a domain D if it satisfies the biharmonic equation

$$\Delta^2 u \equiv u_{xxxx} + 2u_{xxyy} + u_{yyyy} = 0 \qquad \text{in } D \tag{4-130}$$

Any function $u(x,y)$ biharmonic in D can be represented in the Goursat form

$$u = \text{Re}\,[z^*\varphi(z) + \chi(z)] \tag{4-131}$$

[1] See L. Kantorovich and V. Krylov, "Approximate Methods of Higher Analysis," pp. 367ff., Interscience Publishers, Inc., New York, 1958.

where φ and χ are analytic (but not necessarily single-valued) functions of $z = x + iy$ in D.

To establish this result, we observe that Δu is harmonic in D and so

$$\Delta u = \text{Re}\, h(z)$$

where $h(z) \equiv P(x,y) + iQ(x,y)$ is an analytic function in D. Hence the function

$$\varphi(z) \equiv p + iq = \tfrac{1}{4}\int h(z)\, dz$$

is also analytic and

$$\frac{\partial p}{\partial x} = \frac{\partial q}{\partial y} = \tfrac{1}{4}P \qquad \frac{\partial p}{\partial y} = -\frac{\partial q}{\partial x} = -\tfrac{1}{4}Q$$

It is now easily verified that $u - xp - yq \equiv u - \text{Re}\,[z^*\varphi(z)]$ is harmonic. Therefore $u - \text{Re}\,[z^*\varphi(z)] = \text{Re}\,\chi(z)$ for some analytic function $\chi(z)$, and the Goursat representation is established.

We note that, when u is given, $\varphi(z)$ is determined to within an arbitrary additive linear function $iaz + b$ (a real, b complex); $\chi(z)$ is then determined to within an arbitrary additive constant ic (c real). The Goursat representation of a biharmonic function is analogous to the representation of a harmonic function as the real part of an analytic function.

By means of the Goursat representation, a boundary-value problem for the biharmonic equation $\Delta^2 u = 0$ in a domain D is reduced to the problem of finding two analytic functions $\varphi(z)$ and $\chi(z)$ that satisfy certain conditions on the boundary Γ of D. (These boundary conditions are usually supplemented by the requirement that certain functions associated with φ and χ be single-valued in D.) In elasticity theory, for example, a typical form of boundary condition is

$$\varphi(z) + z[\varphi'(z)]^* + [\psi(z)]^* = g(s) \qquad \text{on } \Gamma \tag{4-132}$$

where $\psi(z) \equiv \chi'(z)$ and where $g(s) \equiv g_1(s) + ig_2(s)$ is a prescribed function of a position parameter s on Γ.

Problems of this kind are easily solved for domains of certain standard shapes, e.g., the interior or the exterior of a circle (see Exercises 3 and 4). It is evident that conformal-mapping techniques can be used to solve boundary-value problems of the biharmonic equation. Thus, for example, let D be the domain interior to a simple closed contour Γ in the z plane, and let $z = f(\zeta)$ be a transformation that maps the circular domain $|\zeta| < 1$ in the ζ plane conformally onto D. Then, with the notation

$$\Phi(\zeta) = \varphi[f(\zeta)] \qquad \Psi(\zeta) = \psi[f(\zeta)]$$

the boundary condition (4-132) transforms into the condition

$$\Phi(\zeta) + f(\zeta)\left[\frac{\Phi'(\zeta)}{f'(\zeta)}\right]^* + [\Psi(\zeta)]^* = G(\theta) \qquad \text{on } |\zeta| = 1$$

where $G(\theta)$ is a known function of position on the circle $\zeta = e^{i\theta}$.

Boundary-value problems of the biharmonic equation are important in fluid mechanics and in elasticity theory. The application of complex-variable techniques in elasticity theory has been extensively discussed by Muskhelishvili.[1]

EXERCISES

1. (*a*) Show that, under the transformation $z = f(\zeta)$, the biharmonic equation $\Delta^2 u = 0$ transforms into $\Delta_1[\Delta_1 U/|f'(\zeta)|^2] = 0$, where $\Delta_1 \equiv \partial^2/\partial\xi^2 + \partial^2/\partial\eta^2$.

(*b*) Can conformal mapping be useful in the equation $\Delta\varphi + k^2\varphi = 0$? In what kinds of equations would conformal mapping be useful?

2. Show that the biharmonic equation $\Delta^2 u = 0$ can be written in the form $\partial^4 u/(\partial z^2\, \partial z^{*2}) = 0$ (where z and z^* are regarded as independent variables), and hence deduce the Goursat representation for u.

3. D is the domain $|z| < 1$. Find the functions $\varphi(z)$ and $\psi(z)$, single-valued in D such that $\varphi(0) = 0$, $\text{Im}\ \varphi'(0) = 0$, and

$$\varphi(z) + z[\varphi'(z)]^* + [\psi'(z)]^* = \sum_{n=-\infty}^{+\infty} A_n e^{in\theta} \qquad \text{on } z = e^{i\theta}$$

where the A_n are complex constants [Fourier coefficients of $g(\theta)$].

4. D is the domain $|z| > 1$. Find the functions $\varphi(z)$ and $\psi(z)$, single-valued in D such that

$$\varphi(z) + z[\varphi'(z)]^* + [\psi(z)]^* = 0 \qquad \text{on } |z| = 1$$

and $\varphi(z) \to az$, $\psi \to -2az$ as $z \to \infty$ $(a > 0)$.

5. Γ is the ellipse $x^2/4a^2 + y^2/4b^2 = 1$, and D is the domain exterior to Γ. Find $\varphi(z)$ and $\psi(z)$, single-valued in D, such that

$$\varphi(z) + z[\varphi'(z)]^* + [\psi(z)]^* = 0 \qquad \text{on } \Gamma$$

and $\varphi(z) \to kz$, $\psi(z) \to 2kz$ as $z \to \infty$ $(k > 0)$.

6. Discuss the use of conformal mapping to solve a boundary-value problem for a biharmonic function u in which u and its normal derivative $\partial\varphi/\partial n$ are prescribed on the boundary.

[1] N. I. Muskhelishvili, "Some Basic Problems of the Mathematical Theory of Elasticity," Erven P. Noordhoff, NV, Groningen, Netherlands, 1953.

5

special functions

5-1 The Gamma Function

The *gamma function* $\Gamma(z)$ is a function of a complex variable which, for positive integers n, has the property $\Gamma(n + 1) = n!$ We first define $\Gamma(z)$ for $\operatorname{Re} z > 0$ by Euler's formula

$$\Gamma(z) = \int_0^\infty e^{-t} t^{z-1}\, dt \tag{5-1}$$

where the principal value of t^{z-1} is to be taken. Although this integral converges only when $\operatorname{Re} z > 0$, we shall subsequently find that $\Gamma(z)$ can be continued analytically onto the left-hand plane. Differentiation with respect to z can be carried out by the usual process, and we find

$$\Gamma'(z) = \int_0^\infty e^{-t} t^{z-1} \ln t\, dt$$

This integral also converges (only) for $\operatorname{Re} z > 0$; hence, $\Gamma(z)$ is analytic when $\operatorname{Re} z > 0$.

Repeated integration by parts of Eq. (5-1), using $\Gamma(1) = 1$, shows that $\Gamma(n + 1) = n!$, so that $\Gamma(z)$ is a generalized factorial function.[1]

[1] The notation $z!$ is sometimes used for $\Gamma(z + 1)$.

The definition (5-1) cannot be used when $\operatorname{Re} z \le 0$. An alternative definition, due to Hankel, is

$$\Gamma(z) = \frac{1}{2i \sin \pi z} \int_C e^t t^{z-1}\, dt \tag{5-2}$$

where the path of integration C starts at $-\infty$ on the real axis, circles the origin once in the positive direction, and returns to $-\infty$; the initial and final arguments of t are to be $-\pi$ and π, respectively. Observe that e^t, and not e^{-t}, appears in the integrand. Equation (5-2) defines an analytic function for all z other than $0, \pm 1, \pm 2, \ldots$. The reader can show that, for the special case $\operatorname{Re} z > 0$ (using L'Hospital's rule at $z = 1, 2, \ldots$), Eqs. (5-2) and (5-1) have identical implications. Thus, Eq. (5-2) provides the analytic continuation of $\Gamma(z)$ onto the left half-plane.

For z equal to zero or a negative integer, the integral in Eq. (5-2) is finite, but the denominator vanishes to the first order, so that at such points $\Gamma(z)$ has a simple pole. The residue at $z = -n$ is easily seen to be $(-1)^n/n!$

Thus, $\Gamma(z)$ is analytic everywhere, except at the points $z = 0, -1, -2, \ldots$.

There are two infinite-product expressions for $\Gamma(z)$ which are useful in deriving some of its properties. To obtain the first, use Eq. (5-1) with

$$e^{-t} = \lim_{n\to\infty} \left(1 - \frac{t}{n}\right)^n$$

to give

$$\Gamma(z) = \lim_{n\to\infty} \int_0^n \left(1 - \frac{t}{n}\right)^n t^{z-1}\, dt$$

$$= \lim_{n\to\infty} n^z \int_0^1 (1-\tau)^n \tau^{z-1}\, d\tau$$

For integral n, integration by parts now gives

$$\int_0^1 (1-\tau)^n \tau^{z-1}\, d\tau = \left[\frac{1}{z}\tau^z(1-\tau)^n\right]_0^1 + \frac{n}{z}\int_0^1 (1-\tau)^{n-1}\tau^z\, d\tau$$

$$= \cdots = \frac{n(n-1)\cdots(1)}{z(z+1)\cdots(z+n)}$$

so that

$$\Gamma(z) = \lim_{n\to\infty} \frac{n!\,n^z}{z(z+1)\cdots(z+n)}$$

$$= \frac{1}{z}\prod_1^\infty \left[\left(1+\frac{1}{n}\right)^z \left(1+\frac{z}{n}\right)^{-1}\right] \tag{5-3}$$

This infinite product is convergent and represents an analytic function for all z different from zero or a negative integer, so that the restriction $\operatorname{Re} z > 0$ used in the derivation can now be removed. Equation (5-3) was used by Gauss as the definition of $\Gamma(z)$.

Before obtaining the second infinite-product representation for $\Gamma(z)$, we digress to define the Euler-Mascheroni constant γ by

$$\gamma = \lim_{n\to\infty}\left[\left(1 + \frac{1}{2} + \frac{1}{3} + \cdots + \frac{1}{n}\right) - \ln n\right]$$
$$= 0.5772156649 \cdots \tag{5-4}$$

We can prove that this limit exists and also obtain an integral representation for γ by writing

$$1 + \frac{1}{2} + \frac{1}{3} + \cdots + \frac{1}{n} = \int_0^\infty (e^{-t} + e^{-2t} + \cdots + e^{-nt})\, dt$$
$$= \int_0^\infty \frac{e^{-t} - e^{-(n+1)t}}{1 - e^{-t}}\, dt$$
$$\ln n = \int_1^n dx \int_0^\infty e^{-xt}\, dt = \int_0^\infty \frac{e^{-t} - e^{-nt}}{t}\, dt$$

Subtracting and letting $n \to \infty$, we obtain the limiting value as the convergent integral

$$\gamma = \int_0^\infty e^{-t}\left(\frac{1}{1 - e^{-t}} - \frac{1}{t}\right) dt \tag{5-5}$$

Return now to Eq. (5-3), and write

$$\frac{1}{\Gamma(z)} = \lim_{n\to\infty}\left[z\left(1 + \frac{z}{1}\right)\left(1 + \frac{z}{2}\right) \cdots \left(1 + \frac{z}{n}\right)e^{-z\ln n}\right]$$
$$= \lim_{n\to\infty}\left[z\left(1 + \frac{z}{1}\right)e^{-z}\left(1 + \frac{z}{2}\right)e^{-\frac{1}{2}z} \cdots \left(1 + \frac{z}{n}\right)e^{-(1/n)z}e^{[1+\frac{1}{2}+\cdots+(1/n)-\ln n]z}\right]$$
$$= ze^{\gamma z}\prod_1^\infty\left[\left(1 + \frac{z}{n}\right)e^{-(1/n)z}\right] \tag{5-6}$$

This infinite product is uniformly convergent in any bounded region of the z plane, so that $1/\Gamma(z)$ is an entire function. It has a first-order zero only where $\Gamma(z)$ has a pole. Since $1/\Gamma(z)$ has no poles, $\Gamma(z)$ can have no zeros. Equation (5-6) was used by Weierstrass as the definition of $\Gamma(z)$.

Exercises

1. Show that other expressions for γ are:

(*a*) $\gamma = \int_0^1 \frac{1 - e^{-t}}{t}\, dt - \int_1^\infty \frac{e^{-t}}{t}\, dt$

(*b*) $\gamma = \int_0^1 \frac{1 - e^{-t} - e^{-1/t}}{t}\, dt$

(*c*) $\gamma = \int_0^\infty \left(\frac{1}{1 + t} - e^{-t}\right)\frac{dt}{t}$

(*d*) $\gamma = -\int_0^\infty e^{-t} \ln t \, dt$

(*e*) $\gamma = -\int_0^1 \ln\left(\ln \frac{1}{t}\right) dt$

(*f*) $\gamma = 1 - \frac{1}{2}\sum_2^\infty \frac{1}{n^2} - \frac{1}{3}\sum_2^\infty \frac{1}{n^3} \cdots$

2. Show that $\Gamma(1) = 1$, $\Gamma'(1) = -\gamma$, $\Gamma(\frac{1}{2}) = \sqrt{\pi}$, and that, for $\operatorname{Re} z > 0$,

$$\Gamma(z) = \int_0^1 \left(\ln \frac{1}{t}\right)^{z-1} dt$$

3. If $p > 1$, show that

$$\int_0^\infty \cos t^p \, dt = \frac{1}{p} \cos \frac{\pi}{2p} \Gamma\left(\frac{1}{p}\right)$$

$$\int_0^\infty \sin t^p \, dt = \frac{1}{p} \sin \frac{\pi}{2p} \Gamma\left(\frac{1}{p}\right)$$

Properties of $\Gamma(z)$

We have already seen that $\Gamma(z)$ is a single-valued meromorphic function of z, having poles only at $z = 0, -1, -2, \ldots, -n, \ldots$; each such pole is of first order, with residue $(-1)^n/n!$ We have also seen that $1/\Gamma(z)$ is entire.

From Eq. (5-6), it follows that $\Gamma(z)$ is real for real values of z and also that $\Gamma(z^*) = \Gamma^*(z)$ for general values of z. For $z = iy$, where y is real, Eq. (5-6) gives

$$\frac{1}{\Gamma(iy)\Gamma^*(iy)} = y^2 \prod_1^\infty \left(1 + \frac{y^2}{n^2}\right)$$

so that

$$|\Gamma(iy)| = \left(\frac{\pi}{y \sinh \pi y}\right)^{1/2} \tag{5-7}$$

which implies that $\Gamma(z)$ decays exponentially along the imaginary axis.

Either from an integration by parts of Eq. (5-2) or by use of Eq. (5-3), we find that $\Gamma(z)$ satisfies the basic difference equation

$$\Gamma(z + 1) = z\Gamma(z) \tag{5-8}$$

which holds for all $z \neq 0, -1, -2, \ldots$. For z equal to a real positive integer n, we again obtain $\Gamma(n + 1) = n!$

Next, we use Eq. (5-6) to form the product

$$\frac{1}{\Gamma(z)} \frac{1}{\Gamma(-z)} = -z^2 \prod_1^\infty \left(1 - \frac{z^2}{n^2}\right) = -\frac{z}{\pi} \sin \pi z$$

Since $\Gamma(-z) = -(1/z)\Gamma(1-z)$ from Eq. (5-8), we have

$$\Gamma(z)\Gamma(1-z) = \frac{\pi}{\sin \pi z} \tag{5-9}$$

In a sense, this formula may be thought of as a kind of reflection principle, the reflection being with respect to the point $z = \frac{1}{2}$. In fact, replacing z by $\zeta + \frac{1}{2}$, we obtain the more symmetrical formula

$$\Gamma\left(\frac{1}{2} + \zeta\right)\Gamma\left(\frac{1}{2} - \zeta\right) = \frac{\pi}{\cos \pi\zeta} \tag{5-10}$$

Setting $\zeta = 0$ in Eq. (5-10), we find $\Gamma(\frac{1}{2}) = \pi^{\frac{1}{2}}$, in agreement with Exercise 2 above.

The use of Eqs. (5-2) and (5-9) now leads to a simple contour-integral expression for $1/\Gamma(z)$,

$$\frac{1}{\Gamma(z)} = \frac{1}{2\pi i}\int_C e^t t^{-z}\,dt \tag{5-11}$$

where $|\arg t| \le \pi$ and where C is the Hankel contour of Eq. (5-2).

Consider next the product $\Gamma(z)\Gamma(z + \frac{1}{2})$. From Eq. (5-3), we have

$$\begin{aligned}
\Gamma(z)\Gamma&\left(z + \frac{1}{2}\right) \\
&= \lim_{n\to\infty} \frac{(n!)^2 n^{2z+\frac{1}{2}}}{[z(z+1)\cdots(z+n)][(z+\frac{1}{2})(z+\frac{3}{2})\cdots(z+n+\frac{1}{2})]} \\
&= \lim_{n\to\infty} \frac{n^{2z}n^{\frac{1}{2}}(n!)^2 2^{2n+2}}{2z(2z+1)(2z+2)\cdots(2z+2n)(2z+2n+1)} \\
&= 2^{-2z}\Gamma(2z)\lim_{n\to\infty}\frac{n^{\frac{1}{2}}(n!)^2 2^{2n+2}}{(2n+1)!}
\end{aligned}$$

The final limit does not involve z and so may be obtained by setting $z = \frac{1}{2}$ in the other terms in this equation. The final result (due to Legendre) is

$$2^{2z-1}\Gamma(z)\Gamma(z + \tfrac{1}{2}) = \pi^{\frac{1}{2}}\Gamma(2z) \tag{5-12}$$

which is called the *duplication formula*. By a similar argument, the reader may obtain the following generalization (due to Gauss) of Eq. (5-12),

$$\Gamma(z)\Gamma\left(z + \frac{1}{n}\right)\Gamma\left(z + \frac{2}{n}\right)\cdots\Gamma\left(z + \frac{n-1}{n}\right) = (2\pi)^{\frac{1}{2}(n-1)} n^{\frac{1}{2}-nz}\Gamma(nz) \tag{5-13}$$

[in evaluating the corresponding limit in this case, set $z = 1/n$, and use Eq. (5-9)]. Notice that Eq. (5-13) can be used to relate the value of $\Gamma(z)$ at a point far from the origin to its values at a set of points close to the origin.

Finally, we shall state (with the proof deferred to Chap. 6) that, for large values of $|z|$, a good approximation to $\Gamma(z)$ is given by the following asymptotic representation, valid for $|\arg z| < \pi$:

$$\ln \Gamma(z) \sim \left(z - \frac{1}{2}\right) \ln z - z + \frac{1}{2} \ln 2\pi + \sum_{n=1}^{p} \frac{B_{2n}}{(2n-1)(2n)z^{2n-1}} + 0\left(\frac{1}{z^{2p+1}}\right) \quad (5\text{-}14)$$

Here the B_n are the Bernoulli numbers. Writing out the first few terms, we have

$$\Gamma(z) \sim e^{-z} e^{(z-\frac{1}{2}) \ln z} (2\pi)^{\frac{1}{2}} \left(1 + \frac{1}{12z} + \frac{1}{288z^2} + \cdots\right)$$

The familiar Stirling formula,

$$n! \sim (2\pi n)^{\frac{1}{2}} \left(\frac{n}{e}\right)^n$$

is easily proved from Eq. (5-14). The reader may verify that Eq. (5-14) is in agreement with Eq. (5-7) and moreover that a consequence of Eq. (5-14) is that, for real x and y, $|\Gamma(x + iy)|$ decays exponentially as $|y| \to \infty$ with x fixed—i.e., along any line parallel to the imaginary axis. To determine the asymptotic behavior of $\Gamma(z)$ for $-\pi < \arg(-z) < \pi$, we would first use Eq. (5-9) and then Eq. (5-14).

Exercises

4. Show that $|\Gamma(x + iy)| < \Gamma(x)$ for x, y real and $x > 0$.

5. Prove that, if n is a positive integer,

$$\Gamma\left(n + \frac{1}{2}\right) = \frac{(2n-1)!\pi^{\frac{1}{2}}}{2^{2n-1}(n-1)!}$$

$$\Gamma\left(-n + \frac{1}{2}\right) = \frac{(-1)^n 2^{2n-1}(n-1)!\pi^{\frac{1}{2}}}{(2n-1)!}$$

6. If α is an arbitrary complex number and $|z| < 1$, show that

$$(1 + z)^\alpha = \sum_{n=0}^{\infty} \frac{\Gamma(\alpha + 1)}{n!\Gamma(\alpha - n + 1)} z^n$$

7. Show that, if $0 < \operatorname{Re} z < 1$,

$$\int_0^\infty y^{-z} \cos\left(\frac{\pi}{2} z - y\right) dy = \frac{\pi}{\Gamma(z)} \quad (5\text{-}15)$$

8. Show that, if $\operatorname{Re} z < 0$,

$$\int_0^\infty \left[e^{-t} - 1 + t - \frac{t^2}{2!} \cdots + (-1)^{p+1} \frac{t^p}{p!}\right] t^{z-1}\, dt \quad (5\text{-}16)$$

converges to $\Gamma(z)$ for $\operatorname{Re} z > -(p + 1)$.

9. If α, β, ρ, δ are complex numbers satisfying $\alpha + \beta + \rho + \delta = 0$, show that (apart from the obvious exceptional cases)

$$\prod_1^\infty \frac{(1 + \alpha/n)(1 + \beta/n)}{(1 - \rho/n)(1 - \delta/n)} = \frac{\delta\rho\Gamma(-\rho)\Gamma(-\delta)}{\alpha\beta\Gamma(\alpha)\Gamma(\beta)}$$

and use this formula to evaluate

$$\prod_1^\infty \left(1 + \frac{1}{2n}\right)\left(1 - \frac{1}{2n + 2}\right)$$

10. Prove that a generalization of Eq. (5-9) is

$$\prod_1^\infty \left[1 - \left(\frac{z}{n}\right)^p\right] = -\frac{1}{z^p}\left[\prod_{k=1}^p \Gamma(-ze^{2\pi ik/p})\right]^{-1}$$

where $p = 2, 3, \ldots$.

The ψ Function

The function $\psi(z)$ is the derivative of the logarithm of $\Gamma(z)$, that is,

$$\psi(z) = \frac{\Gamma'(z)}{\Gamma(z)} = \frac{d}{dz}[\ln \Gamma(z)] \tag{5-17}$$

It is meromorphic, with simple poles at $z = 0, -1, -2, \ldots$; the residue of $\psi(z)$ at each pole is -1. Taking the logarithm of both sides of Eq. (5-6) and differentiating, we obtain

$$\psi(z) = -\gamma - \frac{1}{z} + \sum_1^\infty \frac{z}{n(z + n)} \tag{5-18}$$

from which $\psi(1) = -\gamma$. Also, Eq. (5-18) shows that $\psi(z)$ is real on the real axis and $\psi(z^*) = \psi^*(z)$. By using Eq. (5-4), Eq. (5-18) becomes

$$\psi(z) = \lim_{n\to\infty}\left(\ln n - \frac{1}{z} - \frac{1}{z + 1} - \cdots - \frac{1}{z + n}\right) \tag{5-19}$$

which is a generalization of Eq. (5-4). The function $\psi(z)$ also satisfies a basic difference equation, which may be obtained from Eq. (5-8) by direct differentiation,

$$\psi(z + 1) = \frac{1}{z} + \psi(z) \tag{5-20}$$

Equation (5-20) holds if $z \neq 0, -1, -2, \ldots$. For a positive integer n, Eq. (5-20) reduces to

$$\psi(n + 1) = \left(1 + \frac{1}{2} + \frac{1}{3} + \cdots + \frac{1}{n}\right) - \gamma \tag{5-21}$$

so that $\psi(n) \sim \ln n$ as $n \to \infty$.

Differentiation of Eq. (5-10) yields

$$\psi(\tfrac{1}{2} + z) - \psi(\tfrac{1}{2} - z) = \pi \tan \pi z \tag{5-22}$$

and we observe the special case of this for an integral value of z. Equation (5-12) may also be differentiated to give a duplication formula for $\psi(z)$,

$$\psi(2z) = \ln 2 + \tfrac{1}{2}\psi(z) + \tfrac{1}{2}\psi(z + \tfrac{1}{2})$$

A more general result can of course be obtained by differentiating Eq. (5-13).

There are several interesting integral representations for $\psi(z)$. A particularly simple one, valid for $\text{Re } z > 0$, is

$$\psi(z) = -\gamma + \int_0^1 \frac{1 - t^{z-1}}{1 - t} \, dt \tag{5-23}$$

as may be verified by expanding $(1 - t)^{-1}$ in a series and integrating term by term. The replacement of t by $e^{-\tau}$ gives

$$\psi(z) = -\gamma + \int_0^\infty \frac{e^{-t} - e^{-tz}}{1 - e^{-t}} \, dt \tag{5-24}$$

again valid for $\text{Re } z > 0$. We now use Eq. (5-5) to obtain *Gauss's formula*,

$$\psi(z) = \int_0^\infty \left(\frac{e^{-t}}{t} - \frac{e^{-tz}}{1 - e^{-t}} \right) dt \tag{5-25}$$

All these integrals are valid only for $\text{Re } z > 0$; representations for $\text{Re } z < 0$ may be obtained by use of Eq. (5-22).

Exercises

11. Show that

$$\frac{1}{(1 + z)^2} + \frac{1}{(2 + z)^2} + \cdots = \psi'(1 + z)$$

and $$\frac{1}{1 + z} + \frac{1}{2 + z} + \cdots + \frac{1}{n + z} = \psi(z + n + 1) - \psi(z + 1)$$

12. Derive *Dirichlet's formula*, valid for $\text{Re } z > 0$,

$$\psi(z) = \int_0^\infty \left[\frac{e^{-t}}{t} - \frac{1}{t(1 + t)^z} \right] dt \tag{5-26}$$

and use this result to show that

$$\gamma = -\int_0^\infty \left(\cos t - \frac{1}{1 + t^2} \right) \frac{dt}{t}$$

13. Express

$$\int_0^\infty e^{-at} \frac{\sinh bt}{\sinh ct} \, dt$$

in terms of ψ functions, and state the range of variables a, b, c for which the result is valid.

14. Show from Eq. (5-14) that, for large z with $|\arg z| < \pi$,

$$\psi(z) \sim \ln z - \frac{1}{2z} - \sum_{n=1}^{p} \frac{B_{2n}}{(2n)z^{2n}} + 0\left(\frac{1}{z^{2p+2}}\right) \tag{5-27}$$

The Beta Function

We define the *beta function* $\mathrm{B}(p,q)$ by

$$\mathrm{B}(p,q) = \int_0^1 t^{p-1}(1 - t)^{q-1}\, dt \tag{5-28}$$

where $\mathrm{Re}\, p > 0$ and $\mathrm{Re}\, q > 0$; the principal values of the various powers are to be taken.

Replacing t by $1 - \tau$, we see that $\mathrm{B}(p,q) = \mathrm{B}(q,p)$. Setting $t = \sin^2 \theta$ in Eq. (5-28), we obtain a useful trigonometric integral,

$$\mathrm{B}(p,q) = 2 \int_0^{\pi/2} (\sin \theta)^{2p-1}(\cos \theta)^{2q-1}\, d\theta \tag{5-29}$$

In Eq. (5-28), we set $t = \tau/(1 + \tau)$, to give

$$\mathrm{B}(p,q) = \int_0^\infty \frac{t^{p-1}}{(1 + t)^{p+q}}\, dt \tag{5-30}$$

It is possible to express $\mathrm{B}(p,q)$ in terms of gamma functions. For $\mathrm{Re}\, p > 0$ and $\mathrm{Re}\, q > 0$, we have

$$\begin{aligned}
\Gamma(p)\Gamma(q) &= \int_0^\infty e^{-t}t^{p-1}\, dt \int_0^\infty e^{-\tau}\tau^{q-1}\, d\tau \\
&= \int_0^\infty e^{-t}t^{p-1}\, dt \left(t^q \int_0^\infty e^{-tx}x^{q-1}\, dx\right) \\
&= \int_0^\infty x^{q-1}\, dx \int_0^\infty e^{-t(x+1)}t^{p+q-1}\, dt \\
&= \int_0^\infty \frac{x^{q-1}}{(1 + x)^{p+q}}\, dx \int_0^\infty e^{-\tau}\tau^{p+q-1}\, d\tau
\end{aligned}$$

It follows from Eq. (5-30) that

$$\mathrm{B}(p,q) = \frac{\Gamma(p)\Gamma(q)}{\Gamma(p + q)} \tag{5-31}$$

Although $\mathrm{B}(p,q)$ was originally defined only for $\mathrm{Re}\, p > 0$ and $\mathrm{Re}\, q > 0$, Eq. (5-31) provides the analytic continuation of $\mathrm{B}(p,q)$ onto the left halves of the p and q planes.

An integral representation for $\mathrm{B}(p,q)$, valid for all nonintegral values of p and q is given by *Pochhammer's integral*,

$$\mathrm{B}(p,q) = \frac{-e^{-i\pi(p+q)}}{4 \sin \pi p \sin q\pi} \int_C t^{p-1}(1 - t)^{q-1}\, dt \tag{5-32}$$

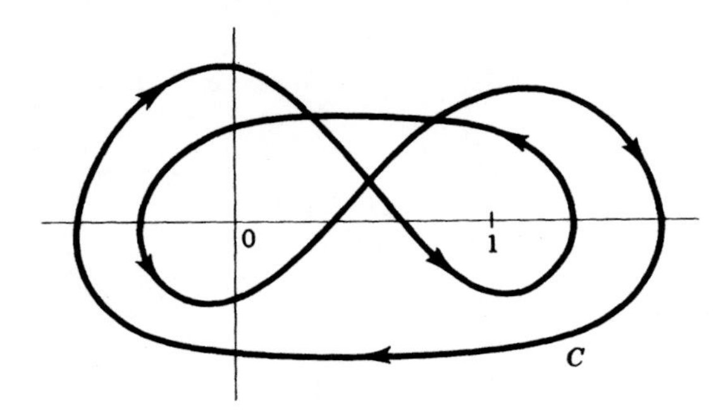

Fig. 5-1 Contour for integral of Eq. (5-32).

where the contour C is that shown in Fig. 5-1. The proof of this formula is left as an exercise. The fact that the integrand in Eq. (5-32) returns to its initial value after a complete traversal of C is advantageous in any manipulations of Eq. (5-32) that involve integration by parts.

EXERCISES

15. If all relevant quantities are real, with $\alpha > 0$ and $\beta > 0$, and if the range of integration is over all positive values of x and y satisfying $x + y \leq 1$, show that (Dirichlet)

$$\iint f(x + y)x^{\alpha-1}y^{\beta-1}\,dx\,dy = \frac{\Gamma(\alpha)\Gamma(\beta)}{\Gamma(\alpha+\beta)}\int_0^1 f(x)x^{\alpha+\beta-1}\,dx$$

Can you think of a problem in physics or in probability theory where this result would be useful? Can you generalize the formula so as to hold for three or more variables of integration?

16. For an appropriate range of the parameters, evaluate in terms of beta functions the integrals

$$\int_0^\infty (\sinh t)^\alpha (\cosh t)^\beta\,dt$$

$$\int_a^b \frac{(t-a)^\alpha (b-t)^\beta}{(t-c)^{\alpha+\beta+2}}\,dt \qquad \text{with } a, b, c \text{ real; } c < a < b$$

Zeta Function

Another member of the gamma-function family is the *generalized zeta function*, defined for Re $s > 1$ and for $a \neq 0, -1, -2, \ldots$ by

$$\zeta(s,a) = \sum_{n=0}^{\infty} \frac{1}{(a+n)^s} \tag{5-33}$$

The special case $\zeta(s,1)$ is denoted by $\zeta(s)$ and is called *Riemann's zeta function*.

The series (5-33) converges uniformly in any region Re $s > 1 + \epsilon$ and so is analytic for Re $s > 1$. For Re $a > 0$ and Re $s > 1$, we can use

$$\frac{\Gamma(s)}{(a+n)^s} = \int_0^\infty t^{s-1}e^{-(n+a)t}\,dt$$

to show easily that

$$\zeta(s,a) = \frac{1}{\Gamma(s)}\int_0^\infty \frac{t^{s-1}e^{-at}}{1-e^{-t}}\,dt \tag{5-34}$$

Thus, in particular,

$$\zeta(s) = \frac{1}{\Gamma(s)}\int_0^\infty \frac{t^{s-1}}{e^t - 1}\,dt$$

for Re $s > 1$ [which is reminiscent of the definition of $\Gamma(s)$]. Just as for $\Gamma(z)$, a Hankel-type contour integral can be found for Re $a > 0$ [Eq. (5-36) in Exercise 17 below] which reduces to Eq. (5-34) for Re $s > 1$ but which defines an analytic function for more general values of s. For Re $a > 0$, Eq. (5-36) shows $\zeta(s,a)$ to be analytic over the entire s plane, except at $s = 1$, where it has a simple pole of residue unity. By manipulation of the contour of integration in Eq. (5-35)—we omit the details—it can be shown that $\zeta(s)$ satisfies the functional equation

$$(2\pi)^s\zeta(1-s) = 2\Gamma(s)\cos\frac{\pi s}{2}\,\zeta(s) \tag{5-35}$$

The function $\zeta(s)$ has an intriguing relationship to the prime-number sequence, first discovered by Euler. Since

$$\zeta(s) = 1 + \frac{1}{2^s} + \frac{1}{3^s} + \frac{1}{4^s} + \frac{1}{5^s} + \cdots$$

we have

$$\left(1 - \frac{1}{2^s}\right)\zeta(s) = 1 + \frac{1}{3^s} + \frac{1}{5^s} + \frac{1}{7^s} + \frac{1}{9^s} + \frac{1}{11^s} + \cdots$$

$$\left(1 - \frac{1}{3^s}\right)\left(1 - \frac{1}{2^s}\right)\zeta(s) = 1 + \frac{1}{5^s} + \frac{1}{7^s} + \frac{1}{11^s} + \cdots$$

$$\left(1 - \frac{1}{5^s}\right)\left(1 - \frac{1}{3^s}\right)\left(1 - \frac{1}{2^s}\right)\zeta(s) = 1 + \frac{1}{7^s} + \frac{1}{11^s} + \cdots$$

A continuation of this process gives

$$\prod\left(1 - \frac{1}{p^s}\right)\zeta(s) = 1$$

where the product is taken over all prime numbers p. [The product clearly converges for Re $s > 1$, since $\prod_{2}^{\infty} (1 - 1/n^s)$ does.] Thus, for Re $s > 1$,

$$\frac{1}{\zeta(s)} = \prod \left(1 - \frac{1}{p^s}\right)$$

which provides a remarkable relationship between the analytic function $\zeta(s)$ and the sequence of prime numbers. It is almost unfortunate that $\zeta(s)$ is only infrequently encountered in applied mathematics.

Finally, we remark that, since $\zeta(1 - s)$ has a pole only at $s = 0$, Eq. (5-35) shows that $\zeta(s)$ must vanish at $s = -2, -4, -6, \ldots$; Riemann conjectured that all other zeros of $\zeta(s)$ are to be found on the line Re $s = \frac{1}{2}$, but this conjecture is as yet unproved.

EXERCISE

17. Show that for Re $a > 0$

$$\zeta(s,a) = \frac{-1}{2\pi i} \Gamma(1 - s) \int_C (-t)^{s-1} \frac{e^{-at}}{1 - e^{-t}} dt \tag{5-36}$$

where C is a contour starting at $+\infty$, circling the origin in the positive direction, and returning to $+\infty$; the argument of t satisfies $|\arg(-t)| \leq \pi$.

Show that $\zeta(s)$ is nonzero for Re $s > 1$ or for Re $s < 0$, except at $s = -2, -4, \ldots$.

5-2 *Differential Equations*

Although differential equations frequently arise within the context of real functions, it is advantageous to study the solutions of such equations in the complex domain. One reason for this is that series expansions are often used, and we have previously seen that power series are most naturally discussed in the complex domain. The continuation into the complex domain is also useful when solution functions are encountered in the integrands of contour integrals.

In this section, we shall consider the nth-order linear homogeneous differential equation

$$w^{(n)} = p_{n-1}(z)w^{(n-1)} + \cdots + p_1(z)w' + p_0(z)w \tag{5-37}$$

where the $p_j(z)$ are functions analytic within some common circle centered on the point z_0—which without loss of generality will be taken to be the origin. We shall show that a solution $w(z)$ exists which is analytic inside this circle and that, if $w(0), w'(0), \ldots, w^{(n-1)}(0)$ are specified, then this solution is unique. To simplify the algebra, we shall give the proof for

$n = 2$; there is no difficulty in extending it to hold for general values of n.

We consider, then, the equation

$$w'' = p_1 w' + p_0 w \tag{5-38}$$

where p_1 and p_0 are analytic for $|z| < R$. Let $|p_1| < M_1$ and $|p_0| < M_0$ for $|z| \le r < R$. Define the functions

$$\psi_1(z) = \frac{M_1}{1 - z/r} \qquad \psi_0(z) = \frac{M_0}{1 - z/r}$$

Both ψ_1 and ψ_0 have positive real derivatives at the origin, and, by Cauchy's inequality,

$$|p_1^{(n)}(0)| \le \psi_1^{(n)}(0) \qquad |p_0^{(n)}(0)| \le \psi_0^{(n)}(0)$$

Consider now, concurrently with Eq. (5-38), the equation

$$W'' = \psi_1 W' + \psi_0 W \tag{5-39}$$

Let $w(0)$ and $w'(0)$ be specified arbitrarily. We choose

$$W(0) = |w(0)| \qquad W'(0) = |w'(0)|$$

as initial conditions for Eq. (5-39).

Now, if an analytic solution of Eq. (5-38) exists, the coefficients of the power series—or, equivalently, the values of the derivatives at the origin—are [with Eq. (5-38) repeatedly differentiated]

$$\begin{aligned} w''(0) &= p_1(0)w'(0) + p_0(0)w(0) \\ w'''(0) &= p_1'(0)w'(0) + p_1(0)w''(0) + p_0'(0)w(0) + p_0(0)w'(0) \\ &\cdots\cdots\cdots \end{aligned} \tag{5-40}$$

Similarly, if an analytic solution of Eq. (5-39) exists, we have

$$\begin{aligned} W''(0) &= \psi_1(0)W'(0) + \psi_0(0)W(0) \\ W'''(0) &= \psi_1'(0)W'(0) + \psi_1(0)W''(0) + \psi_0'(0)W(0) + \psi_0(0)W'(0) \\ &\cdots\cdots\cdots \end{aligned}$$

Since the formalism is exactly the same in each case, the previous choice for $W(0)$ and $W'(0)$ ensures that

$$|W^{(n)}(0)| \ge |w^{(n)}(0)|$$

for all values of n. It follows that, if the power series for $W(z)$ converges, then so does that for $w(z)$.

We have thus reduced the problem to that of showing that Eq. (5-39) possesses a power-series solution, convergent for $|z| < r$. Writing

$$W = c_0 + c_1 z + c_2 z^2 + \cdots$$

and substituting this series into Eq. (5-39), we obtain the recursion formula

$$c_{n+2} = \frac{(n/r) + M_1}{n+2} c_{n+1} + \frac{M_0}{(n+1)(n+2)} c_n \tag{5-41}$$

The c_n are all real and > 0. We can take $M_1 > 2/r$, so that

$$\frac{c_{n+2}}{c_{n+1}} > \frac{1}{r}$$

Upon dividing Eq. (5-41) through by c_{n+1}, it follows that

$$\frac{c_{n+2}}{c_{n+1}} \to \frac{1}{r}$$

as $n \to \infty$. Consequently the series for W converges for $|z| < r$, and the proof is complete.

We have thus proved that, for an arbitrary choice of $w(0)$ and $w'(0)$, there exists a solution of Eq. (5-39) which is analytic in the region $|z| \leq r < R$. Moreover, Eqs. (5-40) show that the solution is unique. In practice, rather than use Eqs. (5-40), it may of course be simpler to substitute a power series with undetermined coefficients for the function $w(z)$ in Eq. (5-38), to write $p_0(z)$ and $p_1(z)$ also in the form of power series, and to multiply out and collect coefficients of the various powers of z so as to find the coefficients of the power series for $w(z)$.

We observe that, if $w(0)$ and $w'(0)$ are zero, then $w(z)$ is identically zero. Also, if we define $w_0(z)$ to be that solution of Eq. (5-38) for which $w_0(0) = 1$, $w_0'(0) = 0$, and if we define $w_1(z)$ to be that solution for which $w_1(0) = 0$, $w_1'(0) = 1$, then *any* solution of Eq. (5-39) can be written as a linear combination of this pair of solutions. In fact, if a solution $w(z)$ is to have $w(0) = a_0$ and $w'(0) = a_1$, then

$$w(z) = a_0 w_0(z) + a_1 w_1(z)$$

This result implies that there are two and only two linearly independent solutions of Eq. (5-38). Any pair of linearly independent solutions is said to form a *fundamental set* of solutions.

The Wronskian

Let $w_1(z)$ and $w_2(z)$ be any two solutions of Eq. (5-38), linearly independent or not. Define their *wronskian* $W(z)$ as the determinant

$$W(z) = \begin{vmatrix} w_1(z) & w_2(z) \\ w_1'(z) & w_2'(z) \end{vmatrix} \tag{5-42}$$

Direct differentiation, by using Eq. (5-38), shows that

$$W'(z) = p_1(z) W(z)$$

so that

$$W(z) = C \exp\left[\int p_1(z)\, dz\right] \tag{5-43}$$

where C is a constant. Since the exponential factor in Eq. (5-43) cannot vanish for $|z| < R$, $W(z)$ cannot vanish for any one value of z in $|z| < R$, unless it vanishes for all values of z.

A necessary and sufficient condition that $W \equiv 0$ is that w_1 and w_2 are linearly dependent. For if constants c_1 and c_2 exist, not both zero, such that $c_1w_1(z) + c_2w_2(z) \equiv 0$, then $c_1w_1'(z) + c_2w_2'(z) \equiv 0$ and so $W \equiv 0$. Conversely, if $W \equiv 0$, then quantities c_1 and c_2, not both zero, can be found such that $c_1w_1(0) + c_2w_2(0) = 0$ and $c_1w_1'(0) + c_2w_2'(0) = 0$. Define $w_3 = c_1w_1(z) + c_2w_2(z)$; then $w_3(0) = w_3'(0) = 0$, so that by the uniqueness theorem of the last section we must have $w_3(z) \equiv 0$. Consequently, $c_1w_1(z) + c_2w_2(z) \equiv 0$, which is the condition for linear dependence.

If one solution $w_1(z)$ of Eq. (5-38) is known, then a linearly independent solution $w_2(z)$ can be found by use of Eq. (5-43). Write Eq. (5-43) as

$$w_2'(z)w_1(z) - w_2(z)w_1'(z) = C \exp\left[\int p_1(z)\,dz\right]$$

This first-order differential equation for $w_2(z)$ can be solved at once to give

$$w_2(z) = Cw_1(z) \int \frac{dz}{w_1{}^2(z)} \exp\left[\int p_1(z)\,dz\right] + Kw_1(z) \tag{5-44}$$

where both C and K are arbitrary constants.

Just as in the real-variable case, a knowledge of a pair of linearly independent solutions $w_1(z)$ and $w_2(z)$ of Eq. (5-38) can be used to generate the solution of the nonhomogeneous equation

$$w'' - p_1w' - p_0w = f(z) \tag{5-45}$$

We write $w = a(z)w_1 + b(z)w_2$, where $a(z)$ and $b(z)$ are as yet undetermined functions. Then

$$w' = a'w_1 + b'w_2 + aw_1' + bw_2'$$

and we now restrict a and b so that

$$a'w_1 + b'w_2 = 0 \tag{5-46}$$

Then

$$w'' = a'w_1' + b'w_2' + aw_1'' + bw_2''$$

and substituting for w, w', and w'' into Eq. (5-45) (taking account of the fact that w_1 and w_2 satisfy the homogeneous equation), we obtain

$$a'w_1' + b'w_2' = f \tag{5-47}$$

Equations (5-46) and (5-47) can be solved as a pair of simultaneous linear equations in the unknowns $a'(z)$ and $b'(z)$ (note that the coefficient determinant is the wronskian, which here cannot vanish); one integration then gives $a(z)$ and $b(z)$. Using the fact that the difference of any two solutions of Eq. (5-45) is a solution of the homogeneous equation, the most general solution of Eq. (5-45) is therefore

$$w = -w_1 \int \frac{fw_2}{W}\,dz + w_2 \int \frac{fw_1}{W}\,dz + Aw_1 + Bw_2$$

where A and B are arbitrary constants. It is clear, of course, that any solution of Eq. (5-45) with prescribed values of $w(0)$ and $w'(0)$ must be unique.

For an nth-order linear homogeneous differential equation, the wronskian of n solutions, $w_1(z)$, $w_2(z)$, . . . , $w_n(z)$, is defined similarly,

$$W = \begin{vmatrix} w_1 & w_2 & \cdots & w_n \\ w_1' & w_2' & \cdots & w_n' \\ \cdots & \cdots & \cdots & \cdots \\ w_1^{(n-1)} & \cdots & \cdots & w_n^{(n-1)} \end{vmatrix}$$

The reader will have no difficulty in showing that W satisfies a simple first-order differential equation, that $W \equiv 0$ is a necessary and sufficient condition that the $w_j(z)$ be linearly dependent, and that a set of n linearly independent solutions of the homogeneous equation can be used to construct the solution of the nonhomogeneous equation.

EXERCISES

1. (*a*) Show that the substitution $w(z) = y(z) \exp\left[\frac{1}{2}\int p_1(z)\,dz\right]$ reduces Eq. (5-38) to the form

$$y''(z) = P_0(z)y(z)$$

Can anything similar be done for a third-order equation?

(*b*) Show that an nth-order differential equation can be written as a set of n first-order equations.

2. One solution of

$$zw'' + 2w' + zw = 0$$

is $(\sin z)/z$; find a linearly independent solution, and explain why its behavior near $z = 0$ does not contradict the theorem of this section.

Singularities of Coefficient Functions

So far, we have confined our attention to functions $p_1(z)$ and $p_0(z)$ which are analytic in the region of interest. In many of the differential equations of applied mathematics, $p_1(z)$ and $p_0(z)$ turn out to have poles at one or more points z_0. We now turn our attention to second-order differential equations for which $p_1(z)$ has at most a simple pole and $p_0(z)$ at most a pole of order 2. Such a point z_0 is said (rather confusingly) to be a *regular singular point*[1] of the differential equation, and a well-developed theory (due largely to Fuchs and Frobenius) exists.

[1] In contrast to an *ordinary point*, at which $p_1(z)$ and $p_0(z)$ are both analytic.

Again, we shall take $z_0 = 0$. Let $S(z) = zp_1(z)$ and $T(z) = z^2p_0(z)$ be analytic in $|z| < R$, with the expansions

$$\begin{aligned} S(z) &= s_0 + s_1z + s_2z^2 + \cdots \\ T(z) &= t_0 + t_1z + t_2z^2 + \cdots \end{aligned} \tag{5-48}$$

The differential equation (5-38) can be written

$$z^2w'' = zS(z)w' + T(z)w \tag{5-49}$$

If we try to solve Eq. (5-49) by a conventional power series for w, we are soon led to contradictory recurrence relations, so that, except in very special cases, no such power-series solution can exist. We therefore try, formally, the more general expression

$$w = z^\rho(a_0 + a_1z + a_2z^2 + \cdots) \tag{5-50}$$

where $a_0 \neq 0$ and where the constant ρ is an as yet undetermined complex number. We substitute this expression into Eq. (5-49) and ask for what ρ and a_j, if any, w is a solution. Subsequently, we inquire whether the series so obtained converges. Substitution of Eq. (5-50) into Eq. (5-49) gives

$$\begin{aligned} a_0[\rho(\rho - 1) - \rho s_0 - t_0] &= 0 \\ a_1[(\rho + 1)\rho - (\rho + 1)s_0 - t_0] &= (\rho s_1 + t_1)a_0 \\ a_2[(\rho + 2)(\rho + 1) - (\rho + 2)s_0 - t_0] &= (\rho s_2 + t_2)a_0 \\ &\quad + [(\rho + 1)s_1 + t_1]a_1 \\ \cdots\cdots\cdots\cdots &\cdots\cdots\cdots\cdots \\ a_n[(\rho + n)(\rho + n - 1) - (\rho + n)s_0 - t_0] &= (\rho s_n + t_n)a_0 + \cdots \\ &\quad + [(\rho + n - 1)s_1 + t_1]a_{n-1} \\ \cdots\cdots\cdots\cdots &\cdots\cdots\cdots\cdots \end{aligned} \tag{5-51}$$

The first of Eqs. (5-51) shows that, since $a_0 \neq 0$, the constant ρ must be a root of the *indicial equation*

$$\rho(\rho - 1) - \rho s_0 - t_0 = 0 \tag{5-52}$$

This equation has two roots ρ_1 and ρ_2. Normally, we can expect to choose either one of these and then determine $a_1, a_2, \ldots$ as multiples of a_0 by the use of Eqs. (5-51). In this way, we would obtain two different formal solutions of Eq. (5-49). However, this process would fail if the coefficient

$$(\rho + n)(\rho + n - 1) - (\rho + n)s_0 - t_0$$

were to vanish for some $n > 0$. Since each such coefficient is obtained from the left-hand side of the indicial equation by replacing ρ by $\rho + n$, such an awkward situation occurs only when ρ_1 and ρ_2 differ by an integer. We temporarily set this case aside and assume that $\rho_2 - \rho_1$ is nonintegral.

We shall now show that each of the two series so obtained is convergent. Consider the choice $\rho = \rho_1$; the case $\rho = \rho_2$ is of course similar. Let $S(z)$

and $T(z)$ be analytic for $|z| \leq r < R$; then there is a constant $M > 1$ such that

$$|s_n| < \frac{M}{r^n} \qquad |t_n| < \frac{M}{r^n} \qquad |\rho_1 s_n + t_n| < \frac{M}{r^n}$$

Since a_0 is simply a multiplicative constant, choose it so that $|a_0| < 1$. Let $K > 1$ be an upper bound of the sequence

$$|1 + (\rho_1 - \rho_2)|^{-1} \qquad |1 + \tfrac{1}{2}(\rho_1 - \rho_2)|^{-1} \qquad |1 + \tfrac{1}{3}(\rho_1 - \rho_2)|^{-1}, \cdots$$

This upper bound exists since $\rho_1 - \rho_2$ is not a negative integer. Then, upon using

$$(\rho_1 + n)(\rho_1 + n - 1) - (\rho_1 + n)s_0 - t_0 = n(\rho_1 - \rho_2 + n)$$

Eqs. (5-51) give

$$\begin{aligned} |a_1| &< K\frac{M}{r} \\ |a_2| &< \frac{K}{2^2}\left[\frac{M}{r^2} + \left(\frac{M}{r} + \frac{M}{r}\right)K\frac{M}{r}\right] \\ &< \frac{K^2M^2}{r^2} \end{aligned}$$

and a continuation shows that $|a_n| < K^n M^n / r^n$ for all values of n. Thus, the series

$$a_0 + a_1 z + a_2 z^2 + \cdots$$

converges uniformly for $|z| < r/MK$ and represents an analytic function.

Consequently, if $\rho_2 - \rho_1$ is nonintegral, two solutions $w_1(z)$ and $w_2(z)$ of the form (5-50) do exist, corresponding to $\rho = \rho_1$ and $\rho = \rho_2$, respectively. Moreover (since $\rho_2 \neq \rho_1$), w_1 and w_2 are linearly independent. At the origin, each solution has a branch point, with $w_1(ze^{i2\pi}) = \exp(2\pi i \rho_1) w_1(z)$ and $w_2(ze^{i2\pi}) = \exp(i2\pi\rho_2) w_2(z)$. In any simply connected region Ω contained within $|z| = r/MK$, but not containing the origin, the two solutions are still valid; here they are free from branch points but of course are still linearly independent. Thus, from our previous results, any solution of the differential equation is expressible within Ω as a linear combination of the two solutions. In this sense, our two solutions are complete; there are no others.

For special values of ρ_1 or ρ_2, the solution may not have a branch point at the origin. If ρ_1 is an integer, $w_1(z)$ is single-valued, and if ρ_1 is a positive integer, $w_1(z)$ is analytic at the origin.

EXERCISES

3. What can be said about the solutions of Eq. (5-39) at a regular point if $p_0(z)$ has a simple pole there?

4. Will the method of this section work if both $p_0(z)$ and $p_1(z)$ have second-order poles at z_0?

5. Show that if $\rho_2 = \rho_1 + 1$, with $\rho_1 s_1 + t_1 = 0$, then two linearly independent solutions are obtained by setting $\rho = \rho_1$ in Eqs. (5-51). Must one of these be related to that obtained by setting $\rho = \rho_2$? Generalize to $\rho_2 = \rho_1 + n$; what accessory relations are required? Discuss Exercise 2 above in this connection.

6. Let $s_0 = t_0 = t_1 = 0$, so that z_0 is an ordinary point. Show in detail that Eqs. (5-51) then lead to the expected results. In what sense is the previous theorem, involving analytic $p_1(z)$ and $p_0(z)$, more powerful than the theorem that could be obtained in this way?

Indicial Roots Separated by an Integer

We index the roots of the indicial equation so that $\rho_1 - \rho_2 = n$, where n is now to be a positive integer or zero. Then, since none of the left-hand coefficients multiplying a_1, a_2, etc., vanishes if we set $\rho = \rho_1$ in Eq. (5-51), we still obtain one solution $w_1(z)$ of Eq. (5-49). With the exception noted in Exercise 5 above, setting $\rho = \rho_2$ does not provide a solution, since the particular a_j multiplied by a vanishing coefficient in Eq. (5-51) would be undefined. [If all previous a_j are set equal to zero, we simply recover $w_1(z)$.] In other words, the form of solution given by Eq. (5-50) is not appropriate for $w_2(z)$. However, since we do have one solution $w_1(z)$, we can find another by use of Eq. (5-44),

$$w_2(z) = Cw_1(z) \int \frac{dz}{w_1^2(z)} \exp\left[\int \frac{S(z)}{z}\,dz\right] \tag{5-53}$$

But except for a constant of integration which can be absorbed into C,

$$\int \frac{S(z)}{z}\,dz = s_0 \ln z + s_1 z + \tfrac{1}{2} s_2 z^2 + \cdots$$

and

$$\frac{1}{w_1^2(z)} = z^{-2\rho_1} \frac{1}{a_0^2}\left(1 - 2\frac{a_1}{a_0} z + \cdots\right)$$

so that Eq. (5-53) gives (with a_0^2 absorbed into C also)

$$w_2(z) = Cw_1(z) \int z^{-2\rho_1 + s_0} g(z)\,dz \tag{5-54}$$

where $g(z)$ is analytic at $z = 0$ [and, in fact, $g(0) = 1$]. Now, from Eq. (5-52), $\rho_1 + \rho_2 = s_0 + 1$, and, by assumption, $\rho_1 - \rho_2 = n$, so that the exponent $-2\rho_1 + s_0$ is equal to $-(n + 1)$. Upon writing

$$g(z) = 1 + g_1 z + g_2 z^2 + \cdots$$

Eq. (5-54) now gives (after we discard an additive multiple of w_1, as unnecessary in the expression for w_2)

$$\begin{aligned} &(a)\ \text{For } n = 0 \\ &w_2(z) = Cw_1(z)[\ln z + h(z)] \qquad \text{where } h(z) \text{ is analytic at } z = 0 \\ &(b)\ \text{For } n \neq 0 \\ &w_2(z) = C[w_1(z)g_n \ln z + z^{\rho_2}h(z)] \qquad \text{where } h(z) \text{ is analytic at } z = 0 \end{aligned} \tag{5-55}$$

Notice that, in case b, the second part of $w_2(z)$ is that which would be expected if $\rho_1 - \rho_2$ were nonintegral. In special cases (cf. Exercise 5 above), the coefficient g_n may vanish, and $w_2(z)$ does then not involve a logarithmic term.

In actual practice, the solution $w_2(z)$ would usually be obtained by assuming it to have the form of Eq. (5-55), with $h(z)$ expressed as $h_0 + h_1z + h_2z^2 + \cdots$, and then substituting it into Eq. (5-49) so as to determine the h_j.

Behavior of Solutions in the Large

In general, the coefficient functions $p_1(z)$ and $p_0(z)$ of Eq. (5-38) will be analytic except at certain points. In the neighborhood of any ordinary point—i.e., a point where both p_1 and p_0 are analytic—we can always find a pair of linearly independent analytic solutions; by analytic continuation, these can be extended to any point in the plane at which p_0 and p_1 are not singular. If we make a circuit around an isolated singularity, the two solutions normally will not return to their initial values after a complete circuit; however, the two new functions will still be linearly independent, for otherwise some linear combination of them must vanish. However, since the zero function is a single-valued solution, the same linear combination of the original functions would also have to vanish.

If the singularity in question is a regular point, then the way in which the two solutions become modified after a circuit of this point is particularly simple, for each solution is a linear combination of a pair of functions with branch points like z^{ρ_1} and z^{ρ_2}, respectively (or with logarithmic branch points, in special cases).

An extensive discussion of further topics in differential-equation theory—including singularities which are not regular points, singularities of higher-order differential equations, criteria for the existence of certain kinds of solution, etc., will be found in the literature.[1]

5-3 *Hypergeometric Functions*

In Eq. (5-38), let $p_1(z)$ and $p_0(z)$ be analytic everywhere except for the three distinct points (a,b,c) (one of which may be the point at ∞; otherwise

[1] A. R. Forsyth, "Theory of Differential Equations," vol. 4, Cambridge University Press, New York, 1906; Dover reprint, 1959.

p_1 and p_0 are to be analytic at ∞). Let these three points be regular points of Eq. (5-38). Many differential equations of physical interest are of this form, and it is useful to derive some general results concerning such equations and their solutions.

Consider first the case in which none of the points (a,b,c) is at ∞. Then $p_1(z)$ and $p_2(z)$ must have the form

$$p_1(z) = \frac{S(z)}{(z-a)(z-b)(z-c)} \qquad p_0(z) = \frac{T(z)}{(z-a)^2(z-b)^2(z-c)^2}$$

where $S(z)$ and $T(z)$ are entire functions. Using the transformation $z = 1/\xi$, Eq. (5-38) is replaced by

$$w_{\xi\xi} = -\left[\frac{S(1/\xi)}{\xi^2(1/\xi - a)(1/\xi - b)(1/\xi - c)} + \frac{2}{\xi}\right] w_\xi + \left[\frac{T(1/\xi)}{\xi^4(1/\xi - a)^2(1/\xi - b)^2(1/\xi - c)^2}\right] w$$

Since $\xi = 0$ is an ordinary point, we can conclude that $T(z)$ is at most a second-order polynomial and that $S(z)$ is precisely a second-order polynomial with the coefficient -2 for the term z^2. This means that we can write $p_1(z)$ and $p_0(z)$ in the forms

$$p_1(z) = \frac{A}{z-a} + \frac{B}{z-b} + \frac{C}{z-c}$$
$$p_0(z) = \left(\frac{A'}{z-a} + \frac{B'}{z-b} + \frac{C'}{z-c}\right)\frac{1}{(z-a)(z-b)(z-c)}$$

where A, A', etc., are constants. These constants are arbitrary, except that $A + B + C = -2$. At the point $z = a$, let the roots of the indicial equation (called the *exponents* at a, for an obvious reason) be α and α'. Then since the indicial equation for an expansion about $z = a$ is

$$\rho^2 - \rho(A+1) - \frac{A'}{(a-b)(a-c)} = 0$$

we have $\qquad A = \alpha + \alpha' - 1 \qquad A' = (a-b)(c-a)\alpha\alpha'$

Upon denoting similarly the exponents at b and c by β, β' and γ, γ', respectively, the differential equation thus becomes

$$w'' = \left(\frac{\alpha+\alpha'-1}{z-a} + \frac{\beta+\beta'-1}{z-b} + \frac{\gamma+\gamma'-1}{z-c}\right) w' + \left[\frac{\alpha\alpha'(a-b)(c-a)}{z-a} + \frac{\beta\beta'(b-c)(a-b)}{z-b} + \frac{\gamma\gamma'(c-a)(b-c)}{z-c}\right] \frac{w}{(z-a)(z-b)(z-c)} \tag{5-56}$$

Since $A + B + C = -2$, the exponents must satisfy the condition

$$\alpha + \alpha' + \beta + \beta' + \gamma + \gamma' = 1 \tag{5-57}$$

Thus, if the only singularities (in the extended plane) of Eq. (5-38) are those corresponding to three distinct regular points, none of which is at ∞, then the equation must have the form of Eq. (5-56), called *Riemann's P equation*. The equation is completely determined by stating the locations of the three points and the exponents at those points—which within the limitations of Eq. (5-57) can be chosen quite arbitrarily. The fact that a function w satisfies Eq. (5-56) is conventionally written

$$w = P \left\{ \begin{matrix} a & b & c & \\ \alpha & \beta & \gamma & z \\ \alpha' & \beta' & \gamma' & \end{matrix} \right\} \tag{5-58}$$

Suppose next that one of the regular points, say c, is at ∞. Let the exponents at $c = \infty$ continue to be denoted by γ, γ'. Then using the same method, the reader may show that the differential equation becomes

$$w'' = \left(\frac{\alpha + \alpha' - 1}{z - a} + \frac{\beta + \beta' - 1}{z - b} \right) w' + \left[\frac{\alpha\alpha'(b - a)}{z - a} + \frac{\beta\beta'(a - b)}{z - b} - \gamma\gamma' \right] \frac{w}{(z - a)(z - b)} \tag{5-59}$$

and that Eq. (5-57) must continue to hold. Since this result can be derived from Eq. (5-56) by letting $c \to \infty$, we conclude that Eq. (5-56) includes the special case in which one of the regular points is at ∞.

Another special case is obtained by allowing two of the regular points to coalesce; the resulting equation is described as being *confluent*. We shall subsequently see that Bessel's equation provides an example of this case.

Exercise

1. If Eq. (5-38) has only two singularities, which are regular points at a and b with exponents α, α' and β, β', respectively (the point at ∞ being an ordinary point), show that $\alpha = -\beta$, $\alpha' = -\beta'$ and that the most general form of the equation is

$$w'' = \left(\frac{\alpha + \alpha' - 1}{z - a} - \frac{\alpha + \alpha' + 1}{z - b} \right) w' - \frac{\alpha\alpha'(a - b)^2}{(z - a)^2(z - b)^2} w \tag{5-60}$$

Simplify this equation by use of the substitution $\xi = (z - a)/(z - b)$, and thus show that the general solution of the equation is

$$w = A \left(\frac{z - a}{z - b} \right)^{\alpha} + B \left(\frac{z - a}{z - b} \right)^{\alpha'} \tag{5-61}$$

where A and B are arbitrary constants. Discuss the special case $\alpha = \alpha'$.

Transformation of the P Equation

First, let none of a, b, c be at ∞, and let

$$w = P\begin{Bmatrix} a & b & c & \\ \alpha & \beta & \gamma & z \\ \alpha' & \beta' & \gamma' & \end{Bmatrix}$$

Define a function $w_1(z)$ by

$$w_1 = \frac{(z-a)^\lambda (z-b)^\mu}{(z-c)^{\lambda+\mu}} w \tag{5-62}$$

where λ and μ are arbitrary complex numbers. The equation for w_1, obtained by substituting for w in Eq. (5-56), is clearly a linear second-order differential equation, with its solutions well behaved except at the points (a,b,c). At a, $w(z)$ has branches that behave like $(z-a)^\alpha$, $(z-a)^{\alpha'}$; from Eq. (5-62), $w_1(z)$ must have branches that behave like $(z-a)^{\alpha+\lambda}$, $(z-a)^{\alpha'+\lambda}$. Thus, the point at a is still a regular point, and the exponents of w_1 at $z = a$ must be $(\alpha + \lambda, \alpha' + \lambda)$. Considering the other two points similarly shows that

$$w_1' = P\begin{Bmatrix} a & b & c & \\ \alpha+\lambda & \beta+\mu & \gamma-\lambda-\mu & z \\ \alpha'+\lambda & \beta'+\mu & \gamma'-\lambda-\mu & \end{Bmatrix} \tag{5-63}$$

We note that the sum of the exponents is still unity—otherwise, the analyticity at ∞ would have been destroyed. Thus, if none of (a,b,c) is at ∞, the transformation of dependent variable described by Eq. (5-62) results in an alteration of the exponents at the regular points, but not in an alteration of the locations of those points. The case where one of the points—say c—is at ∞ can be obtained from Eq. (5-62) by multiplying by the constant $-c^{\lambda+\mu}$ (note that any constant times a solution is still a solution) and then allowing $c \to \infty$; the appropriate transformation is thus

$$w_1 = (z-a)^\lambda (z-b)^\mu w \tag{5-64}$$

for $c = \infty$. In other words, to use Eq. (5-62) for the case in which one of the points is at ∞, simply leave out the term corresponding to that point.

It is also possible to make a change in independent variable. If we use a bilinear transformation

$$z = \frac{A_0 z_1 + A_1}{B_0 z_1 + B_1} \tag{5-65}$$

where A_0, A_1, B_0, B_1 are arbitrary complex constants, then it is easily verified that Eq. (5-56) becomes another P equation for w in terms of z_1, with exactly three regular points (a_1,b_1,c_1), which are the images under the transformation (5-65) of the points (a,b,c), and with no alteration of exponents at those points. The constants in Eq. (5-65) may be chosen so as to make (a,b,c) into any desired set of points (a_1,b_1,c_1); as a special case, one of these points may be the point at ∞.

Hypergeometric Equation

The hypergeometric equation is defined to be

$$z(1 - z)w'' + [C - (A + B + 1)z]w' - ABw = 0 \tag{5-66}$$

where A, B, C are arbitrary complex constants. It is a P equation, with regular points at $(0,1,\infty)$, and can be described by the scheme

$$w = P\left\{\begin{matrix} 0 & 1 & \infty & \\ 0 & 0 & A & z \\ 1 - C & C - A - B & B & \end{matrix}\right\} \tag{5-67}$$

In a sense, the hypergeometric equation can be considered to be a standard form for the P equation, since any equation of the form (5-56) can be put into the form (5-66) by a suitable change of dependent and independent variables. To do this, we first use a transformation of the form (5-62), with $\lambda = -\alpha$ and $\mu = -\beta$, and then use the transformation (5-65) to move the regular points (a,b,c) to $(0,1,\infty)$, respectively. Thus, if $\Phi(A,B,C,z)$ is any solution of Eq. (5-66), then a solution of Eq. (5-56) must be

$$w = \frac{(z - a)^\alpha (z - b)^\beta}{(z - c)^{\alpha+\beta}} \Phi\left[\alpha + \beta + \gamma, \alpha + \beta + \gamma', 1 + \alpha - \alpha', \frac{(z - a)(b - c)}{(z - c)(b - a)}\right] \tag{5-68}$$

This is not the only way of constructing a solution of Eq. (5-56) from the function Φ. There are 6 different ways of mapping (a,b,c) into $(0,1,\infty)$, depending on which point is mapped where; also, since α and α', or β and β', can be interchanged with each other in Eq. (5-56) without altering that equation, this gives 4 possible combinations to be considered with each of the previous 6. Thus, there are 24 ways of obtaining a solution $w(z)$ for Eq. (5-56) in terms of a given solution Φ of Eq. (5-66). Of course, not more than 2 of these can be linearly independent in any region.

The Hypergeometric Series

As in Exercise 5 of Sec. 2-6, we define the *hypergeometric series* or *hypergeometric function* to be

$$F(A,B;C;z) = 1 + \frac{A \cdot B}{1 \cdot C} z + \frac{A(A + 1)B(B + 1)}{1 \cdot 2 \cdot C(C + 1)} z^2 + \frac{A(A + 1)(A + 2)B(B + 1)(B + 2)}{1 \cdot 2 \cdot 3 \cdot C(C + 1)(C + 2)} z^3 + \cdots \tag{5-69}$$

It is convergent for $|z| < 1$ and divergent for $|z| > 1$; direct substitution shows that it is a solution of Eq. (5-66). Thus, it is suitable for use as the Φ function of Eq. (5-68).

Equation (5-66) is itself a P equation, so that solutions of Eq. (5-66) can be obtained by use of Eq. (5-68), with Φ replaced by F. For example, leaving the locations of the three regular points unaltered but interchanging α and α' gives one solution of Eq. (5-66) as

$$z^{1-C}F(1 + A - C, 1 + B - C; 2 - C; z) \tag{5-70}$$

which, together with $F(A,B;C;z)$, would ordinarily provide a set of fundamental solutions of Eq. (5-66), valid in the vicinity of the origin. Similarly, interchanging β and β' shows that another solution of Eq. (5-66) is

$$(z - 1)^{C-A-B}F(C - B, C - A; C; z) \tag{5-71}$$

and interchanging both of α, α' and β, β' gives still another solution as

$$z^{1-C}(z - 1)^{C-A-B}F(1 - B, 1 - A; 2 - C; z) \tag{5-72}$$

Together with $F(A, B; C; z)$ itself, these provide a set of four solutions for Eq. (5-66), valid at least near the origin. Since not more than two of them can be linearly independent, there must be a linear relationship between any three of them; the coefficients in this relationship can be found by examining the behavior near the origin. The reader may show, for example, that the solution given by Eq. (5-71) is simply $(-1)^{C-A-B}F(A,B;C;z)$ and that given by Eq. (5-72) is equal to $(-1)^{C-A-B}$ times that given by Eq. (5-70).

Another set of four solutions can be obtained by interchanging the points 0 and 1, by the transformation $z_1 = 1 - z$. A typical member of this set is the solution

$$F(A, B; 1 + A + B - C; 1 - z) \tag{5-73}$$

Similarly, we could interchange the points 0 and ∞ by the transformation $z_1 = 1/z$; a typical member of the new set [put $a = \infty$, $b = 1$, $c = 0$ in Eq. (5-68), omitting the factor $(z - a)^\alpha$] is

$$z^{-A}F\left(A, A - C + 1; 1 + A - B; \frac{1}{z}\right) \tag{5-74}$$

Note that the series in the solution (5-73) converges for $|z - 1| < 1$, while the series in (5-74) converges for $|z| > 1$. Continuing in this way, just as in the discussion following Eq. (5-68), we obtain a total of 24 solutions for the hypergeometric equation, each expressed in terms of F. The variable will be one of the quantities z, $1 - z$, $1/z$, $(1 - z)^{-1}$, $1 - (1/z)$, $z/(z - 1)$. In any domain, not more than two can be linearly independent. A knowledge of the relationships between the various solutions is useful in constructing analytic continuations of solutions of the hypergeometric equation; a list of all 24 solutions, and the relations between them, is given in Erdélyi, Magnus, Oberhettinger, and Tricomi.[1]

Properties of the $F(A,B;C;z)$

The function F is clearly symmetric in the two variables A and B. It is a polynomial if either A or B is a negative integer, but is undefined if C is a negative integer. The functions $F(A \pm 1, B; C; z)$, $F(A, B \pm 1; C; z)$, and $F(A, B; C \pm 1; z)$ are said to be *contiguous* to $F(A, B; C; z)$; Gauss proved that any three such functions are linearly related, the coefficients in this linear relationship being themselves linear in z. Typical formulas are

$$\begin{aligned}(B - A)F + AF(A + 1) &= BF(B + 1)\\ C(1 - z)F + (C - B)zF(C + 1) &= CF(A - 1)\end{aligned}$$

where the notation has an obvious interpretation. They can be proved by direct comparison of series expansions (note that, in the first example, it is sufficient to check only the first two terms, since each side is a solution of the same second-order equation). A complete list is given in Erdélyi.[2]

Term-by-term comparison also shows that

$$\frac{d}{dz} F(A,B;C;z) = \frac{AB}{C} F(A + 1, B + 1; C + 1; z) \tag{5-75}$$

and repetitive application of this result gives a formula for the nth derivative; by use of the relations between contiguous functions, the result can be exhibited in a variety of forms.

The regular points of the hypergeometric equation are at 0, 1, ∞, so that these are the only points at which F can have singularities. Since F is analytic at $z = 0$, the only singularity of F in the finite plane is at $z = 1$; normally, this is an algebraic branch point, but for special values of the parameters it may be a logarithmic branch point or may not be a singularity at all. In any event, it must be possible to continue the function F defined for $|z| < 1$ by Eq. (5-69) onto the z plane cut from $+1$ to ∞ (say along the real positive axis), and usually when we refer to F we mean the function defined by Eq. (5-69) plus its analytic continuation to the cut plane.

Contour integration is very useful in finding expressions for F which converge for $|z| > 1$. The simplest such integral is one due to Euler, but it is valid only for Re $C >$ Re $B > 0$. It is

$$F(A,B;C;z) = \frac{\Gamma(C)}{\Gamma(B)\Gamma(C - B)} \int_0^1 t^{B-1}(1 - t)^{C-B-1}(1 - tz)^{-A}\, dt \tag{5-76}$$

where z is not real and ≥ 1 (otherwise $1 - tz$ vanishes on the path of integration) and where the principal values of each of the three terms in the inte-

[1] A. Erdélyi, W. Magnus, F. Oberhettinger, and F. Tricomi, "Higher Transcendental Functions," vol. 1, pp. 105–108, McGraw-Hill Book Company, New York, 1953.

[2] *Loc. cit.* p. 103.

grand are to be taken. This integral converges at its two end points if Re $C >$ Re $B > 0$. To prove that it coincides with Eq. (5-69) for $|z| < 1$, we need only expand $(1 - tz)^{-A}$ by the binomial theorem and integrate term by term, using Eq. (5-28).

A more general contour integral is due to Barnes,

$$F(A,B;C;z) = \frac{1}{2\pi i}\frac{\Gamma(C)}{\Gamma(A)\Gamma(B)}\int_{-i\infty}^{i\infty}\frac{\Gamma(A+s)\Gamma(B+s)\Gamma(-s)}{\Gamma(C+s)}(-z)^s\,ds \tag{5-77}$$

where $|\arg(-z)| < \pi$ and where the path of integration is indented as necessary to make the poles in the s plane of $\Gamma(A+s)\Gamma(B+s)$ lie to the left of the path and those of $\Gamma(-s)$ lie to the right of the path. Equation (5-77) may be proved by closing the contour in the right half-plane and examining the asymptotic behavior of the Γ functions as $|s| \to \infty$; we leave the details to the reader. Equation (5-77) having been established, the contour may now be closed in the left-hand plane, and the sum of the residues now gives two descending series in z, which converge for $|z| > 1$ and which must represent the analytic continuation of F onto the part of the plane $|z| > 1$. Again with details omitted, the result is that, for $|z| > 1$, the function F given by Eq. (5-77) has the series representation (if $B - A$ is not an integer)

$$\begin{aligned}\frac{\Gamma(A)\Gamma(B)}{\Gamma(C)}&F(A,B;C;z)\\ &= \frac{\Gamma(A)\Gamma(B-A)}{\Gamma(C-A)}(-z)^{-A}F(A,\,1-C+A;\,1-B+A;\,z^{-1})\\ &\quad+\frac{\Gamma(B)\Gamma(A-B)}{\Gamma(C-B)}(-z)^{-B}F(B,\,1-C+B;\,1-A+B;\,z^{-1})\end{aligned} \tag{5-78}$$

Thus, this is one method by means of which the previously described relationships among the 24 various solutions of the hypergeometric equation could be obtained.

One of the uses of the relations among the 24 solutions of the hypergeometric equation is to determine the behavior of a particular solution near a singularity. For example, how does F itself behave as $z \to 1$? Consulting a table of these relations, we find that (except when $C - A - B$ is an integer)

$$\begin{aligned}F(A,B;C;z) &= \frac{\Gamma(C)\Gamma(C-A-B)}{\Gamma(C-A)\Gamma(C-B)}F(A,\,B;\,A+B+1-C;\,1-z)\\ &\quad+\frac{\Gamma(C)\Gamma(A+B-C)}{\Gamma(A)\Gamma(B)}\\ &\quad(1-z)^{C-A-B}F(C-A,\,C-B;\,C+1-A-B;\,1-z)\end{aligned}$$

Thus, if Re $(C - A - B) < 0$, then, as $z \to 1$,

$$F(A,B;C;z) \sim \frac{\Gamma(C)\Gamma(A+B-C)}{\Gamma(A)\Gamma(B)}(1-z)^{C-A-B} \tag{5-79}$$

whereas, if Re $(C - A - B) > 0$, then, as $z \to 1$, we obtain

$$F(A,B;C;z) \sim \frac{\Gamma(C)\Gamma(C - A - B)}{\Gamma(C - A)\Gamma(C - B)} \tag{5-80}$$

(note that the series for F converges for this case). If $C - A - B$ is an integer, then the singularity at $z = 1$ is logarithmic and somewhat different formulas result. For example, if $C = A + B$, then, as $z \to 1$,

$$F(A,B;C;z) \sim -\frac{\Gamma(A + B)}{\Gamma(A)\Gamma(B)} \ln(1 - z) \tag{5-81}$$

EXERCISES

2. Use Eq. (5-76) and the expression obtained in Exercise 16 of Sec. 5-1 for $\int_0^\infty (\sinh t)^a (\cosh t)^b \, dt$ in terms of beta functions (with the transformation $e^{-t} = \tau$) to show that

$$F(2\alpha, 2\beta; \alpha + \beta + \tfrac{1}{2}; \tfrac{1}{2}) = \frac{\sqrt{\pi}\,\Gamma(\alpha + \beta + \tfrac{1}{2})}{\Gamma(\alpha + \tfrac{1}{2})\Gamma(\beta + \tfrac{1}{2})}$$

3. Legendre's equation reads

$$(1 - z^2)w'' - 2zw' + \nu(\nu + 1)w = 0$$

where ν is a complex constant. Describe the location and nature of its singularities. Express in P-equation form, and verify Eq. (5-57). Use the method of Eq. (5-65) to move the point at ∞ to the origin, leaving the other two regular points unaltered, and verify the result by direct substitution. Finally, express the solution of Legendre's equation in terms of the hypergeometric series. Discuss the influence of ν on the singularities at $z = \pm 1$.

5-4 Legendre Functions

Legendre's differential equation arises when the method of separation of variables is applied to the laplacian in spherical polar coordinates. Let these coordinates be (r,θ,φ), where the element of distance ds is given by

$$(ds)^2 = (dr)^2 + (r\,d\theta)^2 + (r \sin\theta\, d\varphi)^2$$

Using subscripts to denote partial differentiation, Laplace's equation for a function $V(r,\theta,\varphi)$ can be written

$$\Delta V = \frac{1}{r^2}\left[(r^2V_r)_r + \frac{1}{\sin\theta}(\sin\theta\, V_\theta)_\theta + \frac{1}{\sin^2\theta} V_{\varphi\varphi}\right] = 0 \tag{5-82}$$

In the method of separation of variables, one ordinarily seeks a solution of

Eq. (5-82) having the form

$$V = R(r)\Theta(\theta)\Phi(\varphi) \tag{5-83}$$

and hopes that a linear combination of such solutions can be used to solve the particular boundary-value problem that has been posed. Substituting Eq. (5-83) into Eq. (5-82) and dividing through by V, we obtain

$$\frac{1}{R}(r^2R_r)_r + \frac{1}{\Theta \sin\theta}(\sin\theta\,\Theta_\theta)_\theta + \frac{1}{\Phi\sin^2\theta}\Phi_{\varphi\varphi} = 0 \tag{5-84}$$

This equation cannot possibly be satisfied unless the first term, which depends only on r, is a constant—for otherwise its value could be varied independently of the other terms. For future convenience, we write this constant in the form $\nu(\nu+1)$, where ν is an arbitrary complex number.

Solving the equation

$$\frac{1}{R}(r^2R_r)_r = \nu(\nu+1) \tag{5-85}$$

we obtain

$$R = Ar^\nu + Br^{-\nu-1} \tag{5-86}$$

where A and B are arbitrary constants. Equation (5-84) becomes

$$\nu(\nu+1)\sin^2\theta + \frac{\sin\theta}{\Theta}(\sin\theta\,\Theta_\theta)_\theta + \frac{1}{\Phi}\Phi_{\varphi\varphi} = 0 \tag{5-87}$$

By a similar argument, the last term must now also be a constant, say $-m^2$, where m is an arbitrary complex number. Then

$$\Phi = C\cos m\varphi + D\sin m\varphi \tag{5-88}$$

for any numbers C and D, and Eq. (5-87) now becomes

$$\sin^2\theta\,\Theta_{\theta\theta} + \sin\theta\cos\theta\,\Theta_\theta + [\nu(\nu+1)\sin^2\theta - m^2]\Theta = 0$$

Thus, if we set $\cos\theta = x$,

$$(1-x^2)\frac{d^2\Theta}{dx^2} - 2x\frac{d\Theta}{dx} + \left[\nu(\nu+1) - \frac{m^2}{1-x^2}\right]\Theta = 0 \tag{5-89}$$

We have now determined the conditions that the three functions of Eq. (5-83) must satisfy. The only equation that involves any difficulty is Eq. (5-89), for it cannot be solved in terms of elementary functions. We have seen in Sec. 5-2 that it is advantageous to consider solutions in the complex plane, and we are therefore led to the following final form of what is called *Legendre's associated equation:*

$$(1-z^2)w'' - 2zw' + \left[\nu(\nu+1) - \frac{m^2}{1-z^2}\right]w = 0 \tag{5-90}$$

If the problem is radially symmetric, we set $m = 0$ so as to remove any dependence on φ; Eq. (5-90) then reduces to *Legendre's equation,*

$$(1 - z^2)w'' - 2zw' + \nu(\nu + 1)w = 0 \tag{5-91}$$

Solutions of Eq. (5-91) are called *Legendre functions;* solutions of Eq. (5-90) are called *associated Legendre functions.*

Legendre Functions of the First Kind

Equation (5-91) has three regular points, at $+1$, -1, and ∞. In the finite plane, therefore, solutions of Eq. (5-91) are analytic (but not necessarily single-valued) everywhere except at ± 1, where they will usually have branch points. The exponents at each of ± 1 are (0,0) so that the branch points are logarithmic in character.

In the neighborhood of the origin, we know from Sec. 5-2 that we can find a pair of linearly independent analytic solutions. Writing

$$w = a_0 + a_1 z + a_2 z^2 + \cdots \tag{5-92}$$

and substituting Eq. (5-92) into Eq. (5-91), we find that we can take w_1 and w_2 as

$$\begin{aligned} w_1 &= 1 - \frac{\nu(\nu+1)}{2!} z^2 + \frac{\nu(\nu-2)(\nu+1)(\nu+3)}{4!} z^4 \\ &\qquad - \frac{\nu(\nu-2)(\nu-4)(\nu+1)(\nu+3)(\nu+5)}{6!} z^6 + \cdots \\ &= F\left(-\frac{\nu}{2}, \frac{\nu+1}{2}; \frac{1}{2}; z^2\right) \end{aligned} \tag{5-93a}$$

$$\begin{aligned} w_2 &= z\left[1 - \frac{(\nu-1)(\nu+2)}{3!} z^2 + \frac{(\nu-1)(\nu-3)(\nu+2)(\nu+4)}{5!} z^4 \right. \\ &\qquad \left. - \frac{(\nu-1)(\nu-3)(\nu-5)(\nu+2)(\nu+4)(\nu+6)}{7!} z^6 + \cdots \right] \\ &= zF\left(\frac{1-\nu}{2}, \frac{\nu+2}{2}; \frac{3}{2}; z^2\right) \end{aligned} \tag{5-93b}$$

where F is the hypergeometric series of Eq. (5-69). The general solution for $|z| < 1$ is therefore

$$AF\left(-\frac{\nu}{2}, \frac{\nu+1}{2}; \frac{1}{2}; z^2\right) + BzF\left(\frac{1-\nu}{2}, \frac{\nu+2}{2}; \frac{3}{2}; z^2\right)$$

where A and B are arbitrary constants. Equation (5-81) implies that, as $z \to \pm 1$, this solution behaves like

$$-\ln(1 - z^2)\left\{A \frac{\Gamma(\frac{1}{2})}{\Gamma(-\nu/2)\Gamma[(\nu+1)/2]} \pm B \frac{\Gamma(\frac{3}{2})}{\Gamma[(1-\nu)/2]\Gamma[(\nu+2)/2]}\right\}$$

If ν is not a real integer, then neither of the expressions multiplied by A or B can vanish. It is therefore impossible to find constants A and B (not both zero) such that the solution is finite at both $z = +1$ and $z = -1$. Remembering that in Legendre's equation the variable z corresponds to $\cos \theta$, this means that no solution of the form (5-83) can be finite at both $\theta = 0$ and $\theta = \pi$ if ν is nonintegral. Thus, for example, if we want solutions of the form (5-83) that are valid in the whole interior of a sphere, ν must be integral.

If ν is integral, then w_1 reduces to a polynomial if ν is positive even or negative odd and w_2 reduces to a polynomial if ν is positive odd or negative even. Since a polynomial is certainly finite at $z = \pm 1$, we conclude that solutions finite at both ± 1 exist if and only if ν is integral. Actually, we need consider only nonnegative integers, for if $\nu = -n$, where $n > 0$, then the product $\nu(\nu + 1)$ in Eq. (5-91) [or in Eq. (5-90), for that matter] is equal to $(n - 1)n$, so that the solutions for $\nu = -n$ are exactly the same as those for $\nu = n - 1$. Thus, in considering $\nu = n$, there is no loss in generality in requiring $n \geq 0$, and in the future we shall do so. Then, if $n \geq 0$ is even, w_1 is an nth-order polynomial, and if n is odd, w_2 is an nth-order polynomial. We now claim that, within a multiplicative constant, the polynomials so obtained are identical with the Legendre polynomials $P_n(z)$ of Sec. 3-3. This follows from the fact that, for any choice of n, each of the two polynomials is a solution of Legendre's equation and that each is odd if n is odd or even if n is even, so that one initial condition at $z = 0$ must be enough completely to specify that solution.

Choosing w_1 and w_2 as a fundamental set of solutions has the disadvantage that when ν is integral we shall be interested in one or the other of them, according to whether $\nu = n$ is even or odd. It is advantageous to find a linear combination of them, $P_\nu(z)$, which reduces to $P_n(z)$ when n is an integer. Rather than trying directly to find the coefficients of such a linear combination, it is simpler to recollect the remark made following Eq. (3-75) to the effect that the proof there given showed that Schäfli's integral was a solution of Legendre's equation even for nonintegral values of the index n. We therefore define the *Legendre function of the first kind* by

$$P_\nu(z) = \frac{1}{2\pi i} \int_C \frac{(t^2 - 1)^\nu \, dt}{2^\nu (t - z)^{\nu+1}} \tag{5-94}$$

where the simple closed contour C (Fig. 5-2) encloses the singularities $t = z$ and $t = +1$ of the integrand, but not the singularity at $t = -1$. To remove any ambiguity from the definition, we shall make a cut in the t plane along the real axis from -1 to $-\infty$ (otherwise C is not unique—see dotted curve in Fig. 5-2), and we shall also specify that, on the real axis to the right of the point $t = 1$ (and also to the right of z, if z is real) $\arg (t - 1) = 0$, $\arg (t + 1) = 0$, $|\arg (t - z)| < \pi$. Since we would obtain an entirely different contour C by allowing z to traverse once a closed curve which encircles the point -1, it follows that, to make $P_\nu(z)$ single-valued, we must also cut the z plane along the real axis from -1 to $-\infty$. Apart from this

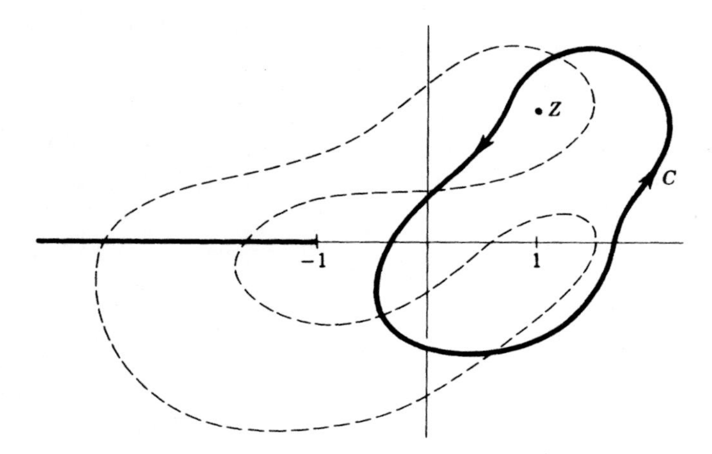

Fig. 5-2 Contour for integral of Eq. (5-94).

cut, which is introduced to identify a particular branch of $P_\nu(z)$, the definition (5-94) shows that P_ν is analytic everywhere except perhaps at $z = -1$, this point being excluded because of the nature of the contour C. For integral values of ν, Eq. (5-94) becomes identical with Eq. (3-74), so that P_ν does then indeed coincide with P_n.

At the point $z = 1$, the integrand in Eq. (5-94) does not have a branch point at $t = 1$ [the same remark applies to the formula for the pth derivative $P_\nu^{(p)}(z)$] so that we expect the coefficients in a series expansion of $P_\nu(z)$ in powers of $z - 1$ to be particularly simple. In fact,

$$P_\nu^{(p)}(1) = \frac{(\nu+1)(\nu+2)\cdots(\nu+p)}{2\pi i}\int_C \frac{(t+1)^\nu\,dt}{2^\nu(t-1)^{p+1}}$$
$$= \frac{(\nu+1)(\nu+2)\cdots(\nu+p)\nu(\nu-1)\cdots(\nu-p+1)}{p!2^p}$$

Thus
$$P_\nu(z) = 1 + \frac{(\nu+1)\nu}{(1!)^2\cdot 2}(z-1) + \frac{(\nu+1)(\nu+2)\nu(\nu-1)}{(2!)^2\cdot 2^2}(z-1)^2 + \cdots$$
$$= F\left(\nu+1, -\nu; 1; \frac{1-z}{2}\right) \tag{5-95}$$

Notice that the branch cuts in P_ν and F are mutually compatible. One consequence of Eq. (5-95) is that, for any value of ν, $P_\nu(z) = P_{-\nu-1}(z)$. We have already mentioned that, for $|z| < 1$, P_ν must be a linear combination of w_1 and w_2 as given by Eq. (5-93); the coefficients of this linear combination are clearly $F(\nu+1, -\nu; 1; \frac{1}{2})$ and $-\frac{1}{2}F'(\nu+1, -\nu; 1; \frac{1}{2})$, and the result of Exercise 2 of Sec. 5-3 now shows that

$$P_\nu(z) = \Gamma\left(\frac{1}{2}\right)\left[\frac{w_1(z)}{\Gamma(1+\nu/2)\Gamma(\frac{1}{2}-\nu/2)} + \frac{\nu w_2(z)}{\Gamma(\frac{1}{2}+\nu/2)\Gamma(1-\nu/2)}\right] \quad (5\text{-}96)$$

Equation (5-96) gives a series representation for $P_\nu(z)$ that is valid for $|z| < 1$, whereas that of Eq. (5-95) is valid for $|z - 1| < 2$. Although the range of validity of the former is contained within that of the latter, the former would of course be more convenient for computing $P_\nu(z)$ for small values of $|z|$. If expressions for $P_\nu(z)$ useful for large z are desired, one would use one of the expressions for P_ν in terms of hypergeometric functions and would consult a table of analytic continuations of the kind discussed in Sec. 5-4.

Finally, we remark that the recursion formulas and various integral expressions for $P_n(z)$ given in Sec. 3-3, derived from manipulations of Schläfli's integral, must also hold for $P_\nu(z)$, where ν is not necessarily an integer. An appropriate choice of branches must be made whenever the replacement of n by ν leads to a multiple-valued expression. A list of results is given by Erdélyi, Magnus, Oberhettinger, and Tricomi.[1]

Exercises

1. Transform Legendre's equation into a hypergeometric equation, and thus (consulting a table of analytic continuations of F), express its solution in terms of F functions for a number of ranges of z. Compare your results with those given in tabular form in Erdélyi.[2]

2. Why does the analyticity of $P_\nu(z)$ at $z = 1$ not contradict the previous result to the effect that P_ν could be finite at ± 1 only if ν were integral? Describe analytically the behavior of P_ν near $z = -1$.

3. Prove in detail that

$$P_\nu(z) = \frac{1}{2\pi}\int_{-\pi}^{\pi} [z + (z^2 - 1)^{1/2}\cos\theta]^\nu \, d\theta$$

for arbitrary values of ν and for $|\arg z| < \frac{1}{2}\pi$. Specify how the branch of the integrand is to be chosen.

4. If Eqs. (5-95) and (5-96) are compared, we obtain a certain identity in F functions. Exhibit this identity in as simple a form as possible, and consult a table of "quadratic transformations" of hypergeometric functions to show that this identity is a special case of such a transformation.

Legendre Function of Second Kind

There are many ways of choosing a second solution of Legendre's equation, to be used together with $P_\nu(z)$ to provide a fundamental set of solutions.

[1] *Op. cit.*, vol. 1, pp. 121ff.

[2] *Ibid.*, vol. 1, pp. 125ff.

It is natural to use a definition as analogous as possible to that for $P_\nu(z)$; this suggests the use of Schläfli's integral again, but this time with a different contour. [We recollect that Schläfli's integral, Eq. (3-74), gives a solution of Legendre's equation provided that the integrand returns to its original value when the contour of integration is traversed.] We define the *Legendre function of the second kind*, $Q_\nu(z)$, by

$$Q_\nu(z) = \frac{1}{4i \sin \nu\pi} \int_C \frac{(t^2 - 1)^\nu \, dt}{2^\nu (z - t)^{\nu+1}} \tag{5-97}$$

where C is the figure-of-eight contour shown in Fig. 5-3; z is not to be a real number in the range $(-1,1)$, and C is to be so chosen that z lies entirely outside C. To complete the definition, let $\arg (t + 1) = \arg (t - 1) = 0$ on the real axis to the right of the point $t = 1$; at $t = 0$, let $\arg (z - t) = \arg z$, with $|\arg z| \leq \pi$. Note that the integrands in Eqs. (5-97) and (5-94) are not identical. If ν is an integer, the definition (5-97) becomes indeterminate. To evaluate it for this case, we first assume $\text{Re } \nu > -1$ and deform the path of integration so as to coincide with the real axis between $(-1,1)$, with small circles around ± 1; this gives

$$Q_\nu(z) = \frac{1}{2^{\nu+1}} \int_{-1}^{1} \frac{(1 - t^2)^\nu \, dt}{(z - t)^{\nu+1}} \tag{5-98}$$

for $\text{Re } (\nu + 1) > 0$. Here the quantity $(1 - t^2)^\nu$ is to be given its principal value. There is now no difficulty in letting $\nu \to n$, where n is an integer ≥ 0, so that Eq. (5-98) can be used to define $Q_n(z)$ for integral $n \geq 0$. If $\nu = -n$, where n is an integer >0, then since, as previously remarked, the differential equation for $\nu = -n$ is the same as the differential equation for $\nu = n - 1$, we can take the solutions as P_{n-1}, Q_{n-1}.

Let us now find a series expansion for $Q_\nu(z)$, in powers of $1/z$. We assume first that $\text{Re } (\nu + 1) > 0$ so that Eq. (5-98) may be used. Writing

$$\frac{1}{(z - t)^{\nu+1}} = \frac{1}{z^{\nu+1}} \left(1 - \frac{t}{z}\right)^{-\nu-1}$$

expanding by the binomial theorem, and integrating term by term, we

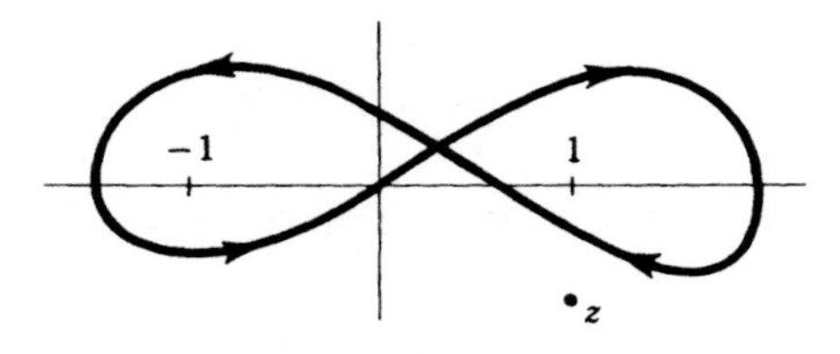

Fig. 5-3 Contour for integral of Eq. (5-97).

obtain

$$Q_\nu(z) = \frac{\sqrt{\pi}\,\Gamma(\nu+1)}{(2z)^{\nu+1}\Gamma(\nu+\frac{3}{2})} F\left(\frac{\nu+1}{2}, \frac{\nu+2}{2}; \nu+\frac{3}{2}; \frac{1}{z^2}\right) \tag{5-99}$$

Since each side is analytic in ν, this result, obtained for Re $(\nu + 1) > 0$, must hold for all values of ν, except the already excluded negative integers. We can make Q_ν single-valued by cutting the z plane from $+1$ to $-\infty$, and this is conventionally done.

As in the case of $P_\nu(z)$, the definition of $Q_\nu(z)$ by an integral of the Schläfli type means that the various recurrence formulas of Sec. 3-3 continue to hold for $Q_\nu(z)$. Also, an integral similar to the Laplace integral for $P_\nu(z)$ may be derived from Eq. (5-97); the reader may show that, for Re $(\nu + 1) > 0$,

$$Q_\nu(z) = \int_0^\infty [z + (z^2-1)^{1/2}\cosh\theta]^{-\nu-1}\,d\theta \tag{5-100}$$

For the case in which ν is a nonnegative integer n, Eq. (5-98) can be integrated directly. Thus, for $n = 0$ and $n = 1$, we obtain

$$\begin{aligned} Q_0(z) &= \frac{1}{2}\ln\frac{z+1}{z-1} = \tfrac{1}{2}P_0(z)\ln\frac{z+1}{z-1} \\ Q_1(z) &= \frac{z}{2}\ln\frac{z+1}{z-1} - 1 = \tfrac{1}{2}P_1(z)\ln\frac{z+1}{z-1} - 1 \end{aligned} \tag{5-101}$$

Using the recurrence relation (cf. Sec. 3-3),

$$(n+1)Q_{n+1}(z) = (2n+1)zQ_n(z) - nQ_{n-1}(z)$$

together with Eq. (5-101), we see that $Q_n(z)$ must have the form

$$Q_n(z) = \tfrac{1}{2}P_n(z)\ln\frac{z+1}{z-1} - W_{n-1}(z) \tag{5-102}$$

where $W_{n-1}(z)$ is a polynomial of order $n - 1$, having real coefficients. Thus $Q_n(z)$ is single-valued in the z plane cut along the line $(-1,1)$. Notice that Eq. (5-102) describes the behavior of $Q_n(z)$ over the whole (cut) plane and so is more general than the result that could be obtained from Eq. (5-55).

Consider now a point x on the real axis in the interval $(-1,1)$. As z approaches this point from above, Eq. (5-102) gives

$$Q_n(x+i0) = \tfrac{1}{2}P_n(x)\left(\ln\frac{1+x}{1-x} - i\pi\right) - W_{n-1}(x)$$

(where the principal value of the ln is taken), whereas an approach from below gives

$$Q_n(x-i0) = \tfrac{1}{2}P_n(x)\left(\ln\frac{1+x}{1-x} + i\pi\right) - W_{n-1}(x)$$

On the cut itself, $Q_n(z)$ is not defined; it is convenient to choose as a definition

$$\begin{aligned} Q_n(x) &= \tfrac{1}{2}Q_n(x + i0) + \tfrac{1}{2}Q_n(x - i0) \\ &= \tfrac{1}{2}P_n(x) \ln \frac{1+x}{1-x} - W_{n-1}(x) \end{aligned} \tag{5-103}$$

for $-1 < x < 1$. More generally, we define

$$Q_\nu(x) = \tfrac{1}{2}Q_\nu(x + i0) + \tfrac{1}{2}Q_\nu(x - i0)$$

for arbitrary values of ν and for $-1 < x < 1$. The function $Q_\nu(x)$ so defined clearly satisfies Legendre's equation for real x.

EXERCISES

5. For integral $n \geq 0$, show that

(a) $Q_n(z) = \dfrac{1}{2}\displaystyle\int_{-1}^{1} P_n(y)\,\dfrac{dy}{z-y}$ *Neumann's formula*

(b) $Q_n(z) = 2^n n! \displaystyle\int_z^\infty \int_t^\infty \int_t^\infty \cdots \int_t^\infty (t^2 - 1)^{-n-1}(dt)^{n+1}$

This last result is an analog of Rodrigues' formula.

6. The combination of the two contours C_1 and C_2 of Fig. 5-4 is equivalent to a single contour C_3 which surrounds -1, surrounds z, and excludes $+1$. Using the same integrand that was used in the definition of $P_\nu(z)$ and making the transformation $\tau = -t$ for the integral involving C_3, show that

$$\frac{2}{\pi}(\sin \pi\nu)Q_\nu(z) = e^{\pm\nu\pi i}P_\nu(z) - P_\nu(-z)$$

where the sign $\pm$ is taken according to whether Im $z < 0$ or >0, respectively.

7. Evaluate the wronskian of P_ν and Q_ν.

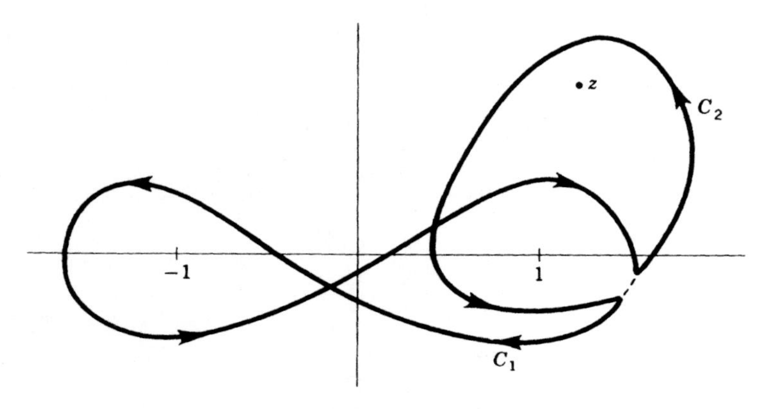

Fig. 5-4 Contour for Exercise 6 of Sec. 5-4.

Associated Legendre Functions

The Legendre functions P_ν and Q_ν are solutions of Eq. (5-91), which was derived from Eq. (5-90) by setting $m = 0$. For $m \neq 0$, we must seek solutions of the more general equation (5-90). Equation (5-90) can be simplified by the transformation

$$w = (z^2 - 1)^{\frac{1}{2}m} y \tag{5-104}$$

which gives

$$(1 - z^2)y'' - 2(m + 1)zy' + (\nu - m)(\nu + m + 1)y = 0 \tag{5-105}$$

This equation has the form

$$y = P\begin{Bmatrix} 1 & -1 & \infty & \\ 0 & 0 & m + \nu + 1 & z \\ -m & -m & m - \nu & \end{Bmatrix} \tag{5-106}$$

which, in conjunction with Eq. (5-104), tells us the nature of the branch points of the solution at the regular points of the equation. We can also make use of Eq. (5-106) to express the solutions of Eq. (5-105) in terms of hypergeometric functions, using Eq. (5-68). Just as in the discussion of Legendre functions, however, we shall not use this approach directly; rather we shall try to find a pair of linearly independent solutions which are closely analogous to $P_\nu(z)$ and $Q_\nu(z)$ and which reduce to them for $m = 0$.

The most important case is that in which m is integral; only then, for example, can Eq. (5-88) hold for all values of the angular coordinate φ [since $\Phi(\varphi + 2\pi)$ must then equal $\Phi(\varphi)$]. Also, the requirement that m be integral is not as restrictive as might at first appear, in view of the possibility of writing a more general trigonometric function of φ as a Fourier series in $\cos m\varphi$ and $\sin m\varphi$, with m integral. In any event, we shall in future assume that m is integral, and from Eq. (5-90) it is clear that there is no loss in generality in requiring $m \geq 0$.

Let us now differentiate Eq. (5-91) m times. This gives

$$(1 - z^2)w^{(m+2)} - 2(m + 1)zw^{(m+1)} + (\nu - m)(\nu + m + 1)w^{(m)} = 0$$

(where by $w^{(m)}$ we mean $d^m w/dz^m$, etc.); comparing this with Eqs. (5-104) and (5-105), we see that the functions

$$\begin{aligned} P_\nu{}^m(z) &= (z^2 - 1)^{\frac{1}{2}m} \frac{d^m}{dz^m}[P_\nu(z)] \\ Q_\nu{}^m(z) &= (z^2 - 1)^{\frac{1}{2}m} \frac{d^m}{dz^m}[Q_\nu(z)] \end{aligned} \tag{5-107}$$

can be chosen as a fundamental set of solutions for Eq. (5-90). For definiteness we require in Eqs. (5-107) that the arguments of $z + 1$ and $z - 1$ have their principal values. This is equivalent to making a cut in the z plane, running from $+1$ to $-\infty$. The functions P_ν^m and Q_ν^m are referred to as *associated Legendre functions of the first and second kinds*, respectively.

The cut from $+1$ to $-\infty$ includes the interval $-1 < x < 1$, so that for real values of z in this range we need a different definition. The one usually chosen is the natural one [although the notation is not always consistent; sometimes a factor $(-1)^m$ is incorporated],

$$
\begin{aligned}
P_\nu^m(x) &= (1 - x^2)^{\frac{1}{2}m} \frac{d^m}{dx^m} [P_\nu(x)] \\
Q_\nu^m(x) &= (1 - x^2)^{\frac{1}{2}m} \frac{d^m}{dx^m} [Q_\nu(x)]
\end{aligned}
\tag{5-108}
$$

for $-1 < x < 1$, where the principal value of $(1 - x^2)^{\frac{1}{2}m}$ is to be taken.

As before, to obtain a solution that is finite at both $x = +1$ and $x = -1$ we must choose the P_ν^m function, with ν integral. The first few of these (with $x = \cos\theta$) are

$$
\begin{array}{lll}
P_1^1 = \sin\theta & P_2^1 = \frac{3}{2}\sin 2\theta & P_3^1 = \frac{3}{8}(\sin\theta + 5\sin 3\theta) \\
P_1^2 = 0 & P_2^2 = \frac{3}{2}(1 - \cos 2\theta) & P_3^2 = \frac{15}{4}(\cos\theta - \cos 3\theta)
\end{array}
$$

We observe, of course, that $P_n^m = 0$ if $m > n$.

There is no difficulty in obtaining recursion relations and integral representations for the associated Legendre functions analogous to those previously obtained for the Legendre functions. A generalization to nonintegral values of m may also be made. There are no new methods involved here; so we shall merely refer the reader to the extensive list of results given by Erdélyi.[1] A thorough discussion of Legendre functions is given by Hobson.[2]

EXERCISES

8. If ν is integral in Eq. (5-107), what branch cuts are necessary for P_ν^m and Q_ν^m? For general values of ν, relate the values of $P_\nu^m(z)$ and $Q_\nu^m(z)$ on the two sides of $(-1,1)$ to $P_\nu^m(x)$ and $Q_\nu^m(x)$.

9. Determine the number of zeros of P_n^m in $(-1,1)$, and evaluate

$$\int_{-1}^{1} P_n^m(x) P_r^m(x)\, dx$$

for positive integers m, n, r, with $r > m$ and $n > m$.

5-5 Bessel Functions

In Sec. 5-4, we were led to Legendre's equation when we "separated variables" in Laplace's equation (5-82), as expressed in spherical polar

[1] *Ibid.*, vol. 1, p. 120.

[2] E. W. Hobson, "The Theory of Spherical and Ellipsoidal Harmonics," Cambridge University Press, New York, 1931, reprinted by Chelsea Publishing Company, New York, 1955.

coordinates (r,θ,φ). If, instead, we consider the Helmholtz equation (the scalar wave equation) for a function V,

$$\frac{1}{r^2}\left[(r^2V_r)_r + \frac{1}{\sin\theta}(\sin\theta \cdot V_\theta)_\theta + \frac{1}{\sin^2\theta}V_{\varphi\varphi}\right] + k^2V = 0 \qquad (5\text{-}109)$$

and set $V = R(r)\Theta(\theta)\Phi(\varphi)$, then we obtain as the equation for R

$$\frac{1}{R}(r^2R_r)_r + k^2r^2 = \nu(\nu + 1) \qquad (5\text{-}110)$$

where ν, as before, is an arbitrary complex constant. Again, the equations for Θ and Φ are Eqs. (5-88) and (5-89). Writing $R = r^{-\frac{1}{2}}w(r)$ in Eq. (5-110), we obtain

$$r^2w_{rr} + rw_r + [k^2r^2 - (\nu + \tfrac{1}{2})^2]w = 0 \qquad (5\text{-}111)$$

which is *Bessel's equation*.

Alternatively, consider the Helmholtz equation in cylindrical coordinates (r,θ,z), with the element of distance ds given by

$$(ds)^2 = (dr)^2 + (r\,d\theta)^2 + (dz)^2$$

Then the equation

$$\frac{1}{r}\left[(rV_r)_r + \frac{1}{r}V_{\theta\theta}\right] + V_{zz} + k^2V = 0 \qquad (5\text{-}112)$$

has solutions of the form $R(r)\Theta(\theta)Z(z)$ provided that

$$Z = A\cos\alpha z + B\sin\alpha z$$
$$\Theta = C\cos m\theta + D\sin m\theta$$

where α, m are arbitrary complex constants, and provided that

$$r^2R'' + rR' + [(k^2 - \alpha^2)r^2 - m^2]R = 0 \qquad (5\text{-}113)$$

Again, this is Bessel's equation.

We shall take the standard form of Bessel's equation to be as given by Eq. (5-114) below. If we denote a solution of this equation by $C_\nu(z)$, then a solution of Eq. (5-111) is given by $C_{\nu+\frac{1}{2}}(kr)$, and a solution of Eq. (5-113) is given by $C_m(r\sqrt{k^2 - \alpha^2})$.

Solutions of Bessel's Equation

Bessel's equation is

$$z^2w'' + zw' + (z^2 - \nu^2)w = 0 \qquad (5\text{-}114)$$

where ν is an arbitrary complex constant. From the general discussion of Sec. 5-2, it is seen that this equation has two singularities—a regular point at the origin and an irregular point at ∞. Thus, it is not a P equation; it can, however, be exhibited (cf. Exercise 1 below) as the limiting case of a P equation in which two of the regular points coalesce (become *confluent*) at ∞.

The only singularity in the finite plane is at the origin, so that solutions can be expected to be analytic everywhere except (possibly) for a branch point at the origin. To find the solutions, we write

$$w = z^\rho(a_0 + a_1 z + a_2 z^2 + \cdots)$$

and apply Eq. (5-51) to obtain

$$\begin{aligned} a_0(\rho^2 - \nu^2) &= 0 \\ a_1[(\rho+1)^2 - \nu^2] &= 0 \\ a_2[(\rho+2)^2 - \nu^2] &= -a_0 \\ a_3[(\rho+3)^2 - \nu^2] &= -a_1 \\ \cdots\cdots & \cdots\cdots \\ a_n[(\rho+n)^2 - \nu^2] &= -a_{n-2} \\ \cdots\cdots & \cdots\cdots \end{aligned} \tag{5-115}$$

The first of these equations requires that $\rho = \pm\nu$; if the two square roots of ν^2 have unequal real parts, we shall always denote that root with the larger real part by $+\nu$ and the other one by $-\nu$. Then, when we choose $\rho = +\nu$ in the sequence (5-115), none of the coefficients multiplying the a_j on the left-hand side vanishes and we always obtain one solution of the form

$$a_0 z^\nu \left[1 - \frac{1}{1\cdot(\nu+1)}\left(\frac{z}{2}\right)^2 + \frac{1}{1\cdot 2\cdot(\nu+1)(\nu+2)}\left(\frac{z}{2}\right)^4 - \cdots\right]$$

It is conventional to set $a_0 = 1/[2^\nu\Gamma(\nu+1)]$. Then one solution of Bessel's equation is always given by the function

$$J_\nu(z) = \left(\frac{z}{2}\right)^\nu \sum_{m=0}^{\infty} \frac{(-1)^m}{m!\Gamma(\nu+m+1)}\left(\frac{z}{2}\right)^{2m} \tag{5-116}$$

called the *Bessel function of first kind, of order* ν. Apart from the multiplicative factor $(z/2)^\nu$, $J_\nu(z)$ involves only even powers of $z/2$. The radius of convergence of the series is ∞, so that $(z/2)^{-\nu}J_\nu(z)$ is an entire function of z. If ν is a real integer $n \geq 0$, then

$$J_n(z) = \left(\frac{z}{2}\right)^n \left[\frac{1}{0!n!} - \frac{1}{1!(n+1)!}\left(\frac{z}{2}\right)^2 + \frac{1}{2!(n+2)!}\left(\frac{z}{2}\right)^4 - \cdots\right] \tag{5-117}$$

is called the *Bessel coefficient of order n*. Except for the additional factorial in each denominator, the series (5-117) is the same as that for $\exp[-(z/2)^2]$; in any event, $J_n(z)$ shares with the exponential function the property of being an entire function.

If the difference between the indicial roots, 2ν, is not an integer, then the choice $\rho = -\nu$ in Eq. (5-115) leads to another solution, $J_\nu(z)$. Since the leading terms of J_ν and $J_{-\nu}$ involve z^ν and $z^{-\nu}$, respectively, these two solutions must be linearly independent. Hence, if 2ν is not an integer, J_ν and $J_{-\nu}$ form a fundamental set of solutions of Bessel's equation.

Even when 2ν is an odd integer, $J_\nu(z)$ and $J_{-\nu}(z)$ are linearly independent solutions of Bessel's equation. For if $2\nu = 2p + 1$, where p is an integer ≥ 0, the choice $\rho = -\nu$ in Eqs. (5-115) will yield

$$\begin{aligned} a_0(0) &= 0 \\ a_1[(-p + \tfrac{1}{2})^2 - (p + \tfrac{1}{2})^2] &= 0 \\ a_2[(-p + \tfrac{3}{2})^2 - (p + \tfrac{1}{2})^2] &= -a_0 \\ a_3[(-p + \tfrac{5}{2})^2 - (p + \tfrac{1}{2})^2] &= -a_1 \\ \cdots\cdots\cdots\cdots & \\ a_{2p+1}(0) &= -a_{2p-1} \end{aligned}$$

Thus, each of $a_{2p-1}, a_{2p-3}, \ldots, a_1$ must vanish; however, a_{2p+1} can be chosen arbitrarily. Thus, we obtain two arbitrary constants a_0 and a_{2p+1}; moreover, the reader may verify that the series multiplied by a_0 and a_{2p+1} are proportional, respectively, to $J_{-\nu}$ and J_ν [as defined by Eq. (5-116)]. Hence the only peculiarity about the case $2\nu = 2p + 1$ is that both solutions J_ν and $J_{-\nu}$ can be obtained from the choice $\rho = -\nu$ in Eqs. (5-115); these solutions, of course, are still linearly independent (consider the leading terms).

Thus, *whenever ν is not an integer, J_ν and $J_{-\nu}$ provide a fundamental set of solutions for Bessel's equation.*

Consider now the case $\nu = n$, where $n \geq 0$ is an integer. One solution is still J_n, as given by Eq. (5-117); however, the choice $\rho = -n$ in Eq. (5-115) merely leads again to the result that J_n is a solution. This result is consistent with the fact that the choice $\nu = -n$ in Eq. (5-116) [with $1/\Gamma(-m) = 0$ for integral $m \geq 0$] leads to

$$J_{-n}(z) = (-1)^n J_n(z) \tag{5-118}$$

To obtain a second solution, we could use the general formula (5-55); however, a more compact definition of a second solution can be obtained by use of a method due to Hankel. Since the linear combination

$$J_\nu(z) - (-1)^n J_{-\nu}(z)$$

$\rightarrow 0$ as $\nu \rightarrow n$, we might form the quotient

$$\lim_{\nu \rightarrow n} \frac{J_\nu(z) - (-1)^n J_{-\nu}(z)}{\nu - n} \tag{5-119}$$

and ask whether or not it has any interesting properties. Now each of J_ν and $J_{-\nu}$ is differentiable with respect to ν (the two defining series are uniformly convergent), so that the limit must exist and have the value

$$\lim_{\nu \rightarrow n} \frac{(J_\nu - J_n) - (-1)^n (J_{-\nu} - J_{-n})}{\nu - n} = \left[\frac{\partial J_\nu}{\partial \nu} - (-1)^n \frac{\partial J_{-\nu}}{\partial \nu}\right]_{\nu=n}$$

In this form, it is easy to verify that the limit (5-119) is a solution of Bessel's equation—merely differentiate Eq. (5-114) with respect to ν. Moreover, a series expansion (which we shall obtain shortly) shows that this new solution of Bessel's equation is linearly independent of $J_n(z)$.

Having found in this way a second solution for integral values of ν and desiring to define a Bessel function of the second kind in a way which would hold whether or not ν was integral, Hankel suggested that the pair of basic solutions of Bessel's equation be taken as J_ν and the expression

$$2\pi e^{\nu\pi i} \frac{J_\nu(z) \cos \nu\pi - J_{-\nu}(z)}{\sin 2\pi\nu}$$

This definition fails if $\nu = n + \frac{1}{2}$, where n is integral, so that this case is excluded. For all other ν, this expression (or its limiting value if $\nu = n$) is a solution of Bessel's equation and is linearly independent of J_ν. To avoid the nuisance of having the definition fail for ν equal to half an odd integer, the standard definition (suggested by Weber and Schläfli as a modification of Hankel's definition) of the *Bessel function of the second kind* is now generally written

$$Y_\nu(z) = \frac{J_\nu(z) \cos \nu\pi - J_{-\nu}(z)}{\sin \nu\pi} \tag{5-120}$$

(or its limit, if ν is integral).

When ν is nonintegral, Y_ν is clearly independent of J_ν. We must now show that this is true also for integral values of ν. To do this, we write

$$Y_n(z) = \frac{1}{\pi} \left[\frac{\partial J_\nu}{\partial \nu} - (-1)^n \frac{\partial J_{-\nu}}{\partial \nu} \right]_{\nu=n} \tag{5-121}$$

Direct differentiation of Eq. (5-116) can now be used to evaluate $Y_n(z)$; the only point that deserves notice is that, if p is a nonnegative integer,

$$\lim_{z \to -p} \frac{\Gamma'(z)}{\Gamma^2(z)} = (-1)^{p+1} p!$$

which follows at once from the fact that the residue of $\Gamma(z)$ at $z = -p$ is $(-1)^p/p!$ The final result is (for $n \geq 0$)

$$Y_n(z) = \frac{2}{\pi} (\gamma + \ln \tfrac{1}{2}z) J_n(z) - \frac{1}{\pi} (\tfrac{1}{2}z)^{-n} \sum_{m=0}^{n-1} \frac{(n-m-1)!}{m!} (\tfrac{1}{2}z)^{2m}$$
$$- \frac{1}{\pi} (\tfrac{1}{2}z)^n \sum_{m=0}^{\infty} \frac{(-1)^m (\tfrac{1}{2}z)^{2m}}{m!(m+n)!} (h_m + h_{m+n}) \tag{5-122}$$

where
$$h_m = \begin{cases} 0 & \text{if } m = 0 \\ 1 + \frac{1}{2} + \frac{1}{3} + \cdots + \frac{1}{m} & \text{if } m \neq 0 \end{cases}$$

[If $n = 0$, the finite sum in Eq. (5-122) is omitted.] Thus, in general, the singularity of Y_n at the origin is a combination of an nth-order pole and a logarithmic branch point; Y_0 has only a logarithmic singularity. Except at the origin, Y_ν is analytic in z. Both J_ν and Y_ν are real for real ν and real positive z.

We shall show in Chap. 6 that the leading terms in the asymptotic expansions of J_ν and Y_ν are

$$J_\nu(z) \sim \left(\frac{2}{\pi z}\right)^{1/2} \cos\left(z - \tfrac{1}{2}\nu\pi - \tfrac{1}{4}\pi\right)$$
$$Y_\nu(z) \sim \left(\frac{2}{\pi z}\right)^{1/2} \sin\left(z - \tfrac{1}{2}\nu\pi - \tfrac{1}{4}\pi\right)$$

These expressions are valid for large $|z|$, with $|\arg z| < \pi$. It is often convenient to have a solution of Bessel's equation which tends exponentially to zero in a half-plane, as $|z| \to \infty$; it is clear from the asymptotic formulas that the combinations

$$J_\nu \pm iY_\nu \sim \left(\frac{2}{\pi z}\right)^{1/2} e^{\pm i(z - \frac{1}{2}\nu\pi - \frac{1}{4}\pi)} \tag{5-123}$$

have this property. We define *the Bessel functions of the third kind*, or *the Hankel functions of the first and second kinds*, by

$$\begin{aligned} H_\nu^{(1)}(z) &= J_\nu(z) + iY_\nu(z) \\ H_\nu^{(2)}(z) &= J_\nu(z) - iY_\nu(z) \end{aligned} \tag{5-124}$$

It is clear that $H_\nu^{(1)}$ and $H_\nu^{(2)}$ form a fundamental set of solutions, the first vanishing exponentially (i.e., as e^{iz}) in the upper half-plane and the second in the lower half-plane. Like J_ν and Y_ν, they are single-valued in the plane cut from 0 to $-\infty$.

In physical problems, Bessel's equation sometimes arises in the form

$$z^2w'' + zw' - (z^2 + \nu^2)w = 0$$

For nonintegral ν, two independent solutions are $J_\nu(iz)$ and $J_{-\nu}(iz)$; however, the special notation

$$I_\nu(z) = e^{-i\nu\pi/2}J_\nu(iz) \tag{5-125}$$

is ordinarily used. The function $I_\nu(z)$ is called *the modified Bessel function of the first kind.* For nonintegral ν, I_ν and $I_{-\nu}$ form a fundamental set of solutions; the reader may verify that each is real for real ν and real positive z. If ν is integral, we need an independent second solution. For convenience, we would like a solution which, like $I_\nu(z)$, is real for real positive z. From Eq. (5-122), we see that no multiple of $Y_n(iz)$ satisfies this requirement but that the function

$$\begin{aligned} K_n(z) &= \tfrac{1}{2}\pi i e^{in\pi/2}[J_n(iz) + iY_n(iz)] \\ &= \tfrac{1}{2}\pi i e^{in\pi/2}H_n^{(1)}(iz) \end{aligned} \tag{5-126}$$

is real for real positive z. It also vanishes exponentially as $z \to \infty$ along the real axis [cf. Eq. (5-123)]. More generally, we could define

$$K_\nu(z) = \tfrac{1}{2}\pi i e^{i\nu\pi/2} H_\nu^{(1)}(iz) \\ = -\tfrac{1}{2}\pi i e^{-i\nu\pi/2} H_\nu^{(2)}(-iz) \tag{5-127}$$

and I_ν and K_ν are a fundamental set of solutions of the modified Bessel's equation, valid for arbitrary values of ν. The function $K_\nu(z)$ is termed *the modified Bessel function of the third kind.*

Finally, we define *Kelvin's functions* by

$$\text{ber } z \pm i \text{ bei } z = I_0(ze^{\pm i\pi/4})$$
$$\text{ker } z \pm i \text{ kei } z = K_0(ze^{\pm i\pi/4})$$

where each of these new functions is real for real positive z.

Exercises

1. Show that

$$(a)\ J_\nu(z) = \lim_{\substack{\alpha\to\infty \\ \beta\to\infty}} \frac{(\frac{1}{2}z)^\nu}{\Gamma(\nu+1)} F\left(\alpha, \beta; \nu+1; -\frac{z^2}{4\alpha\beta}\right)$$

$$(b)\ J_0(z) = \lim_{n\to\infty} P_n\left(\cos\frac{z}{n}\right)$$

[*Hint:* See Eq. (5-95).]

2. Show that

(a) $Y_{-n}(z) = (-1)^n Y_n(z)$
(b) $[J_\nu(z)]^* = J_{\nu^*}(z^*)$
(c) $[H_\nu^{(1)}(z)]^* = H_{\nu^*}^{(2)}(z^*)$

3. A solution of the wave equation

$$c^2\,\Delta\varphi = \frac{\partial^2\varphi}{\partial t^2}$$

in spherical polar coordinates that represents monochromatic spherical waves emanating from the origin is (apart from a complex multiplicative constant)

$$\varphi = \frac{1}{r}\, e^{ik(r-ct)}$$

Explain why an analogous solution in cylindrical coordinates, for a two-dimensional problem, is

$$\varphi = H_0^{(1)}(kr)e^{-ikct}$$

rather than, say, $J_0(kr)e^{-ikct}$, $Y_0(kr)e^{-ikct}$, or $H_0^{(2)}(kr)e^{-ikct}$.

4. Show that the wronskian (Sec. 5-2) of $J_\nu(z)$ and $J_{-\nu}(z)$ is equal to $-2(\sin\nu\pi)/\pi z$. (Observe that it vanishes only for integral values of ν.) Find the wronskian of $J_\nu(z)$ and $Y_\nu(z)$. Find simple indefinite integrals of $[zJ_\nu^2(z)]^{-1}$ and $[zJ_\nu(z)Y_\nu(z)]^{-1}$.

Elementary Properties of the Bessel Functions

If we multiply Eq. (5-116) by z^ν or $z^{-\nu}$ and differentiate term by term, we obtain, respectively,

$$\frac{d}{dz}[z^\nu J_\nu(z)] = z^\nu J_{\nu-1}(z)$$
$$\frac{d}{dz}[z^{-\nu} J_\nu(z)] = -z^{-\nu} J_{\nu+1}(z) \tag{5-128}$$

These equations may be written

$$zJ'_\nu(z) + \nu J_\nu(z) = zJ_{\nu-1}(z)$$
$$zJ'_\nu(z) - \nu J_\nu(z) = -zJ_{\nu+1}(z) \tag{5-129}$$

Addition and subtraction gives

$$\frac{2\nu}{z} J_\nu(z) = J_{\nu-1}(z) + J_{\nu+1}(z)$$
$$2J'_\nu(z) = J_{\nu-1}(z) - J_{\nu+1}(z) \tag{5-130}$$

These three sets of equations provide various forms of the basic *recurrence formulas* for the Bessel functions.

Using the definition of Y_ν given by Eq. (5-120), and Eqs. (5-128) both as they stand and also with ν replaced by $-\nu$, it follows easily that the functions Y_ν also satisfy these various recurrence relations. Although Eq. (5-120) is valid only for nonintegral ν, the functions Y_ν and their derivatives are continuous in ν so that the recurrence relations hold also for integral ν. It also follows that the same recurrence relations apply to the functions $H_\nu^{(1)}(z)$ and $H_\nu^{(2)}(z)$. Recurrence formulas for the modified functions I_ν and K_ν differ slightly from the above equations; they may be easily obtained by replacing z by iz in the above formulas and using Eqs. (5-125) and (5-127).

In particular, the recurrence relations apply to the Bessel coefficients $J_n(z)$; from these, we can obtain such useful special relations as $J'_0(z) = -J_1(z)$. We can also use the recurrence relations to construct a generating function for the $J_n(z)$. Define

$$\varphi(z,t) = \sum_{n=-\infty}^{\infty} J_n(z)t^n \tag{5-131}$$

and assume tentatively that this series converges. Then

$$\frac{\partial}{\partial z}\varphi(z,t) = \sum_{n=-\infty}^{\infty} [\tfrac{1}{2}J_{n-1}(z) - \tfrac{1}{2}J_{n+1}(z)]t^n$$
$$= \frac{1}{2}\left(t - \frac{1}{t}\right)\varphi(z,t)$$

so that

$$\varphi(z,t) = f(t)e^{\frac{1}{2}z(t-1/t)}$$

Setting $z = 0$, we see that $f(t) = 1$, so that, if there exists a generating function of the form (5-131), it must be

$$e^{\frac{1}{2}z(t-1/t)} = \sum_{n=-\infty}^{\infty} J_n(z)t^n \tag{5-132}$$

To prove that Eq. (5-132) is indeed valid, we need observe only that each of $e^{\frac{1}{2}zt}$ and $e^{-\frac{1}{2}z/t}$ can be expanded into absolutely convergent series of powers of t; their product is again absolutely convergent, so that the terms can be rearranged. Collecting powers of t and comparing the coefficient of t^n with the definition of $J_n(z)$, Eq. (5-132) is established.

Equation (5-132) allows us to obtain very easily certain properties of the $J_n(z)$. As a first example, setting $t = e^{i\theta}$ and also $t = -e^{i\theta}$ and combining the two results so obtained, we find

$$\begin{aligned} \cos(z \sin \theta) &= J_0(z) + 2 \sum_{n=1}^{\infty} J_{2n}(z) \cos 2n\theta \\ \sin(z \sin \theta) &= 2 \sum_{n=0}^{\infty} J_{2n+1}(z) \sin(2n+1)\theta \end{aligned} \tag{5-133}$$

The replacement of θ by $\pi/2 - \theta$ would give similar Fourier expansions for $\cos(z \cos \theta)$ and $\sin(z \cos \theta)$. The reader should note the results of the special choices $\theta = 0$ and $\theta = \pi/2$.

Second, we observe that Eq. (5-132) provides a Laurent expansion for $e^{(z/2)(t-1/t)}$. But the coefficients of such an expansion are obtainable in terms of contour integrals of the function itself, so that

$$J_n(z) = \frac{1}{2\pi i} \int_C t^{-n-1} e^{\frac{1}{2}z(t-1/t)}\, dt$$

Choosing C to be the unit circle, with $t = e^{i\theta}$, we obtain

$$J_n(z) = \frac{1}{\pi} \int_0^{\pi} \cos(n\theta - z \sin \theta)\, d\theta \tag{5-134}$$

which is called *Bessel's integral;* it is valid only for integral n. One consequence of Eq. (5-134) is that $|J_n(x)| \leq 1$ for all real x.

Equation (5-134) can be generalized so as to include the case of nonintegral index. In Sec. 3-3, we obtained a solution of a difference equation which becomes Bessel's equation in the limit as the parameter $h \to 0$. If, therefore, we let $h \to 0$ in Eq. (3-100), we obtain *Schläfli's generalization of Bessel's integral* as

$$J_\nu(z) = \frac{1}{\pi} \int_0^{\pi} \cos(\nu\theta - z \sin \theta)\, d\theta - \frac{\sin \nu\pi}{\pi} \int_0^{\infty} e^{-\nu t - z \sinh t}\, dt \tag{5-135}$$

where we have replaced the real argument by z, justifying this by analytic continuation. We require $|\arg z| < \pi/2$ in Eq. (5-135) in order that the second integral converge. We shall subsequently rederive Eq. (5-135) by a different method.

Another integral expression for J_ν was obtained by Poisson for integral values of ν and generalized by Lommel to arbitrary values of ν satisfying $\operatorname{Re} \nu > -\frac{1}{2}$. *Poisson's integral* is

$$J_\nu(z) = \frac{(\frac{1}{2}z)^\nu}{\Gamma(\nu + \frac{1}{2})\Gamma(\frac{1}{2})} \int_0^\pi \cos (z \cos \theta) \sin^{2\nu} \theta \, d\theta \tag{5-136}$$

To prove this, we write the general term in the series expansion for J_ν as

$$\begin{aligned} \frac{(-1)^m(\frac{1}{2}z)^{\nu+2m}}{m!\Gamma(m + \nu + 1)} &= \frac{(\frac{1}{2}z)^\nu}{\Gamma(\nu + \frac{1}{2})\Gamma(\frac{1}{2})} \left[\frac{(-1)^m z^{2m}}{(2m)!} \frac{\Gamma(\nu + \frac{1}{2})\Gamma(m + \frac{1}{2})}{\Gamma(\nu + m + 1)} \right] \\ &= \frac{(\frac{1}{2}z)^\nu}{\Gamma(\nu + \frac{1}{2})\Gamma(\frac{1}{2})} \frac{(-1)^m z^{2m}}{(2m)!} \int_0^1 t^{\nu-\frac{1}{2}} (1 - t)^{m-\frac{1}{2}} \, dt \end{aligned}$$

by use of Eq. (5-28). Summing this series, we verify Eq. (5-136).

For the special case in which $\nu = n + \frac{1}{2}$, where $n \geq 0$ is a real integer, Eq. (5-136) can be written

$$J_{n+\frac{1}{2}}(z) = \frac{(\frac{1}{2}z)^{n+\frac{1}{2}}}{n!\pi^{\frac{1}{2}}} \int_{-1}^1 e^{izt}(1 - t^2)^n \, dt$$

A first integration by parts gives

$$J_{n+\frac{1}{2}}(z) = \frac{(\frac{1}{2}z)^{n+\frac{1}{2}}}{n!\pi^{\frac{1}{2}}} \left\{ - \int_{-1}^1 \frac{e^{izt}}{iz} \frac{d}{dt} [(1 - t^2)^n] \, dt \right\}$$

and it is clear that $2n - 1$ further integrations by parts will lead to an expression containing a finite number of terms involving e^{iz}, e^{-iz}, and powers of z. Thus, $J_{n+\frac{1}{2}}(z)$ can be expressed as the sum of a finite number of algebraic and trigonometric functions of z. The recurrence relations now enable us to extend this result to $J_{-n+\frac{1}{2}}$, so that we have the general theorem that any J_ν can be so expressed whenever $\nu = n + \frac{1}{2}$. It has been proved by Liouville[1] that, when $\nu \neq n + \frac{1}{2}$, no (nonzero) solution of Bessel's equation can be expressed in terms of a finite number of logarithmic, exponential, or integral operations on algebraic functions of z.

EXERCISES

5. Prove that no Bessel function can have a double zero at any point other than the origin. Show also that, if ν is real and positive, the real positive zeros of $J_\nu(x)$ and $J_{\nu+1}(x)$ interlace one another.

6. By term-by-term comparison of the series expansions, show that

$$|J_n(z)| \leq \frac{|\frac{1}{2}z|^n}{n!} \exp \left(\frac{\frac{1}{4}|z|^2}{n + 1} \right)$$

[1] See G. N. Watson, "Theory of Bessel Functions," pp. 111ff., Cambridge University Press, New York, 1948.

Obtain a similar inequality for $J_\nu(z)$, from Poisson's integral.

7. Show that

(a) $K_\nu(z)I_{\nu+1}(z) + K_{\nu+1}(z)I_\nu(z) = \dfrac{1}{z}$

(b) $J_{½+\nu}(z)J_{½-\nu}(z) + J_{-½+\nu}(z)J_{-½-\nu}(z) = \dfrac{2\cos\nu\pi}{\pi z}$

8. Show that, if n is an odd positive integer,

$$J_n(z) = (-1)^{½(n-1)}\frac{2}{\pi}\int_0^{\pi/2}\cos n\theta \sin(z\cos\theta)\,d\theta$$

9. Prove that

$$\int_0^\pi e^{\alpha\cos\theta}\cos(\beta\sin\theta)\,d\theta = \pi J_0[(\beta^2-\alpha^2)^{½}]$$

10. Show that

$$J_{½}(z) = \left(\frac{2}{\pi z}\right)^{½}\sin z \qquad J_{3/2}(z) = \left(\frac{2}{\pi z}\right)^{½}\left(\frac{\sin z}{z} - \cos z\right)$$

$$J_{-½}(z) = \left(\frac{2}{\pi z}\right)^{½}\cos z \qquad J_{-3/2}(z) = \left(\frac{2}{\pi z}\right)^{½}\left(-\frac{\cos z}{z} - \sin z\right)$$

11. Prove the *addition formula*

$$J_n(\alpha+\beta) = \sum_{m=-\infty}^{\infty} J_m(\alpha)J_{n-m}(\beta)$$

12. Prove that any positive integral power of z may be expanded in a series of Bessel functions, via the formula

$$z^n = 2^n\sum_{m=0}^{\infty}\frac{(n+2m)(m+n-1)!}{m!}J_{n+2m}(z)$$

13. Find the generating function for

$$\sum_{n=-\infty}^{\infty} I_n(z)t^n$$

and show that

$$\sum_{m=-\infty}^{\infty} I_{n-m}(z)J_m(z) = \begin{cases}\dfrac{z^n}{n!} & \text{for } n \geq 0\\ 0 & \text{for } n < 0\end{cases}$$

14. J. C. Miller[1] has shown that the recurrence formula for the I_n

[1] See "Mathematical Tables," vol. X, pt. II, Bessel Functions, British Association for the Advancement of Science, Cambridge University Press, New York, 1960. These tables contain a general discussion of computational methods for Bessel functions.

functions may be used "backward" so as to generate I_p, I_{p-1}, I_{p-2}, . . . , where p is some chosen number. Let p and x be real and > 0. Then the procedure is to define a sequence of functions

$$\varphi_{r-1}(x) = \frac{2r}{x}\varphi_r(x) + \varphi_{r+1}(x)$$

starting with $\varphi_{p+n+1}(x) = 0$, $\varphi_{p+n}(x) = 1$.

Show that this process yields

$$\varphi_p(x) = \alpha\left[I_p(x) + (-1)^n \frac{I_{p+n+1}(x)}{K_{p+n+1}(x)} K_p(x)\right]$$
$$\varphi_{p-1}(x) = \alpha\left[I_{p-1}(x) + (-1)^{n+1} \frac{I_{p+n+1}(x)}{K_{p+n+1}(x)} K_{p-1}(x)\right]$$
$$\cdots\cdots\cdots\cdots\cdots\cdots\cdots\cdots$$

and find a compact expression for the common multiplier α. From Chap. 6, the asymptotic behavior of $I_n(x)$ and $K_n(x)$ is such that the ratio $I_{p+n+1}(x)/K_{p+n+1}(x)$ can be made as small as desired by choosing n sufficiently large. The functions $\varphi_p(x)$, $\varphi_{p-1}(x)$, . . . are then effectively multiples of $I_p(x)$, $I_{p-1}(x)$, . . . , and the multiplier α can be found either by comparing $\varphi_0(x)$ with tabulated values of $I_0(x)$ or by the use of some such formula as $1 = I_0(x) - 2I_2(x) + 2I_4(x) - \cdots$. The method is particularly effective for evaluating a sequence of $I_p(x)$ functions for large values of p and x; it is not necessary to use different formulas depending on the relative sizes of p and x. The method has also been used for complex values of argument and for other kinds of Bessel functions, as well as for associated Legendre functions and repeated error integrals.

Contour Integral Representation

One often needs an integral representation for a function $w(z)$ defined by a differential equation. A technique that is frequently useful is to write

$$w(z) = \int_C K(z,t)f(t)\,dt$$

where $K(z,t)$, $f(t)$, and the contour C in the complex t plane are so chosen that the differential equation for $w(z)$ is satisfied.

In the case of Bessel functions, several representations of this form have been found. We begin with the choice

$$w(z) = z^\nu \int_C e^{izt}f(t)\,dt \tag{5-137}$$

where the factor z^ν can be anticipated because of its occurrence in the series expansion for $J_\nu(z)$. Substituting Eq. (5-137) into Eq. (5-114), we obtain

$$z^{\nu+2} \int_C e^{izt}(1 - t^2) f(t)\, dt + (2\nu + 1) i z^{\nu+1} \int_C e^{izt} t f(t)\, dt = 0$$

or $$[(t^2 - 1)e^{izt} f(t)]_a^b + \int_C e^{izt} \left\{ (2\nu + 1) t f(t) - \frac{d}{dt}[(t^2 - 1) f(t)] \right\} dt = 0$$

after an integration by parts, where a and b are the two ends of the contour C. Thus we do indeed obtain a solution of Bessel's equation, provided that

$$\frac{d}{dt}[(t^2 - 1) f(t)] - (2\nu + 1) t f(t) = 0$$

and $$[(t^2 - 1) e^{izt} f(t)]_a^b = 0$$

The first of these requires that $f(t)$ be some constant times $(t^2 - 1)^{\nu - \frac{1}{2}}$; the second then requires that $e^{izt}(t^2 - 1)^{\nu + \frac{1}{2}}$ have the same value at the two ends of the contour. This latter requirement could be met by using a closed contour for which $(t^2 - 1)^{\nu+\frac{1}{2}}$ returns to its initial value after a complete traversal, or (if Re $z > 0$) by letting the two ends of C be at $+i\infty$, where e^{izt} vanishes.

Define C_1 to be a figure-of-eight contour encircling $t = 1$ in the positive direction and $t = -1$ in the negative direction, as shown in Fig. 5-5. Then $e^{izt}(t^2 - 1)^{\nu+\frac{1}{2}}$ does return to its initial value after a complete traversal, so that

$$w_1(z) = z^\nu \int_{C_1} e^{izt}(t^2 - 1)^{\nu - \frac{1}{2}}\, dt \tag{5-138}$$

is a solution of Bessel's equation. For definiteness, the arguments of $t - 1$ and $t + 1$ are taken as zero for that point where the contour cuts the real axis to the right of $t = 1$. We exclude the case $\nu = n + \frac{1}{2}$, n integral and ≥ 0, for then the right-hand side of Eq. (5-138) vanishes identically.

Since $w_1(z)$ is a solution of Bessel's equation, it can be expressed in terms of $J_\nu(z)$ and $Y_\nu(z)$. Consider first the case Re $\nu > -\frac{1}{2}$, and shrink C_1 onto the line segment $(-1,1)$ except for small circles around the points ± 1. The restriction on ν ensures that the integrals around the circles centered on

Fig. 5-5 **Contour for integral of Eq. (5-138).**

± 1 vanish in the limit of small radius, so that the result is

$$w_1(z) = 2iz^\nu \cos \nu\pi \int_{-1}^{1} e^{izt}(1 - t^2)^{\nu-\frac{1}{2}}\, dt \tag{5-139}$$

Comparison with Eq. (5-136) shows that, except for a constant factor, the right-hand side of Eq. (5-139) is identical with Poisson's integral for $J_\nu(z)$. Inserting this factor, we therefore have

$$J_\nu(z) = \frac{\Gamma(\frac{1}{2} - \nu)}{2\pi i \Gamma(\frac{1}{2})} \left(\frac{z}{2}\right)^\nu \int_{C_1} e^{izt}(t^2 - 1)^{\nu-\frac{1}{2}}\, dt \tag{5-140}$$

Although this result was proved only for Re $\nu > -\frac{1}{2}$, each side is analytic in ν so that the result must hold for general values of ν. The case in which $\nu + \frac{1}{2}$ is a positive integer is still excepted, since the right-hand side of Eq. (5-140) then becomes indeterminate (however, the limit could be taken in the usual way). This representation of $J_\nu(z)$, due to Hankel, is a generalization of Poisson's integral, in that the restriction Re $\nu > -\frac{1}{2}$ to which Eq. (5-136) is subject is here unnecessary.

Consider next the function

$$w_2(z) = z^\nu \int_{C_2} e^{izt}(t^2 - 1)^{\nu-\frac{1}{2}}\, dt \tag{5-141}$$

where C_2 is the contour of Fig. 5-6. At its two ends, C_2 is asymptotic to a line making an angle φ with the positive real axis, as shown. Then for $-\varphi < \arg z < \pi - \varphi$, the condition that $e^{izt}(t^2 - 1)^{\nu+\frac{1}{2}}$ vanish at both ends of the path is fulfilled, and Eq. (5-141) is a solution of Bessel's equation. Again, we exclude positive integral values of $\nu + \frac{1}{2}$. We choose the initial

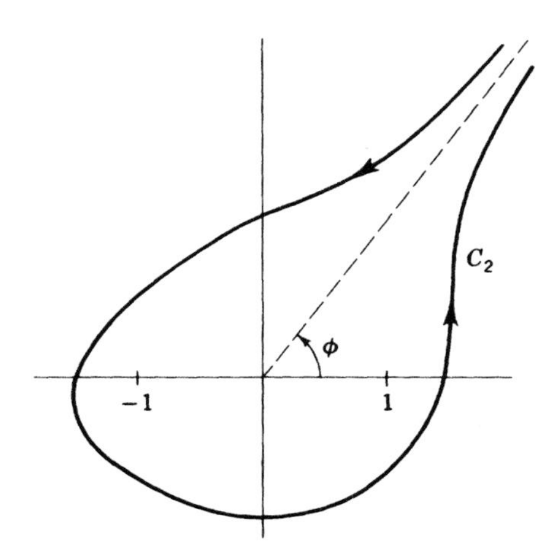

Fig. 5-6 Contour for Eq. (5-141).

and final arguments of each of $t+1$, $t-1$, to be φ and $2\pi+\varphi$. To find what Bessel function Eq. (5-141) represents, choose a contour outside the unit circle, and write

$$(t^2-1)^{\nu-\frac{1}{2}} = t^{2\nu-1}\left(1-\frac{1}{t^2}\right)^{\nu-\frac{1}{2}}$$

where the principal value of the second factor is to be taken and where the initial and final values of $\arg t$ are φ and $2\pi+\varphi$. Expansion by the binomial theorem and term-by-term integration then give

$$w_2(z) = z^\nu\Gamma\left(\nu+\frac{1}{2}\right)\sum_{m=0}^{\infty}\int_{C_2}\frac{(-1)^m e^{izt}}{m!\Gamma(\nu-m+\frac{1}{2})}t^{2\nu-2m-1}\,dt$$

To evaluate

$$\int_{C_2} e^{izt}t^{2\nu-2m-1}\,dt$$

we make the change of variable $\tau = izt$; the path of integration in the τ plane is asymptotic to a line radiating from the origin into the left-hand plane. The initial argument of τ is $\pi/2+\arg z+\varphi$, and the final argument is $5\pi/2+\arg z+\varphi$. The asymptote may be rotated into coincidence with the negative real axis; the initial and final arguments of τ are now π and 3π. Taking account of this difference in arguments from those used in the definition of Eq. (5-11), we can now apply Eq. (5-11) to give

$$\int_{C_2} e^{izt}t^{2\nu-2m-1}\,dt = e^{3\pi i(\nu-m)}z^{-2\nu+2m}\frac{2\pi i}{\Gamma(-2\nu+2m+1)}$$

By Eq. (5-12),

$$\Gamma(-2\nu+2m+1) = 2^{-2\nu+2m}\pi^{-\frac{1}{2}}\Gamma(-\nu+m+\tfrac{1}{2})\Gamma(-\nu+m+1)$$

so that, putting everything together [changing the sign of ν and using Eq. (5-10)], we can finally write Eq. (5-141) in the form

$$J_\nu(z) = \frac{\Gamma(\nu+\frac{1}{2})e^{3\pi i\nu}}{2\pi^{\frac{3}{2}}i}\left(\frac{z}{2}\right)^{-\nu}\int_{C_2} e^{izt}(t^2-1)^{-\nu-\frac{1}{2}}\,dt \qquad (5\text{-}142)$$

where the initial argument of $t\pm 1$ is φ and the final argument is $\varphi+2\pi$; also, $-\varphi < \arg z < \pi-\varphi$. This representation is also due to Hankel.

Hankel's representations were obtained by starting with an assumed form for $w(z)$ as given by Eq. (5-137). A similar form, suggested by Whittaker, leads to contour integrals involving Legendre functions. Write

$$w(z) = z^{\frac{1}{2}}\int_C e^{izt}f(t)\,dt \qquad (5\text{-}143)$$

Substitution into Eq. (5-114) shows that this representation does provide a solution of Bessel's equation, for suitable contours C, if $f(t)$ is a solution of

Legendre's equation for functions of order $\nu - \frac{1}{2}$. A typical result is

$$J_\nu(z) = \frac{(\frac{1}{2}z)^{\frac{1}{2}}e^{-\frac{1}{2}(\nu+\frac{1}{2})\pi i}}{\pi^{\frac{3}{2}}} \int_{C_2} e^{izt}Q_{\nu-\frac{1}{2}}(t)\,dt \tag{5-144}$$

where C_2 is the same as the contour of Fig. 5-6, where $-\varphi < \arg z < \pi - \varphi$ and where the phase of t is zero at the point where C_2 cuts the real axis to the right of $t = 1$. Verification of Eq. (5-144) is left as an exercise.

A quite different representation is suggested by the form of the generating function for the Bessel coefficients. Write

$$w(z) = \int_C e^{\frac{1}{2}z\left(t-\frac{1}{t}\right)} f(t)\,dt \tag{5-145}$$

and we find easily that Bessel's equation is satisfied if $f(t) = t^{-\nu-1}$ and if the two ends of the contour in the t plane are at infinity, with $e^{\frac{1}{2}zt}$ vanishing at those two ends. Let $|\arg z| < \frac{1}{2}\pi$; then a suitable contour C_3 is one which starts at $-\infty$, encircles the origin once in the positive direction, and then returns to $-\infty$. We can evaluate

$$\int_{C_3} e^{\frac{1}{2}z\left(t-\frac{1}{t}\right)} t^{-\nu-1}\,dt$$

by expansion of the function $e^{-\frac{1}{2}z/t}$ in powers of t, followed by term-by-term integration; we thus find that the desired integral representation is (Schläfli)

$$J_\nu(z) = \frac{1}{2\pi i}\int_{C_3} e^{\frac{1}{2}z\left(t-\frac{1}{t}\right)} t^{-\nu-1}\,dt \tag{5-146}$$

where the argument of t starts as $-\pi$ and ends as π. It is clear that Eq. (5-146) represents a generalization of Bessel's integral (5-134). Two other forms of Eq. (5-146) are of interest. The first is obtained by setting $t = 2\tau/z$, which gives

$$J_\nu(z) = \frac{1}{2\pi i}\left(\frac{z}{2}\right)^\nu \int_{C_3} \exp\left(\tau - \frac{z^2}{4\tau}\right)\tau^{-\nu-1}\,d\tau \tag{5-147}$$

There is now no restriction on $\arg z$. The second is obtained by writing $t = e^u$, which gives

$$J_\nu(z) = \frac{1}{2\pi i}\int_{\infty - i\pi}^{\infty + i\pi} e^{z \sinh u - \nu u}\,du \tag{5-148}$$

This representation is valid for $\operatorname{Re} z > 0$.

Exercise

15. A possible contour for use with Eq. (5-137) is one which encloses only one of the points $+1$, -1 and which has its two ends at $+i\infty$.

Show that

$$H_\nu^{(1)}(z) = \frac{\Gamma(\frac{1}{2} - \nu)(\frac{1}{2}z)^\nu}{\pi i \Gamma(\frac{1}{2})} \int_{1+i\infty}^{(1+)} e^{izt}(t^2 - 1)^{\nu - \frac{1}{2}}\, dt \qquad (5\text{-}149)$$

$$H_\nu^{(2)}(z) = \frac{\Gamma(\frac{1}{2} - \nu)(\frac{1}{2}z)^\nu}{\pi i \Gamma(\frac{1}{2})} \int_{-1+i\infty}^{(-1-)} e^{izt}(t^2 - 1)^{\nu - \frac{1}{2}}\, dt \qquad (5\text{-}150)$$

where the notation $\int_{1+i\infty}^{(1+)}$, for example, indicates that the contour begins and ends at $1 + i\infty$ and encircles $t = 1$ in the $+$ direction. It is assumed that $\operatorname{Re} z > 0$. Also, the phase of $t^2 - 1$ is to be zero at a point on the real axis to the right of $+1$ in Eq. (5-149) and is to be π at such a point to the right of -1 in Eq. (5-150).

[*Hint:* Use Eq. (5-140) for J_ν and Eq. (5-142) for $J_{-\nu}$, with each contour deformed so as to coincide as nearly as possible with the two half-lines $z = \pm 1 + iy$, $y > 0$.]

Line Integral Representations

Take the contour of Eq. (5-146) to consist of the real axis from $-\infty$ to -1, the unit circle in the positive direction around the origin, and the real axis from -1 to $-\infty$; then a simple change of variables in the result gives Schläfli's generalization of Bessel's integral,

$$\pi J_\nu(z) = \int_0^\pi \cos(z \sin\theta - \nu\theta)\, d\theta - \sin\nu\pi \int_0^\infty e^{-(z\sinh\varphi + \nu\varphi)}\, d\varphi \qquad (5\text{-}151)$$

which is valid for $\operatorname{Re} z > 0$. This agrees with the result (5-135) as previously obtained. Use of Eq. (5-120) also gives a simple representation for $Y_\nu(z)$, that is,

$$\pi Y_\nu(z) = \int_0^\pi \sin(z\sin\theta - \nu\theta)\, d\theta - \int_0^\infty (e^{\nu\varphi} + e^{-\nu\varphi}\cos\nu\pi)e^{-z\sinh\varphi}\, d\varphi \qquad (5\text{-}152)$$

again valid for $\operatorname{Re} z > 0$. Notice that Eq. (5-152) is valid even for integral values of ν.

Consider next Eq. (5-142), with φ taken as $\pi/2$ and with the contour shrunk down onto the strip $(-1,1)$ and the positive imaginary axis. If $\operatorname{Re}\nu < \frac{1}{2}$, the contributions due to the small circles surrounding the points ± 1 vanish and we obtain

$$J_\nu(z) = \frac{2}{\pi^{\frac{1}{2}}\Gamma(\frac{1}{2} - \nu)} \left(\frac{z}{2}\right)^{-\nu} \left[\int_0^1 (1 - t^2)^{-\nu - \frac{1}{2}} \cos(zt - \nu\pi)\, dt - \sin\nu\pi \int_0^\infty (1 + t^2)^{-\nu - \frac{1}{2}} e^{-zt}\, dt\right] \qquad (5\text{-}153)$$

valid for $\operatorname{Re}\nu < \frac{1}{2}$ and $\operatorname{Re} z > 0$. This equation is similar to Poisson's integral, but it holds for a different range of ν. We can replace ν by $-\nu$ in Eq. (5-153) to obtain an expression for $J_{-\nu}$ valid for $\operatorname{Re}\nu > -\frac{1}{2}$; if this result is then combined with Poisson's integral for J_ν [Eq. (5-136)], we obtain

$$Y_\nu(z) = \frac{2}{\pi^{1/2}\Gamma(\frac{1}{2}+\nu)}\left(\frac{z}{2}\right)^\nu \left[\int_0^1 (1-t^2)^{\nu-1/2}\sin(zt)\,dt - \int_0^\infty (1+t^2)^{\nu-1/2}e^{-zt}\,dt\right] \tag{5-154}$$

valid for $\operatorname{Re}\nu > -\frac{1}{2}$ and $\operatorname{Re} z > 0$.

Equations (5-153) and (5-154) were obtained by taking φ equal to $\pi/2$ in Eq. (5-142). If instead we set $\varphi = 0$, then replacing ν by $-\nu$ can be shown by the reader to lead to a formula for $I_{-\nu}(z)$ valid for $\operatorname{Re}\nu > -\frac{1}{2}$; using the formula for $I_\nu(z)$ as obtained from Poisson's integral then gives the simple result

$$K_\nu(z) = \frac{\pi^{1/2}}{\Gamma(\frac{1}{2}+\nu)}\left(\frac{z}{2}\right)^\nu \int_1^\infty e^{-zt}(t^2-1)^{\nu-1/2}\,dt \tag{5-155}$$

valid for $\operatorname{Re}\nu > -\frac{1}{2}$ and $\operatorname{Re} z > 0$.

From Eqs. (5-151) and (5-152), some interesting contour integrals for $H_\nu^{(1)}(z)$ and $H_\nu^{(2)}(z)$ can be obtained. Following Sommerfeld, consider the integral

$$w(z) = \int_{C_1} e^{iz\cos t + i\nu(t-\pi/2)}\,dt$$

where C_1 is a contour which proceeds along a vertical line from $-\frac{1}{2}\pi + i\infty$ to $-\frac{1}{2}\pi$, along the real axis from $-\frac{1}{2}\pi$ to $\frac{1}{2}\pi$, and along a vertical line from $\frac{1}{2}\pi$ to $\frac{1}{2}\pi - i\infty$. Evaluating the integrals along the various parts of this contour separately and using Eqs. (5-151) and (5-152), we obtain $w(z) = \pi H_\nu^{(1)}(z)$, so that

$$H_\nu^{(1)}(z) = \frac{1}{\pi}\int_{C_1} e^{iz\cos t + i\nu(t-\pi/2)}\,dt \tag{5-156}$$

for $\operatorname{Re} z > 0$. Similarly,

$$H_\nu^{(2)}(z) = \frac{1}{\pi}\int_{C_2} e^{iz\cos t + i\nu(t-\pi/2)}\,dt \tag{5-157}$$

for $\operatorname{Re} z > 0$, where C_2 is a contour which starts at $\pi/2 - i\infty$, proceeds along a vertical line to $\pi/2$, along the real axis from $\pi/2$ to $\frac{3}{2}\pi$, and along a vertical line from $\frac{3}{2}\pi$ to $\frac{3}{2}\pi + i\infty$.

Actually, C_1 and C_2 may be any curves lying in the shaded regions in Fig. 5-7, as the reader may show by considering the behavior of the integrand in these regions. Also, putting the contours C_1 and C_2 together to form C_3 (Fig. 5-7), we see that

$$J_\nu(z) = \frac{1}{2\pi}\int_{C_3} e^{iz\cos t + i\nu(t-\pi/2)}\,dt \tag{5-158}$$

As special cases, we obtain useful line integrals by taking C_1 to be the vertical line $\operatorname{Re} t = 0$ and taking C_2 to be a rectilinear contour from $0 - i\infty$

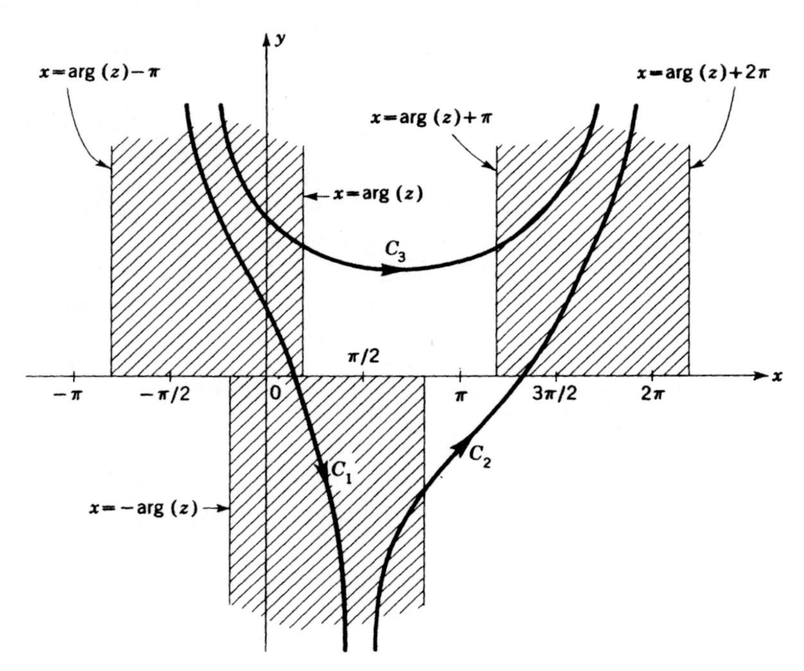

Fig. 5-7 Sommerfeld contours.

to $2\pi + i\infty$,

$$H_\nu^{(1)}(z) = -\frac{i}{\pi} e^{-i\nu\pi/2} \int_{-\infty}^{\infty} e^{iz\cosh t - \nu t}\, dt \tag{5-159}$$

$$H_\nu^{(2)}(z) = \frac{2}{\pi} e^{i\nu\pi/2} \left[\int_0^\pi e^{-iz\cos t} \cos(\nu t)\, dt + i \int_0^\infty e^{iz\cosh t} \cosh(\nu t - i\nu\pi)\, dt \right] \tag{5-160}$$

each of which is valid for $0 < \arg z < \pi$. Similar expressions can be obtained for $-\pi < \arg z < 0$, by choice of appropriate rectilinear contours C_1 and C_2; in particular,

$$H_\nu^{(2)}(z) = \frac{ie^{i\nu\pi/2}}{\pi} \int_{-\infty}^{\infty} e^{-iz\cosh t - \nu t}\, dt \tag{5-161}$$

for $-\pi < \arg z < 0$. If $\arg z = 0$, then the integrals in Eqs. (5-159) and (5-161) still converge, provided that $-1 < \operatorname{Re} \nu < 1$ and we obtain

$$\begin{aligned} J_\nu(x) &= \frac{2}{\pi} \int_0^\infty \sin(x\cosh t - \tfrac{1}{2}\nu\pi) \cosh(\nu t)\, dt \\ Y_\nu(x) &= \frac{-2}{\pi} \int_0^\infty \cos(x\cosh t - \tfrac{1}{2}\nu\pi) \cosh(\nu t)\, dt \end{aligned} \tag{5-162}$$

each valid for $x > 0$ and $-1 < \operatorname{Re} \nu < 1$. As special cases,

$$J_0(x) = \frac{2}{\pi} \int_1^\infty \frac{\sin xt}{(t^2 - 1)^{1/2}} \, dt$$

$$Y_0(x) = \frac{-2}{\pi} \int_1^\infty \frac{\cos xt}{(t^2 - 1)^{1/2}} \, dt$$

Asymptotic Behavior of Bessel Functions

It is often useful to have simple formulas available which describe the behavior of the general Bessel function $C_\nu(z)$ as $|z|$ becomes large, or as $|\nu|$ becomes large, or as both $|z|$ and $|\nu|$ become large together. We shall derive several such formulas in Chap. 6; here we shall list only the behavior for large $|z|$,

$$J_\nu(z) \sim \left(\frac{2}{\pi z}\right)^{1/2} \left[\cos\left(z - \frac{\nu\pi}{2} - \frac{\pi}{4}\right) - \frac{\nu^2 - 1/4}{2z} \sin\left(z - \frac{\nu\pi}{2} - \frac{\pi}{4}\right) + 0\left(\frac{1}{z^2}\right)\right] \tag{5-163}$$

$$Y_\nu(z) \sim \left(\frac{2}{\pi z}\right)^{1/2} \left[\sin\left(z - \frac{\nu\pi}{2} - \frac{\pi}{4}\right) + \frac{\nu^2 - 1/4}{2z} \cos\left(z - \frac{\nu\pi}{2} - \frac{\pi}{4}\right) + 0\left(\frac{1}{z^2}\right)\right] \tag{5-164}$$

each of which is valid for $-\pi < \arg z < \pi$,

$$H_\nu^{(1)}(z) \sim \left(\frac{2}{\pi z}\right)^{1/2} e^{i(z - \nu\pi/2 - \pi/4)} \left[1 + \frac{i(\nu^2 - 1/4)}{2z} + 0\left(\frac{1}{z^2}\right)\right] \tag{5-165}$$

for $-\pi < \arg z < 2\pi$, and

$$H_\nu^{(2)}(z) \sim \left(\frac{2}{\pi z}\right)^{1/2} e^{-i(z - \nu\pi/2 - \pi/4)} \left[1 - \frac{i(\nu^2 - 1/4)}{2z} + 0\left(\frac{1}{z^2}\right)\right] \tag{5-166}$$

for $-2\pi < \arg z < \pi$. There is no restriction on ν in these equations.

The basic power series for $C_\nu(z)$ can often be used to determine the behavior of $C_\nu(z)$ for large $|\nu|$. For example, if ν is real and >0, then the first few terms of Eq. (5-116) adequately describe the behavior of $J_\nu(z)$ as $\nu \to \infty$, with z fixed. If both ν and z are allowed to grow, then we shall see in Chap. 6 that the results will depend on the relative rate of growth of these two variables. The case of real ν and z is there treated in detail; for compex values of z and ν, a table of results is given by Watson.[1]

An interesting example in which one needs asymptotic formulas for $C_\nu(z)$ for large complex values of ν arises in *Watson's transformation.*[2] The problem is to evaluate a series typified by $\sum_{n=1}^{\infty} c_n(-1)^n H_n(r)$, where r is so large

[1] G. N. Watson, "Theory of Bessel Functions," 2d ed., chaps. 7 and 8, Cambridge University Press, 1944. A variety of asymptotic expansion is also listed in Erdélyi, Magnus, Oberhettinger, and Tricomi, *op. cit.*, vol. 2, pp. 85ff.

[2] G. N. Watson, *Proc. Roy. Soc.* (*London*), (A)**95:** 83 (1918).

that the series converges very slowly. The method is to replace the sum of the series by a contour integral in the ν plane, with integrand

$$\frac{c_\nu H_\nu(r)}{\sin \nu\pi}$$

The contour here encloses the positive real axis, but it may be deformed into a new contour enclosing various poles lying outside the original contour; the integral along the new contour vanishes, and the sum of residues at the "outside" poles provides a new series which converges rapidly for large values of r. A general discussion of the condition for vanishing of the integral along the new contour has been given by Pflumm.[1]

[1] E. Pflumm, *N.Y. Univ. Inst. Math. Sci. Rept.* BR-35, October, 1960.

6
asymptotic methods

6-1 *The Nature of an Asymptotic Expansion*

A function $f(z)$ which is analytic at $z = \infty$ can be represented by a power series in $1/z$ converging absolutely in some region $|z| > R$. If $s_n(z)$ denotes the sum of the first n terms, then we know that $|f(z) - s_n(z)|$ tends to zero, not only as $n \to \infty$ for fixed z (with $|z| > R$), but also for fixed n as $|z| \to \infty$. If $f(z)$ is not analytic at ∞, no such convergent power series exists. Nevertheless, it is a remarkable fact that, for large $|z|$, the behavior of such an $f(z)$ can often be described with enormous accuracy by the partial sums of a different kind of series expansion, called an *asymptotic expansion.* The great utility of an asymptotic expansion derives from the fact that, although it may not converge, $|f(z) - s_n(z)|$ does tend to zero as $|z| \to \infty$ for fixed n and for a restricted range of arg z. Before making these remarks precise, let us consider a simple example in which we restrict our attention to a real function of a real variable.

The exponential integral $Ei(z)$ was defined on page 91. For real positive x, it has the form

$$Ei(x) = \int_x^\infty \frac{e^{-t}}{t}\,dt$$

where the path of integration is along the real axis. We wish to evaluate $Ei(x)$ for large positive values of x. A convergent power series in $1/x$ is not possible, because of the nonanalyticity at ∞. However, let us integrate by parts to obtain

$$Ei(x) = \frac{e^{-x}}{x} - \int_x^\infty \frac{e^{-t}}{t^2}\,dt = \frac{e^{-x}}{x} - \frac{e^{-x}}{x^2} + 2\int_x^\infty \frac{e^{-t}}{t^3}\,dt$$

and after n repetitions,

$$Ei(x) = S_n(x) + R_n(x) \tag{6-1}$$

where
$$S_n(x) = e^{-x}\left\{\frac{1}{x} - \frac{1}{x^2} + \frac{2!}{x^3} - \cdots + \frac{(-1)^{n+1}(n-1)!}{x^n}\right\}$$
$$R_n(x) = (-1)^n n! \int_x^\infty \frac{e^{-t}}{t^{n+1}}\,dt \tag{6-2}$$

For fixed x, the terms in $S_n(x)$ eventually grow without limit as n increases; thus, $S_\infty(x)$ does not provide a convergent-series representation for $Ei(x)$. Since Eq. (6-1) is exact, $R_n(x)$ must also grow without limit.

Nevertheless, Eq. (6-1) can be used to evaluate $Ei(x)$ for large values of x. Since $R_n(x)$ satisfies the inequality

$$|R_n(x)| < \frac{n!}{x^{n+1}} \int_x^\infty e^{-t}\,dt = e^{-x}\frac{n!}{x^{n+1}}$$

it follows that, *for fixed n*, $R_n(x)$ is of order e^{-x}/x^{n+1} as $x \to \infty$. Thus, for n fixed and x increasing, $R_n(x)$ tends to zero more rapidly than the final term of $S_n(x)$. Hence, $S_n(x)$ gives an accurate approximation to $Ei(x)$ for a sufficiently large value of x. As a numerical example, take $n = 4$ and $x = 10$. Then $S_4(10) = 0.0914e^{-10}$ and $|R_4(10)| < 0.00024e^{-10}$, so that $|S_4(10) - Ei(10)| < 3 \times 10^{-3} \times Ei(10)$. For larger x, the agreement is still better; thus $|S_4(100) - Ei(100)| < 3 \times 10^{-7} \times Ei(100)$.

The standard notation used to indicate that $Ei(x)$ is more and more closely approximated by $S_4(x)$ as $x \to \infty$ is

$$\begin{aligned} Ei(x) &\sim S_4(x) \\ &\sim e^{-x}\left(\frac{1}{x} - \frac{1}{x^2} + \frac{2!}{x^3} - \frac{3!}{x^4}\right) \end{aligned}$$

We say that $Ei(x)$ and $S_4(x)$ are *asymptotic* to each other. An alternative asymptotic approximation for $Ei(x)$ is given by taking a larger number of terms in $S_n(x)$. For example, $|R_4(x)| < e^{-x} \cdot 4!/x^5$, whereas $|R_8(x)| < e^{-x} \cdot 8!/x^9$; thus, for sufficiently large x, $S_8(x)$ provides a better approximation to $Ei(x)$ than does $S_4(x)$. Observe, however, that for x less than about

6, $S_8(x)$ provides a poorer approximation than does $S_4(x)$. For very small x, the asymptotic representation is useless.

To indicate that n may be chosen arbitrarily, we commonly write

$$Ei(x) \sim e^{-x}\left(\frac{1}{x} - \frac{1}{x^2} + \frac{2!}{x^3} - \frac{3!}{x^4} + \cdots\right) \tag{6-3}$$

with the understanding that, before computing $Ei(x)$ for a given x, the user of this asymptotic approximation will terminate the series with the term of whatever order he wishes to include.[1] The complete series is divergent; as n increases, $|R_n(x)|$ may first decrease but must eventually increase without limit as $n \to \infty$. The reader may show that, for a given value of x, the smallest $|R_n(x)|$ in Eq. (6-2) is obtained by taking n equal to the integral part of x. Note incidentally that in Eq. (6-2) the sign of the error alternates as n increases.

We have already remarked that the asymptotic series (6-3) is useless for small values of x. However, one way of evaluating $Ei(x)$ for small x would be to write $Ei(x) = \int_x^{10} (e^{-t}/t)\,dt + Ei(10)$ and to use numerical integration for the first term and the asymptotic formula for the second.

Exercises

1. Obtain the same asymptotic expansion for $Ei(x)$ by writing

$$\begin{aligned} Ei(x) &= \frac{e^{-x}}{x}\int_0^\infty \frac{e^{-t}}{1 + t/x}\,dt \\ &= \frac{e^{-x}}{x}\int_0^\infty e^{-t}\,dt\left[1 - \frac{t}{x} + \frac{t^2}{x^2} - \cdots + \frac{(-1)^{n-1}t^{n-1}}{x^{n-1}} \right. \\ &\qquad \left. + \frac{(-1)^n t^n}{x^n(1 + t/x)}\right] \end{aligned}$$

and integrating term by term. Prove directly that the remainder term is identical with $R_n(x)$ as defined in Eq. (6-2).

2. The cosine integral $Ci(x)$ is defined by

$$Ci(x) = \int_x^\infty \frac{\cos t}{t}\,dt$$

Show that

$$\begin{aligned} Ci(x) \sim (\sin x)&\left(-\frac{1}{x} + \frac{2!}{x^3} - \frac{4!}{x^5} + \cdots\right) \\ &+ (\cos x)\left(\frac{1}{x^2} - \frac{3!}{x^4} + \frac{5!}{x^6} - \cdots\right) \end{aligned}$$

[1] Because of the special form of $R_n(x)$ in this example, the magnitude of the error is less than the magnitude of the first omitted term. Ordinarily, such a criterion is not applicable.

Obtain the remainder term explicitly, and find a bound for its magnitude. Use this series to evaluate $Ci(10)$.

3. Since integration by parts is equally valid in the complex plane, Eq. (6-1) continues to hold if x is replaced by a complex variable z. Using either the form of the remainder as given by Eq. (6-2) or that given in Exercise 1, show that Eq. (6-3) continues to give an asymptotic representation for $Ei(z)$ which is uniform (in what sense?) in the sector $|\arg z| < \pi - \epsilon$, where ϵ is any small fixed positive number. Can the representation be extended to the sector $|\arg z| < \frac{3}{2}\pi - \epsilon$? If so, must the multivalued nature of $Ei(z)$ be considered? Can the result of Exercise 2 now be obtained from Eq. (6-3)?

4. Use integration by parts to obtain asymptotic expansions, valid for large x, for the integrals:

(a) $\displaystyle\int_0^1 (\cos t + t^2)e^{ixt}\,dt$

(b) $\displaystyle\int_0^1 e^{ixt}t^{-1/2}\,dt$

(c) $\displaystyle\int_0^1 e^{ixt}\,(\ln t)\left(\cos\frac{\pi}{2}t\right)dt$

$\left[\right.$*Hint:* Write (b) as $\displaystyle\int_0^\infty - \int_1^\infty$ and, in (c), define $v(t) = \displaystyle\int_{i\infty}^t e^{ixt}\ln t\,dt.\left.\right]$ In general, the asymptotic form of such integrals depends on the behavior of the integrand near the end points and near those points where some derivative of the integrand has a discontinuity; for a further discussion, see van der Corput,[1] Erdélyi,[2] and Lighthill.[3]

Definition and Properties

A sequence of functions $\{\varphi_j(z)\}$ is said to be *an asymptotic sequence*, as z tends to z_0, if $\varphi_{j+1}(z)/\varphi_j(z) \to 0$ as $z \to z_0$, for $j = 1, 2, \ldots$. A function $f(z)$ is said to have the *asymptotic representation*

$$f(z) \sim a_1\varphi_1(z) + a_2\varphi_2(z) + \cdots \tag{6-4}$$

as $z \to z_0$, if, for each fixed choice of n, $|f(z) - s_n(z)|/\varphi_n(z) \to 0$ as $z \to z_0$. Here the a_n are constants, and $s_n = \sum_1^n a_j\varphi_j(z)$.

The case most commonly encountered is that in which (apart from a common multiplicative function) the $\varphi_j(z)$ are simply powers of $1/z$, the point z_0 is the point at ∞, and the approach of z to z_0 is sectorially con-

[1] J. G. van der Corput, *Ned. Akad. Wetensch. Proc.*, ***51:*** 650 (1948).

[2] A. Erdélyi, *J. Soc. Indust. Appl. Math.*, ***3:*** 17 (1955), ***4:*** 38 (1956).

[3] M. J. Lighthill, "Fourier Analysis and Generalised Functions," p. 46, Cambridge University Press, New York, 1958.

strained by some bounds on arg z. The form of Eq. (6-4) in this case is

$$f(z) \sim p(z)\left(a_0 + \frac{a_1}{z} + \frac{a_2}{z^2} + \cdots\right) \tag{6-5}$$

A given function need not possess an asymptotic expansion of a prescribed form. For example, it is clear that if

$$e^{-x} \sim a_0 + \frac{a_1}{x} + \frac{a_2}{x^2} + \cdots$$

as $x \to \infty$ along the real axis, then all a_j are zero. On the other hand, it is also clear that, if an asymptotic expansion of the form (6-4) exists [with prescribed $\varphi_j(z)$], it must be unique. However, a given asymptotic expansion can be valid for two quite different functions—for example, if $f(x)$ has an asymptotic expansion in powers of $1/x$ as $x \to \infty$, then $f(x) + e^{-x}$ has the same expansion.

Asymptotic series are usually divergent; that this is not always the case is exemplified by the usual power-series expansion for $e^{1/x}$.

Asymptotic representations may be combined linearly; if $f(z) \sim \Sigma a_j\varphi_j(z)$ and $g(z) \sim \Sigma b_j\varphi_j(z)$ and if α and β are any complex constants, then the reader will find it easy to show that

$$\alpha f(z) + \beta g(z) \sim \Sigma(\alpha a_j + \beta b_j)\varphi_j(z)$$

If, in Eq. (6-4), $\varphi_n(z) = 1/z^n$, the asymptotic representation is sometimes called an *asymptotic power series*. Such series can be manipulated in much the same way as conventional power series. Let

$$f(z) \sim a_0 + \frac{a_1}{z} + \frac{a_2}{z^2} + \cdots \qquad \text{for } \alpha_1 < \arg z < \alpha_2$$

$$g(z) \sim b_0 + \frac{b_1}{z} + \frac{b_2}{z^2} + \cdots \qquad \text{for } \beta_1 < \arg z < \beta_2$$

as $z \to \infty$. Then, as an exercise, the reader may show that, for appropriate arg z, the two series may be multiplied together term by term to give an asymptotic representation for $f(z)g(z)$ or (if $b_0 \neq 0$) divided by the usual process to give a representation for $f(z)/g(z)$. Also, term- by-term integration is valid, in the sense that

$$\int_z^\infty \left[f(z) - a_0 - \frac{a_1}{z}\right] dz \sim \frac{a_2}{z} + \frac{a_3}{2z^2} + \frac{a_4}{3z^3} + \cdots$$

and finally, if $f'(z)$ posesses an asymptotic representation, it will be obtained by term-by-term differentiation of that for $f(z)$. Similar results hold for the slightly more general case of Eq. (6-5).

Consider now the case of a function $f(z)$ which is single-valued and analytic outside some circle $|z| > R$. Suppose $f(z) \sim a_0 + a_1/z + a_2/z^2 + \cdots$ for all arg z as $z \to \infty$. Then $f(z)$ must be bounded near ∞ and so possess a conventional power series in $1/z$, valid as $z \to \infty$; this power series must coincide with the asymptotic series. Thus, if $f(z)$ is nonanalytic at ∞, it *cannot* possess an asymptotic power series valid for all arg z as $z \to \infty$. In practice, it is usually found that any asymptotic-power-series representation for $f(z)$ is valid only within some sectorial region (cf. Exercise 3 above) and that a different representation must be used outside that region. Thus, even though $f(z)$ is analytic across the boundary of the sector, the formula for its asymptotic representation changes as we cross the boundary. This situation is referred to as *Stokes' phenomenon;* a detailed discussion will be given in Sec. 6-6.

EXERCISES

5. Let $z = f(\zeta) \sim a_0 + a_1/\zeta + a_2/\zeta^2 + \cdots$ as $|\zeta| \to \infty$, for $\alpha < \arg \zeta < \beta$. Under what conditions can this representation be substituted into a convergent power series $g(z)$ and terms of like order collected to give an asymptotic representation for $g[f(\zeta)]$?

6. The form of the differential equation

$$y'' + \left(1 + \frac{1}{x^2}\right) y = 0$$

suggests that $y \sim e^{\pm ix}$ for large real positive x. Assume an asymptotic expansion of the form

$$y \sim e^{\pm ix}\left(a_0 + \frac{a_1}{x} + \frac{a_2}{x^2} + \cdots\right)$$

and determine the coefficients by substitution into the differential equation.

Euler's Summation Formula

Euler's formula, which can be used to obtain asymptotic expansions for certain kinds of functions, arises from a calculation of the difference between the exact integral of a function and the integral of a trapezoidal approximation to that function. Let $f(x)$ be a function of the real variable x, continuously differentiable as many times as required. Then if j, $j + 1$ are two adjacent positive integers, it is easily verified that

$$\tfrac{1}{2}[f(j) + f(j+1)] - \int_j^{j+1} f(x)\,dx = \int_j^{j+1} [x - (j + \tfrac{1}{2})]f'(x)\,dx$$

Adding together the equations of this form obtained by the choices $j = 0, 1, 2, \ldots, n - 1$ gives

$$f(0) + f(1) + f(2) + \cdots + f(n) = \tfrac{1}{2}[f(0) + f(n)] + \int_0^n f(x)\,dx + \int_0^n (x - [x] - \tfrac{1}{2})f'(x)\,dx \qquad (6\text{-}6)$$

which is *Euler's summation formula*. Here $[x]$ denotes the greatest integer $<x$, so that the factor multiplying $f'(x)$ in the last integral is a sawtooth function. Denote this factor by

$$P_1(x) = x - [x] - \tfrac{1}{2}$$

and notice that, in the interval (0,1), $P_1(x)$ coincides with the first Bernoulli polynomial $\varphi_1(x)$ [cf. Eq. (3-70)]. One integration by parts gives

$$f(0) + f(1) + \cdots + f(n) = \tfrac{1}{2}[f(0) + f(n)] + \int_0^n f(x)\,dx + [P_2 f']_0^n - \int_0^n P_2 f''\,dx$$

where $P_2(x)$ is any indefinite integral of $P_1(x)$. We can repeat this integration by parts indefinitely, using any sequence of functions $P_j(x)$ satisfying $P_j'(x) = P_{j-1}(x)$. A convenient sequence is obtained by making $P_j(x)$ periodic in x, with period 1, and setting $P_j(x) = \varphi_j(x)/j!$ for x in (0,1). Here $\varphi_j(x)$ is the jth Bernoulli polynomial; from Eq. (3-70) these polynomials satisfy $\varphi_j'(x) = j\varphi_{j-1}(x)$ and $\varphi_j(0) = \varphi_j(1) = B_j$, the jth Bernoulli number. Then

$$\begin{aligned} f(0) + f(1) + \cdots + f(n) &= \frac{1}{2}[f(0) + f(n)] + \int_0^n f(x)\,dx \\ &\quad + \frac{B_2}{2!}[f'(n) - f'(0)] + \frac{B_4}{4!}[f'''(n) - f'''(0)] + \cdots \\ &\quad + \frac{B_{2m}}{(2m)!}[f^{(2m-1)}(n) - f^{(2m-1)}(0)] + \int_0^n P_{2m+1}(x) f^{(2m+1)}(x)\,dx \end{aligned} \tag{6-7}$$

This is the most useful form of Euler's formula. An indefinite continuation of this process will scarcely be expected to yield a convergent series, because of the very rapid growth of the Bernoulli numbers. An alternative proof of Eq. (6-7) for analytic $f(z)$, using a contour-integration method, has been given by Evgrafov.[1]

As an example of the use of Euler's formula, let us obtain an asymptotic representation for $n!$. Rewrite Eq. (6-7) with n replaced by $n - 1$, and set $f(x) = \ln(1 + x)$. Then

$$\begin{aligned} \ln n! = \left(n + \frac{1}{2}\right)\ln n - n + A + \frac{B_2}{1\cdot 2}\frac{1}{n} + \frac{B_4}{3\cdot 4}\frac{1}{n^3} + \cdots \\ + \frac{B_{2m}}{(2m-1)(2m)}\frac{1}{n^{2m-1}} - (2m)!\int_n^\infty \frac{P_{2m+1}(x)}{x^{2m+1}}\,dx \end{aligned} \tag{6-8}$$

where

$$\begin{aligned} A &= 1 - \frac{B_2}{1\cdot 2} - \frac{B_4}{3\cdot 4} - \cdots - \frac{B_{2m}}{(2m-1)(2m)} + (2m)!\int_1^\infty \frac{P_{2m+1}(x)}{x^{2m+1}}\,dx \\ &= 1 + \int_1^\infty \frac{P_1(x)}{x}\,dx \end{aligned} \tag{6-9}$$

[1] M. A. Evgrafov, "Asymptotic Estimates and Entire Functions," p. 9, Gordon and Breach, New York, 1961.

The integrals in Eqs. (6-8) and (6-9) converge, because of the bounded (periodic) behavior of $P_j(x)$. The constant A is most easily evaluated by obtaining the first term of the asymptotic expansion by an entirely different method; this is done in Sec. 6-2 as an example of a Laplace integral, and it is found that $A = \ln \sqrt{2\pi}$. An alternative method is given in Exercise 7 below. Thus

$$\ln n! \sim \left(n + \frac{1}{2}\right) \ln n - n + \ln \sqrt{2\pi} + \frac{B_2}{1 \cdot 2} \frac{1}{n} + \frac{B_4}{3 \cdot 4} \frac{1}{n^3} + \cdots \tag{6-10}$$

That this is indeed an asymptotic expansion as $n \to \infty$ follows from the fact that (for a fixed m) the remainder term in Eq. (6-8) is of order $1/n^{2m}$, since P_{2m+1} is bounded. Equation (6-10) was obtained by Stirling in 1730.

To obtain Stirling's formula for complex values of argument, we use Eq. (5-3), which states that

$$\ln \Gamma(z) = \lim_{n \to \infty} [\ln n! + z \ln n - \ln z - \ln (z + 1) \cdots - \ln (z + n)] \tag{6-11}$$

Now setting $f(x) = \ln (x + z)$ in Eq. (6-7) gives

$$\begin{aligned} \ln z + \ln (z + 1) + \cdots + \ln (z + n) &= \left(z + n + \frac{1}{2}\right) \ln (z + n) \\ &\quad - \left(z - \frac{1}{2}\right) \ln z - n + \frac{B_2}{1 \cdot 2} \frac{1}{z + n} + \frac{B_4}{3 \cdot 4} \frac{1}{(z + n)^3} + \cdots \\ &\quad + \frac{B_{2m}}{(2m - 1)(2m)} \frac{1}{(z + n)^{2m-1}} \\ &\quad - (2m)! \int_n^\infty P_{2m+1}(x) \frac{dx}{(x + z)^{2m+1}} - C(z) \end{aligned} \tag{6-12}$$

where

$$\begin{aligned} C(z) = \frac{B_2}{1 \cdot 2} \frac{1}{z} + \frac{B_4}{3 \cdot 4} \frac{1}{z^3} + \cdots + \frac{B_{2m}}{(2m - 1)(2m)} \frac{1}{z^{2m-1}} \\ - (2m)! \int_0^\infty P_{2m+1}(x) \frac{dx}{(x + z)^{2m+1}} \end{aligned}$$

and where $\arg z \neq \pi$. Holding m fixed, and substituting Eq. (6-12) into Eq. (6-11) [with use of Eq. (6-8)], we obtain

$$\ln \Gamma(z) = (z - \tfrac{1}{2}) \ln z - z + A + C(z) \tag{6-13}$$

Just as in the case of Eq. (6-8), this is an exact equation. We obtain an asymptotic expansion by dropping the remainder term in $C(z)$,

$$\ln \Gamma(z) \sim \left(z + \frac{1}{2}\right) \ln z - z + \ln \sqrt{2\pi} + \frac{B_2}{1 \cdot 2} \frac{1}{z} + \frac{B_4}{3 \cdot 4} \frac{1}{z^3} + \cdots \tag{6-14}$$

If we stop with the term in $1/z^{2m-1}$, then the remainder term in $C(z)$ is of order $1/z^{2m}$. As previously remarked, $\arg (z) \neq \pi$; thus the range of validity of Eq. (6-14) is $|\arg z| < \pi$. The value of $\ln \Gamma(z)$ as given in Eq. (6-14) is not necessarily the principal value of the logarithm [which is not surprising, since in Eq. (6-12) we are adding together many logarithms]; however, the continuity of the right-hand side (and its reality if z is real and positive) shows that Eq. (6-12) represents the analytic continuation of $\ln \Gamma(z)$ in the plane cut along the negative real axis, starting from real values on the positive real axis.

EXERCISES

7. Obtain the value of A by combining Eq. (6-8) with Wallis's formula [Eq. (2-42)ff.]. Obtain A in still another way by combining Eqs. (6-13) and (5-12).

8. Prove that the right-hand side of Eq. (6-14) is a divergent series. Combine Eqs. (6-14) and (5-8) to obtain an accurate value of $\Gamma(i)$.

9. Use Euler's summation formula to obtain an asymptotic expansion for the derivative of $\Psi(z)$ defined by Eq. (5-18). Compare the result with that which would be obtained by use of Eq. (6-14).

10. Use Euler's summation formula, with $f(x) = 1/(x + 1)$, to obtain an asymptotic representation for the Euler-Mascheroni constant defined by Eq. (5-4). Use this representation to evaluate γ to five decimal places, and compare the labor with that required by Eq. (5-4). [*Hint:* Obtain an asymptotic representation for $\gamma - (1 + \frac{1}{2} + \cdots + 1/n - \ln n)$.]

6-2 *Laplace's Method*

Laplace's method provides an asymptotic representation for certain real integrals of the form $\int_a^b \varphi(x,t)\, dt$, as $x \to \infty$. The basic idea is to approximate the integral by taking into account only that portion of the range of integration where $\varphi(x,t)$ obtains its largest values; the method is useful if this largest contribution becomes more and more dominant as $x \to \infty$.

As an example, let us find an approximate expression, valid for large positive x, for the function

$$\Psi(x) = \int_{-\infty}^{\infty} e^{-x \cosh \theta}\, d\theta \tag{6-15}$$

The integrand attains its maximum value at $\theta = 0$; here $\cosh \theta$ has value unity and slope zero. For any other value of θ, $\cosh \theta > 1$, so that, as x becomes large, the value of the integrand at any $\theta \neq 0$ tends to become exponentially small in comparison with its value at $\theta = 0$. A set of graphs of the integrand (multiplied by e^x, for scaling purposes) for $x = 1, 10, 100$ is given in Fig. 6-1. It is seen that, as x grows, the value of the integrand tends

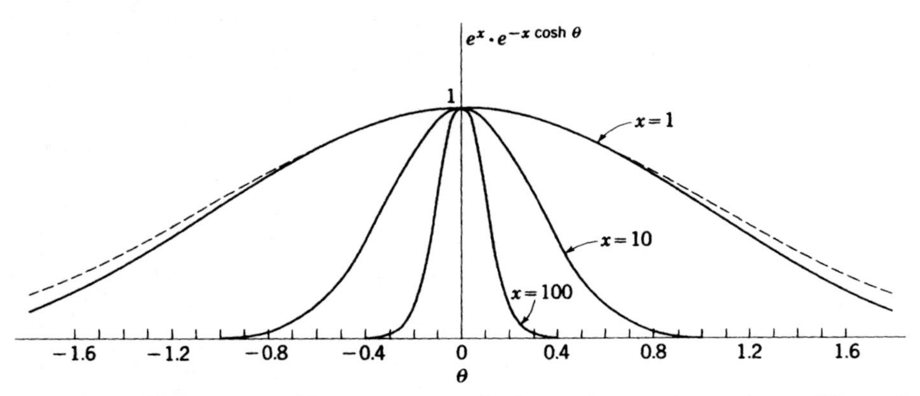

Fig. 6-1 Plots of $e^{x(1-\cosh\theta)}$ for various values of x. (Plots of $\exp(-x\theta^2/2)$ are shown dotted.)

to depend more and more completely on those values of θ near $\theta = 0$. This makes it plausible that, in computing $\Psi(x)$ for large x, we can replace $\cosh\theta$ by its approximate value, $1 + \frac{1}{2}\theta^2$, valid near the origin; thus,

$$\Psi(x) \sim \int_{-\infty}^{\infty} \exp\left[-x(1 + \tfrac{1}{2}\theta^2)\right] d\theta = \sqrt{\frac{2\pi}{x}}\, e^{-x}$$

In Fig. (6-1), the dotted curves represent $\exp(-\frac{1}{2}x\theta^2)$; for $x = 10$ and $x = 100$, they essentially coincide with the curves for $e^{x(1-\cosh\theta)}$. The method we have used is called *Laplace's method.* We observe that, if the integrand of Eq. (6-15) contained a factor $f(\theta)$, so that

$$\Psi_1(x) = \int_{-\infty}^{\infty} f(\theta) e^{-x\cosh\theta}\, d\theta \tag{6-16}$$

then [if $f(\theta)$ is continuous and nonzero at $\theta = 0$] a similar argument would lead to

$$\Psi_1(x) \sim \sqrt{\frac{2\pi}{x}}\, f(0) e^{-x}$$

To obtain further terms in the asymptotic expansion of $\Psi(x)$ as defined by Eq. (6-15), we can either expand $\cosh\theta$ to higher-order terms in θ or, alternatively, make the exact change in variable $\cosh\theta - 1 = t^2$, which leads to

$$\begin{aligned}\Psi(x) &= 2\sqrt{2} \int_0^{\infty} \exp\left[-x(1 + t^2)\right] \left(1 + \frac{t^2}{2}\right)^{-\frac{1}{2}} dt \\ &= 2\sqrt{2}\, e^{-x} \int_0^{\infty} \exp(-xt^2) \left(1 - \frac{t^2}{4} + \frac{3t^4}{32} - \cdots\right) dt \\ &\sim \sqrt{2\pi}\, e^{-x} \left(\frac{1}{\sqrt{x}} - \frac{1}{8x^{3/2}} + \frac{9}{128x^{5/2}} - \cdots\right)\end{aligned} \tag{6-17}$$

For suitable functions $f(\theta)$, a similar process can clearly be carried out for the more general function $\Psi_1(x)$ of Eq. (6-16).

We shall shortly prove that the expansions obtained by Laplace's method are indeed asymptotic. Before doing this, however, let us consider a second example—the evaluation of $\Gamma(x)$ for large positive real x [cf. Eq. (6-10)]. We have

$$\Gamma(x + 1) = \int_0^\infty e^{-t}t^x \, dt = \int_0^\infty e^{-t+x \ln t} \, dt$$

The exponent attains its maximum value at $t = x$, and as t departs from this value, the exponent decreases monotonically to $-\infty$ at each end of the interval of integration. Thus the major contribution of the integrand must arise from a region near $t = x$; however, the peak of the exponent becomes flatter as x increases, so that we do not have the kind of condensation process envisioned in Laplace's method. We therefore make the change in variable $t = x\tau$, which leads to

$$\Gamma(x + 1) = x^{x+1} \int_0^\infty e^{x(-\tau + \ln \tau)} \, d\tau \tag{6-18}$$

The exponent has a peak at $\tau = 1$; as x increases, this peak becomes sharper, so that the effect of the integrand tends to become more and more concentrated near $\tau = 1$. Near $\tau = 1$, the exponent has the behavior

$$x(-\tau + \ln \tau) = x[-1 - \tfrac{1}{2}(\tau - 1)^2 + \cdots]$$

so that, as x becomes large,

$$\Gamma(x + 1) \sim x^{x+1}e^{-x} \int_{-\infty}^\infty \exp\,[-\tfrac{1}{2}x(\tau - 1)^2] \, d\tau \tag{6-19}$$

We have set the lower limit of integration equal to $-\infty$, on the supposition that the only important contribution to the integral arises from a region near $\tau = 1$; this contribution is unaltered by the change in limit of integration. This change in limit of integration is a useful device, for the integral can now be evaluated explicitly, and Eq. (6-19) becomes

$$\Gamma(x + 1) \sim \sqrt{2\pi x}\, x^x e^{-x}$$

To determine the higher-order terms in the asymptotic expansion, we make the change in variable

$$-\tau + \ln \tau = -1 - \xi^2$$

from which [e.g., by differentiation, followed by substitution of a power series for $\tau(\xi)$, involving undetermined coefficients]

$$\tau = 1 + \sqrt{2}\xi + \tfrac{2}{3}\xi^2 + \frac{1}{9\sqrt{2}}\xi^3 - \tfrac{2}{135}\xi^4 + \cdots$$

As in the process leading to Eq. (6-17), we now obtain

$$\Gamma(x+1) \sim \sqrt{2\pi x}\, e^{-x} x^x \left(1 + \frac{1}{12x} + \frac{1}{288x^2} + \cdots\right) \tag{6-20}$$

Exercises

1. Show that the expansion for $\Gamma(x+1)$ as given in Eq. (6-20) is compatible with that for $\ln \Gamma(x)$ as given by Eq. (6-14).

2. The complementary error function erfc x is defined by

$$\begin{aligned}\operatorname{erfc} x &= \frac{2}{\sqrt{\pi}} \int_x^\infty \exp(-t^2)\, dt \\ &= \frac{2}{\sqrt{\pi}} \exp(-x^2) \int_0^\infty \exp(-2tx) \exp(-t^2)\, dt \qquad (6\text{-}21)\end{aligned}$$

The form of the integrand is different from that of Eq. (6-15) or (6-18) in that the maximum occurs at an end point rather than at an interior point; also, the slope of the exponent function is not zero at this maximum point. Nevertheless, the contribution of the integrand tends to be confined, in a more and more concentrated fashion, to the neighborhood of the maximum point as x (positive, real) increases; thus, for an asymptotic description of erfc x, it is necessary in Eq. (6-21) to consider only the immediate neighborhood of the origin. Expanding $\exp(-t^2)$ in a power series around the origin, integrate Eq. (6-21) term by term to obtain an asymptotic expansion for large values of x. Compare the asymptotic results with the exact values erfc $2 = 0.004677735 \cdots$, erfc $4 = 0.00000\,00154\,173 \cdots$. Is the asymptotic series convergent?

General Case

Each of the integrals in Eqs. (6-15), (6-18), and (6-21) is of the form

$$f(x) = \int_a^b e^{xh(t)} q(t)\, dt \tag{6-22}$$

where all quantities are real. In using Laplace's method to obtain an asymptotic expansion for $f(x)$ as $x \to \infty$, attention is concentrated on the neighborhood of that point (or points) where $h(t)$ attains its maximum value—whether this is at an end point or at an interior point. Although the basic idea of the Laplace method is more widely applicable than to integrals of the form (6-22), it is usually worthwhile to try to put a given integral into a form as close to (6-22) as possible, in order to utilize the rapid decay of the exponential factor as one moves away from the point of maximum $h(t)$—a decay which, moreover, becomes more rapid as x increases.

The case in which $h(t)$ has a maximum at an interior point in (a,b) is not essentially different from that in which it has a maximum at an end point, for the range of integration can always be subdivided and the right-hand side of Eq. (6-22) expressed as a sum of integrals in each of which $h(t)$ has a maximum at an end point only.

At these new end points, $h(t)$ may or may not have a vanishing derivative (to any order). Whether it does or not, we can make a change of variable so that each integral has the new form (within a multiplicative function of x)

$$f(x) = \int_0^T e^{-xt} t^\lambda g(t)\, dt \tag{6-23}$$

where the factor t^λ includes whatever singularity results, via the change in variable, from the vanishing of one or more derivatives of h. The general case (6-22) can therefore be reduced to this form; we now wish to prove (*Watson's lemma*) that the application of Laplace's method to Eq. (6-23) will yield a true asymptotic expansion for $f(x)$ as $x \to \infty$—provided, of course, that $g(t)$ satisfies certain conditions.

We require λ to be real and > -1 [so that the integral (6-23) converges at its lower limit]. Let constants K and b exist so that $|g(t)| < Ke^{bt}$ in $(0,T)$; this condition is important only if $T = \infty$, when it guarantees existence of the integral for sufficiently large values of x. Let us suppose that, in some interval $(0,A)$, the function $g(t)$ can be written as

$$g(t) = a_0 + a_1 t + a_2 t^2 + \cdots a_m t^m + R_{m+1}(t)$$

where m is some nonnegative integer and where some constant C exists such that

$$|R_{m+1}(t)| < Ct^{m+1}$$

for t in $(0,A)$. [These latter conditions on $g(t)$ are, for example, certainly satisfied if $g(t)$ is analytic at the origin.] Then

$$f(x) = \int_0^A e^{-xt} t^\lambda (a_0 + a_1 t + \cdots + a_m t^m)\, dt + \int_0^A e^{-xt} t^\lambda R_{m+1}(t)\, dt + \int_A^T e^{-xt} t^\lambda g(t)\, dt \tag{6-24}$$

The first integral in Eq. (6-24) can be written as

$$\int_0^\infty e^{-xt} t^\lambda (a_0 + a_1 t + \cdots + a_m t^m)\, dt - \int_A^\infty e^{-xt} t^\lambda (a_0 + a_1 t + \cdots + a_m t^m)\, dt$$
$$= a_0 \frac{\Gamma(\lambda + 1)}{x^{\lambda+1}} + a_1 \frac{\Gamma(\lambda + 2)}{x^{\lambda+2}} + \cdots + a_m \frac{\Gamma(\lambda + m + 1)}{x^{\lambda+m+1}} + 0(e^{-Ax})$$

where the symbol $0(e^{-Ax})$ means that the last term becomes less in magnitude than some constant times e^{-Ax} as $x \to \infty$. The second integral in Eq. (6-24) is less in magnitude than

$$C \int_0^A e^{-xt} t^\lambda t^{m+1} \, dt = 0\left(\frac{1}{x^{\lambda+m+2}}\right)$$

and the third integral in Eq. (6-24) is clearly exponentially small as $x \to \infty$. Altogether, then, we get, as $x \to \infty$,

$$f(x) \sim a_0 \frac{\Gamma(\lambda+1)}{x^{\lambda+1}} + a_1 \frac{\Gamma(\lambda+2)}{x^{\lambda+2}} + \cdots + a_m \frac{\Gamma(\lambda+m+1)}{x^{\lambda+m+1}} + 0\left(\frac{1}{x^{\lambda+m+2}}\right) \quad (6\text{-}25)$$

which completes the proof of Watson's lemma.

We notice the interesting fact that all the contribution to the asymptotic expansion comes from the interval near $t = 0$, irrespective of how rapidly (via large values of λ) the integrand may vanish at the origin. The upper limit T does not even occur in Eq. (6-25).

EXERCISES

3. In Eq. (6-22), let $h(t)$ attain its maximum at the interior point t_0. Let $h''(t_0) < 0$. Show that the first term in the asymptotic expansion is

$$q(t_0) \exp [xh(t_0)] \left[\frac{-2\pi}{xh''(t_0)}\right]^{1/2}$$

and write out the next few terms.

4. Use Laplace's method to obtain the first few terms of the asymptotic expansion of

$$(a) \int_0^1 e^{-xt} \sin t \, dt \qquad (b) \int_0^1 e^{-xt} t^{-1/2} \, dt$$

and compare with, in the case of (a), the exact result and, in the case of (b), the expansion as obtained by use of the erfc function of Eq. (6-21).

5. Obtain the first few terms of the asymptotic expansions, as $x \to \infty$, of

$$(a) \int_0^\pi \exp(-xt^2) t^{-1/3} \cos t \, dt \qquad (b) \int_0^{5(\pi/2)} e^{x \cos t} J_0(t) \sin 2t \, dt$$

$$(c) \int_0^1 t^x \sin^2 t \, dt \qquad (d) \int_0^\infty e^{xt} t^{-t} \, dt$$

[under (d), obtain the result for $x \to -\infty$ also].

6. Under certain conditions, the Laplace method may be used to find an asymptotic expansion of the somewhat more general expression

$$f(x) = \int_a^b e^{xh(x,t)} q(x,t)\, dt$$

although those theorems which are known are rather awkward to apply.[1] For the special case

$$f(x) = \int_0^\infty \exp\,(-xt + x^\alpha \ln t)\, dt$$

where α is a real positive or negative constant, show that a direct application of the Laplace method gives the correct asymptotic expansion [which can be obtained by first expressing $f(x)$ as a Γ function] only for $\alpha > 0$. Sketch the behavior of the exponent for various values of α and x.

Fourier-type Integrals

A class of integrals frequently encountered has the form

$$I(\beta) = \int_{-\infty}^{\infty} e^{i\beta z} f(z)\, dz$$

where β is real and where the path of integration passes below singularities located at $z_1, z_2, \ldots, z_n$ and above singularities located at $\zeta_1, \zeta_2, \ldots, \zeta_m$, as shown in Fig. 6-2. The asymptotic behavior of $I(\beta)$ for large $|\beta|$ can often be found by use of Laplace's method.

[1] *Ibid.*, p. 20.

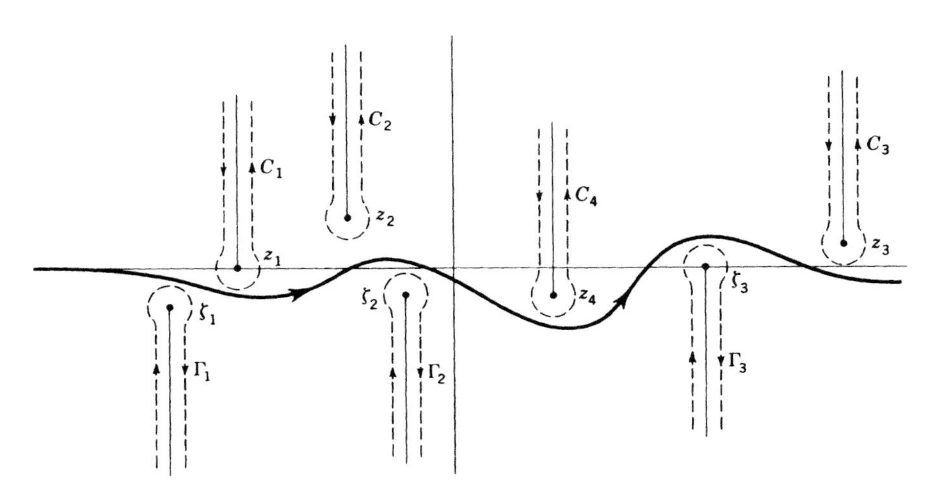

Fig. 6-2

We consider the case in which the singularities are branch points (if poles are present, residue contributions are easily incorporated in our subsequent formulas). Let $|f(z)| \to 0$ uniformly as $|z| \to \infty$ so that the use of Cauchy's theorem and Jordan's lemma leads to

$$I(\beta) = \begin{cases} \int_{C_1+C_2+\cdots+C_n} e^{i\beta z} f(z)\,dz & \text{for } \beta > 0 \\ \int_{\Gamma_1+\Gamma_2+\cdots+\Gamma_m} e^{i\beta z} f(z)\,dz & \text{for } \beta < 0 \end{cases}$$

The C_j and Γ_j are depicted in Fig. 6-2. If $\beta > 0$ and if $f(z)$ near z_j can be written

$$f(z) = (z - z_j)^\nu \sum_{p=0}^{\infty} a_p (z - z_j)^p$$

(with Re $\nu > -1$), then the contribution from C_j to I is

$$\begin{aligned} I_{C_j} &= \int_{C_j} e^{i\beta z}(z - z_j)^\nu \left[\sum_{p=0}^{N} a_p (z - z_j)^p + R_N(z) \right] dz \\ &= -2 \sin \pi\nu \exp\left(i\beta z_j - i\frac{\pi}{2}\nu \right) \int_0^\infty e^{-\beta r} r^\nu \left[\sum_{p=0}^{N} a_p (ir)^p + R_N(z_j + ir) \right] dr \end{aligned}$$

Asymptotically,

$$I_{C_j} \sim -\frac{2 \sin \pi\nu}{\beta^{\nu+1}} \exp\left(i\beta z_j - i\frac{\pi}{2}\nu \right) \sum_{p=0}^{\infty} \frac{a_p i^p \Gamma(\nu + p + 1)}{\beta^p}$$

[Clearly, this result can be extended to include functions $f(z)$ whose behavior at z_j exhibits rather pathological collections of algebraic and logarithmic singularities.] When the asymptotic contributions from the other z_j are added, so as to give an asymptotic expression for I itself, it is clear that the behavior, for large $\beta > 0$, of I is dominated by that particular I_{C_j} associated with the z_j of smallest imaginary part. The asymptotic calculation for $\beta < 0$ is analogous.

As an illustration, consider

$$F(\beta,\epsilon) = \int_{-i\infty}^{i\infty} e^{\beta z}(z^2 + \epsilon \sqrt{z} + 1)^{-1}\,dz$$

where $0 < \epsilon \ll 1$, $\beta > 0$, and $|\arg z| < \pi$. The integrand has two poles located (approximately) at $i[1 + (\epsilon/2)\sqrt{i}]$ and at $-i[1 + (\epsilon/2)\sqrt{-i}]$ (and two other poles in the other Riemann sheet). For large β, the residue terms give $2\pi i \sin \beta \exp[-\epsilon\beta/(2\sqrt{2})] + \cdots$, and we obtain the other portion of the asymptotic expression by integrating along the two sides of a branch cut coinciding with the negative real axis. The result is

$$F(\beta,\epsilon) \sim 2\pi i \sin \beta \exp\left(-\frac{\epsilon\beta}{2\sqrt{2}} \right) + \cdots + \frac{2i\epsilon}{\beta^{3/2}} \Gamma\left(\frac{3}{2}\right) - \frac{2i\epsilon^3}{\beta^{5/2}} \Gamma\left(\frac{5}{2}\right) + \cdots$$

Thus, for sufficiently large β,

$$F(\beta,\epsilon) \sim \frac{2i\epsilon}{\beta^{3/2}}\,\Gamma\left(\frac{3}{2}\right) - \frac{2i\epsilon^3}{\beta^{5/2}}\,\Gamma\left(\frac{5}{2}\right) + \cdots$$

We should observe, however, that there is a range of "moderately large" β in which the discarded ($\sin\beta$) term can be important: note that this term is oscillatory, whereas the final asymptotic formula is monotone-decreasing.

6-3 *Method of Steepest Descents*

The method of steepest descents is basically an extension of Laplace's method to integrals in the complex plane. It consists in deforming the contour of integration in such a way that the major contribution to the integral arises from a small portion of the new path of integration, this major contribution becoming more and more dominant as the parameter of interest grows.

As an example of the method, let us determine the asymptotic behavior as $x \to \infty$ of the integral

$$\varphi(x,\xi) = \frac{1}{2\pi i}\int_{\gamma - i\infty}^{\gamma + i\infty} \frac{\exp\,[x(s\xi - \sqrt{s})]}{s}\,ds \tag{6-26}$$

where ξ is a real positive parameter and where the path of integration is along a vertical line C_1 a distance γ to the right of the origin (Fig. 6-3). The

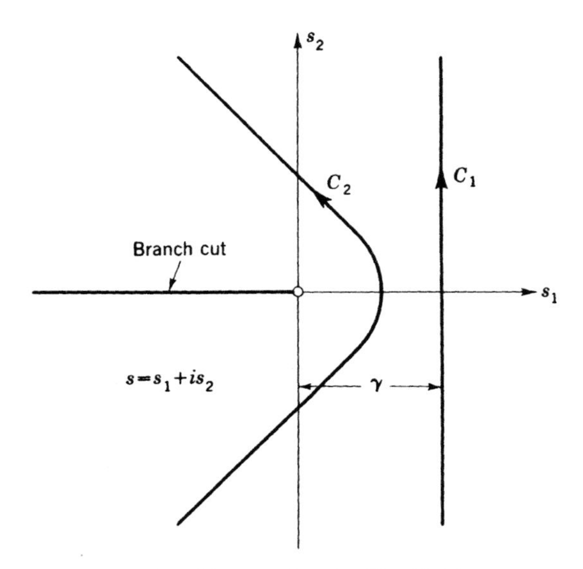

Fig. 6-3 Paths of integration for Eq. (6-26).

quantity $\sqrt{s}$ is given its principal value. (In Chap. 7, we shall encounter integrals of this form in solving diffusion problems.)

The value of the integral is clearly independent of the choice of γ, provided that $\gamma > 0$. Also, the path of integration can be deformed into paths such as C_2 in Fig. 6-3. For any chosen path of integration, the exponential function may well have a maximum modulus at some point s_0 on this path, and this raises the possibility of applying Laplace's method. There is, however, a complication. In general, for points on the path near s_0, the quantity $s\xi - \sqrt{s}$ will be complex, and

$$\exp [x(s\xi - \sqrt{s})] = \exp [x \operatorname{Re} (s\xi - \sqrt{s})] \exp [ix \operatorname{Im} (s\xi - \sqrt{s})]$$

Since the modulus is a maximum at s_0, we can expect $\operatorname{Re} (s\xi - \sqrt{s})$ to decrease as we move along the path away from s_0; moreover, this decrease will be accentuated as x becomes larger. But the value of $\operatorname{Im} (s\xi - \sqrt{s})$ can also be expected to alter as we move along the path away from s_0; with increasing x, the phase of the complex number $\exp [x(s\xi - \sqrt{s})]$ will therefore oscillate more and more rapidly in the neighborhood of s_0. Thus, we *cannot* deduce that the portion of the path of integration near s_0 is dominant, for as x increases, the increasingly dense oscillations in phase may negate the relative increase in real part near s_0.

This suggests that it is desirable to use a path of integration on which $\operatorname{Im} (s\xi - \sqrt{s})$ is constant near s_0; fortunately, the flexibility in choice of path of integration permits this to be done, at least in the present example. Figure 6-4 depicts the contour lines for the real and imaginary parts of $s\xi - \sqrt{s}$, shown as solid and dotted curves, respectively. Arrows on the dotted curves indicate the direction in which $\operatorname{Re} (s\xi - \sqrt{s})$ *increases*. An examination of the figure shows that there is one path of integration, marked C, which satisfies our various requirements. It is a curve of constant $\operatorname{Im} (s\xi - \sqrt{s})$, there is one point on it where $\operatorname{Re} (s\xi - \sqrt{s})$ attains a maximum, and moreover it is similar to C_2 in Fig. 6-3, so that the original path of integration can be deformed into it. [The path C is, of course, to be traversed from lower left to upper left; the arrows on it denote the direction of increasing $\operatorname{Re} (s\xi - \sqrt{s})$, and not necessarily the direction of traversal.]

The curve C is here a parabola; if $s = \alpha + i\beta$, its equation is

$$\alpha = \frac{1}{4\xi^2} - \xi^2\beta^2 \tag{6-27}$$

It cuts the real axis at the point $s_0 = 1/4\xi^2$, the point on C at which $\operatorname{Re} (s\xi - \sqrt{s})$ has its maximum value. A relief map of $\operatorname{Re} (s\xi - \sqrt{s})$ would be saddle-shaped near s_0, with the elevation rising as we move away from s_0 to the left or right but falling as we move away from s_0 in either of the two directions parallel to the imaginary axis. For obvious reasons, the point s_0 is termed a *saddle point*. We shall again remark, for purposes of emphasis, that we *could* choose a path of integration that went "higher" on the saddle—say through the point A in Fig. 6-4. However, even though Re $(s\xi - \sqrt{s})$ would attain a higher maximum on such a path, the dense oscillations in phase preclude the assertion that the integral is dominated by the contribution from a small neighborhood of A.

Starting at s_0, the curve C also gives the direction in which Re $(s\xi - \sqrt{s})$ decreases as rapidly as possible. It is thus a curve of steepest descent, and hence the name of the method. Where the paths of steepest descent involve a saddle point, as in the present example, the method is alternatively called a *saddle-point method.*

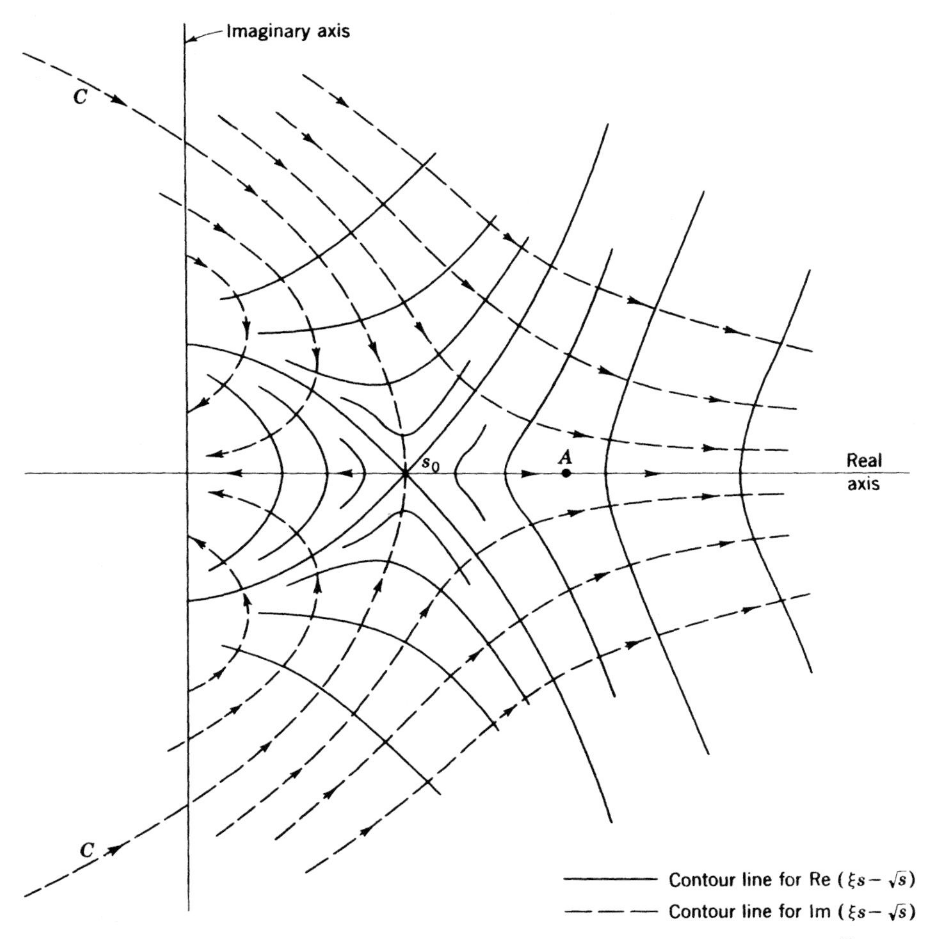

Fig. 6-4 Contour lines for real and imaginary parts of $(\xi s - \sqrt{s})$.

As in Laplace's method, we are primarily concerned with the behavior of the exponent near s_0, as we integrate along C. A power-series expansion for the exponent gives

$$x(s\xi - \sqrt{s}) = -\frac{x}{4\xi} + x\xi^3(s - s_0)^2 + \cdots \tag{6-28}$$

and since C cuts the real axis orthogonally, we can make the local change in variable $s - s_0 = it$, where t is real. We then have

$$\varphi(x) \sim \frac{1}{2\pi i} e^{-x/4\xi}(4\xi^2) \int_{-\infty}^{\infty} i \exp(-x\xi^3 t^2)\, dt$$

where the integration limits on t have been taken as $\pm\infty$ and where the factor s in the integrand has been replaced by s_0, just as in the previous application of Laplace's method to real integrals. Integration gives

$$\varphi(x) \sim 2\sqrt{\frac{\xi}{\pi x}}\, e^{-x/4\xi} \tag{6-29}$$

For many purposes, one needs only this first term of the asymptotic expansion for $\varphi(x)$, as x becomes large. However, one can also compute the higher-order terms. To do this, we observe that Im $(s\xi - \sqrt{s}) \equiv 0$ on C, and moreover it has a zero derivative at s_0. This suggests the *exact* change in variable

$$s\xi - \sqrt{s} = -\frac{1}{4\xi} - \tau^2 \tag{6-30}$$

where τ is real on C. For uniqueness, we require τ to be negative on the lower half of C and so positive on the upper half. For points s not on C, τ will be complex; it is clear that, at least near s_0, $s - s_0$ and τ are single-valued analytic functions of each other. Equation (6-26) now becomes

$$\varphi(x) = \frac{1}{2\pi i} e^{-x/4\xi} \int_{-\infty}^{\infty} \frac{1}{s(\tau)} \exp(-x\tau^2) \frac{ds}{d\tau}\, d\tau \tag{6-31}$$

We need the power-series expansion for $\dfrac{1}{s(\tau)} \dfrac{ds}{d\tau}$ about the point $\tau = 0$. Since Eq. (6-30) is equivalent to

$$s = \left(\frac{1}{2\xi} + \frac{i\tau}{\sqrt{\xi}}\right)^2$$

this series is easily obtained; substitution into Eq. (6-31), followed by term-by-term integration, then yields

$$\varphi(x) \sim 2\sqrt{\frac{\xi}{\pi x}}\, e^{-x/4\xi}\left(1 - 2\frac{\xi}{x} + 12\frac{\xi^2}{x^2} - \cdots\right) \tag{6-32}$$

It will be observed that, on each half of the path C, Watson's lemma can be applied, so as to guarantee that the expansion (6-32) is indeed asymptotic.

An interesting modification of Eq. (6-26) results from an alteration of the denominator. Define

$$\Psi(x) = \frac{1}{2\pi i}\int_{\gamma-i\infty}^{\gamma+i\infty} \frac{\exp\,[x(s\xi - \sqrt{s})]}{s^2 + \omega^2}\,ds \tag{6-33}$$

where ω is a real constant and where the path of integration is again C_1 of Fig. 6-3. As before, the path of integration may be deformed into the steepest descent path C of Fig. 6-4, so as to obtain the asymptotic behavior of $\Psi(x)$ for large x. However, this integral differs from that of Eq. (6-26) in that there are poles at $\pm i\omega$; depending on the size of the parameter ξ, these poles will lie either to the right or to the left of C. If they lie to the right of C, the residue contributions must be added to the saddle-point contributions (since the poles lie between C_1 and C).

We turn now to an example in which the steepest-descent paths do not pass through a saddle point. Let

$$f(x) = \int_0^1 \frac{1}{\sqrt{t}} \exp\,[ix(t + t^2)]\,dt \tag{6-34}$$

where x is real and positive and where the path of integration is along the real axis. We seek the asymptotic behavior of $f(x)$ for large x. Upon

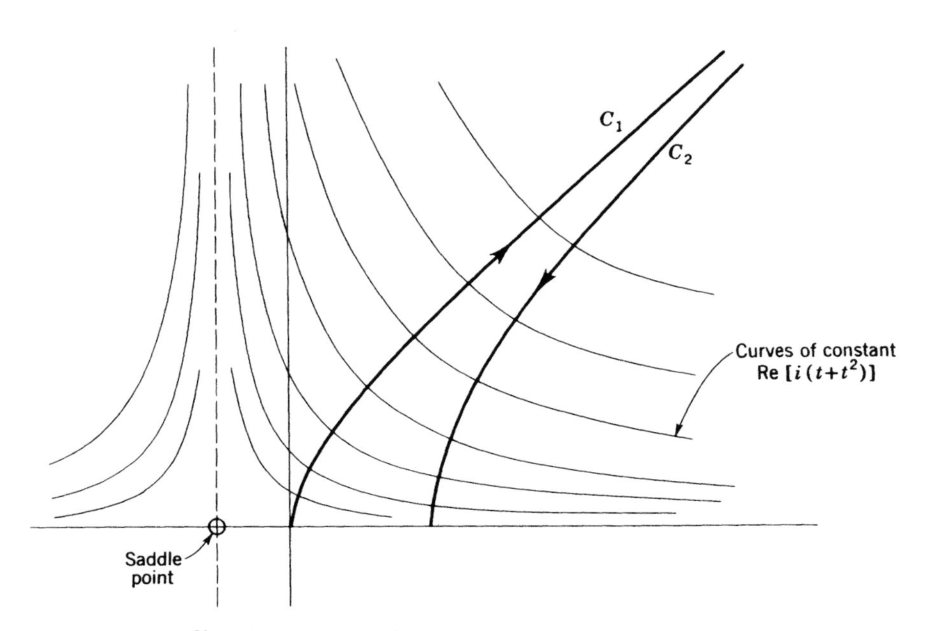

Fig. 6-5 Steepest-descent paths for Eq. (6-34).

writing $t = t_1 + it_2$, the contour lines of Re $[i(t + t^2)]$ satisfy the equation $t_2(1 + 2t_1) = \text{const}$; they are sketched in Fig. 6-5. The path of integration can be deformed into the pair of steepest-descent contours C_1 and C_2, traversed in the directions shown. Making the changes in variable, $i\tau = t + t^2$, $2 + i\eta = t + t^2$, in the C_1, C_2 integrals, respectively, we obtain

$$f(x) = \int_0^\infty \frac{e^{-x\tau}i\,d\tau}{(-\tfrac{1}{2} + \tfrac{1}{2}\sqrt{1 + 4i\tau})^{1/2}\sqrt{1 + 4i\tau}} - \int_0^\infty \frac{e^{i2x}e^{-x\eta}i\,d\eta}{(-\tfrac{1}{2} + \tfrac{1}{2}\sqrt{9 + 4i\eta})^{1/2}\sqrt{9 + 4i\eta}}$$

By use of Watson's lemma, power-series expansions now lead to

$$f(x) \sim \sqrt{\frac{\pi}{x}}\, e^{i(\pi/4)}\left(1 - \frac{3i}{4x} - \frac{105}{32x^2} + \cdots\right) + \frac{i}{3}e^{2ix}\left(-\frac{1}{x} + \frac{7i}{18x^2} + \frac{111}{324x^3} + \cdots\right) \tag{6-35}$$

EXERCISE

1. (*a*) Obtain the asymptotic expansion, valid for large positive x, of $\Psi(x)$ as defined by Eq. (6-33). Consider each of the two cases $\omega \lessgtr 1/(2\xi^2)$.

(*b*) Consider the case in which the steepest-descent path passes to the left of the poles, as shown in Fig. 6-6. Instead of utilizing the entire steepest-descent path, choose a path of integration C which coincides with the steepest-descent path only near the saddle point; the remainder of C stays to the right of the imaginary axis. It can now be argued that the real part of the exponent decreases monotoni-

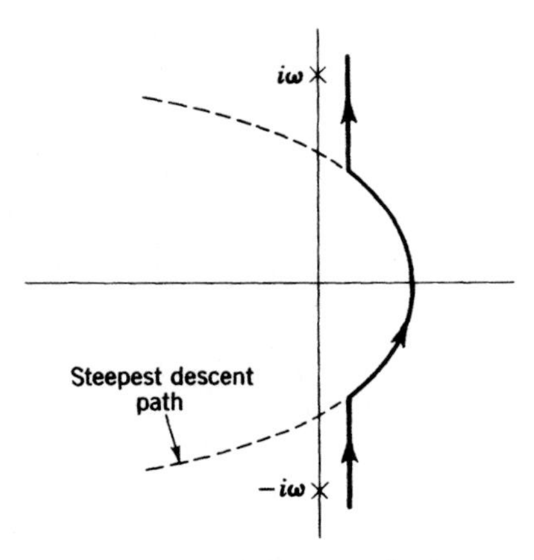

Fig. 6-6 Integration path for Exercise 7*b*.

cally as we move along C in either direction away from the saddle point, so that, asymptotically for large x, the contributions from the vertical portion of C can be neglected, and we are left with the saddle-point contribution alone. Since there is now apparently no contribution from the poles, can the argument be correct? If so, under what conditions are the residue terms of interest?

(c) The reader should be careful not to infer that saddle-point contributions always dominate contributions from poles. Evaluate, for example, the asymptotic behavior of

$$\varphi(x,\alpha) = \int_{-\infty}^{\infty} \frac{\exp\left(-x\sqrt{\xi^2 - 1^2} + i\alpha\xi\right)}{\xi\sqrt{\xi^2 - 1^2}}\, d\xi$$

where the path of integration is indented above $\zeta = -1$ and below $\xi = 0, 1$. The parameter α is real. Obtain an asymptotic description of $\varphi(x,\alpha)$ as $x \to +\infty$, for each of the cases $\alpha > 0$, $\alpha < 0$.

Saddle Points

We consider now the problem of finding an asymptotic expansion, valid as $x \to \infty$, of

$$f(x) = \int_C e^{xh(t)} g(t)\, dt \tag{6-36}$$

where the path of integration is indented above $\xi = -1$ and below $\xi = 0, 1$. The parameter α is real. Obtain an asymptotic description of $\varphi(x,\alpha)$ as $x \to +\infty$, for each of the cases $\alpha > 0$, $\alpha < 0$.

$$f(z) = \int_C e^{zh(t)} g(t)\, dt$$

as $z \to \infty$ along the ray $z = re^{i\alpha}$ (with α fixed), the factor $e^{i\alpha}$ can be absorbed into $h(t)$ to give the form of Eq. (6-36). Both $h(t)$ and $g(t)$ in Eq. (6-36) are to be analytic functions of t.

In accordance with the saddle-point discussion of the preceding section, the general method is to modify C, as necessary, so as to pass through saddle points of Re $h(t)$ (if this can be done), in such a way that the exponential factor decreases as rapidly as possible as we move along the revised path away from each saddle point. These paths of steepest descent away from saddle points are simultaneously paths of constant phase, since the gradient vector of the harmonic function Re $h(t)$ is tangent to the contour curves of the conjugate function Im $h(t)$.

Writing $t = \tau + i\sigma$, we observe that at a saddle point of Re $h(t)$ the first partial derivatives with respect to τ and σ of Re $h(t)$ vanish. The Cauchy-Riemann conditions then imply that the criterion for a saddle point is that $h'(t) = 0$. Thus it is appropriate to speak of the saddle points of $h(t)$, rather than merely of its real part. It may happen that the first m derivatives of $h(t)$ vanish at the saddle point, in which case it is said to be *a saddle point of order m* (such a saddle point can be thought of as a confluence of lower-order saddle points).

Let $t_0 = \tau_0 + i\sigma_0$ be a saddle point. Then, for points t sufficiently close to t_0, a power-series expansion of $h(t)$ gives

$$h(t) = h(t_0) + [a_2 \exp (i\alpha_2)](t - t_0)^2 + [a_3 \exp (i\alpha_3)](t - t_0)^3 + \cdots$$

where we have used the fact that $h'(t_0) = 0$ and where we write $(1/n!)h^{(n)}(t_0)$ in the form $a_n \exp (i\alpha_n)$ with α_n and $a_n(>0)$ real. Then, if $t - t_0 = re^{i\theta}$,

$$\text{Re } h(t) = \text{Re } h(t_0) + a_2r^2 \cos (2\theta + \alpha_2) + a_3r^3 \cos (3\theta + \alpha_3) + \cdots$$

Thus, if $h''(t_0) \neq 0$, the function Re $h(t)$ has two "valleys" and two "peaks" near the point t_0, corresponding to the behavior of $\cos (2\theta + \alpha_2)$ as θ goes through the range $(0,2\pi)$. The contour curves of Re $h(t)$ in the (τ,σ) plane intersect orthogonally at the point (τ_0,σ_0), and there are two curves on which Im $h(t) =$ const; these bisect the angles of intersection of the contour curves. One of these curves climbs the peaks on the two sides of the saddle point; the other descends into the two valleys and is the path of steepest descent associated with the saddle point at t_0. More generally, if all derivatives of $h(t)$ up to the mth derivative vanish, then

$$\text{Re } h(t) = \text{Re } h(t_0) + a_{m+1}r^{m+1} \cos [(m + 1)\theta + \alpha_{m+1}] + \cdots$$

and there will be $m + 1$ contour curves intersecting at equal angles at (τ_0,σ_0); again, $m + 1$ curves of constant phase will bisect these angles, and the path of integration would presumably utilize valley rather than mountain curves, although now the two parts of such a path need not be collinear. We shall subsequently encounter examples in which such a higher-order saddle point does indeed occur.

If the function $h(t)$ is complicated, it may be troublesome to find an appropriate new variable of integration for use along a steepest-descent path; a useful tool is Lagrange's formula [Eq. (2-35)]. Let t_0 be a saddle point, away from which a path C_1 is proceeding (i.e., the two halves C_1, C_2 of the path are considered separately). The contribution of C_1 to the value of $f(x)$ in Eq. (6-36) is

$$f_1(x) = \exp [xh(t_0)] \int_{C_1} \exp \{-x[h(t_0) - h(t)]\}g(t)\, dt \qquad (6\text{-}37)$$

Since Im $h(t)$ is constant along C_1, the quantity $h(t_0) - h(t)$ is purely real (and positive) along C_1. We make the change in variable

$$h(t_0) - h(t) = u \qquad (6\text{-}38)$$

and now, to evaluate $g[t(u)]$ and dt/du for use in Eq. (6-37), we need the inverse of Eq. (6-38)—that is, t in terms of u. If all of $h'(t_0)$, $h''(t_0)$, . . . , $h^{(m)}(t_0)$ vanish, Eq. (6-38) will read

$$-\frac{1}{(m+1)!} h^{(m+1)}(t_0)(t - t_0)^{m+1} + \cdots = u$$

and it follows that $t - t_0$ will be an analytic function of $u^{1/(m+1)}$ (an appropriate branch must be chosen). We then treat the equation

$$u^{1/(m+1)} = [h(t_0) - h(t)]^{1/(m+1)}$$

as the equation to be inverted so as to give a power-series representation for t in powers of $u^{1/(m+1)}$. The coefficients in this representation can be found, for example, by use of Lagrange's formula [Eq. (2-35)]. Equation (6-37) now becomes

$$f_1(x) = \exp[xh(t_0)] \int_0^\infty e^{-xu} g[t(u)] \frac{dt}{du} du$$

Even though the image of C_1 in the u plane may extend only from 0 to L, the integration is taken as from 0 to ∞ in accordance with the spirit of Laplace's method. The reader will find it useful to carry out the details for an example such as that of Exercise 2 below.

In some cases the asymptotic expansion resulting from the steepest-descent method may not be sufficiently accurate for the values of x of interest—that is, x may not be large enough. It might then be worth considering an actual numerical integration along the steepest-descent path; oscillation troubles are thus avoided.[1] Also, if there is a pole of $g(t)$ close to t_0, so that for moderately large x the saddle-point contribution does not completely dominate the contribution from the rest of the steepest-descent path, then it is often possible to subtract off the pole of $g(t)$ and to write the right-hand side of Eq. (6-36) as the sum of two integrals. One of these will have a simple form, and the other has no pole near t_0. A formal discussion has been given by van der Waerden.[2]

Historically, the method of steepest descents was first suggested by Riemann,[3] in a problem involving the asymptotic behavior of a hypergeometric function. Debye developed the method independently[4] and used it to determine the asymptotic behavior of Bessel functions of large order. We shall next consider one particularly instructive such example.

Exercises

2. An integral (associated with the name of Airy) which arises in optical problems is

$$I(x) = \int_{-\infty}^{\infty} \exp[ix(\tfrac{1}{3}t^3 + t)]\, dt \tag{6-39}$$

It is required to find an asymptotic expansion valid for large positive real values of x. Find the appropriate saddle-point path, and show

[1] For an example of this, see R. Esch, *J. Fluid Mech.*, **3:** 289 (1957).

[2] B. L. van der Waerden, *Appl. Sci. Research*, **B2:** 33 (1950).

[3] B. Riemann, "Gesammelte Mathematische Werke," p. 428, Dover Publications, Inc., New York, 1953.

[4] P. Debye, *Math. Ann.*, 1900, p. 535; "Collected Papers," p. 583, Interscience Publishers, Inc., New York, 1954.

that the original path of integration (along the real axis) can be deformed into it. Carry out the change in variable of Eq. (6-38), and show by use of Lagrange's formula [Eq. (2-35)] that

$$t - i = \sum_{n=1}^{\infty} \frac{\Gamma(\frac{3}{2}n - 1)}{n!\Gamma(n/2)} \left(\frac{i}{3}\right)^{n-1} (\sqrt{u})^n$$

where $\sqrt{u}$ is positive on one side of $t = i$ and negative on the other side. Show finally that

$$I(x) \sim \frac{1}{\sqrt{x}} e^{-\frac{2}{3}x} \sum_{m=0}^{\infty} \frac{\Gamma(3m + \frac{1}{2})}{(2m)!} (-9x)^{-m}$$

3. By the change in variable $xt = \tau$, the integrand of Eq. (6-39) takes the form $\exp(i\tau^3/3x^2)\exp(i\tau)$. Use the saddle-point method on the resulting integral [with $h(\tau) = i\tau^3$] to obtain an asymptotic representation for $I(x)$ valid for *small* positive values of x. Is the resulting series convergent? Find a second-order differential equation satisfied by $I(x)$, and relate it to Bessel's equation.

4. An integral representation for the modified Bessel function of integral order, $I_n(x)$, is

$$I_n(x) = \frac{1}{2\pi i} e^{-in\pi/2} \int_C t^{-n-1} e^{(ix/2)\left(t - \frac{1}{t}\right)} dt$$

where C is a closed contour around the origin. Use steepest descents to determine the asymptotic behavior of $I_n(x)$ for large real positive values of x.

Bessel Functions

We shall apply the method of steepest descents to Eq. (5-157) so as to obtain an asymptotic representation for the function $H_\nu^{(2)}(z)$. The matter of immediate interest is the way in which the saddle-point technique depends on the size of the parameter ν; we shall therefore avoid the minor complication associated with allowing ν and z to be complex[1] and shall confine our attention to real positive z and ν.

Equation (5-157) may be written

$$H_\nu^{(2)}(x) = \frac{1}{\pi} \int_C e^{-ix \sin t + i\nu t} dt \tag{6-40}$$

[1] A discussion of the corresponding steepest-descent paths for complex values of z and ν is given in G. N. Watson, "A Treatise on the Theory of Bessel Functions," 2d ed., p. 262, Cambridge University Press, New York, 1948. We shall discuss the case of complex z in connection with Stokes' phenomenon, in Sec. 6-6.

where the curve C starts at $0 - i\infty$ and ends at $\pi + i\infty$. We consider first the case in which ν is held fixed and $x \to \infty$; although the result is valid for any choice of ν, it will turn out to be of practical utility only for $x \gg \nu$, so that the case of large ν will require a modification in the technique.

We first choose the function $h(t)$ of Eq. (6-36) to be $-i \sin t$. The appropriate saddle point is at $t = \pi/2$, and (with $t = t_1 + it_2$) the curves of steepest descent satisfy

$$\sin t_1 \cosh t_2 = 1$$

This curve has two branches through the point $t = \pi/2$, intersecting the t_1 axis at $\pm 45°$, as shown in Fig. 6-7. On curve B, the real part of $h(t)$ increases most steeply as we move in either direction away from the point $t = \pi/2$; on curve A, this real part decreases most steeply. Moreover, the path of integration in Eq. (6-40) can be deformed into branch A, so that this is the appropriate path for the steepest-descent method. In the equation

$$H_\nu^{(2)}(x) = \frac{e^{-ix}}{\pi} \int_A e^{-x[i(\sin t - 1)]} e^{i\nu t}\, dt$$

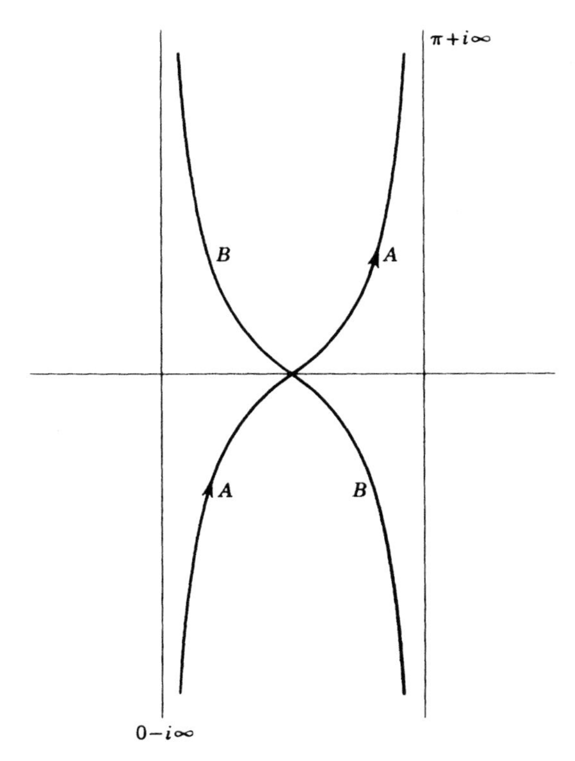

Fig. 6-7 Steepest-descent paths for $H_\nu^{(2)}(x)$ (small ν).

we now make the change in variable

$$u = i(\sin t - 1)$$

or

$$\frac{1}{2}\left(t - \frac{\pi}{2}\right) = \pm\arcsin\left(\frac{ui}{2}\right)^{1/2}$$

where by $(ui/2)^{1/2}$ we mean the principal value of this expression, where $\arcsin \alpha \to 0$ as $\alpha \to 0$, and where the + and − signs are to be used on the upper and lower halves, respectively, of curve A. Then

$$H_\nu^{(2)}(x) = \frac{4}{\pi} e^{-ix+i\nu\pi/2} \int_0^\infty e^{-xu} \cos\left[2\nu \arcsin\left(\frac{ui}{2}\right)^{1/2}\right] \frac{d}{du}\left[\arcsin\left(\frac{ui}{2}\right)^{1/2}\right] du$$

and an integration by parts (considering the product of the second two terms in the integrand as the derivative of a sine function) gives

$$H_\nu^{(2)}(x) = \frac{2x}{\pi\nu} e^{-ix+i\nu\pi/2} \int_0^\infty e^{-xu} \sin\left[2\nu \arcsin\left(\frac{ui}{2}\right)^{1/2}\right] du \qquad (6\text{-}41)$$

The expansion of $\sin 2\nu\theta$ in terms of powers of $\sin\theta$ is given[1] by

$$\sin 2\nu\theta = \sum_{n=0}^{\infty} \frac{(-1)^n 2^{2n+1}\nu\Gamma(\frac{1}{2} + \nu + n)}{(2n+1)!\Gamma(\frac{1}{2} + \nu - n)} \sin^{2n+1}\theta$$

and using this result in Eq. (6-41), we obtain the asymptotic expansion of $H_\nu^{(2)}(x)$ as

$$H_\nu^{(2)}(x) \sim \left(\frac{2i}{\pi x}\right)^{1/2} e^{-ix+i\nu\pi/2} \sum_{n=0}^{\infty} \frac{\Gamma(\frac{1}{2} + \nu + n)}{n!\Gamma(\frac{1}{2} + \nu - n)} \frac{1}{(2ix)^n} \qquad (6\text{-}42)$$

Since $H_\nu^{(1)}(x) = [H_\nu^{(2)}(x)]^*$, the corresponding expression for $H_\nu^{(1)}(x)$ can be obtained by simply replacing i by $-i$ in Eq. (6-42).

Equation (6-42) had been obtained by Hankel, prior to Debye, starting from a different integral representation and using Laplace's method. We have used Debye's method here, since it can be extended in a natural way to the case of large ν.

The coefficients in Eq. (6-42) can be written

$$\frac{\Gamma(\frac{1}{2} + \nu + n)}{n!\Gamma(\frac{1}{2} + \nu - n)} = \frac{(4\nu^2 - 1^2)(4\nu^2 - 3^2) \cdots [4\nu^2 - (2n-1)^2]}{n!2^{2n}}$$

[1] To obtain the coefficients a_n in

$$\sin 2\nu\theta = \sum_0^\infty a_n \sin^n\theta$$

we differentiate twice with respect to θ, so as to obtain the recursion relation

$$a_{n+2} = a_n \frac{n^2 - (2\nu)^2}{(n+1)(n+2)}$$

Using this relation and Eqs. (5-9) and (5-10) gives the desired formula for a_n.

and this form makes it clear that Eq. (6-42) is useful only if $x \gg \nu$. Thus, for ν equal to about 100, we might have to take x as large as 10,000 before Eq. (6-42) could be reliably used; it could certainly not be used for a value of x as small as 200.

We therefore seek an asymptotic formula for $H_\nu^{(2)}(x)$ valid when x and ν are both large. A natural approach is to set $\nu = \xi x$ in Eq. (6-40), where ξ (real and > 0) is to be some fixed parameter; then the exponent in Eq. (6-40) can be written

$$xh(t) = x(-i \sin t + i\xi t) \tag{6-43}$$

and the method of steepest descents used afresh. If for example $\xi = \frac{1}{2}$, then we could reasonably expect the method to give us an accurate asymptotic expression for $H_{100}^{(2)}(200)$.

The saddle points of $h(t)$ in Eq. (6-43) satisfy the equation

$$\cos t = \xi$$

In the strip depicted in Fig. 6-7, there is one saddle point when $\xi < 1$; it lies in the interval $(0,\pi/2)$ of the t_1 axis, and it approaches the point $t = 0$ as $\xi \to 1$. For $\xi > 1$, there are two saddle points on the t_2 axis, equidistant from the origin.

For $\xi < 1$, the procedure is similar to that used in the derivation of Eq. (6-42); the point at which the path A of Fig. 6-7 cuts the t_1 axis is merely shifted to the left. With the analytical details left as an exercise for the reader, the final result is (with $\cos t_0 = \xi = \nu/x$ and $\nu < x$)

$$\begin{aligned} H_\nu^{(2)}(x) \sim \frac{1}{\pi} \exp\left[-ix(\sin t_0 - t_0 \cos t_0)\right] \Bigg\{ & \frac{e^{i\pi/4}\Gamma(\frac{1}{2})}{[(x/2)\sin t_0]^{1/2}} \\ & + \frac{(\frac{1}{8} + \frac{5}{24}\cot^2 t_0)e^{i3\pi/4}\Gamma(\frac{3}{2})}{[(x/2)\sin t_0]^{3/2}} \\ + \left(\frac{3}{128} + \frac{7}{576}\cot^2 t_0 + \frac{385}{3{,}456}\cot^4 t_0\right) & \frac{e^{i5\pi/4}\Gamma(\frac{5}{2})}{[(x/2)\sin t_0]^{5/2}} + \cdots \Bigg\} \end{aligned} \tag{6-44}$$

As was to be expected, the first term of Eq. (6-44) approaches the first term of Eq. (6-42) as $\nu \to 0$. It may be of interest to write the first term of Eq. (6-44) explicitly,

$$H_\nu^{(2)}(x) \sim \left(\frac{2}{\pi}\right)^{1/2} (x^2 - \nu^2)^{-1/4} \exp\left[-i(x^2 - \nu^2)^{1/2} + i\nu \arccos\frac{\nu}{x} + \frac{i\pi}{4}\right]$$

so as to observe the marked difference in form between Eqs. (6-42) and (6-44). The latter representation is useful when x and ν are both large, with $\nu < x$; the previously mentioned example of $\nu = 100$, $x = 200$ would now present no difficulty. If, however, $\nu \approx x$ with $\nu < x$, then the term $\sin t_0$ in each denominator of Eq. (6-44) is close to zero and the representation breaks down. We shall defer this case, and we turn now to the case $\nu > x$.

For $\nu > x$, there are two saddle points, at $\pm t_0$, where $t_0 = i \operatorname{arccosh} \xi$. The appropriate path of integration is shown in Fig. 6-8. Starting at

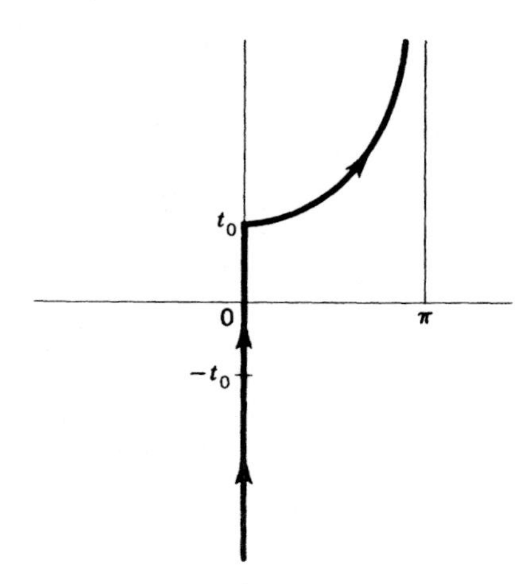

Fig. 6-8 Path of integration for $\xi > 1$.

$0 - i\infty$, the real part of the function $h(t)$ given by Eq. (6-43) increases as rapidly as possible, as we move along the imaginary axis to $-t_0$; continuing along the imaginary axis, this real part now decreases as rapidly as possible. At the branch point t_0, the path of most rapid decrease departs from the imaginary axis, and becomes the curve shown. We need not compute the asymptotic series corresponding to t_0, since it is clearly exponentially small compared with that which arises from $-t_0$. Define α to be that positive number satisfying $\cosh \alpha = \xi$; then the final result is found to be (necessarily) of exactly the same form as Eq. (6-44), with, however, t_0 replaced by $-i\alpha$. Alternatively, it can be written

$$H_\nu^{(2)}(x) \sim \frac{i}{\pi} e^{-x \sinh \alpha + \nu\alpha} \left\{ \frac{\Gamma(\frac{1}{2})}{[(x/2) \sinh \alpha]^{\frac{1}{2}}} - \frac{\Gamma(\frac{3}{2})}{[(x/2) \sinh \alpha]^{\frac{3}{2}}} \left(\frac{1}{8} - \frac{5}{24} \coth^2 \alpha \right) + \cdots \right\} \tag{6-45}$$

for $\nu > x$. Again, this equation fails if $\nu \cong x$.

Suppose next that $\nu \cong x$; set $\nu = x + \epsilon$, and take $h(t) = -i \sin t + it$, with

$$H_\nu^{(2)}(x) = \frac{1}{\pi} \int_C e^{xh(t)} e^{i\epsilon t}\, dt$$

The saddle point at $t = 0$ is a double one; the reader may verify that the path of integration is that shown in Fig. 6-9 and that the final result is

$$H_\nu^{(2)}(x) \sim \frac{1}{\pi\sqrt{3}} \left[e^{\frac{1}{3}\pi i} \frac{\Gamma(\frac{1}{3})}{(x/6)^{\frac{1}{3}}} + e^{\frac{1}{3}\pi i} \epsilon \frac{\Gamma(\frac{2}{3})}{(x/6)^{\frac{2}{3}}} + \cdots \right] \tag{6-46}$$

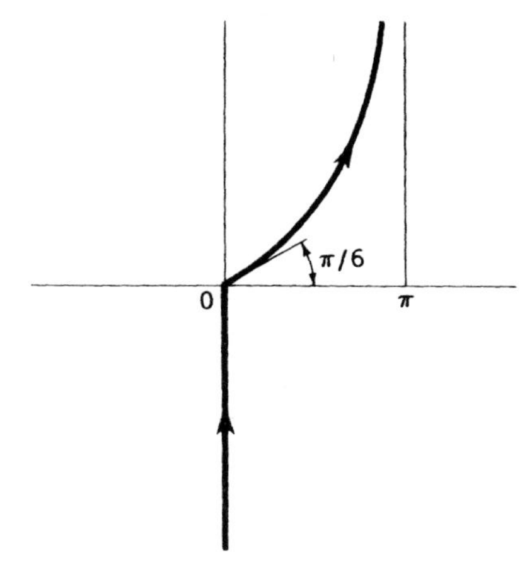

Fig. 6-9 Path of integration for $\xi \cong 1$.

It may be observed that, for $\xi < 1$, the asymptotic representation for $J_\nu(x)$ can be obtained from Eq. (6-42) or Eq. (6-44) [by using the fact that $H_\nu^{(1)}(x)$ is the complex conjugate of $H_\nu^{(2)}(x)$, for real x and ν]. For $\xi > 1$, however, the formula $J_\nu(x) = \frac{1}{2}H_\nu^{(1)}(x) + \frac{1}{2}H_\nu^{(2)}(x)$ gives zero when applied to Eq. (6-45). The reason for this is that the asymptotic expansion for $J_\nu(\nu/\xi)$ (which can be obtained by use of an appropriate integral representation) is transcendentally small in ν as $\nu \to \infty$.

There is no difficulty in extending these results to complex values of x; Fig. 5-7 shows how the end points of the path of integration depend on arg x. It is clear that, as arg x alters, then Eq. (6-42), say, will continue to be valid—until that value of arg z is reached beyond which we must switch to a different saddle point. Thus we can expect the asymptotic expansion to alter abruptly as arg z passes certain critical values—an example of Stokes' phenomenon. We shall return to this matter in Sec. 6-6. In Sec. 6-7, we shall discuss the WKB method of obtaining asymptotic expansions—a method which is based on the differential equation satisfied by the function rather than on an integral representation for it. In terms of the WKB method, the necessity for treating separately the cases $\nu > x$ and $\nu < x$ is related to the different signs of the coefficient $\nu^2 - x^2$ in Bessel's equation; it may be mentioned here that by Langer's modification (Sec. 6-7) of the WKB method a single representation for the Bessel functions can be obtained which is uniformly valid for large ν and x.

Exercises

5. Verify Eqs. (6-44), (6-45), and (6-46).

6. With ν real and >0, determine the range of arg x over which Eq. (6-42) remains valid.

6-4 Method of Stationary Phase

The method of stationary phase differs from the method of steepest descents in that it makes use of the self-canceling oscillation, rather than the decay, of an exponential factor in an integrand, so as to allow the contribution of the integrand to be neglected everywhere except near one or more critical points. It was first used by Stokes[1] to obtain an asymptotic representation for the function

$$f(x) = \int_0^\infty \cos x(\omega^3 - \omega)\, d\omega \tag{6-47}$$

valid for large positive real values of x. Denote the "phase" of the cosine function by $\varphi = x(\omega^3 - \omega)$. Then a plot of φ versus ω (Fig. 6-10) shows that, for large values of x, φ is a rapidly increasing or a rapidly decreasing function of ω—except near the point $\omega_0 = (\frac{1}{3})^{\frac{1}{2}}$, where $\partial\varphi/\partial\omega = 0$, so that the phase is there "stationary." Thus, the cosine function alternates rapidly in sign as we move along the ω axis; everywhere except in the immediate vicinity of ω_0, the positive and negative loops of the cosine function become

[1] G. G. Stokes, "Mathematical and Physical Papers," vol. 2, p. 329, Cambridge University Press, 1883. See also W. Thomson, "Mathematical and Physical Papers," vol. 4, p. 303, Cambridge University Press, New York, 1910, and H. Poincaré, *Proc. Roy. Soc. London*, *72:* 48 (1904).

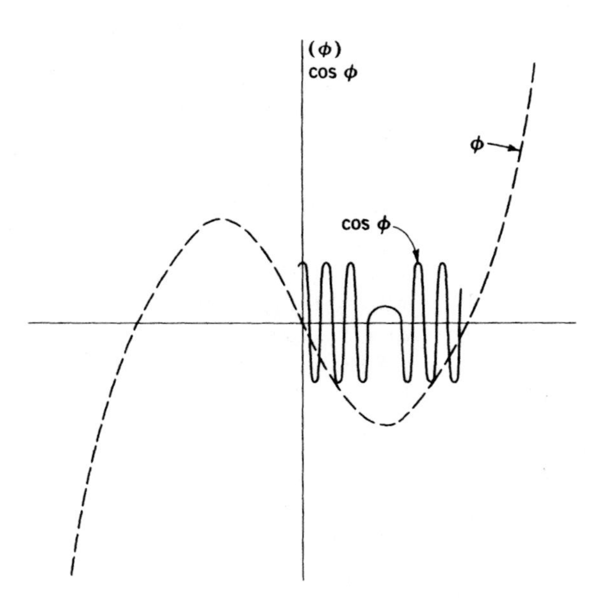

Fig. 6-10 Plot of ϕ and $\cos\phi$ for $\phi = x(\omega^3 - \omega)$.

more and more closely packed as x grows. Stokes' argument was that the contribution of these positive and negative loops to the integral should therefore tend to be mutually canceling, as $x \to \infty$, so that the major contribution to the integral would be that arising from the vicinity of ω_0. Since a power-series expansion yields

$$\omega^3 - \omega \cong (\omega_0{}^3 - \omega_0) + 3\omega_0(\omega - \omega_0)^2$$

for the approximate behavior of the phase function near ω_0, we have

$$f(x) \sim \int_{-\infty}^{\infty} \cos \left[x(\omega_0{}^3 - \omega_0) + 3x\omega_0(\omega - \omega_0)^2\right] d\omega$$

where the lower limit has been changed to $-\infty$ for ease of integration (a procedure that is suggested by the same rapid-oscillation argument). Integration gives

$$f(x) \sim \left(\frac{\pi}{3x\omega_0}\right)^{1/2} \cos \left[x(\omega_0{}^3 - \omega_0) + \frac{\pi}{4}\right]$$

As a second example, consider the Bessel-function integral

$$J_0(x) = \frac{2}{\pi} \int_0^{\pi/2} \cos (x \cos \theta)\, d\theta$$

where again x is to be real and >0. Here the point of stationary phase is at one end ($\theta = 0$) of the interval. Near this point,

$$\cos (x \cos \theta) \cong \cos [x(1 - \tfrac{1}{2}\theta^2)]$$

so that

$$J_0(x) \sim \frac{2}{\pi} \int_0^{\infty} \cos [x(1 - \tfrac{1}{2}\theta^2)]\, d\theta$$

where the upper limit of integration has been taken as ∞, on the supposition that only the region of integration near the origin is important. Integration then gives

$$J_0(x) \sim \left(\frac{2}{\pi x}\right)^{1/2} \cos \left(x - \frac{\pi}{4}\right) \tag{6-48}$$

General Case

Consider now the problem of finding an asymptotic representation, valid for large real positive x, of the function

$$f(x) = \int_a^b e^{ixh(t)} g(t)\, dt \tag{6-49}$$

Here $h(t)$ and $g(t)$ are real functions (no analyticity is necessary), and the interval of integration (a,b) is along the real axis. Let us suppose that there is one point t_0 in the interior of (a,b), having the property that $h'(t_0) = 0$; also, let $h''(t_0) \neq 0$. In accordance with the idea of the method of stationary phase, we assume that only the neighborhood of the point t_0 is of significance, and we write

$$ixh(t) \cong ix[h(t_0) + \tfrac{1}{2}h''(t_0)(t - t_0)^2]$$

Then

$$f(x) \sim \int_{-\infty}^{\infty} g(t_0) \exp \{ix[h(t_0) + \tfrac{1}{2}h''(t_0)(t - t_0)^2]\} \, dt$$

$$\sim \left[\frac{2\pi}{x|h''(t_0)|}\right]^{\frac{1}{2}} g(t_0) \exp \left[ixh(t_0) \pm \frac{i\pi}{4}\right] \qquad (6\text{-}50)$$

where the + or − sign is chosen for $h''(t_0) > 0$ or <0, respectively.

There is no difficulty in obtaining a similar formula for the case in which the first nonvanishing derivative of $h(t)$ at t_0 is of higher order than the second, or for the case in which t_0 is an end point. The function $g(t)$ may also have certain singularities at the end points; a formal theorem covering several of these cases is given by Erdélyi.[1] Some such situations arise in the exercises below.

If the functions $g(t)$ and $h(t)$ are analytic, then we recognize the point of stationary phase, t_0, as a saddle point in the complex t plane. The situation is then somewhat as shown in Fig. 6-11, which indicates the contour lines of the function Re $ixh(t)$ over the t plane. Here a shaded region denotes a valley, and an unshaded region a hill [the orientation of hills and valleys shown corresponds to $h''(t_0) > 0$]. The path of integration, from a to b along the real axis, is itself a contour line; we remark in passing that one way of looking at the method of stationary phase is to observe that the self-canceling of oscillations is a weaker decay mechanism than the possible exponential decay of the exponential factor in the integrand, and so can be used only for the borderline case in which the exponential factor has constant absolute magnitude over the path of integration.

[1] A. Erdélyi, "Asymptotic Expansions," p. 52, Dover Publications, Inc., New York, 1956.

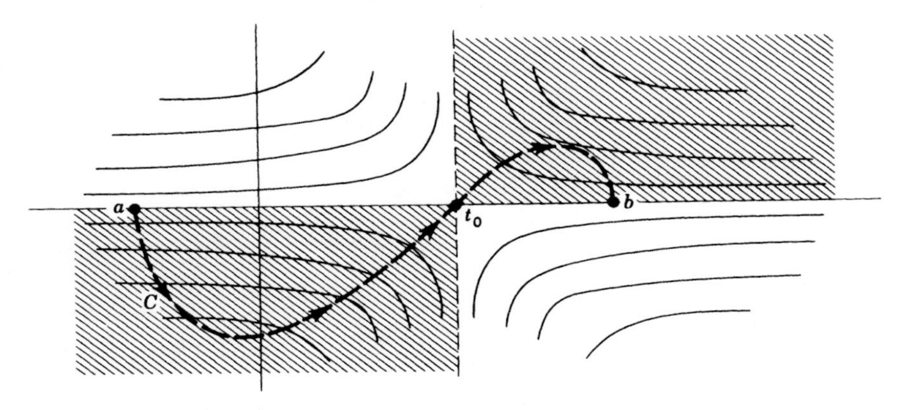

Fig. 6-11 Contour lines of Re $[ixh(t)]$ in method of stationary phase.

In the present case, we can use the method of steepest descents, by deforming the path of integration in some such way as shown in Fig. 6-11 by curve C. In the simple case shown [only one saddle point, and well-behaved $g(t)$], contributions to the asymptotic behavior of $f(x)$ arise from the two end points a, b and from the saddle point; straightforward computation gives the first terms of these contributions as

$$\frac{-e^{ixh(a)}g(a)}{x|h'(a)|} \qquad \frac{ie^{ixh(b)}g(b)}{xh'(b)} \qquad \left(\frac{2\pi}{xh''(t_0)}\right)^{\frac{1}{2}} g(t_0) \exp\left[ixh(t_0) + i\frac{\pi}{4}\right]$$

respectively. [For the case shown, with $h''(t_0) > 0$, we have $h'(a) < 0$ and $h'(b) > 0$.] Since the three exponential factors here have modulus unity, the relative asymptotic behavior depends only on the power of x in the denominator and it follows that the saddle-point contribution is dominant.

EXERCISES

1. Use the method of stationary phase to find an asymptotic representation, valid as $x \to \infty$, of

(a) $\int_0^1 \cos(xt^p)\,dt$, with p real and >1

(b) $\int_0^{\pi/2} \left(1 - \frac{2\theta}{\pi}\right)^{-\frac{1}{2}} \cos(x\cos\theta)\,d\theta$

6-5 *Phase, Group, and Signal Velocities*

A single harmonic wave propagating in the direction of increasing x can be written as Re $Ae^{ik(x-ct)}$, where t is time, A is the complex amplitude, k the wave parameter ($k = 2\pi/\lambda$, λ = wavelength), and c the so-called *phase velocity*. We are interested in those situations (termed *dispersive*) in which c is a function of k. (For the physical genesis of one such example, see Exercise 1 below.)

A continuous superposition of simple harmonic waves can be represented by

$$\varphi(x,t) = \text{Re} \int_{-\infty}^{\infty} a(k)e^{ik[x-c(k)t]}\,dk \tag{6-51}$$

where $a(k)$ is some complex function of k. The method of stationary phase can be used to compute an approximate value for $\varphi(x,t)$, where x and t are large. It is convenient to describe the result in terms of t and $\xi = x/t$. Equation (6-51) can now be written

$$\varphi(x,t) = \text{Re} \int_{-\infty}^{\infty} a(k)e^{it(k\xi-kc)}\,dk$$

and, for sufficiently large t, the principle of stationary phase suggests that the dominant contribution to the integral arises from the neighborhood of that value of k for which

$$\frac{d}{dk}(k\xi - kc) = 0$$

or

$$\xi - \frac{d}{dk}(kc) = 0 \tag{6-52}$$

Conversely, we can say that the contribution to the integral arising from a chosen $(k, k + dk)$ interval will be of predominating importance for those large x and t for which $\xi = x/t$ satisfies Eq. (6-52). This speed ξ is called the *group velocity* associated with the wave parameter k.

In a sense, then, a group $(k, k + dk)$ of waves tends to travel with the group velocity ξ; it is perhaps plausible to expect the group velocity to correspond in some way with the velocity of signal propagation and also with the velocity with which the energy contained within a traveling group of waves would be propagated. For some cases (cf. Exercise 1), interpretations such as these are indeed possible; however, there are other cases—particularly those in which there is absorption—where these interpretations are not at all correct. A famous example is that of the group velocity in an optical medium, in the neighborhood of a region of strong anomalous dispersion (c decreases as λ increases); here ξ as calculated by Eq. (6-52) can easily be greater than the velocity of light in free space. This fact led Sommerfeld and Brillouin[1] to investigate more carefully the idea of signal velocity for electromagnetic waves; in particular, Sommerfeld proved that the first (extremely small) disturbance was always propagated with the velocity of light in free space, irrespective of the value of the index of refraction, and Brillouin showed how the saddle-point method could be used to give a criterion for the velocity with which appreciable signal energy would be propagated. The optical case discussed by Sommerfeld and Brillouin is complicated by the character of the function describing the complex index of refraction in terms of frequency; the basic ideas involved will therefore be discussed in the next section in the analytically simpler case of signal propagation along a transmission line.

EXERCISE

1. Let acoustic waves be transmitted along a rectangular tube, one edge of which is the z axis. Let the sides of the tube be at $x = 0$, $x = l_1$ and $y = 0$, $y = l_2$. With subscripts to indicate partial derivatives, the velocity potential φ must satisfy $\varphi_x = 0$ at $x = 0, l_1$ and $\varphi_y = 0$ at $y = 0, l_2$. Prove that the wave equation

$$\varphi_{xx} + \varphi_{yy} + \varphi_{zz} = (1/c_0^2)\varphi_{tt}$$

[1] A. Sommerfeld, *Ann. Phys.*, (4) *44:* 177 (1914), and L. Brillouin, *Ann. Phys.*, (4) *44:* 203 (1914), both translated and reprinted in L. Brillouin, "Wave Propagation and Group Velocity," Academic Press, Inc., New York, 1960.

(where c_0 is the acoustic velocity) and the boundary conditions are satisfied by the particular (m,n) "mode"

$$\varphi = \cos \frac{m\pi x}{l_1} \cos \frac{n\pi y}{l_2} \psi(z,t)$$

and find the differential equation satisfied by ψ. Setting

$$\psi = A \sin \omega \left(t - \frac{z}{c}\right)$$

find the phase velocity c in terms of m/l_1, n/l_2, ω, and c_0; show that $c > c_0$. Prove that the group velocity u is given by $u = c_0^2/c$, so that here $u < c_0$. (Observe that there is a cutoff frequency which limits the value of ω.)

Interpret the phase velocity physically, by means of superposition of plane acoustic waves traveling with velocity c_0, so oriented as to make the net normal derivative of φ vanish at the tube surfaces (i.e., so that the tube can be removed altogether; reflection is replaced by superposition).

Computing the acoustic velocity from the gradient of φ and the pressure from $-\rho\varphi_t$, where ρ is base density, show that the average rate of energy transport per unit area over the whole cross section is $A^2\rho\omega^2/8c$. Show that the average energy density (kinetic plus potential), when divided into this average rate of energy transport, again gives the group velocity.

Signal Velocity

We shall use an electrical transmission line as an example of the way in which the motion of a path of steepest descent can be used to give a definition of signal velocity. Consider a transmission line with resistance R, inductance L, and capacity C, all per unit length; then the voltage E satisfies the equation

$$E_{xx} = LCE_{tt} + RCE_t \tag{6-53}$$

where x is distance along the line $(x \geq 0)$, t is time, and subscripts indicate partial differentiation.

A solution of Eq. (6-53) that represents a sinusoidal wave of angular frequency ω traveling along the line is given by

$$E = Ae^{-\sigma x} \sin \omega \left(t - \frac{x}{c}\right) \tag{6-54}$$

where A is an arbitrary constant, provided that

$$\sigma = RC\,\frac{c}{2}$$

$$\omega^2 \left(\frac{1}{c^2} - \frac{1}{c_0^2}\right) = \sigma^2 \tag{6-55}$$

where $c_0^2 = 1/LC$. Equation (6-55) shows that the phase velocity c is less than c_0 (we shall subsequently find that c_0 is the maximum velocity of signal transmission) and that $c \to c_0$, as $\omega \to \infty$. Using Eq. (6-52), we obtain the group velocity u as

$$u = 2c - \frac{c^3}{c_0^2} \tag{6-56}$$

and straightforward algebra shows that, for values of RC and ω in the range

$$\frac{RCc_0^2}{2\omega} < (2 + \sqrt{5})^{1/2}$$

the group velocity u will be greater than c_0. But, as previously mentioned, no signal can travel faster than c_0; thus the conventional interpretation of group velocity as a signal velocity cannot hold here.

To clarify the situation, let us write down the exact solution of Eq. (6-53), corresponding to the application at $x = 0$ of a voltage which is zero for $t < 0$ and which for $t > 0$ has the form

$$E(0,t) = E_0 \sin \omega t$$

where E_0 is a constant. By the Laplace-transform technique (Chap. 7) the solution is

$$E(x,t) = \frac{E_0\omega}{2\pi i} \int_{\gamma - i\infty}^{\gamma + i\infty} \frac{\exp \{st - (x/c_0)[s^2 + (R/L)s]^{1/2}\}}{s^2 + \omega^2}\, ds \tag{6-57}$$

where the path of integration is any vertical line to the right of the origin and where the square-root expression is to be given its principal value on this path. We shall take the interval $(-R/L, 0)$ of the real axis in the s plane as a branch cut; for large $|s|$, the square-root expression then behaves like s itself. The integral can be evaluated in terms of Bessel functions [cf. Eq. (6-64)], but for our present purposes it is best left in the form (6-57).

We observe first that, if $t < x/c_0$, the path of integration may be closed by a large semicircle in the right half-plane, and it follows easily that $E = 0$; thus the signal cannot propagate faster than c_0. Exactly the same result would hold for any kind of applied signal, since the exponential function in Eq. (6-57) depends only on the properties of the line and not on those of the input signal. [This result could also have been deduced from the theory of characteristics as applied to Eq. (6-53).] We shall next show that any point x on the line will begin to experience a disturbance at the time $t = x/c_0$, so that c_0 is indeed the velocity with which the initial disturbance travels. Write

$$t = \frac{x}{c_0} + \tau$$

where τ is small and > 0, and close the path of integration in Eq. (6-57) by a large semicircle in the left-hand plane. The whole path of integration can now be deformed into a large circle centered on the origin. We make the change in variable $s\tau = z$, so that Eq. (6-57) becomes

$$E\left(x, \frac{x}{c_0} + \tau\right) = \frac{E_0\omega\tau}{2\pi i} \int \frac{\exp\{z + (x/c_0\tau)[z - z(1 + R\tau/Lz)^{1/2}]\}}{z^2 + \omega^2\tau^2}\, dz \tag{6-58}$$

where we shall take the path of integration in the z plane as being the unit circle. Now $|z| = 1$ on this circle, so that, if $\omega\tau < 1$ and if $R\tau/L < 1$, the integrand can be expanded to give

$$E\left(x, \frac{x}{c_0} + \tau\right) = \frac{E_0\omega\tau}{2\pi i} \exp\left(-\frac{xR}{2Lc_0}\right) \int \frac{1}{z^2}\left(1 - \frac{\omega^2\tau^2}{z^2} + \cdots\right) \exp\left(z + \frac{xR^2\tau}{8c_0L^2z} + \cdots\right) dz$$

For small values of τ, we can discard everything except the first term, to obtain

$$E\left(x, \frac{x}{c_0} + \tau\right) \cong E_0\omega\tau \exp\left(-\frac{xR}{2Lc_0}\right) \tag{6-59}$$

which is nonzero for $\tau > 0$, so that a signal does indeed arrive with velocity c_0. Actually, we can obtain a somewhat more accurate expression by keeping also the second term in the exponent; the result is recognized as a Bessel integral, and we have

$$E\left(x, \frac{x}{c_0} + \tau\right) \cong \frac{E_0\omega L}{R}\left(\frac{8c_0\tau}{x}\right)^{1/2} \exp\left(-\frac{xR}{2Lc_0}\right) I_1\left[\frac{2R}{L}\left(\frac{x\tau}{8c_0}\right)^{1/2}\right] \tag{6-60}$$

where I_1 is the modified Bessel function of the first kind of order unity. From the expansion process used in Eq. (6-58), it is seen that the conditions for validity of Eq. (6-59) or Eq. (6-60) are that $\tau \ll 1/\omega$ and $\tau \ll L/R$. We see from Eq. (6-59) that, as x increases, the initial signal becomes exponentially small (of course, so does the steady-state signal; compare the decay rates).

Let us now use the saddle-point method to examine the character of the signal for large values of x. What we are really after is to see whether or not some recognizable signal, of relatively large amplitude and of frequency ω, travels with a velocity less than c_0. To this end, set $t = \xi x/c_0$, with ξ fixed and > 1. The exponent in Eq. (6-57) becomes

$$\frac{x}{c_0}\left[\xi s - \left(s^2 + \frac{R}{L}s\right)^{1/2}\right]$$

which has a saddle point at

$$s = \frac{R}{2L}\left[-1 \pm \frac{\xi}{(\xi^2 - 1)^{1/2}}\right]$$

The two saddle points are denoted by α, β in Fig. 6-12. The path of steepest descent through α may be verified to be an ellipse, passing also through β,

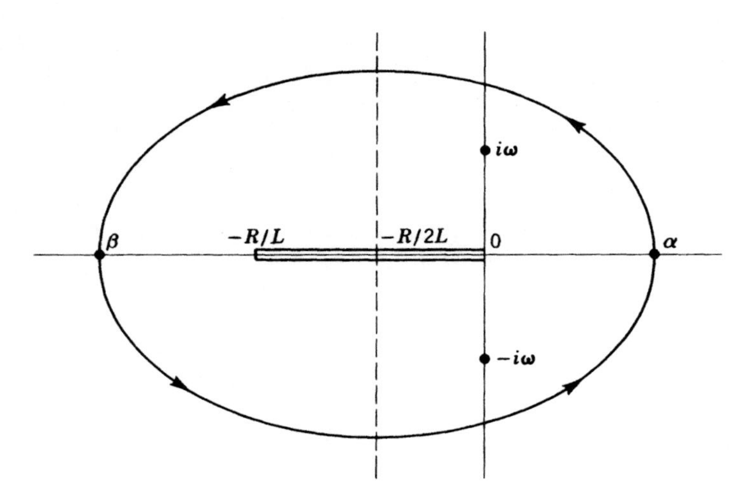

Fig. 6-12 Steepest descents for transmission-line problem.

and with foci at $(0, -R/L)$. On this ellipse, the real part of the exponent has a maximum at α and decreases monotonically as we move along the curve in either direction from α toward β. The ellipse cuts the imaginary axis at

$$s = \pm \frac{iR}{2L\xi\sqrt{\xi^2 - 1}}$$

For values of ξ close to unity—i.e., for the ratio of t to x/c_0 only slightly greater than 1, α and β are far from the origin, and the ellipse contains the poles at $\pm i\omega$ in its interior, as shown in Fig. 6-12. The original path of integration in Eq. (6-57) may then be replaced by the ellipse, and the only contribution for large x is that arising from the saddle point at α. As ξ increases, the saddle point α moves along the real axis toward the origin and eventually the ellipse will exclude the poles at $\pm i\omega$ from its interior. The integral of Eq. (6-57) is then equal to the integral along the ellipse (again obtainable asymptotically by considering only the saddle point), plus, now, a contribution from the residues at the two poles. The value of ξ at which the ellipse just cuts the poles is given by

$$\xi_1 = \left[\frac{1}{2} + \frac{1}{2}\left(1 + \frac{R^2}{\omega^2 L^2}\right)^{\frac{1}{2}}\right]^{\frac{1}{2}} \tag{6-61}$$

Computation now gives, for large values of x, the contribution of the saddle point as

$$E\left(x, \xi\frac{x}{c_0}\right) \sim \frac{E_0}{2\omega}\left(\frac{c_0 R}{\pi L x}\right)^{\frac{1}{2}} \frac{(\xi^2 - 1)^{\frac{1}{4}} \exp\left[-(Rx/2Lc_0)(\xi - \sqrt{\xi^2 - 1})\right]}{(\xi^2 - 1) + (R/2\omega L)^2(\xi - \sqrt{\xi^2 - 1})^2} \tag{6-62}$$

For $\xi < \xi_1$, this is (asymptotically) all there is; for $\xi > \xi_1$, however, we must add to the right-hand side of Eq. (6-62) the residue term, which after some algebra is seen to have the form

$$E_0 e^{-\sigma x} \sin\left(\frac{x\omega\xi}{c_0} - \frac{x\omega}{c}\right)$$

where σ and c are as defined in Eq. (6-55). Since this residue term should be included for all values of $\xi > \xi_1$, we can write it as

$$E_0 e^{-\sigma x} \sin \omega\left(t - \frac{x}{c}\right) \tag{6-63}$$

where $t > \xi_1 x/c_0$. We note that this term has the same form as the steady-state solution given by Eq. (6-54).

Thus, for large values of x, the steady-state solution appears to propagate along the line with velocity c_0/ξ_1, where ξ_1 is given by Eq. (6-61). It is only when this steady-state solution "arrives" at a point that the sinusoidal time variation, of angular frequency ω, is felt. Thus it would be reasonable to define signal velocity here as c_0/ξ_1 (which is $< c_0$); on the other hand, an adequately sensitive instrument would detect the "forerunner," or "precursor," described by whichever of Eq. (6-59) or (6-62) is appropriate. We observe that these two latter equations are different in character, in that the first holds for small values of $t - x/c_0$, whereas the second holds for (fixed) values of $c_0 t/x$ satisfying $1 < c_0 t/x < \xi_1$. [For $\xi > 1$, but close to 1, Eq. (6-62) does indeed hold—but only for very large values of x, since the second derivative at the saddle point is small for ξ close to 1. Thus $t - x/c_0$ would be large before Eq. (6-62) would be applicable, so that we do not necessarily expect to be able to deduce Eq. (6-59) from Eq. (6-62).]

We observe that the "signal speed," as defined here, depends on the angular frequency of the sinusoidal-input signal. Nonsinusoidal inputs would lead to quite different formulas for signal speed; in some cases, the concept would not have a clear meaning. For reference, we give the exact solution of Eq. (6-53), corresponding to an input signal which is zero for $t < 0$ and $E(0,t)$ for $t > 0$,

$$\begin{aligned} E(x,t) &= \exp\left(-\frac{Rx}{2Lc_0}\right) E\left(0, t - \frac{x}{c_0}\right) \\ &\quad + \frac{Rx}{2Lc_0} \int_{x/c_0}^{t} e^{-(R/2L)\tau} E(0, t - \tau) \frac{I_1[(R/2L)\sqrt{\tau^2 - x^2/c_0^2}]}{\sqrt{\tau^2 - x^2/c_0^2}}\, d\tau \end{aligned} \tag{6-64}$$

where I_1 is the modified Bessel function of the first kind. Equation (6-64) holds, of course, only for $t > x/c_0$. We might notice that if we wanted to find an asymptotic representation of a convolution integral such as that given in Eq. (6-64), valid for large x and t, a useful procedure might be to begin by taking a Laplace transform; this would lead to an equation similar to Eq. (6-57).

EXERCISES

2. If $t > x/c_0$, the integral in Eq. (6-57) may be written as the sum of the residues at the two poles, plus an integral whose path of integration has been shrunk onto the branch cut. The residue contribution, given by Eq. (6-63), is the steady-state solution; consequently the branch-cut integral must represent a transient. Find a compact formulation for this transient contribution.

3. The "large-s" method was used to find the behavior of $E(x, x/c_0 + \tau)$ for small values of τ, leading to Eq. (6-59). In the Bessel-function discussion of Sec. 6-3, the analogous problem of finding an asymptotic expansion for $H_\nu^{(2)}(x)$ was solved by setting $\nu = x + \epsilon$ and using the saddle-point method. Could anything similar have been done here, by writing $t = x/c_0 + \tau$ and keeping τ fixed as $x \to \infty$? Can anything be done for small τ with the branch-cut integral of Exercise 2?

4. To clarify the large-s method, consider the easily verified equation

$$\sin \omega t = \frac{1}{2\pi i} \int_{\gamma - i\infty}^{\gamma + i\infty} \frac{\omega e^{st}}{s^2 + \omega^2}\, ds$$

where the path of integration is any vertical line to the right of the origin. Evaluate the integral, approximately, for small t, by the method used in the discussion of Eq. (6-58); now, however, let the path of integration in the z plane be a circle of radius R. Find a bound on the error, and discuss the effect of R. Can the approximation be improved by including more terms?

5. Use steepest descents to discuss the idea of signal velocity in the acoustic problem of Exercise 1. Here the response to an input signal (in the (m,n) mode) of

$$\psi(0,t) = \begin{cases} 0 & \text{for } t < 0 \\ \psi_0 \sin \omega t & \text{for } t > 0 \end{cases}$$

is given by

$$\psi(z,t) = \frac{\psi_0 \omega}{2\pi i} \int_{\gamma - i\infty}^{\gamma + i\infty} \frac{\exp\left[-z(k^2 + s^2/c_0^2)^{1/2} + st\right]}{s^2 + \omega^2}\, ds$$

where

$$k^2 = \frac{m^2\pi^2}{l_1^2} + \frac{n^2\pi^2}{l_2^2}$$

What happens if ω is below the cutoff frequency?

6. Rework the transmission-line problem treated in the text for the case of an input signal which is zero for $t < 0$ and which for $t > 0$ satisfies

$$E(0,t) = E_0 \sin(\omega t + \alpha)$$

where α is some constant. In Eq. (6-57), the factor $\omega/(s^2 + \omega^2)$ is now replaced by $(s \sin \alpha + \omega \cos \alpha)/(s^2 + \omega^2)$.

6-6 *Differential-equation Methods*

With the exception of the Euler-summation-formula method, the asymptotic-expansion methods discussed in the preceding sections are based on an integral representation of the function in question. We turn now to another group of methods which make use of the differential equation satisfied by the function. The general idea of most differential-equation methods is to rearrange the differential equation so that some terms become negligible as the independent variable (or a parameter) becomes large; when these terms are discarded, the solution of the resulting simplified equation can be expected to provide an asymptotic approximation to the solution of the original equation.

As an example, consider Bessel's equation (5-114),

$$z^2w'' + zw' + (z^2 - \nu^2)w = 0 \tag{6-65}$$

The substitution $w = z^{-\frac{1}{2}}y$ gives

$$y'' + \left(1 - \frac{\nu^2 - \frac{1}{4}}{z^2}\right)y = 0 \tag{6-66}$$

For a fixed choice of ν, it now seems reasonable to neglect the term $(\nu^2 - \frac{1}{4})/z^2$ for large values of $|z|$. The resulting equation would have solutions $e^{\pm iz}$; so we expect an asymptotic solution of Eq. (6-66) to be

$$y \sim Ae^{iz} + Be^{-iz} \tag{6-67}$$

where A and B are arbitrary constants. In terms of w, we would have

$$w \sim Az^{-\frac{1}{2}}e^{iz} + Bz^{-\frac{1}{2}}e^{-iz} \tag{6-68}$$

and this is indeed compatible with our previous formulas for the asymptotic behavior of the Bessel functions.

Of course, the fact that the differential equation (6-66) is "close" to the equation

$$y'' + y = 0 \tag{6-69}$$

for large values of z does not in itself *prove* that the solutions of these two equations must also be close, but it does make this very plausible. In the next section, in connection with a discussion of the behavior in the complex plane of asymptotic solutions, we shall prove that the expansion whose first terms are given by Eq. (6-68) is asymptotic to an exact solution of Eq. (6-65) for an appropriate range of values of arg z.

Consider first, however, the problem of improving the representation in, say, Eq. (6-67). A method that at once suggests itself is to use successive approximations for the last term in Eq. (6-69); i.e., we solve in sequence the equations

$$\begin{aligned} y_a'' + y_a &= 0 \\ y_b'' + y_b &= \frac{\nu^2 - \frac{1}{4}}{z^2}\, y_a \\ y_c'' + y_c &= \frac{\nu^2 - \frac{1}{4}}{z^2}\, y_b \\ &\cdots\cdots\cdots\cdots \end{aligned}$$

Choosing y_a as e^{iz}, for example, would lead us to take as a suitable solution of the second equation

$$\begin{aligned} y_b &= e^{iz} + \frac{\nu^2 - \frac{1}{4}}{2i} \int_{i\infty}^{z} (e^{iz} - e^{-iz+2it})\, \frac{dt}{t^2} \\ &= e^{iz}\left[1 - \frac{1}{2i}\, \frac{\nu^2 - \frac{1}{4}}{z} + 0\left(\frac{1}{z^2}\right)\right] \end{aligned} \tag{6-70}$$

(a discussion of this equation will be given in the next section). We could proceed indefinitely in this way. Alternatively, we could decide that this looks like the beginning of an asymptotic series of the form

$$y \sim e^{iz}\left(1 + \frac{a_1}{z} + \frac{a_2}{z^2} + \cdots\right) \tag{6-71}$$

and simply substitute this series into Eq. (6-66) so as to determine the constants a_j. In the same way, an improved series with initial factor e^{-iz} could be obtained. The reader may verify that the two series so obtained, when the factor $z^{-1/2}$ associated with Eq. (6-65) is included, are proportional to the previous asymptotic series for $H_\nu^{(1)}(z)$ and $H_\nu^{(2)}(z)$.

Here we were led fairly naturally to the form (6-71) by the relationship of Eq. (6-66) to Eq. (6-69) and by the successive-approximation method leading to Eq. (6-70). We could equally well have started by assuming—as Stokes did—that an asymptotic solution of Eq. (6-65) has the form

$$w \sim e^{\pm iz}(az^\alpha + bz^\beta + cz^\gamma + \cdots) \tag{6-72}$$

where $a, b, c, \ldots, \alpha, \beta, \gamma, \ldots$ are unknown constants, to be determined by substitution of this series into Eq. (6-65).

For a second-order differential equation, any such series-substitution method usually provides us with a pair of linearly independent asymptotic representations; the problem then arises of finding that linear combination of these two representations which corresponds to a particular exact solution. For example, how would one find A and B in Eq. (6-68) so as to give the asymptotic behavior of $J_\nu(z)$? In a sense, the problem here is that of matching the behavior of a solution near the origin with its behavior near ∞.

This can sometimes be done by finding the first term of the asymptotic expansion by some other method—e.g., stationary phase—but in some cases it may be necessary to try to match the solutions numerically.

The transformation that reduced Eq. (6-65) to Eq. (6-66) is an example of the transformation

$$w = y \exp\left[-\int \tfrac{1}{2} a_1(z)\, dz\right] \tag{6-73}$$

that can be applied to the general second-order equation

$$w'' + a_1(z)w' + a_0(z)w = 0$$

so as to remove the term in the first derivative. Thus, without loss of generality, most discussions of asymptotic solutions found in the literature are concerned with equations of the form

$$y'' + f(z)y = 0$$

Exercises

1. Substitute the series (6-72) into Eq. (6-65) so as to determine the unknown constants. Obtaining the two coefficients of the linear combination by the method of stationary phase, write down the first few terms of the asymptotic representation for $J_1(z)$.

2. The differential equation

$$y' = \left(1 + \frac{1}{z}\right) y$$

has the exact solution ze^z. Disregarding this fact, try to find an asymptotic solution by first discarding the term in $1/z$ and then by improving this term by successive substitutions. Consider also the use of an assumed asymptotic representation of a form similar to that of Eq. (6-72). Consider next the nonhomogeneous equation

$$y' + \left(1 + \frac{1}{z}\right) y = \frac{1}{z}$$

How could you proceed if you did not know the exact solution and wanted an asymptotic solution?

3. Find asymptotic solutions of

$$zy'' - (z + 2)y = 0$$

valid for large z, and compare with the exact solutions (one of which is ze^z).

4. If

$$w'' = p_1 w' + p_0 w$$

where p_1 and p_0 are certain functions of z, analytic at the point at ∞, find conditions that the coefficients of the expansions of p_1 and p_0 in powers of $1/z$ must fulfill if w is to possess a *convergent* expansion of the form

$$z^{-\rho}\left(w_0 + \frac{w_1}{z} + \frac{w_2}{z^2} + \cdots\right)$$

[If p_1 and p_0 have, at worst, poles at ∞, the singularity there is said to be of *finite rank;* it may then be proved[1] that asymptotic solutions of the more general form

$$e^{\varphi(z)}z^{-\rho}\left(w_0 + \frac{w_1}{z} + \frac{w_2}{z^2} + \cdots\right)$$

exist. We shall subsequently find that this is also the form obtained by use of the WKB method.]

Behavior in the Complex Plane

We shall continue with the Bessel function example, which led to Eq. (6-66) and to the asymptotic solution (6-71). Let the coefficients a_i of this series be obtained by substitution into Eq. (6-66). Denote the sum of the first $n + 1$ terms by y_1,

$$y_1 = e^{iz}\left(1 + \frac{a_1}{z} + \frac{a_2}{z^2} + \cdots + \frac{a_n}{z^n}\right) \tag{6-74}$$

When y_1 is substituted into the left-hand side of Eq. (6-66), we obtain

$$\begin{aligned} y_1'' + \left(1 - \frac{\nu^2 - \frac{1}{4}}{z^2}\right)y_1 &= a_n e^{iz} z^{-n-2}\left[n(n+1) - \left(\nu^2 - \frac{1}{4}\right)\right] \\ &= y_1 \cdot 0(z^{-n-2}) \end{aligned}$$

Thus $y_1(z)$ satisfies *exactly* an equation of the form

$$y_1'' + \left[1 - \frac{\nu^2 - \frac{1}{4}}{z^2} + \alpha(z)\right]y_1 = 0 \tag{6-75}$$

where $\alpha(z)$ is a known function which is $0(z^{-n-2})$. Similarly, we define y_2 as the sum of the first $n + 1$ terms of the second asymptotic solution (with an exponent of opposite sign),

$$y_2 = e^{-iz}\left(1 + \frac{b_1}{z} + \frac{b_2}{z^2} + \cdots + \frac{b^n}{z^n}\right) \tag{6-76}$$

Then

$$y_2'' + \left[1 - \frac{\nu^2 - \frac{1}{4}}{z^2} + \beta(z)\right]y_2 = 0 \tag{6-77}$$

where $\beta(z)$ is a known function which is $0(z^{-n-2})$. Since Eqs. (6-75) and (6-77) are close to Eq. (6-69), in that they differ from it only by terms of $0(z^{-n-2})$, we intuitively expect that, for large z, y_1 and y_2 will be close to a

[1] E. L. Ince, "Ordinary Differential Equations," chaps. 17 and 18, Dover Publications, Inc., New York, 1956.

pair of exact solutions of Eq. (6-65). The following formal proof of this fact will serve the second purpose of delineating the region of the z plane over which the result holds true.

We first change the notation slightly. Define $w_1(z) = y_1(z)$, and $w_2(z) = y_2(z) + \gamma(z)$, where $\gamma(z)$ is to be a function of the form $e^{-iz}0(z^{-n-2})$ so chosen[1] that $w_2(z)$ satisfies Eq. (6-75). Then we have two known functions w_1 and w_2 which are independent exact solutions of Eq. (6-75) and which coincide with the two truncated asymptotic series y_1 and y_2 within terms of higher order. Rewrite Eq. (6-66) in the form

$$y'' + \left[1 - \frac{\nu^2 - \frac{1}{4}}{z^2} + \alpha(z)\right] y = \alpha(z)y \tag{6-78}$$

This equation has the form of Eq. (5-54), with $f(z) = \alpha(z)y$, so that its general solution can be written

$$y = Aw_1(z) + Bw_2(z) + \frac{1}{W}\int [w_1(t)w_2(z) - w_1(z)w_2(t)]\alpha(t)y(t)\,dt \tag{6-79}$$

where A and B are arbitrary constants and where W is the (constant) wronskian of w_1 and w_2. Since the unknown function $y(t)$ occurs in the integrand, this is an integral equation for $y(z)$, which may be solved by iteration. We want first to find a solution which is close to $w_1(z)$ [and so to $y_1(z)$]; we shall show that this can be done by choosing $A = 1$, $B = 0$, by setting the lower limit of integration equal to $i\infty$, and by suitably restricting the path of integration in the t plane. To solve

$$y = w_1(z) + \int_{i\infty}^{z} [w_1(t)w_2(z) - w_1(z)w_2(t)]\alpha(t)y(t)\,dt \tag{6-80}$$

by successive substitutions, we obtain a first approximation for $y(z)$ by replacing $y(t)$ in the integrand by $w_1(t)$, a second approximation by replacing $y(t)$ in the integrand by the first approximation just obtained, and so on. The function $\alpha(t)$ is analytic at ∞, so its singularities lie within some bounded region centered on the origin. Take the path of integration in the t plane so as to circumvent this region and also so that Im t decreases monotonically along the path (Fig. 6-13). For the first approximation, the integrand of Eq. (6-80) can be written as

$$[w_1(t)w_2(z) - w_1(z)w_2(t)]w_1(t)\alpha(t) = -w_1(z)(1 - e^{2i(t-z)} + \cdots)\alpha(t)$$

and since Im $(t - z) > 0$ everywhere on the path, the right-hand side can be written as $w_1(z)0(t^{-n-2})$. After integration, we therefore obtain a term

[1] We can determine $\gamma(z)$ by observing that it must satisfy

$$\gamma'' + \left(1 - \frac{\nu^2 - \frac{1}{4}}{z^2} + \alpha\right)\gamma = (\beta - \alpha)y_2$$

which can be solved since we know one solution, namely, y_1, of the corresponding homogeneous equation.

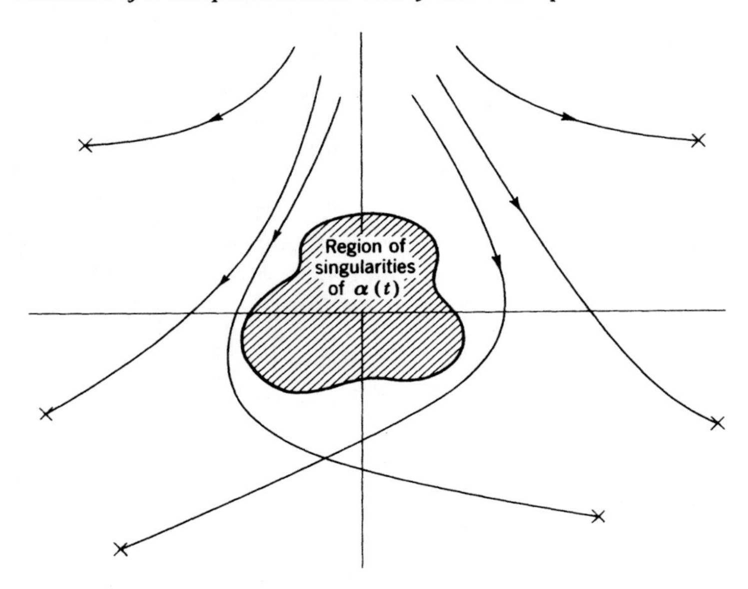

Fig. 6-13 Possible paths of integration in t plane.

$w_1(z)0(z^{-n-1})$ as a correction to $w_1(z)$. Continuing in this way gives a series of correction terms; bounds on the magnitudes of these correction terms are easily computed, and the series is seen to be (uniformly) convergent. The sum is an exact solution of Eq. (6-80) and so of Eq. (6-66), which differs from y_1 only by terms of $0(z^{-n-1})$. Moreover, this result is valid for any point z which can be reached by a path of integration of the kind described—i.e., for any point z for which $-\pi < \arg z < 2\pi$. In the same way, choosing $B = 1$, $A = 0$ and taking the lower limit of integration as $-i\infty$ in Eq. (6-79) will lead to an exact solution of Eq. (6-66) which differs from y_2 only by terms of $0(z^{-n-1})$ throughout the range $-2\pi < \arg z < \pi$.

As previously remarked, the asymptotic series which we have been discussing are essentially those for $H_\nu^{(1)}(z)$ and $H_\nu^{(2)}(z)$; we have now proved that these asymptotic series are indeed valid for those ranges of arg z which were specified for Eqs. (5-165) and (5-166). If we try to extend the limits of arg z, then we can no longer find a path of integration for which Im t changes monotonically and the "correction" terms in the integral equation can become very large as $|z|$ grows. Thus we cannot necessarily expect the asymptotic series to continue to represent $H_\nu^{(1)}(z)$ or $H_\nu^{(2)}(z)$ for values of arg z outside the ranges stated; in fact, we shall shortly show that the representation does break down. What this means is that, if we start with z on the positive imaginary axis, say, and allow z to circle the origin in the positive direction, with z always large, then the series (5-165) will give a good representation for $H_\nu^{(1)}(z)$ until arg z approaches 2π; at and beyond this limit, the representation is no longer valid, despite the fact that $H_\nu^{(1)}(z)$ itself varies perfectly smoothly and continuously as we cross this *critical ray*. This rather surprising situation was discovered by Stokes[1] and is called the *Stokes phenomenon*.

We can point up the situation still further by writing $z = re^{i\theta}$ and observing that $H_\nu^{(1)}(re^{i(\theta+2\pi)})$ can be expressed as a linear combination of $H_\nu^{(1)}(re^{i\theta})$ and $H_\nu^{(2)}(re^{i\theta})$ (since these latter two functions are a pair of independent solutions of Bessel's equation). For θ in the range $(-\pi,\pi)$, these latter two functions are asymptotically represented by the series (5-165) and (5-166), so that $H_\nu^{(1)}(re^{i(\theta+2\pi)})$ must have an asymptotic representation given by a linear combination of these two series. Thus an asymptotic representation for $H_\nu^{(1)}(z)$, as arg z increases past 2π, still involves the two series (5-165) and (5-166), but the coefficients of the linear combination are no longer (1,0). There is a discontinuous change of coefficients (which we shall shortly calculate), despite the continuous variation of $H_\nu^{(1)}(z)$. This discontinuous change in the coefficients of a linear combination of two asymptotic series, which must be made in order to continue to represent a particular exact solution, is characteristic of the Stokes phenomenon.

Of the two asymptotic series for $H_\nu^{(1)}(z)$ and $H_\nu^{(2)}(z)$, the first is exponentially small compared with the second in the upper half-plane, but exponentially large in the lower; thus the first series is said to be *dominant* in the lower half-plane and *subdominant*, or *recessive*, in the upper half-plane. Along any ray, any solution of Bessel's equation is some linear combination of $H_\nu^{(1)}(z)$ and $H_\nu^{(2)}(z)$ and so must be expressible asymptotically by the same linear combination of the series (5-165) and (5-166). Except near the real axis, one of these series is exponentially small compared with the other and so may be discarded; conversely, a knowledge of only the dominant portion of the asymptotic series for a particular exact solution can tell us nothing about the asymptotic behavior in a region where the roles of dominance and subdominance have become reversed.

We have seen that Eq. (6-66) possesses an exact solution, say $\varphi_1(z)$, which is asymptotically represented by the series $y_1(z)$ for $-\pi < \arg z < 2\pi$. Similarly, it possesses an exact solution $\varphi_2(z)$ which is asymptotically represented by $y_2(z)$ for $-2\pi < \arg z < \pi$. Let us now find an asymptotic representation for $\varphi_1(z)$ which is valid for $\arg z > 2\pi$. Although this could be done by making use of known relations between Hankel functions (cf. the second to last paragraph), it will be instructive to use the differential equation (6-66) directly. The indicial roots of Eq. (6-66) at the origin are $\frac{1}{2} \pm \nu$, so that any exact solution—and in particular $\varphi_1(z)$—can be written as

$$\varphi_1(z) = z^{\frac{1}{2}-\nu}p(z^2) + z^{\frac{1}{2}+\nu}q(z^2)$$

[1] G. G. Stokes, "Mathematical and Physical Papers," vol. 4, pp. 77 and 293, Cambridge University Press, New York, 1904.

where p and q are power series in z^2. Set $z = re^{i\theta}$. Now φ_1 is asymptotically represented by y_1 for the particular choices $\theta = 0$ and $\theta = \pi$, so that, as $r \to \infty$,

$$r^{\frac{1}{2}-\nu}p(r^2) + r^{\frac{1}{2}+\nu}q(r^2) \sim e^{ir}$$
$$r^{\frac{1}{2}-\nu}e^{i\pi(\frac{1}{2}-\nu)}p(r^2) + r^{\frac{1}{2}+\nu}e^{i\pi(\frac{1}{2}+\nu)}q(r^2) \sim e^{-ir}$$

where we have taken only the first term of the asymptotic behavior. When $\theta = 2\pi$, $\varphi_1(z)$ has the form

$$r^{\frac{1}{2}-\nu}e^{i2\pi(\frac{1}{2}-\nu)}p(r^2) + r^{\frac{1}{2}+\nu}e^{i2\pi(\frac{1}{2}+\nu)}q(r^2)$$

which may be expressed as a linear combination of the left-hand sides of the preceding two equations, from which it is seen to be asymptotic to $e^{ir} + (2i \cos \nu\pi)e^{-ir}$.

Thus, for $\theta = 2\pi$,

$$\varphi_1(z) \sim e^{iz} + (2i \cos \nu\pi)e^{-iz} \tag{6-81}$$

For $-\pi < \theta < \pi$, all of $\varphi_1(re^{i\theta})$, $\varphi_2(re^{i\theta})$, and $\varphi_1(re^{i(\theta+2\pi)})$ are exact solutions of Eq. (6-66); the first two are independent (cf. their asymptotic behavior), so that the last can be expressed as a linear combination of them. But its asymptotic representation must then be given by the same linear combination of the asymptotic representations of $\varphi_1(re^{i\theta})$ and $\varphi_2(re^{i\theta})$—that is, of $y_1(z)$ and $y_2(z)$, known to be valid representations for this range of θ. Thus for $-\pi < \theta < \pi$

$$\varphi_1(re^{i(\theta+2\pi)}) \sim Ay_1(z) + By_2(z)$$

Choosing $\theta = 0$ and using Eq. (6-81) give A as 1 and B as $2i \cos \nu\pi$. Consequently,

$$\varphi_1(z) \sim y_1(z) + (2i \cos \nu\pi)y_2(z) \tag{6-82}$$

for $\pi < \arg z < 3\pi$. Comparing this with our previous result

$$\varphi_1(z) \sim y_1(z) \tag{6-83}$$

for $-\pi < \arg z < 2\pi$, we see that in the range $\pi < \arg z < 2\pi$ we have our choice of two formulas; either may be used, since the second term in Eq. (6-82) is here negligible. However, as $\arg z$ reaches or passes the value 2π, Eq. (6-82) *must* be used; Eq. (6-83) would be incorrect. Actually, the transition from Eq. (6-83) to Eq. (6-82) could, of course, be made to take place at any chosen value of θ in $(\pi, 2\pi)$.

EXERCISE

5. Using Eqs. (5-165) and (5-166), obtain the asymptotic expansion for $J_\nu(z)$, and state its range of validity. Is the result consistent with Eq. (5-163)? Obtain also an asymptotic representation for $J_\nu(z)$ valid for the range $0 < \arg z < 2\pi$, first by use of an extended formula for $H_\nu^{(2)}(z)$ and second by using directly the fact that $J_\nu(z) = z^\nu f(z)$. For which values of arg z is $H_\nu^{(2)}(z)$ exponentially small in the lower half-plane?

6-7 *WKB Method*

Continuing with our study of differential-equation methods, we now turn our attention to the WKB method. Its purpose is to obtain an asymptotic representation for the solution of a differential equation containing a parameter; the representation is to be valid for small values of the parameter. As a standard form of differential equation for discussion, we choose

$$y'' + \frac{1}{h^2} f(x)y = 0 \tag{6-84}$$

where $f(x)$ is some given function of x. We require an asymptotic solution valid as $h \to 0$. This kind of problem arose in connection with Schrödinger's wave equation (h is then Planck's constant) and was attacked independently about 1926 by Wentzel, Kramers, and Brillouin, whose initials form the name of the method. It actually has a long previous history, with which the names of Liouville, Green, Horn, Rayleigh, Gans, and Jeffreys are associated; a useful historical survey has been given by Heading.[1] We might observe that Eq. (6-84) is not so special as might at first appear, since any term in y' in an antecedent differential equation can be removed by the transformation (6-73); also, it is often easy to make a simple change of variable or parameter that gives the form (6-84). Even if the dependence on the parameter is more complicated than that given in Eq. (6-84), it will become clear that the WKB method may often be used with but little modification.

We begin by motivating the WKB method physically, by assuming that Eq. (6-85) has resulted from a one-dimensional wave-motion problem. Let $\varphi(x,t)$ satisfy the wave equation

$$\varphi_{xx} = \frac{1}{c^2} \varphi_{tt} \tag{6-85}$$

where subscripts denote partial derivatives with respect to position x or time t and where c is the x-dependent local rate at which a disturbance propagates. To find the steady-state response to an excitation of angular frequency ω, we write

$$\varphi = \operatorname{Re}\,[\psi(x)e^{i\omega t}] \tag{6-86}$$

[1] J. Heading, "An Introduction to Phase Integral Methods," chap. 1, Methuen & Co., Ltd., London, 1962.

where ψ is a complex amplitude function. Since the expression obtained by making a time translation of $\pi/2\omega$ in Eq. (6-86) must also be a solution, we can drop the Re symbol; substitution into Eq. (6-85) then gives

$$\psi'' + \frac{\omega^2}{c^2}\psi = 0 \tag{6-87}$$

which has the form of Eq. (6-84). Let the average value of c over some range of x be denoted by c_0. If c had the constant value c_0, solutions of Eq. (6-87) would have the form $\exp(\pm i\omega x/c_0)$, which in combination with the previous $e^{i\omega t}$ factor represents traveling waves. Guided by this, we write

$$\psi = \alpha(x) \exp\left[-\frac{i}{c_0}\omega\beta(x)\right] \tag{6-88}$$

where α and β are real functions of x. Account being taken of the factor $e^{i\omega t}$, a point of constant phase moves with a local velocity of c_0/β'. It is physically plausible (and in fact corresponds to Huygens' principle in optics) to expect this velocity to be approximately equal to the local velocity $c(x)$, so that

$$\beta' \cong \pm\frac{c_0}{c} \tag{6-89}$$

To determine the conditions under which Eq. (6-89) is valid, let us obtain the exact equation satisfied by β, by substituting Eq. (6-88) into Eq. (6-87). Doing this and equating separately the real and imaginary parts, we obtain

$$(\beta')^2 = \left(\frac{c_0}{c}\right)^2 + \left(\frac{c_0}{\omega}\right)^2\frac{\alpha''}{\alpha} \tag{6-90}$$

$$2\alpha'\beta' + \alpha\beta'' = 0 \tag{6-91}$$

Equation (6-90) shows that Eq. (6-89) is a good approximation if

$$\left(\frac{c}{\omega}\right)^2\frac{\alpha''}{\alpha} \ll 1 \tag{6-92}$$

(where we have used $c_0 \cong c$). If we define the local wavelength λ by $\lambda = 2\pi c/\omega$, then the inequality (6-92) can be interpreted as requiring the amplitude α not to fluctuate too much[1] over a distance of a wavelength. Thus, just as in optics, we can expect Eq. (6-89) to become more and more accurate as $\lambda \to 0$; in this sense, Eq. (6-89) is an asymptotic version of Eq. (6-90). Once β has been determined, Eq. (6-91) gives $\alpha = \text{const}\cdot(1/c)^{-1/2}$. This value of α could then be substituted into the second term of Eq. (6-90) so as to give an improved value for β.

In optics, the equation

$$(\beta')^2 = \left(\frac{c_0}{c}\right)^2 \tag{6-93}$$

[1] More exactly, not to depart very much from a linear rate of change.

which results from omitting the second term of Eq. (6-90) is known as the *eikonal equation.* (Besides being the approximate equation satisfied by the phase function, it is the exact equation satisfied by Hamilton's characteristic function; it also arises in the method of characteristics.)

The result then is that, as $\lambda \to 0$, we expect solutions of Eq. (6-87) to be given asymptotically by

$$\psi \sim \text{const} \times \left(\frac{1}{c(x)}\right)^{-1/2} \exp\left[\pm i\omega \int \frac{dx}{c(x)}\right]$$

In terms of Eq. (6-84), we analogously anticipate

$$y \sim \text{const} \times [(f(x)]^{-1/4} \exp\left[\pm \frac{i}{h} \int (f(x))^{1/2}\, dx\right] \tag{6-94}$$

These two asymptotic solutions, corresponding to the two choices of sign, are the WKB solutions of Eq. (6-84). The most general asymptotic solution is then an arbitrary linear combination of them. From the foregoing discussion, we expect them to improve in accuracy as $h \to 0$.

Let us now obtain Eq. (6-94) in a different and more formal manner. Thinking of all the factors in Eq. (6-94) as being incorporated into the exponent suggests that we might begin by assuming an asymptotic expansion for Eq. (6-84) of the form

$$y \sim \exp\left[\frac{1}{h} \int (\varphi_0 + h\varphi_1 + h^2\varphi_2 + \cdots)\, dx\right] \tag{6-95}$$

where the φ_i are certain functions of x. Substituting into Eq. (6-84) and collecting powers of h, we obtain a sequence of defining relations for the φ_i,

$$\begin{aligned} \varphi_0^2 + f &= 0 \\ 2\varphi_0\varphi_1 + \varphi_0' &= 0 \\ \varphi_1^2 + 2\varphi_0\varphi_2 + \varphi_1' &= 0 \\ \cdots\cdots\cdots\cdots& \end{aligned} \tag{6-96}$$

The reader may verify that, if we stop after the first two of these equations (as is often done), we recover Eq. (6-94).

Although the physical motivation of Eq. (6-94) leads us to expect it to be a good approximation to an exact solution of Eq. (6-84) for sufficiently small h, there are two properties of Eq. (6-94) which result in a modification of this expectation. Suppose that $f(x)$ is well behaved over some region but vanishes at a point x_0. Then any exact solution of Eq. (6-84) is also well behaved at x_0, but Eq. (6-94) is not, because of the factor $f^{-1/4}$. Thus the asymptotic property of Eq. (6-94) must break down near a zero of $f(x)$. Second, if we consider now complex values of x, an exact solution of Eq. (6-84) is single-valued insofar as circuits around a zero of $f(x)$ are concerned, but Eq. (6-94) is not. Thus, for complex values of x, an asymptotic solution of the form (6-94) cannot be valid in any region which completely surrounds a zero of $f(x)$, even if a subregion around that zero has been deleted. We shall return to this matter in the next section.

One way of investigating the accuracy of Eq. (6-94) is to observe that the function y_1 defined by

$$y_1 = f^{-1/4} \exp\left(\pm \frac{i}{h} \int f^{1/2}\, dx\right) \tag{6-97}$$

is an exact solution of

$$y_1'' + y_1 \left\{\frac{1}{h^2} f + \left[\frac{1}{4}\frac{f''}{f} - \frac{5}{16}\left(\frac{f'}{f}\right)^2\right]\right\} = 0 \tag{6-98}$$

The term in square brackets does not depend on h and so becomes negligible in comparison with the term f/h^2 as $h \to 0$, again provided that $f(x)$ is not close to zero. Thus, as $h \to 0$, the differential equation satisfied by y_1 becomes closer and closer to Eq. (6-84), and we can expect corresponding solutions also to approach one another. Rather simple error bounds for the WKB method have been obtained by Blumenthal; an extended and more complete account is given by Olver.[1]

Exercises

1. Airy's equation

$$y'' + zy$$

is essentially a special case of Bessel's equation. Ignoring this fact, use the change in variable $z = h^{-2/3}\xi$ to obtain an equation to which the WKB method can be applied. Use this method to find an asymptotic representation for solutions of Airy's equation, valid for large z.

2. If we set $z = (\nu^2 - 1/4)^{1/2}\xi$ in Eq. (6-66), we obtain

$$y_{\xi\xi} + \left(\nu^2 - \frac{1}{4}\right)\left(1 - \frac{1}{\xi^2}\right) y = 0 \tag{6-99}$$

For each of the two cases $\xi < 1$, $\xi > 1$, apply the WKB method to Eq. (6-99) so as to obtain the asymptotic behavior of $H_\nu^{(2)}(x)$ for large ν; compare the results with Eqs. (6-44) and (6-45).

3. Replace f in Eq. (6-84) by $f_0 + hf_1$. What equations now replace Eqs. (6-96)?

Consider next the differential equation

$$y'' + \frac{1}{h} f(x) y' + \frac{1}{h^2} g(x) y = 0$$

See whether or not a direct substitution of Eq. (6-95) into this equation will give the same result as such a substitution following a preliminary removal of the second term.

[1] F. W. J. Olver, *Proc. Cambridge Phil. Soc.*, *57:* 790 (1961).

4. Schrödinger's wave equation for the one-dimensional oscillator reads

$$y'' + \frac{8\pi^2 m[E - V(x)]}{h^2} y = 0$$

where m is particle mass, h is Planck's constant, $V(x)$ is potential energy, and E is energy level. Obtain the WKB solutions for the case $E > V(x)$, and compare them with those obtained if E is allowed to depend on h, via $E = kh$, where k is some constant.

Transition Points

It has already been remarked that Eq. (6-94) fails at a zero of $f(x)$. Suppose for the moment that we are dealing with real variables only; let $f(x)$ have a simple zero at x_0, with $f'(x_0) = c > 0$. Then the two exponents in Eq. (6-94) are purely imaginary for $x > x_0$, and the general asymptotic solution is oscillatory in character, whereas the exponents are purely real for $x < x_0$ and the general asymptotic solution then consists of an exponentially increasing and an exponentially decreasing term. The transition between these two kinds of behavior must take place in a region centered on x_0. For this reason x_0 is called a *transition point*, although the expression *turning point* is also used (because of the fact that x_0 sometimes corresponds to the classical limit of motion of a wave-mechanical particle bound by a potential field).

A particular asymptotic solution of Eq. (6-84) will thus be represented by quite differently behaving functions on the two sides of x_0, and it is necessary to find out how to "connect" from one side of x_0 to the other. The provision of such connection formulas is part of the WKB method; we shall here consider several methods of deriving them.

One method, a rather natural one, is to replace Eq. (6-84) by an equation which is approximately equivalent to Eq. (6-84) in a region around x_0 and which can be solved exactly. Without loss of generality, we may set $x_0 = 0$. Then, in the neighborhood of the origin, Eq. (6-84) has the approximate form

$$y'' + \frac{c}{h^2} xy = 0 \tag{6-100}$$

Equation (6-100) is a form of Airy's equation, itself a form of Bessel's equation. Its general solution is

$$y = \left(\frac{\pi}{3h}\right)^{1/2} Ax^{1/2} J_{1/3}\left(\frac{2}{3}\frac{c^{1/2}}{h} x^{3/2}\right) + \left(\frac{\pi}{3h}\right)^{1/2} Bx^{1/2} J_{-1/3}\left(\frac{2}{3}\frac{c^{1/2}}{h} x^{3/2}\right) \tag{6-101}$$

where A and B are arbitrary constants. The factor $(\pi/3h)^{1/2}$ is included for future convenience. The right-hand side of Eq. (6-101) has no singularity at the origin; so the choice of branch is immaterial as long as we use the same branch in each factor. Take $\arg x = 0$ for $x > 0$ and $\arg x = \pi$ for $x < 0$. Then, for $x < 0$, the argument ξ of the Bessel functions in Eq. (6-101) is

$$\xi = \frac{2}{3}\frac{c^{1/2}}{h} e^{i\frac{3}{2}\pi}|x|^{3/2}$$

and its phase angle $\frac{3}{2}\pi$ must be taken into account [via the z^ν behavior of $J_\nu(z)$] before using the usual asymptotic formulas for the $J_\nu(z)$ functions, which are valid only for $-\pi < \arg z < \pi$. [We do want asymptotic expressions for the Bessel functions in Eq. (6-101), since ξ becomes very large as $h \to 0$.] The asymptotic formulas are

(*a*) For $x > 0$

$$y \sim (cx)^{-1/4}\left[A \cos\left(\frac{2}{3}\frac{c^{1/2}}{h}x^{3/2} - \frac{5}{12}\pi\right) + B\cos\left(\frac{2}{3}\frac{c^{1/2}}{h}x^{3/2} - \frac{1}{12}\pi\right)\right]$$

(*b*) For $x < 0$

$$y \sim \frac{1}{2}(c|x|)^{-1/4}\left[(B - A)\exp\left(\frac{2}{3}\frac{c^{1/2}}{h}|x|^{3/2}\right) + (Ae^{i(\pi/6)} + Be^{-i(\pi/6)})\exp\left(-\frac{2}{3}\frac{c^{1/2}}{h}|x|^{3/2}\right)\right]$$

Even if $|x|$ is small, we can choose h sufficiently small so that the WKB solutions given by Eq. (6-94) are valid [cf. Eq. (6-98)]; in order that they be compatible with the results just obtained we must consider $\frac{2}{3}(c^{1/2}/h)x^{3/2}$ and $(1/h)\int_0^x f^{1/2}\,dx$ as becoming equivalent for small x. In terms of the WKB solutions, we therefore have

(*a*) If $B \neq A$, then

$$y \sim f^{-1/4}\left[A\cos\left(\frac{1}{h}\int_0^x f^{1/2}\,dx - \frac{5}{12}\pi\right) + B\cos\left(\frac{1}{h}\int_0^x f^{1/2}\,dx - \frac{1}{12}\pi\right)\right] \tag{6-102}$$

valid for $x > 0$, connects onto

$$y \sim \frac{1}{2}|f|^{-1/4}(B - A)\exp\left(\frac{1}{h}\int_x^0 |f|^{1/2}\,dx\right) \tag{6-103}$$

valid for $x < 0$.

(*b*) If $B = A$, then

$$y \sim \sqrt{3}\,Af^{-1/4}\cos\left(\frac{1}{h}\int_0^x f^{1/2}\,dx - \frac{\pi}{4}\right) \tag{6-104}$$

valid for $x > 0$, connects onto

$$y \sim \frac{\sqrt{3}}{2}A|f|^{-1/4}\exp\left(-\frac{1}{h}\int_x^0 |f|^{1/2}\,dx\right) \tag{6-105}$$

valid for $x < 0$. In either case, the constants A and B are those which occur in Eq. (6-101), which is valid at and near $x = 0$. We have taken the case $c = f'(0) > 0$; the case $f'(0) < 0$ can be handled by simply reversing the direction in which x is positive.

The argument of the cosine function in Eq. (6-104) is often described as behaving as if one-eighth wavelength were used up to the left of the transition point. Observe that, if an asymptotic description of the solution for $x < 0$ were known, then only in the case that it is exponentially decreasing can the WKB solution for $x > 0$ be determined uniquely.

An alternative method of deriving the above connection formulas is to start again with Eq. (6-100) but to obtain its asymptotic solutions directly, rather than via Bessel functions. Setting $y = x^{-1/4}p(\xi)$, with $\xi = 2c^{1/2}x^{3/2}/3h$, in Eq. (6-100), we obtain

$$p_{\xi\xi} + p\left(1 + \frac{5}{36\xi^2}\right) = 0 \tag{6-106}$$

This clearly has asymptotic solutions like $e^{\pm i\xi}$, for large ξ (i.e., for small h); by a discussion analogous to that given for Eq. (6-78), the solution $e^{i\xi}$ is asymptotic to an exact solution for $-\pi < \arg \xi < 2\pi$, and the solution $e^{-i\xi}$ is asymptotic to an exact solution for $-2\pi < \arg \xi < \pi$. This suggests that we divide the complex x plane into three sectors, and in these sectors define the following functions (we take $c > 0$):

$$(a) \qquad 0 < \arg x < \frac{2\pi}{3} \qquad y_1 = x^{-1/4} \exp\left(i\frac{2}{3}\frac{c^{1/2}}{h}x^{3/2}\right)$$

$$(b) \qquad \frac{2\pi}{3} < \arg x < \frac{4\pi}{3} \qquad y_2 = x^{-1/4} \exp\left(-i\frac{2}{3}\frac{c^{1/2}}{h}x^{3/2}\right)$$

$$(c) \qquad -\frac{2\pi}{3} < \arg x < 0 \qquad y_3 = x^{-1/4} \exp\left(-i\frac{2}{3}\frac{c^{1/2}}{h}x^{3/2}\right)$$

The argument ranges of x to be used for computing $x^{3/2}$ for these three functions are to be those given in the definition of the sector. Each function is exponentially decaying in its own sector and from the ξ-plane result carries over its asymptotic relationship to an exact solution into the adjoining two sectors.

There is some exact solution to which y_2 is asymptotic in sector b and also in sectors a and c—but not on the line separating (b) from (c). Let us find the asymptotic behavior of this exact solution on this line. [Incidentally, y_2 does not coincide with y_3 in sector c, since, in moving from (b) to (c), the argument of x must vary continuously.] Solutions y_1 and y_3 are independent, because of their different behavior in sector a; therefore, in sector b we must have

$$y_2 = Ly_1 + My_3 \tag{6-107}$$

where L and M are constants. This relation holds not only between the asymptotic solutions but also between the corresponding exact solutions. Now y_1 and y_3 are dominant, but y_2 is subdominant, in sector b; thus the net coefficient of the dominant term on the right-hand side of Eq. (6-107) must vanish. Upon taking account of the various interpretations of $x^{-1/4}$ and $x^{3/2}$, this condition yields $L + Mi = 0$. It was remarked that Eq. (6-107) holds also for the corresponding exact solutions [which are single-valued, as is seen from the form of Eq. (6-100); see Chap. 5]; we now add that for the exact solutions this relation must hold through all sectors. The same relation must therefore also hold for the asymptotic solutions, within their limits of definition. In particular, it is valid in sector a, and comparing coefficients of dominant terms there gives $M = 1$. Thus Eq. (6-107) becomes

$$y_2 = -iy_1 + y_3$$

Since each of y_1 and y_3 is known on the positive real axis, our object is accomplished; the reader may verify that the result leads again to Eqs. (6-104) and (6-105). Equations (6-102) and (6-103) are even more easily obtained, since we merely start with an arbitrary linear combination of y_1 and y_3 on the positive real axis, and then each function may be followed around in sector b.

The connection methods we have so far described were those used by early writers on the WKB method, particularly Gans, Jeffreys, and Zwaan. They may be extended to problems in which $f(x)$ has a zero of higher order.[1] Formal connection rules for a number of wave-propagation problems have been discussed in detail by Heading.[2] Kemble[3] has given an interesting variation of the second of the above methods, in which he writes the exact solution as a linear combination of the WKB approximations; the coefficients of course are then functions of position rather than constants, and differential equations for these functions may be obtained. We mention also a useful paper by Olver,[4] which presents a survey of the regions of the complex plane over which the WKB solutions (as well as Langer's modification, to be discussed) give uniform asymptotic representations.

A somewhat different approach to the connection-formula problem has been given by Langer,[5] who obtained a single asymptotic formula, valid at the transition point as well as on its two sides. We shall give a simple heuristic derivation of Langer's result for the case in which $f(x)$ has a simple zero at the origin, with $f'(0) > 0$.

[1] S. Goldstein, *Proc. London Math. Soc.*, ***28*:** 81 (1928).

[2] Heading, *op. cit.*

[3] E. C. Kemble, "The Fundamental Principles of Quantum Mechanics," p. 97, Dover Publications, Inc., New York, 1958.

[4] F. W. J. Olver, *Phil. Trans. Roy. Soc. London*, (A)***250*:** 60 (1958).

[5] R. E. Langer, *Trans. Am. Math. Soc.*, ***33*:** 23 (1931), ***34*:** 447 (1932), ***36*:** 90 (1934). See also *Phys. Rev.*, ***51*:** 669 (1937).

In the discussion leading to Eqs. (6-102) to (6-105), it was pointed out that, for small h, the WKB approximations would coalesce with the asymptotic form of Eq. (6-101) for small $|x|$ if the quantities $\frac{2}{3}(c^{1/2}/h)x^{3/2}$ and $(1/h)\int_0^x f^{1/2}\,dx$ were considered as equivalent. We can remove the restriction that $|x|$ be small by replacing $\frac{2}{3}(c^{1/2}/h)x^{3/2}$ in the arguments of $J_{1/3}$ and $J_{-1/3}$ by $(1/h)\int_0^x f^{1/2}\,dx$ and by replacing the coefficient factor $x^{1/2}$ by

$$f^{-1/4}\left(\frac{1}{h}\int_0^x f^{1/2}\,dx\right)^{1/2}$$

This does not alter the situation near $x = 0$; for there $f \cong cx$; also, the asymptotic forms of $J_{1/3}$ and $J_{-1/3}$ now lead to the WKB approximations. Thus a uniform asymptotic representation, valid for all x, is

$$y \sim f^{-1/4}\left(\frac{1}{h}\int_0^x f^{1/2}\,dx\right)^{1/2}\left[KJ_{1/3}\left(\frac{1}{h}\int_0^x f^{1/2}\,dx\right) + LJ_{-1/3}\left(\frac{1}{h}\int_0^x f^{1/2}\,dx\right)\right] \tag{6-108}$$

where K and L are arbitrary constants. The choice of branches may be made in the same way as for Eq. (6-101). The reader may now use Eq. (6-108) to rederive the previous connection formulas.

Equation (6-108) can also be used to provide a uniform asymptotic representation for Bessel functions of large argument and large order [Eq. (6-66) indicates a transition point at $\nu \approx z$]; since Eq. (6-108) itself involves Bessel functions, there is, as remarked by Langer, a sort of poetic justice in this particular application.

Equation (6-108) ceases to be valid in the neighborhood of a second zero of $f(x)$; a generalization of the method to the case of two zeros—involving parabolic cylinder functions—is possible (see Exercise 10).

Exercises

5. Obtain the WKB solutions for Eq. (6-100), compare them with the exact solutions, and determine how close to the origin—in terms of both distance and number of cycles of the oscillatory solution—the WKB solutions remain useful.

6. Discuss by any method the WKB solutions for Eq. (6-84), given that $f(x)$ has a double zero at the origin.

7. Use Eqs. (6-108) and (6-66) to obtain a uniform asymptotic representation for $J_\nu(x)$, x and ν real and positive.

8. The radial-wave equation

$$\frac{d^2\psi}{dr^2} + \left[\frac{8\pi^2 m}{h^2}\left(E + \frac{Ze^2}{r}\right) - \frac{l(l+1)}{r^2}\right]\psi = 0$$

has a singularity at $r = 0$. Here m, E, Z, e, l, h are all constants. Show that the change in variables (Langer) $r = e^x$, $\psi = e^{x/2}u$ gives an equation which is amenable to the WKB method.

9. Use the WKB method to discuss the reflection of an advancing one-dimensional wave train at a point where c of Eq. (6-85) has a simple zero. Investigate amplitudes, energies, and phase changes.

10. (*a*) Provide the details of the following calculation, which leads to a uniform asymptotic representation for solutions $y(x)$ of Eq. (6-84), for the case in which $f(x)$ is >0 for x in (a,b) and is <0 for $x > b$ or for $x < a$ (the zeros at a and b are to be simple). The desired representation is in terms of solutions $w(\xi)$ of the equation

$$w'' + \frac{1}{h^2}(c - \xi^2)w = 0 \tag{6-109}$$

where $c > 0$. Solutions of Eq. (6-109) are parabolic cylinder functions; a survey of their properties will be found in A. Erdélyi.[1]

The transformation $y(x) = f(x)u(\xi)$, where ξ and u are new independent and dependent variables, respectively, transforms Eq. (6-84) into an equation without a term in u' if $f(x) = (\xi_x)^{-1/2}$. To make the resulting equation have a term in $1/h^2$ similar to that of Eq. (6-109), we let

$$\xi_x = f^{1/2}(c - \xi^2)^{-1/2}$$

and obtain

$$u_{\xi\xi} + u\left[\frac{1}{h^2}(c - \xi^2) - s(\xi)\right] = 0 \tag{6-110}$$

where

$$s(\xi) = (x_\xi)^{1/2}[(x_\xi)^{-1/2}]_{\xi\xi}$$

Choosing c so that

$$\int_a^b \sqrt{f}\,dx = \int_{-\sqrt{c}}^{\sqrt{c}} \sqrt{c - \xi^2}\,d\xi$$

makes $s(\xi)$ well behaved at $\xi = \pm\sqrt{c}$; moreover, a calculation of $s(\xi)$ in terms of f shows that $s(\xi)$ is well behaved elsewhere (for appropriate f). Since $s(\xi)$ does not depend on h, solutions of Eqs. (6-84) and (6-110) must coalesce as $h \to 0$. Finally, there are no branch-point ambiguities in the resulting uniform asymptotic representation for $y(x)$ in terms of solutions of Eq. (6-109).

(*b*) Show how a similar method could have been used for the case of a single zero of $f(x)$; using such a method, rederive Eq. (6-108).

[1] A. Erdélyi et al., "Higher Transcendental Functions," vol. 2, p. 115, McGraw-Hill Book Company, New York, 1953.

7
transform methods

7-1 Fourier Transforms

The Fourier transform $F(\lambda)$ of a function $f(t)$ is defined by

$$F(\lambda) = \frac{1}{\sqrt{2\pi}} \int_{-\infty}^{\infty} e^{i\lambda\tau} f(\tau)\, d\tau$$

It is of great value in the treatment of many problems involving differential or integral equations. Intrinsically, $F(\lambda)$ is an analytic function of the complex variable λ, and this fact underlies most of the power of the methods in which it appears. Nevertheless, a natural—but heuristic—way in which to motivate the concept of the Fourier transform and to derive the basic inversion theorem that it satisfies is to consider the limiting case of a real-variable Fourier series whose interval of definition grows without limit.

A function $f(t)$ defined in $-T < t < T$ can be represented in Fourier-series form via

$$f(t) = \tfrac{1}{2}a_0 + \sum_{1}^{\infty} \left(a_n \cos \frac{n\pi t}{T} + b_n \sin \frac{n\pi t}{T} \right) \qquad (7\text{-}1)$$

The coefficients a_n and b_n are given by the formulas

$$a_n = \frac{1}{T}\int_{-T}^{T} f(\tau)\cos\frac{n\pi\tau}{T}\,d\tau \qquad b_n = \frac{1}{T}\int_{-T}^{T} f(\tau)\sin\frac{n\pi\tau}{T}\,d\tau \tag{7-2}$$

Combining Eqs. (7-1) and (7-2), we obtain

$$\begin{aligned} f(t) &= \frac{1}{2T}\int_{-T}^{T} f(\tau)\,d\tau + \sum_{1}^{\infty}\frac{1}{T}\int_{-T}^{T} f(\tau)\cos\frac{n\pi}{T}(\tau - t)\,d\tau \\ &= \sum_{-\infty}^{\infty}\frac{1}{2T}\int_{-T}^{T} f(\tau)\cos\frac{n\pi}{T}(\tau - t)\,d\tau \end{aligned} \tag{7-3}$$

If we set $n\pi/T = \lambda_n$ and $\pi/T = \lambda_{n+1} - \lambda_n = \delta\lambda_n$, then the process $T \to \infty$ formally yields the result

$$f(t) = \frac{1}{2\pi}\int_{-\infty}^{\infty} d\lambda \int_{-\infty}^{\infty} f(\tau)\cos\lambda(\tau - t)\,d\tau \tag{7-4}$$

which is a form of the *Fourier integral theorem*. Clearly, the symbol "=" in Eq. (7-4) can be expected to have as special an interpretation as it had in the Fourier series (7-1) from which Eq. (7-4) was derived.

Since $\cos\lambda(\tau - t)$ is even in λ, Eq. (7-4) can be written

$$f(t) = \frac{1}{\pi}\int_{0}^{\infty} d\lambda \int_{-\infty}^{\infty} f(\tau)\cos\lambda(\tau - t)\,d\tau \tag{7-5}$$

Also, since $\sin\lambda(\tau - t)$ is odd in λ, Eq. (7-4) is equivalent to

$$f(t) = \frac{1}{2\pi}\int_{-\infty}^{\infty} d\lambda \int_{-\infty}^{\infty} f(\tau)e^{i\lambda(\tau - t)}\,d\tau \tag{7-6}$$

The form of Eq. (7-6) suggests the definition of $F(\lambda)$, the *Fourier transform* of $f(t)$, as

$$F(\lambda) = \frac{1}{\sqrt{2\pi}}\int_{-\infty}^{\infty} e^{i\lambda\tau} f(\tau)\,d\tau \tag{7-7}$$

Equation (7-6) then provides the inversion formula

$$f(t) = \frac{1}{\sqrt{2\pi}}\int_{-\infty}^{\infty} e^{-i\lambda t} F(\lambda)\,d\lambda \tag{7-8}$$

Clearly, for a large class of functions $f(t)$, the transform $F(\lambda)$ is an analytic function of the complex variable λ; we shall give a formal proof of this property in Sec. 7-3. When needed, the forms (7-4), (7-5), and (7-6) can readily be recovered from Eqs. (7-7) and (7-8). Notice that Eqs. (7-7) and (7-8) are not quite symmetrical. Alternative definitions for $F(\lambda)$ are sometimes encountered; in fact, one of us ordinarily uses the definitions

$$F(\lambda) = \int_{-\infty}^{\infty} e^{-i\lambda t} f(t)\, dt$$
$$f(t) = \frac{1}{2\pi} \int_{-\infty}^{\infty} e^{i\lambda t} F(\lambda)\, d\lambda$$

but was outvoted insofar as this book is concerned.

Since Eq. (7-4) holds for the real part and for the imaginary part of a complex-valued $f(t)$, Eqs. (7-7) and (7-8) are not restricted to real-valued $f(t)$. For the special case of a real-valued $f(t)$, Eq. (7-7) gives

$$\begin{aligned} \operatorname{Re} F(\lambda) &= \frac{1}{\sqrt{2\pi}} \int_{-\infty}^{\infty} f(\tau) \cos \lambda\tau\, d\tau \\ \operatorname{Im} F(\lambda) &= \frac{1}{\sqrt{2\pi}} \int_{-\infty}^{\infty} f(\tau) \sin \lambda\tau\, d\tau \end{aligned} \tag{7-9}$$

These formulas are analogous to those for a_n and b_n in Eq. (7-2). Thus, in a sense, $F(\lambda)$ is equivalent to the combination $a_n + ib_n$, and $|F(\lambda)|$ corresponds to the amplitude $\sqrt{a_n^2 + b_n^2}$ of a frequency component in the spectral decomposition (7-1).

The foregoing derivation of the Fourier integral theorem is essentially that originally given by Fourier. It is, of course, not rigorous; in fact, we have said nothing at all about the conditions that $f(t)$ must satisfy. In the next section, we shall prove this theorem for certain $f(t)$, but we emphasize immediately that the class of $f(t)$ for which $F(\lambda)$ exists and for which a useful modification of Eq. (7-8) is correct is far more extensive than that covered by any theorem of this type.

Exercises

1. Use Eq. (7-7) to derive the following transform pairs, and verify Eq. (7-8) in each case. Here t, λ, a are real, with $a > 0$.

(*a*) $f(t) = \dfrac{\sin at}{t}$, $F(\lambda) = \sqrt{\dfrac{\pi}{2}}$ for $|\lambda| < a$
$= 0$ for $|\lambda| > a$

(*b*) $f(t) = (a^2 + t^2)^{-1}$, $F(\lambda) = \sqrt{\dfrac{\pi}{2}}\dfrac{e^{-a|\lambda|}}{a}$

(*c*) $f(t) = \exp(-at^2)$, $F(\lambda) = \dfrac{1}{\sqrt{2a}} \exp(-\lambda^2/4a)$

(Observe the special case $a = \frac{1}{2}$.)

(*d*) $f(t) = J_0(at)$, $F(\lambda) = \sqrt{\dfrac{2}{\pi}}\, (a^2 - \lambda^2)^{-\frac{1}{2}}$ for $|\lambda| < a$
$= 0$ for $|\lambda| > a$

2. (*a*) Let $f(t) = 1$ for $0 < t < 1$ and 0 otherwise. Find $F(\lambda)$, and evaluate $f(t)$ for the special cases $t = 0$, $t = 1$. (*Note:* Evaluate $\int_{-\infty}^{\infty}$ as $\lim_{T\to\infty} \int_{-T}^{T}$.) Are the results consistent with those obtained for similar situations in Fourier-series expansions?

(*b*) Interpreting the integral in Eq. (7-8) as $\lim_{T\to\infty} \int_{-T}^{T}$, substitute for $F(\lambda)$ in Eq. (7-8), and invert the order of integration to show heuristically that

$$f(t) = \lim_{T\to\infty} \frac{1}{\pi} \int_{-\infty}^{\infty} f(\tau) \frac{\sin T(t-\tau)}{t-\tau}\, d\tau$$

This result is called *Fourier's single integral formula.*

(*c*) For the function $f(t)$ described in Exercise 2*a*, show that

$$\frac{1}{\sqrt{2\pi}} \int_{-T}^{T} e^{-i\lambda t} F(\lambda)\, d\lambda = 1 - \frac{1}{\pi} \int_{T-Tt}^{\infty} \frac{\sin \xi}{\xi}\, d\xi - \frac{1}{\pi} \int_{tT}^{\infty} \frac{\sin \xi}{\xi}\, d\xi$$

Consider small values of $t > 0$. For large T, the second integral may be discarded; however, no matter how large T is, the last integral is an oscillatory function of t. Deduce that, as we consider any of a sequence of values of T tending to ∞, the above formula always gives an "overshoot" near $t = 0$ and that the magnitude of this overshoot is (essentially) independent of T. (Of course, as $T \to \infty$, the value of t at which the maximum overshoot occurs moves closer and closer to $t = 0$.) As in the case of the sum of a Fourier series near a point of discontinuity, this situation is referred to as *Gibbs's phenomenon.*

3. Let $f(t)$ be an analytic function of the complex variable t which, in particular, is regular on the interval (a,b) of the real axis. Define

$$F(\lambda) = \frac{1}{\sqrt{2\pi}} \int_a^b e^{i\lambda\tau} f(\tau)\, d\tau \qquad I(t) = \frac{1}{\sqrt{2\pi}} \int_{-\infty}^{\infty} e^{-i\lambda t} F(\lambda)\, d\lambda$$

By breaking the path of integration for $I(t)$ into the two halves $(-\infty,0)$ and $(0,\infty)$, inserting the definition of $F(\lambda)$, and replacing the path (a,b) by either an upper or lower half-plane path of integration, as appropriate, use contour integration to show that $I(t) = f(t)$ for $a < t < b$. What do we obtain for $I(t)$ if t is outside the range (a,b)? (The method is due to MacRobert,[1] who used it to derive a number of transform inversion theorems.)

[1] T. M. MacRobert, *Proc. Roy. Soc. Edinburgh*, *51*: 116 (1931).

A Proof of Fourier's Integral Theorem

We shall prove Eq. (7-4) for the case[1] in which $f(t)$ is a real piecewise smooth function, with $|f(t)|$ integrable from $-\infty$ to $+\infty$. ["Piecewise smooth" means that each of $f(t)$, $f'(t)$ is continuous in any finite interval, except at a finite number of points where $f(t)$ and $f'(t)$ may have bounded discontinuities.] Since the real and complex parts of a complex-valued function may be treated individually, the result applies at once to suitable complex $f(t)$. Furthermore, the result can be extended to include such functions $f(t)$ as $t^{-1/2}$, $t^{-1/3}$, etc.

We shall need the identity (cf. Exercise 2)

$$\lim_{L\to\infty} \int_{-T}^{T} f(t+\tau)\frac{\sin L\tau}{\tau}\,d\tau = \frac{\pi}{2}[f(t+0)+f(t-0)] \tag{7-10}$$

Here, T is any finite real positive number, and $f(t+0)$, $f(t-0)$ denote the limits of $f(t+\epsilon)$, $f(t-\epsilon)$, respectively, as $\epsilon\ (>0) \to 0$. To establish Eq. (7-10), we divide the range of integration into the two halves $(-T,0)$ and $(0,T)$ and write, for example,

$$\begin{aligned}\int_0^T f(t+\tau)\frac{\sin L\tau}{\tau}\,d\tau &= \int_0^T [f(t+\tau)-f(t+0)]\frac{\sin L\tau}{\tau}\,d\tau \\ &\quad + f(t+0)\int_0^T \frac{\sin L\tau}{\tau}\,d\tau\end{aligned} \tag{7-11}$$

Since $\int_T^\infty \sin L\tau\ (d\tau/\tau) \to 0$ as $L\to\infty$, we have

$$\lim_{L\to\infty} f(t+0)\int_0^T \frac{\sin L\tau}{\tau}\,d\tau = f(t+0)\int_0^\infty \frac{\sin\tau}{\tau}\,d\tau = \frac{\pi}{2}f(t+0)$$

In the first integral on the right-hand side of Eq. (7-11), the existence of $f'(t+0)$ implies that, given any $\epsilon>0$, a number h can be found such that

$$f(t+\tau)-f(t+0) = \tau[f'(t+0)+g(\tau)]$$

with $|g(\tau)|<\epsilon$ for $0<\tau<h$. Thus

$$\begin{aligned}\int_0^T [f(t+\tau)-f(t+0)]\frac{\sin L\tau}{\tau}\,d\tau &= \int_0^h [f'(t+0)+g(\tau)]\sin L\tau \\ &\quad + \int_h^T [f(t+\tau)-f(t+0)]\frac{\sin L\tau}{\tau}\,d\tau\end{aligned}$$

from which it is clear that the first integral on the right-hand side of Eq.

[1] Proofs of the Fourier integral theorem for functions $f(t)$ satisfying other conditions will be found in E. C. Titchmarsh, "Introduction to the Theory of Fourier Integrals," 2d ed., Oxford University Press, New York, 1948.

(7-11) tends to zero as $L \to \infty$. Thus

$$\lim_{L\to\infty} \int_0^T f(t+\tau) \frac{\sin L\tau}{\tau} \, d\tau = \frac{\pi}{2} f(t+0)$$

and a similar calculation for the $\int_{-T}^{0}$ half of Eq. (7-10) completes the proof.

We shall now prove that

$$\frac{1}{\pi} \lim_{L\to\infty} \int_0^L d\lambda \int_{-\infty}^{\infty} f(t+\tau) \cos \lambda\tau \, d\tau = \frac{1}{2} [f(t+0) + f(t-0)] \quad (7\text{-}12)$$

which is equivalent to Eq. (7-5). For $T_1 > T > 0$,

$$\int_0^L d\lambda \int_{-T_1}^{T_1} f(t+\tau) \cos \lambda\tau \, d\tau - \int_0^L d\lambda \int_{-T}^{T} f(t+\tau) \cos \lambda\tau \, d\tau$$
$$= \int_{-T_1}^{-T} f(t+\tau) \frac{\sin L\tau}{\tau} \, d\tau + \int_T^{T_1} f(t+\tau) \frac{\sin L\tau}{\tau} \, d\tau$$

and since $K = \int_{-\infty}^{\infty} |f(\tau)| \, d\tau$ exists, it follows that

$$\left| \int_0^L \int_{-T_1}^{T_1} - \int_0^L \int_{-T}^{T} \right| < \frac{K}{T}$$

Thus, in the limit $T_1 \to \infty$,

$$\left| \int_0^L \int_{-\infty}^{\infty} - \int_0^L \int_{-T}^{T} \right| < \frac{K}{T} \quad (7\text{-}13)$$

The second expression on the left-hand side of Eq. (7-13) can be written as

$$E = \int_{-T}^{T} f(t+\tau) \frac{\sin L\tau}{\tau} \, d\tau$$

and $E \to (\pi/2)[f(t+0) + f(t-0)]$ as $L \to \infty$, by Eq. (7-10). Thus, as $L \to \infty$, Eq. (7-13) gives

$$\left| \lim_{L\to\infty} \int_0^L d\lambda \int_{-\infty}^{\infty} f(t+\tau) \cos \lambda\tau \, d\tau - \frac{\pi}{2} [f(t+0) + f(t-0)] \right| < \frac{K}{T}$$

Since T is arbitrarily large, the validity of Eq. (7-12) has now been established.

Since $\int_{-\infty}^{\infty} |f(t)| \, dt$ exists and since $\sin \lambda(t-\tau)$ is odd in λ, we can also conclude that Eq. (7-8) is valid, provided, of course, that $f(t)$ is interpreted as $\frac{1}{2}[f(t+0) + f(t-0)]$.

Extension of Fourier's Integral Theorem

An illustration of the fact that the Fourier integral theorem is valid for many functions $g(t)$ which do not fall within the scope of the preceding proof is provided by the function

$$g(t) = \begin{cases} t^{-1/2} & t > 0 \\ 0 & t < 0 \end{cases} \tag{7-14}$$

The transform $G(\lambda)$ as given by Eq. (7-7) is

$$G(\lambda) = \frac{1}{2\sqrt{|\lambda|}}\left(1 + i\frac{\lambda}{|\lambda|}\right) \tag{7-15}$$

for λ real and nonzero. Use of Eq. (7-8) leads again to Eq. (7-14), so that the integral theorem is indeed valid for this $g(t)$.

The extension of the Fourier integral theorem to include certain other functions requires a modification of the inversion formula (7-8). For example, let

$$f(t) = \begin{cases} t^{1/2} & t > 0 \\ 0 & t < 0 \end{cases} \tag{7-16}$$

and note that the defining integral for its transform,

$$F(\lambda) = \frac{1}{\sqrt{2\pi}} \int_0^\infty e^{i\lambda t} t^{1/2}\, dt \tag{7-17}$$

converges only when $\operatorname{Im} \lambda > 0$. For such λ, we have

$$F(\lambda) = \frac{1}{2\sqrt{2}} e^{i\frac{3}{4}\pi} \lambda^{-3/2} \tag{7-18}$$

where $0 < \arg \lambda < \pi$, and by analytic continuation this definition can be extended to include all λ except $\lambda = 0$. If we write $\lambda = \lambda_1 + i\lambda_2$, where λ_1 and λ_2 are real, and hold $\lambda_2 > 0$ fixed, then $F(\lambda)$ is a function of λ_1 alone; define

$$G(\lambda_1) = F(\lambda_1 + i\lambda_2)$$

It is clear that $G(\lambda_1)$ is the transform, in the domain of real λ_1, of the function $g(t) = \exp(-\lambda_2 t)f(t)$. The inversion of $G(\lambda_1)$ is assured by the Fourier integral theorem, which gives

$$g(t) = \frac{1}{\sqrt{2\pi}} \int_{-\infty}^{\infty} \exp(-i\lambda_1 t) G(\lambda_1)\, d\lambda_1$$

Consequently

$$\begin{aligned} f(t) = \exp(\lambda_2 t) g(t) &= \frac{1}{\sqrt{2\pi}} \int_{-\infty}^{\infty} \exp[-i(\lambda_1 + i\lambda_2)t] F(\lambda_1 + i\lambda_2)\, d\lambda_1 \\ &= \frac{1}{\sqrt{2\pi}} \int_{-\infty + i\lambda_2}^{\infty + i\lambda_2} e^{-i\lambda t} F(\lambda)\, d\lambda \end{aligned}$$

Using Cauchy's integral theorem, we can write immediately

$$f(t) = \frac{1}{\sqrt{2\pi}} \int_{\Gamma_1} e^{-i\lambda t} F(\lambda)\, d\lambda \tag{7-19}$$

where Γ_1 is the real axis except for an indentation above the point $\lambda = 0$. The reader should verify that Eq. (7-19) does indeed lead to Eq. (7-16).

In the preceding example, the function $f(t)$ was identically zero for $t < 0$, and this is why the function $g(t) = \exp(-\lambda_2 t) f(t)$ satisfied the conditions under which we derived the Fourier integral theorem. As our next example, we choose a function $h(t)$ which is identically zero for $t > 0$. Let

$$h(t) = \begin{cases} 0 & t > 0 \\ t^{1/3} e^{i\alpha t} & t < 0 \end{cases} \tag{7-20}$$

where α is real and where the principal value of $t^{1/3}$ is to be taken. We have

$$H(\lambda) = \frac{1}{\sqrt{2\pi}} \int_{-\infty}^{0} e^{i(\lambda+\alpha)t} t^{1/3}\, dt \tag{7-21}$$

which converges for $\operatorname{Im} \lambda < 0$. By a similar process, we now establish that

$$h(t) = \frac{1}{\sqrt{2\pi}} \int_{\Gamma_2} e^{-i\lambda t} H(\lambda)\, d\lambda \tag{7-22}$$

where Γ_2 is the line $\operatorname{Im} \lambda = 0$, except for an indentation below the point $\lambda = -\alpha$.

Finally, we put these last two examples together so as to form a function $k(t)$ which is not identically zero in either of $t < 0$, $t > 0$. Define

$$k(t) = \begin{cases} t^{1/2} & t > 0 \\ t^{1/3} e^{i\alpha t} & t < 0 \end{cases} \tag{7-23}$$

The Fourier transform of $k(t)$ is

$$K(\lambda) = F(\lambda) + H(\lambda)$$

and

$$k(t) = \frac{1}{\sqrt{2\pi}} \int_{\Gamma} e^{-i\lambda t} [F(\lambda) + H(\lambda)]\, d\lambda \tag{7-24}$$

where Γ is any path which is asymptotic to $\operatorname{Im} \lambda = 0$ as $|\lambda| \to \infty$ and which passes above the origin and below the point $\lambda = -\alpha$.

It is clear that appropriate inversion formulas could be established rigorously, in much the same way, for a huge class of functions $f(t)$. However, to embark on a comprehensive effort of this sort is a rather sterile exercise. We shall see, in fact, when we encounter examples of the use of Fourier transforms, that the appropriate inversion path will be implicit in the full statement of any well-set problem to which transform methods are applicable. Even before we encounter these examples, however, one type of ambiguity should be identified. Suppose that we are asked to find the function $f(t)$ whose transform is $F(\lambda) = \lambda^{-1/2}$. Even if the inversion path is confined to the real axis, the lack of uniqueness in the specification of $F(\lambda)$ means that we cannot infer a unique function $f(t)$. Thus an observation which is of crucial importance is this: The statement of every well-set problem must include all information needed to allow a unique specification on the inversion path of the values of any apparently multivalued $F(\lambda)$ which arises in the analysis. This point will receive detailed exemplification later in this section. [The reader may be reminded at this point that he has already had an opportunity, in Exercise 26 of Sec. 3-1, to identify several functions and their transforms in cases where the domain of convergence of the definition of $F(\lambda)$ does not include real λ and where the inversion path is not the line $\text{Im}\,\lambda = 0$.]

EXERCISE

4. In Eq. (7-23), let α be complex; describe the new inversion contour for Eq. (7-24). What happens as $\alpha \to 0$?

Sine and Cosine Transforms

In many problems, the domain of interest of $f(t)$ is $t > 0$. In such problems, we are free to extend the definition of $f(t)$ to the range $t < 0$ in such a way that $f(t)$ is even; that is, $f(-t) = f(t)$. Equation (7-7) then shows that $F(\lambda)$ is also even, and Eqs. (7-7) and (7-8) become

$$F(\lambda) = \sqrt{\frac{2}{\pi}} \int_0^\infty (\cos \lambda\tau) f(\tau)\, d\tau$$

$$f(t) = \sqrt{\frac{2}{\pi}} \int_0^\infty (\cos \lambda t) F(\lambda)\, d\lambda$$

Such expressions are known as *Fourier cosine transforms.* We define the Fourier cosine transform[1] $F_c(\lambda)$ of $f(t)$ by

$$F_c(\lambda) = \sqrt{\frac{2}{\pi}} \int_0^\infty (\cos \lambda\tau) f(\tau)\, d\tau \tag{7-25}$$

and then from the result just proved we have the perfectly symmetrical inversion formula

$$f(t) = \sqrt{\frac{2}{\pi}} \int_0^\infty (\cos \lambda t) F_c(\lambda)\, d\lambda \tag{7-26}$$

Equation (7-25) involves values of $f(t)$ only for $t > 0$; Eq. (7-26) gives the same result for $f(-t)$ as for $f(t)$.

[1] In general, we shall try to be consistent in using a capital letter to denote a transform and the corresponding lower-case letter to denote the original function.

Similarly, by extending the definition of a function $f(t)$ given for $t > 0$ so that $f(-t) = -f(t)$, Eqs. (7-7) and (7-8) lead to the definition of the *Fourier sine transform* $F_s(\lambda)$,

$$F_s(\lambda) = \sqrt{\frac{2}{\pi}} \int_0^\infty (\sin \lambda\tau) f(\tau)\, d\tau \tag{7-27}$$

and to the inversion formula

$$f(t) = \sqrt{\frac{2}{\pi}} \int_0^\infty (\sin \lambda t) F_s(\lambda)\, d\lambda \tag{7-28}$$

Finally, if we define $g(t) = f(-t)$ for $t > 0$, then it is readily verified that, for an arbitrary function $f(t)$,

$$F(\lambda) = \frac{1}{2}[F_c(\lambda) + G_c(\lambda)] + \frac{i}{2}[F_s(\lambda) - G_s(\lambda)]$$

Tables of sine and cosine transforms are given by Erdélyi et al.,[1] and tables of Fourier transforms are given by Campbell and Foster.[2] Beware of the notation.

Exercises

5. Can a table of sine and cosine transforms be used to invert Fourier transforms?

6. If

$$f(t) = \begin{cases} (1 - t^2)^{\nu - \frac{1}{2}} & \text{for } t < 1 \\ 0 & \text{for } t > 1 \end{cases}$$

for $\nu > 0$, expand the cosine term in $F_c(\lambda)$, and use term-by-term integration to show that

$$F_c(\lambda) = 2^{\nu - \frac{1}{2}}\Gamma(\nu + \tfrac{1}{2})\lambda^{-\nu} J_\nu(\lambda)$$

Poisson Sum Formula

Let $f(t)$ be a continuous function defined for $t \geq 0$, and consider the series

$$S(a) = \sum_{n=1}^{\infty} f(na)$$

where a is an arbitrary positive constant. We confine our attention to such functions $f(t)$ that the series converges. We shall obtain a formula relating $S(a)$ to the sum of a similar series involving the cosine transform $F_c(\lambda)$ of

[1] A. Erdélyi, W. Magnus, F. Oberhettinger, and F. Tricomi, "Tables of Integral Transforms," vol. 4, McGraw-Hill Book Company, New York, 1954.

[2] G. Campbell and R. Foster, "Fourier Integrals for Practical Applications," D. Van Nostrand Company, Inc., Princeton, N.J., 1948.

$f(t)$. Consider first the partial sum $S_N(a)$,

$$S_N(a) = \sum_{n=1}^{N} f(na) = \sqrt{\frac{2}{\pi}} \sum_{n=1}^{N} \int_0^\infty F_c(\lambda) \cos na\lambda \, d\lambda$$

$$= \sqrt{\frac{2}{\pi}} \int_0^\infty F_c(\lambda) \, d\lambda \sum_{n=1}^{N} \cos na\lambda$$

$$= \sqrt{\frac{2}{\pi}} \int_0^\infty F_c(\lambda) \left[-\frac{1}{2} + \frac{\sin \lambda a(N + \frac{1}{2})}{2 \sin \frac{1}{2}\lambda a} \right] d\lambda$$

Consequently $$\tfrac{1}{2}f(0) + \sum_{n=1}^{N} f(na) = \sqrt{\frac{2}{\pi}} \int_0^\infty F_c(\lambda) \frac{\sin \lambda a(N + \frac{1}{2})}{2 \sin \frac{1}{2}\lambda a} \, d\lambda$$

As $N \to \infty$, the situation becomes that covered by Fourier's single-integral theorem of Exercise 2*b* and Eq. (7-10) (note that the denominator behaves like a linear function in the immediate neighborhood of each of its zeros). Use of the formula of Exercise 2*b*—or Eq. (7-10)—with due regard to the fact that at the origin there is only "half of a contribution," gives now

$$\sqrt{a} \left[\tfrac{1}{2}f(0) + \sum_{n=1}^{\infty} f(na) \right] = \sqrt{b} \left[\tfrac{1}{2}F_c(0) + \sum_{n=1}^{\infty} F_c(nb) \right] \tag{7-29}$$

where $ab = 2\pi$. This result is called the *Poisson sum formula*. It is interesting that only the mesh-point values of $f(t)$ and $F_c(\lambda)$ occur.

If $g(t)$ is a function defined and continuous for the whole interval $-\infty < t < \infty$, then, writing $g(t)$ as the sum of an even and an odd function and using Eq. (7-29), we find that an alternative form of Eq. (7-29) is

$$\sqrt{a} \sum_{n=-\infty}^{\infty} g(na) = \sqrt{b} \sum_{n=-\infty}^{\infty} G(nb) \tag{7-30}$$

where $ab = 2\pi$ as before and where $G(\lambda)$ is the Fourier transform of $g(t)$. Again, we confine our attention to those $g(t)$ for which these series converge.

It is also possible to obtain a formula involving sine transforms; the reader may show that, if $f(t)$ is continuous for $t \geq 0$ and if $F_s(\lambda)$ is the sine transform of $f(t)$, then

$$\sqrt{a} \, [f(a) - f(3a) + f(5a) - \cdots] = \sqrt{b} \, [F_s(b) - F_s(3b) + F_s(5b) - \cdots] \tag{7-31}$$

where now $ab = \pi/2$.

These various formulas are often used for the summation of series; thus, for example, Eq. (7-29) and the result of Exercise 6 can be used to obtain a closed-form expression for the sum of the infinite series $\Sigma(nb)^{-\nu}J_\nu(nb)$. Even if the Poisson sum formula does not produce a series which is more easily summed in finite terms than the original series, it may still happen that the new series converges more rapidly than the original one.

The reader may find it a useful exercise to use the Poisson sum formula to verify some of the series summations obtained by residue methods in Chap. 3.

Transform of a Derivative

Transform methods are most commonly used for solving ordinary and partial linear differential equations; their usefulness arises from the fact that the transform of a derivative is simply related to the transform of the original function. If we confine our attention to functions $f(t)$ which vanish as $|t| \to \infty$ (we shall modify this restriction in a later section), then the Fourier transform of $f'(t)$, which we denote by $F_1(\lambda)$, is given by

$$\begin{aligned} F_1(\lambda) &= \frac{1}{\sqrt{2\pi}} \int_{-\infty}^{\infty} e^{i\lambda t} f'(t)\, dt \\ &= \frac{1}{\sqrt{2\pi}} \left\{ [e^{i\lambda t} f(t)]_{\infty}^{-\infty} - i\lambda \int_{-\infty}^{\infty} e^{i\lambda t} f(t)\, dt \right\} \end{aligned} \tag{7-32}$$

following an integration by parts. Thus

$$F_1(\lambda) = -i\lambda F(\lambda)$$

Similarly, denoting the transform of the nth derivative of $f(t)$ by $F_n(\lambda)$ and assuming that all derivatives up to the $(n-1)$st of $f(t)$ vanish at $\pm\infty$, we obtain

$$F_n(\lambda) = (-i\lambda)^n F(\lambda) \tag{7-33}$$

The following example illustrates the use of this result. The equation for one-dimensional heat conduction along an indefinitely long insulated rod is

$$\varphi_t = \alpha \varphi_{xx} \tag{7-34}$$

in $-\infty < x < \infty$ and $t > 0$. Subscripts indicate partial derivatives, and α is a positive constant. We seek to determine the temperature, $\varphi(x,t)$, when $\varphi(x,0) = g(x)$ is given.

We define the Fourier transform of φ with respect to x by

$$\Phi(\lambda,t) = \frac{1}{\sqrt{2\pi}} \int_{-\infty}^{\infty} e^{i\lambda x} \varphi(x,t)\, dx \tag{7-35}$$

Multiplying both sides of Eq. (7-21) by $(1/\sqrt{2\pi})e^{i\lambda x}$, integrating with respect to x from $-\infty$ to ∞ (in future we shall describe such a process as "taking the transform of the equation with respect to x"), and using Eq. (7-33), we obtain

$$\frac{\partial \Phi}{\partial t} = \alpha(-i\lambda)^2 \Phi \tag{7-36}$$

On the left-hand side, we have interchanged the order of differentiation with respect to t and integration with respect to x. [Since we do not yet know what $\Phi(\lambda,t)$ will be, this interchange of order is necessarily based on optimism rather than on a justifying theorem. It is characteristic of the use of transform methods that such heuristically motivated steps are necessary and that the final justification consists in an a posteriori "checking of the answer." In the case of the present problem—which is a well-posed one with a unique solution—if the function $\varphi(x,t)$ which we finally construct obeys all the constraints of the problem, we have reached our objective and no further justification is necessary.]

The use of the Fourier transform has enabled us to replace a problem in which we must solve a partial differential equation in two variables by one in which, in essence, we are confronted with an ordinary differential equation in which λ plays the role of a parameter. The solution of Eq. (7-36) is

$$\Phi(\lambda,t) = A(\lambda) \exp(-\alpha\lambda^2 t) \tag{7-37}$$

where $A(\lambda)$ is a function of λ to be determined from the initial condition. For $t = 0$, Eq. (7-37) reduces to

$$A(\lambda) = \Phi(\lambda,0) = \frac{1}{\sqrt{2\pi}} \int_{-\infty}^{\infty} e^{i\lambda x} g(x)\, dx \tag{7-38}$$

Thus, from the known initial temperature distribution, $A(\lambda)$ can be determined, and hence $\varphi(x,t)$ can be obtained from an application of the Fourier inversion theorem to Eq. (7-37).

As a particular example, let $g(x) = B \exp(-\beta x^2)$, where B and β are real positive constants. Then Eq. (7-38) gives

$$A(\lambda) = \frac{B}{\sqrt{2\beta}} \exp\left(-\frac{\lambda^2}{4\beta}\right)$$

and we have

$$\begin{aligned}\varphi(x,t) &= \frac{B}{2\sqrt{\beta\pi}} \int_{-\infty}^{\infty} \exp\left(-i\lambda x - \lambda^2\left[\frac{1}{4\beta} + \alpha t\right]\right) d\lambda \\ &= \frac{B}{\sqrt{1 + 4\alpha\beta t}} \exp\left(-\frac{x^2\beta}{1 + 4\alpha\beta t}\right)\end{aligned} \tag{7-39}$$

Direct substitution verifies that this is indeed a solution of Eq. (7-34).

We turn now to the sine and cosine transforms of the nth derivative of $f(t)$ and denote them by $F_{sn}(\lambda)$ and $F_{cn}(\lambda)$, respectively. For functions $f(t)$ for which no contributions arise at $t \to \infty$ from the integrations by parts, we obtain

$$\begin{aligned} F_{s1}(\lambda) &= -\lambda F_c(\lambda) & F_{c1}(\lambda) &= -\sqrt{\frac{2}{\pi}} f(0) + \lambda F_s(\lambda) \\ F_{s2}(\lambda) &= \sqrt{\frac{2}{\pi}}\, \lambda f(0) - \lambda^2 F_s(\lambda) & F_{c2}(\lambda) &= -\sqrt{\frac{2}{\pi}} f'(0) - \lambda^2 F_c(\lambda) \\ &\cdots & &\cdots \end{aligned} \tag{7-40}$$

Since $F_s(\lambda)$ and $F_c(\lambda)$ depend only on values of $f(t)$ for $t > 0$, they are potentially suitable for problems in which one variable has a semi-infinite (rather than a fully infinite) range.

Consider the heat-conduction problem for a semi-infinite rod which occupies the interval $x \geq 0$. In order to have a meaningful physical problem and, hence, a well-set mathematical problem, we must specify not only the initial temperature along the rod but also a boundary condition at $x = 0$. The sine transform of Eq. (7-34) with respect to x is

$$\frac{\partial \Phi_s}{\partial t} = \alpha \sqrt{\frac{2}{\pi}} \lambda \varphi(0,t) - \alpha \lambda^2 \Phi_s \tag{7-41}$$

where
$$\Phi_s(\lambda,t) = \sqrt{\frac{2}{\pi}} \int_{-\infty}^{\infty} (\sin \lambda x) \varphi(x,t)\, dx$$

We can solve Eq. (7-41) for $\Phi_s(\lambda,t)$ only if $\varphi(0,t)$ is prescribed. Thus a sine transform would be useful if the boundary condition at $x = 0$ consisted in a prescription of the temperature $\varphi(0,t)$, for all t. On the other hand, the cosine transform of Eq. (7-34) is

$$\frac{\partial \Phi_c}{\partial t} = -\alpha \sqrt{\frac{2}{\pi}} \varphi_x(0,t) - \alpha \lambda^2 \Phi_c \tag{7-42}$$

Thus a cosine transform is useful if $\varphi_x(0,t)$ is given—i.e., if the boundary condition at $x = 0$ is one which prescribes the thermal gradient (or equivalently the rate of heat efflux from the end of the rod).

As an example of the use of Eq. (7-41), let $\varphi(0,t) = 1$ for $t > 0$ and $\varphi(x,0) = 0$ for $x > 0$. Then Eq. (7-41) leads to

$$\Phi_s = \sqrt{\frac{2}{\pi}} \frac{1 - \exp(-\alpha \lambda^2 t)}{\lambda}$$

where the condition $\Phi_s(\lambda,0) = 0$ has been used. Inversion yields

$$\begin{aligned} \varphi(x,t) &= 1 - \frac{2}{\sqrt{\pi}} \int_0^{x/2\sqrt{\alpha t}} \exp(-\eta^2)\, d\eta \\ &= \operatorname{erfc} \frac{x}{2\sqrt{\alpha t}} \end{aligned} \tag{7-43}$$

Again, substitution into Eq. (7-34) shows that $\varphi(x,t)$ as given by Eq. (7-43) satisfies the differential equation.

A somewhat different example is that of two-dimensional steady-state heat conduction in a metal plate occupying the upper half of the xy plane. The temperature $\varphi(x,y)$ satisfies the equation

$$\varphi_{xx} + \varphi_{yy} = 0 \tag{7-44}$$

in $-\infty < x < \infty$, $y > 0$. We denote the prescribed temperature on $y = 0$ by $\varphi(x,0)$. The Fourier transform in x of Eq. (7-44) is

$$-\lambda^2\Phi(\lambda,y) + \frac{\partial^2}{\partial y^2}\Phi(\lambda,y) = 0 \tag{7-45}$$

which has the solution

$$\Phi(\lambda,y) = A(\lambda)e^{\lambda y} + B(\lambda)e^{-\lambda y} \tag{7-46}$$

where $A(\lambda)$ and $B(\lambda)$ are as yet unknown functions of λ. The edge condition at $y = 0$ is not enough to determine both $A(\lambda)$ and $B(\lambda)$; we must also impose the condition that $\varphi(x,y)$ be bounded as $y \to \infty$ [we require the given edge temperature $\varphi(x,0)$ to be bounded]. The transform of the temperature cannot grow exponentially as $y \to \infty$; since $A(\lambda)$ and $B(\lambda)$ do not depend on y, the only way to avoid such an exponential growth in $\Phi(\lambda,y)$ is to take $A(\lambda) = 0$ for $\lambda > 0$ and $B(\lambda) = 0$ for $\lambda < 0$. Equation (7-46) then takes the form

$$\Phi(\lambda,y) = C(\lambda)e^{-|\lambda|y} \tag{7-47}$$

and we could find the unknown function $C(\lambda)$ as the transform of $\varphi(x,0)$. Fourier inversion then gives $\varphi(x,y)$.

We note that Eq. (7-47) could have been obtained more directly by writing the solution of Eq. (7-45) in the form

$$\Phi(\lambda,y) = C(\lambda)\exp[-(\lambda^2)^{1/2}y] + D(\lambda)\exp[(\lambda^2)^{1/2}y] \tag{7-48}$$

where $(\lambda^2)^{1/2}$ may be interpreted as the multiple-valued function defined by

$$(\lambda^2)^{1/2} = \lim_{\epsilon\to 0}(\lambda^2 + \epsilon^2)^{1/2} \tag{7-49}$$

For definiteness, we take the branch cuts to lie along the imaginary axis portions $|\text{Im}\,\lambda| > \epsilon$ and take $(\lambda^2 + \epsilon^2)^{1/2}$ as > 0 for λ real and > 0 (and so also > 0 for λ real and < 0). Then a consideration of the limit $y \to \infty$ shows that we must set $D(\lambda) = 0$, and we recover Eq. (7-47).

Alternatively, Eq. (7-44) can be solved by taking a sine transform in y, to give

$$\frac{\partial^2}{\partial x^2}\Phi_s(x,\lambda) + \sqrt{\frac{2}{\pi}}\,\lambda\varphi(x,0) - \lambda^2\Phi_s(x,\lambda) = 0 \tag{7-50}$$

The solution of this equation is

$$\Phi_s(x,\lambda) = e^{\lambda x}\left[A(\lambda) - \frac{1}{\sqrt{2\pi}}\int_0^x \varphi(\xi,0)e^{-\lambda\xi}\,d\xi\right] + e^{-\lambda x}\left[B(\lambda) + \frac{1}{\sqrt{2\pi}}\int_0^x \varphi(\xi,0)e^{\lambda\xi}\,d\xi\right] \tag{7-51}$$

Since $\Phi_s(x,\lambda)$ cannot grow exponentially as $|x| \to \infty$, we clearly must choose

$$A(\lambda) = \frac{1}{\sqrt{2\pi}} \int_0^\infty \varphi(\xi,0)e^{-\lambda\xi}\, d\xi \qquad B(\lambda) = -\frac{1}{\sqrt{2\pi}} \int_0^{-\infty} \Phi(\xi,0)e^{\lambda\xi}\, d\xi$$

so that Eq. (7-51) becomes

$$\sqrt{2\pi}\, \Phi_s(x,\lambda) = \int_x^\infty \varphi(\xi,0)e^{\lambda(x-\xi)}\, d\xi + \int_{-\infty}^x \varphi(\xi,0)e^{\lambda(\xi-x)}\, d\xi \tag{7-52}$$

The inversion formula for the sine transform may now be used to find $\varphi(x,y)$. We can in fact do this quite generally; if we multiply Eq. (7-52) through by $\sin \lambda y$ and integrate with respect to λ from 0 to ∞, we obtain

$$\varphi(x,y) = \frac{1}{\pi} \int_{-\infty}^\infty \varphi(\xi,0) \frac{y}{y^2 + (x-\xi)^2}\, d\xi \tag{7-53}$$

as the general solution of the half-plane thermal-conduction problem. As a check, the laplacian of $y/[y^2 + (x-\xi)^2]$ vanishes for $y > 0$; so Eq. (7-44) is satisfied. To verify that the edge condition at $y = 0$ is also satisfied is not quite so easy; the function $y/[y^2 + (x-\xi)^2] \to 0$ as $y \to 0$ except at the point $\xi = x$, where it grows without limit as $y \to 0$. We shall see later that the limit of this function as $y \to 0$ acts like a unit-impulse function (or delta function) and that the right-hand side of Eq. (7-53) does indeed reduce to $\varphi(x,0)$ as $y \to 0$.

Although the operation of taking a Fourier transform is basically a linear one, so that there is no general formula giving the transform of the product of two functions in terms of their individual transforms, it is sometimes possible to circumvent this restriction—particularly if one of the functions is simply a power of x. For example, let the Fourier transform of $y(x)$ be $Y(\lambda)$. Then the transform of $x^n y(x)$ is given by

$$\begin{aligned}\frac{1}{\sqrt{2\pi}} \int_{-\infty}^\infty e^{i\lambda x} x^n y(x)\, dx &= \left(\frac{1}{i}\right)^n \frac{d^n}{d\lambda^n} \left[\frac{1}{\sqrt{2\pi}} \int_{-\infty}^\infty e^{i\lambda x} y(x)\, dx\right] \\ &= (-i)^n \frac{d^n Y(\lambda)}{d\lambda^n}\end{aligned} \tag{7-54}$$

for those functions $y(x)$ for which the above operations exist. As an example of the use of this result, let $y(x)$ satisfy Bessel's equation of order zero, so that

$$xy_{xx} + y_x + xy = 0$$

Taking transforms, we have

$$-i(-\lambda^2 Y)' - i\lambda Y - iY' = 0$$

where a prime denotes $d/d\lambda$. The solution of this first-order differential equation is

$$Y(\lambda) = (1 - \lambda^2)^{-1/2}$$

within an arbitrary multiplicative constant. The inversion formula provides, directly, an integral representation of $J_0(x)$.

EXERCISES

7. (*a*) Let B and β be real positive constants. Show that the solution of Eq. (7-34) in $-\infty < x < \infty, t > 0$, with $\varphi(x,0) = B(x^2 + \beta^2)^{-1}$, is

$$\varphi = \frac{B\sqrt{\pi}}{2\beta\sqrt{\alpha t}} \exp\left(\frac{\beta^2 - x^2}{4\alpha t}\right) \text{Re}\left[\exp\left(\frac{ix\beta}{2\alpha t}\right) \text{erfc}\left(\frac{\beta + ix}{2\sqrt{\alpha t}}\right)\right]$$

and use an asymptotic expression for the erfc function to verify that $\varphi(x,t)$ attains the correct initial value as $t \to 0$.

(*b*) Let $\varphi(x,0) = Be^{-\beta x}$, where B and β are real positive constants. Use sine or cosine transforms to solve Eq. (7-34) in $0 < x < \infty, t > 0$ for each of the two end conditions, (*a*) $\varphi(0,t) = 0$; (*b*) $\varphi_x(0,t) = 0$.

8. Solve the heat-conduction problem for the semi-infinite rod $(0 < x < \infty)$ with the boundary condition

$$p\varphi(0,t) + q\varphi_x(0,t) = g(t)$$

where p, q are constants and $g(t)$ is a given function of t. [*Hint:* Define $\psi(x,t) = p\varphi(x,t) + q\varphi_x(x,t)$, and observe that ψ satisfies Eq. (7-34). An equivalent process is to begin with an appropriate linear combination of sine and cosine transforms. We shall subsequently find, however, that the Laplace-transform method is more direct.]

For the special case $\varphi(x,0) = Be^{-x}$, where B is a positive real constant, and $\varphi(0,t) + \varphi_x(0,t) = 0$, use transforms to show that the solution of Eq. (7-44) is $\varphi = Be^{-x+\alpha t}$.

9. Obtain the result (7-53) from Eq. (7-47). [First express $C(\lambda)$ in terms of $\varphi(\xi,0)$, then multiply both sides of Eq. (7-47) by $e^{-i\lambda x}$, and integrate with respect to λ from $-\infty$ to $+\infty$.]

10. A strip of metal occupies the region $-b \leq y \leq b, -\infty < x < \infty$. An imposed temperature profile $\theta(x)$ (above some reference level) is a function of x alone. As a consequence of this temperature distribution, certain stresses are set up in the strip; the stresses are given by the various second derivatives of a function $\varphi(x,y)$ satisfying the equation[1]

$$\varphi_{xxxx} + 2\varphi_{xxyy} + \varphi_{yyyy} = -E\alpha\theta_{xx}$$

where the constants α and E are the linear coefficient of thermal expansion and the elastic modulus, respectively. The boundary conditions on φ are $\varphi(x, \pm b) = \varphi_y(x, \pm b) = 0$. Let $\theta = \theta_0 e^{\epsilon x}$ for $x < 0$ and $\theta = 0$ for $x > 0$, where ϵ is a small positive constant (eventually

[1] See, for example, B. A. Boley and J. H. Weiner, "Theory of Thermal Stresses," chap. 4, John Wiley & Sons, Inc., New York, 1960.

allowed to approach zero). Take a Fourier transform in x to show that

$$\varphi = -\frac{E\alpha\theta_0}{\pi}\int_{-\infty}^{\infty} e^{-i\lambda x}\frac{1}{i\lambda^3}\left[A\cosh\lambda y + B\lambda y\sinh\lambda y - \frac{1}{2}\right]d\lambda$$

where
$$A = \frac{\sinh b\lambda + \lambda b(\cosh b\lambda)}{2\lambda b + \sinh 2\lambda b}$$

and
$$B = -\frac{\sinh b\lambda}{2\lambda b + \sinh 2\lambda b}$$

Obtain the explicit form of the longitudinal stress given by $\tau = \varphi_{yy}$.

To find numerical values of τ, either we can use direct numerical integration—often a tedious process with transform formulas, because of the oscillating integrands—or we can try to sum the residues. Determine at least approximately where the poles of the denominator are. What curve is asymptotic to their locations? If we approximate the denominator by $\sinh 2\lambda b$ alone, the residues may be summed explicitly; show that this leads to

$$\frac{\tau}{E\alpha\theta_0} = \frac{e^{\pi x/b} + \cos(\pi y/b)}{2\cos(\pi y/b) + 2\cosh(\pi x/b)} - 1 + \frac{1}{\pi}\arctan\frac{2e^{\pi x/2b}\cos(\pi y/2b)}{e^{\pi x/b} - 1} - \frac{y}{2b}\left[\frac{\sinh(\pi x/2b)\sin(\pi y/2b)}{\cos^2(\pi y/2b) + \sinh^2(\pi x/2b)}\right]$$

for the case $x < 0$. (The function arctan is an angle in the second quadrant.) By examining the difference between a typical exact and approximate residue, show how correction terms could be obtained, and comment on their magnitudes.

11. Let $\varphi(x,t)$ satisfy Eq. (7-34) in $0 < x < \infty, 0 < t < \infty$. Equations (7-41) and (7-42) were obtained by taking transforms in x. Since the range of the t variable is also semi-infinite, it is alternatively possible to take sine or cosine transforms in t; do so, and show how to combine the two results so as to solve the problem. In addition to the specification of $\varphi(x,0)$ and a boundary condition at $x = 0$, what condition at $x \to 0$ must be imposed?

The Delta Function

In Eq. (7-39), we saw that the solution of $\varphi_t = \alpha\varphi_{xx}$ in $-\infty < x < \infty$, $t > 0$, satisfying the initial condition

$$\varphi(x,0) = B\exp(-\beta x^2) \tag{7-55}$$

was given by

$$\varphi(x,t) = \frac{B}{\sqrt{1 + 4\alpha\beta t}}\exp\left(-\frac{x^2\beta}{1 + 4\alpha\beta t}\right) \tag{7-56}$$

Let us now set $B = (\beta/\pi)^{1/2}$, so that $\int_{-\infty}^{\infty} \varphi(x,0)\,dx = 1$ for all $\beta > 0$. (Thus, for any choice of β, a unit quantity of heat is initially placed in the bar.) If β is made to increase, the initial temperature distribution becomes more and more concentrated near the origin; as $\beta \to \infty$, this initial temperature distribution becomes a unit "impulse" function, called a *delta function*, $\delta(x)$. Thus,

$$\begin{aligned}\delta(x) &\overset{?}{=} \lim_{\beta\to\infty}\left[\left(\frac{\beta}{\pi}\right)^{1/2}\exp\left(-\beta x^2\right)\right]\\ &\overset{?}{=} \lim_{\epsilon\to 0}\left[(\pi\epsilon)^{-1/2}\exp\left(\frac{-x^2}{\epsilon}\right)\right]\end{aligned} \tag{7-57}$$

One might loosely describe $\delta(x)$ as a function which is zero everywhere except at $x = 0$, where it is infinite, and which has unit area under its graph. Of course, $\delta(x)$ is not a conventional mathematical function, and this fact is indicated by the notation $\overset{?}{=}$ in Eq. (7-57). However, difficulties of definition and usage can be avoided by not proceeding to the limit $\beta = \infty$ or $\epsilon = 0$ until after the completion of whatever mathematical process is to be applied to $\delta(x)$.

In our heat-conduction problem, for example, in order to determine the value of $\varphi(x,t)$ for $t > 0$ corresponding to $\varphi(x,0) = \delta(x)$, we would first obtain the solution (7-56) for $\beta \neq \infty$, and then [with $B = (\beta/\pi)^{1/2}$] require $\beta \to \infty$. This gives

$$\varphi(x,t) = \frac{1}{\sqrt{4\pi\alpha t}}\exp\left(-\frac{x^2}{4\alpha t}\right) \tag{7-58}$$

Similarly, to evaluate an expression such as $\int_{a-b}^{a+b} f(x)\,\delta(x-a)\,dx$ (where a and b are real, with $b > 0$), we first replace $\delta(x-a)$ by $(\pi\epsilon)^{1/2}$ $\exp\left[-(x-a)^2/\epsilon\right]$, carry out the integration, and then allow $\epsilon \to 0$. If $f(x)$ is continuous at $x = a$, it is clear that this process must yield

$$\int_{a-b}^{a+b} f(x)\,\delta(x-a)\,dx = f(a) \tag{7-59}$$

Thus the delta function is here introduced as *a symbolic representation for a deferred limiting process.* As special cases of Eq. (7-59), we have (for $n = 1, 2, 3, \ldots$)

$$\int_{a-b}^{a+b} \delta(x-a)\,dx = 1 \qquad \int_{a-b}^{a+b} x^n\,\delta(x-a)\,dx = a^n$$

Many functions other than the one in Eq. (7-57) may be used for a limiting representation of $\delta(x)$. Other possibilities are

$$\delta(x) \overset{?}{=} \lim_{\epsilon\to 0}\begin{bmatrix}\dfrac{1}{2\epsilon} & |x| < \epsilon\\ 0 & |x| > \epsilon\end{bmatrix} \tag{7-60}$$

$$\delta(x) \overset{?}{=} \lim_{\epsilon\to 0}\left[\frac{\epsilon}{\pi(\epsilon^2 + x^2)}\right] \tag{7-61}$$

For any $\epsilon > 0$, the expressions within brackets in Eqs. (7-57), (7-60), and (7-61) all have unit area under their graphs; as $\epsilon \to 0$, each expression $\to 0$ for $x \neq 0$ and $\to \infty$ for $x = 0$.

The delta function can be discussed within a more elegant framework,[1] but such a discussion would be only a digression here. The simple device of deferring the limiting process $\epsilon \to 0$ until the integration relevant to the problem in question has been performed [and such an integration always arises, at least implicitly, in any meaningful use of $\delta(x)$] will be entirely adequate for our purposes. The reader should prove the following results, in which a and b are any real numbers:

1. $\delta(-t) = \delta(t)$
2. $\delta(at) = \dfrac{1}{|a|}\,\delta(t)$
3. $t\delta(t) = 0$
4. $\delta(t^2 - a^2) = \dfrac{1}{2|a|}\,[\delta(t + a) + \delta(t - a)]$
5. $\int_{-\infty}^{\infty} \delta(t - a)\,\delta(t - b)\,dt = \delta(a - b)$
6. $\delta(t) = -t\delta'(t)$

In (6), $\delta'(t)$ denotes the derivative of $\delta(t)$, defined by an analogous limiting process; for example,

$$\delta'(t) \overset{?}{=} \lim_{\epsilon \to 0} \frac{d}{dt} \frac{\epsilon}{\pi(\epsilon^2 + t^2)} \tag{7-62}$$

By using Eq. (7-59), the Fourier transform of the function

$$g(t) = \delta(t - a)$$

(with a real) is

$$G(\lambda) = \frac{1}{\sqrt{2\pi}}\, e^{i\lambda a} \tag{7-63}$$

If we encounter any difficulties in using Eq. (7-63), then, in accordance with our interpretation of the delta function, we merely go back a step and replace $g(t)$ by (for example)

$$g_\epsilon(t) = \frac{1}{\sqrt{\pi\epsilon}} \exp\left[-\frac{(x - a)^2}{\epsilon}\right]$$

which leads to

$$G_\epsilon(\lambda) = \frac{1}{\sqrt{2\pi}} \exp\left(i\lambda a - \tfrac{1}{2}\epsilon\lambda^2\right) \tag{7-64}$$

[1] See L. Schwarz, "Théorie des distributions," Hermann & Cie, Paris, 1950; N. Wiener, *Acta Math.*, **55:** 117 (1930); M. J. Lighthill, "An Introduction to Fourier Analysis and Generalized Functions," Cambridge University Press, New York, 1959.

We observe that, with this interpretation, we can apply the Fourier inversion formula (7-8) to Eq. (7-63) to obtain

$$\frac{1}{2\pi}\int_{-\infty}^{\infty} e^{i\lambda(a-t)}\, d\lambda = \delta(t-a)$$

$\left[\text{The formula } \int_{-\infty}^{\infty} e^{i\lambda t}\, d\lambda = 2\pi\delta(t) \text{ is a convenient one to remember.}\right]$

Consider again the problem $\varphi_t = \alpha\varphi_{xx}$ in $-\infty < x < \infty$, $t > 0$, with $\varphi(x,0) = \delta(x-a)$. As in the process leading to Eq. (7-37), a Fourier transform in x of the differential equation yields

$$\Phi(\lambda,t) = A(\lambda)\exp(-\alpha\lambda^2 t)$$

where $A(\lambda)$ is now the transform of $\delta(x-a)$, that is, $(1/\sqrt{2\pi})e^{ia\lambda}$. Inversion yields Eq. (7-58) directly; we see that, with suitable precautions where necessary, the use of the delta function and its transform can result in a convenient brevity in calculating the effect of a concentrated source. We might also remark that, since the heat-conduction problem is linear and since any initial temperature distribution $\varphi(x,0)$ can be written as

$$\varphi(x,0) = \int_{-\infty}^{\infty} \varphi(\xi,0)\delta(\xi - x)\, d\xi$$

superposition can be used to express the general solution of the heat-conduction problem as

$$\varphi(x,t) = \int_{-\infty}^{\infty} \varphi(\xi,0)\frac{1}{\sqrt{4\pi\alpha t}}\exp\left[-\frac{(x-\xi)^2}{4\alpha t}\right] d\xi \tag{7-65}$$

[where we have used Eq. (7-58)]. We shall see later that Eq. (7-65) could also be obtained by use of the convolution theorem for Fourier transforms.

Exercises

12. Explain the sense in which one can write

(*a*) $g(t) = t \Rightarrow G(\lambda) = -i\sqrt{2\pi}\,\delta'(\lambda)$

(*b*) $g(t) = \sin\omega t \Rightarrow G(\lambda) = \frac{1}{2i}\sqrt{2\pi}\,[\delta(\omega+\lambda) - \delta(\omega-\lambda)]$

(*c*) $g(t) = \delta'(t) \Rightarrow G(\lambda) = (-i\lambda)$ [transform of $\delta(t)$]

13. Show that, for a value of t at which $f(t)$ is continuous, the Fourier integral theorem

$$\frac{1}{2\pi}\int_{-\infty}^{\infty} e^{-i\lambda t}\, d\lambda \int_{-\infty}^{\infty} e^{i\lambda\tau} f(\tau)\, d\tau = f(t)$$

can be "proved" by changing the order of integration.

A function closely related to $\delta(t)$ is the *Heaviside unit function*[1] $h(t)$, defined by

$$h(t) = \begin{cases} 0 & \text{for } t < 0 \\ \frac{1}{2} & \text{for } t = 0 \\ 1 & \text{for } t > 0 \end{cases} \tag{7-66}$$

The value $h(0) = \frac{1}{2}$ is specified only for completeness; we could equally well leave $h(0)$ undefined. Using when necessary an $\epsilon \to 0$ interpretation, we can write

$$h(t) = \int_{-\infty}^{t} \delta(\tau)\, d\tau \qquad h'(t) = \delta(t) \tag{7-67}$$

The defining integral for the Fourier transform $H(\lambda)$ converges for Im $\lambda > 0$ to

$$H(\lambda) = -\frac{1}{\sqrt{2\pi}}\frac{1}{i\lambda} \tag{7-68}$$

By analytic continuation, Eq. (7-68) provides a description of $H(\lambda)$ valid for all λ except at $\lambda = 0$, where there is a pole. The inversion formula (7-8) reproduces $h(t)$ if the path of integration is indented above $\lambda = 0$.

To obtain the transform $H_1(\lambda)$ of $h'(t)$, we can replace $h(t)$ by a continuously differentiable function $h_\epsilon(t)$ which $\to h(t)$ as $\epsilon \to 0$; since the transform of $h'_\epsilon(t)$ is $-i\lambda$ times the transform of $h_\epsilon(t)$, it is appropriate to write

$$H_1(\lambda) = -i\lambda\left(-\frac{1}{\sqrt{2\pi}}\frac{1}{i\lambda}\right) = \frac{1}{\sqrt{2\pi}} \tag{7-69}$$

Equation (7-69) is also consistent with the result of taking the transform of $\delta(t) = h'(t)$. Observe that had we replaced $h'(t)$ by zero—which is correct for all $t \neq 0$—we would have obtained the incorrect result $H_1(\lambda) = 0$.

Let us return now to the heat-conduction problem

$$\varphi_t = \alpha\varphi_{xx} \tag{7-70}$$

for $-\infty < x < \infty$, $t > 0$, but now with the initial condition

$$\varphi(x,0) = \begin{cases} 2 & x > 0 \\ 0 & x < 0 \end{cases} \tag{7-71}$$

As before, we could find $\varphi(x,t)$ by taking a Fourier transform in x; to bring $\varphi(x,0)$ within the domain of functions whose transforms exist, we could either require Im $\lambda > 0$ or replace the value 2 for $x > 0$ by $2e^{-\epsilon x}$, with $\epsilon \to 0$ in the final solution. However, let us use this problem to furnish a rudimentary example of the way in which one deals with a multivalued transform—and also to clarify a seldom discussed uniqueness question. We begin by arbitrarily setting $\varphi(x,t) = 0$ for $t < 0$; as long as we do not

[1] More common notations are $H(t)$ or 1. We reserve capital letters for transforms.

try to apply the differential equation to the region $t \leq 0$, this device involves no loss in generality. Define

$$\Phi(x,\lambda) = \frac{1}{\sqrt{2\pi}} \int_{-\infty}^{\infty} e^{i\lambda t}\varphi(x,t)\, dt = \frac{1}{\sqrt{2\pi}} \int_{0}^{\infty} e^{i\lambda t}\varphi(x,t)\, dt \qquad (7\text{-}72)$$

An integration by parts shows that, at least for $\text{Im}\, \lambda > 0$,

$$\frac{1}{\sqrt{2\pi}} \int_{0}^{\infty} e^{i\lambda t}\varphi_t(x,t)\, dt = -\frac{1}{\sqrt{2\pi}}\varphi(x,0) - i\lambda\Phi(x,\lambda)$$

Consequently, if we multiply Eq. (7-70) by $(1/\sqrt{2\pi})e^{i\lambda t}$ and integrate with respect to t from 0 to ∞, we obtain

$$\alpha\Phi_{xx} + i\lambda\Phi = -\frac{2}{\sqrt{2\pi}}h(x) \qquad (7\text{-}73)$$

For $x < 0$, we therefore have

$$\Phi = A(\lambda)\exp\left(x\sqrt{-\frac{i\lambda}{\alpha}}\right) + B(\lambda)\exp\left(-x\sqrt{-\frac{i\lambda}{\alpha}}\right) \qquad (7\text{-}74)$$

where for definiteness we shall define $\sqrt{-i\lambda/\alpha}$ to be such that its real part is > 0 for λ real and > 0. Since $x < 0$ in Eq. (7-74), we must set $B(\lambda) = 0$, to avoid an exponential growth in Φ as $x \to -\infty$. Further, $\text{Re}\,(\sqrt{-i\lambda/\alpha})$ must be positive when λ is real and < 0 if the integral of Eq. (7-8) is to converge; this leads us to adopt for $\sqrt{-i\lambda/\alpha}$ a branch line lying below the real λ axis. We could have drawn this conclusion about the branch line independently by noting that, unless $\varphi(x,t)$ is rather pathological, Eq. (7-72) requires Φ to be analytic in some half-plane $\text{Im}\,\lambda > k$. The same reasoning for the case $x > 0$ leads to

$$\Phi(x,\lambda) = C(\lambda)\exp\left(-x\sqrt{-\frac{i\lambda}{\alpha}}\right) - \frac{2}{\sqrt{2\pi}\, i\lambda} \qquad (7\text{-}75)$$

Continuity of Φ and Φ_x at $x = 0$ leads to $A = -C = -1/(\sqrt{2\pi}\, i\lambda)$. Thus, finally,

$$\varphi(x,t) = \frac{1}{2\pi}\int_{\Gamma}\left[\frac{|x|}{i\lambda x}\exp\left(-|x|\sqrt{-\frac{i\lambda}{\alpha}}\right) - \frac{2}{i\lambda}h(x)\right]e^{-i\lambda t}\, d\lambda \qquad (7\text{-}76)$$

where Γ must be so chosen that all constraints on φ are satisfied. In particular, Γ must be such that $\varphi = 0$ for $t < 0$ and such that $\varphi(x,0) = 2h(x)$. The reader may verify that these conditions are satisfied if Γ lies above the origin; the solution is then

$$\varphi(x,t) = 1 + \text{erf}\,\frac{x}{2\sqrt{\alpha t}} \qquad (7\text{-}77)$$

Despite the foregoing, it is evident that we could write an alternative solution of the problem as

$$\varphi_1(x,t) = \varphi(x,t) + A_1\varphi_x(x,t) + B_1\varphi_{xt}(x,t) + \cdots$$

where A_1, B_1, . . . are constants. Each added function vanishes for $t \to 0$ or for $x^2 + t^2 \to \infty$. The fact of the matter is that the problem as stated does not have a unique solution unless one also imposes the requirement that the solution remain bounded as x and t tend to zero along any path in $t > 0$. Such a requirement, which is ordinarily implicit in scientific investigations, renders the solution unique; furthermore, since the time transforms of φ_x, φ_{xt}, etc., do not exist (because of the origin singularity), the transform method automatically legislates against the unacceptable solutions and acts as a very useful "filter."

EXERCISES

14. Let B, b, C, c be real constants such that, for sufficiently large $|t|$, $|f(t)| < Be^{bt}$ for $t > 0$ and $|f(t)| < Ce^{c|t|}$ for $t < 0$. Define

$$F_+(\lambda) = \frac{1}{\sqrt{2\pi}} \int_0^\infty e^{i\lambda t} f(t)\, dt \qquad F_-(\lambda) = \frac{1}{\sqrt{2\pi}} \int_{-\infty}^0 e^{i\lambda t} f(t)\, dt \tag{7-78}$$

Show that

$$f(t) = \frac{1}{\sqrt{2\pi}} \int_{-\infty+ib_1}^{\infty+ib_1} e^{-i\lambda t} F_+(\lambda)\, d\lambda + \frac{1}{\sqrt{2\pi}} \int_{-\infty-ic_1}^{\infty-ic_1} e^{-i\lambda t} F_-(\lambda)\, d\lambda \tag{7-79}$$

where b_1, c_1 are arbitrary except that $b_1 > b$, $c_1 > c$.

Verify these equations for the special case $f(t) = e^{at}$, $a > 0$. What are the half-planes of convergence for $F_+(\lambda)$ and $F_-(\lambda)$? Repeat for $f(t) = |t|$.

15. A very long, thin-walled, circular tube, of wall thickness t and radius r, elastic modulus E and Poisson's ratio ν, is coaxial with the x axis and for practical purposes may be considered to extend from $x = -\infty$ to $x = \infty$. It is subjected to a temperature distribution $T(x)$. The differential equation governing the shear force $\varphi(x)$ per unit circumferential distance is

$$\frac{d^4\varphi}{dx^4} + k^4\varphi = -\frac{Et\alpha}{r}\frac{d^3T}{dx^3}$$

where α is the linear coefficient of thermal expansion and where $k^4 = 12(1 - \nu^2)/r^2t^2$. Use a Fourier-transform method to show that, for the special case $T(x) = T_0 h(x)$, where T_0 is a constant and $h(x)$ is defined by Eq. (7-66), the shear force is given by

$$\varphi = \frac{Et\alpha T_0}{2rk} \exp\left(-\frac{k}{\sqrt{2}}|x|\right) \cos\left(\frac{\pi}{4} + \frac{k}{\sqrt{2}}|x|\right)$$

16. Show that the Fourier transforms of the sequence

$$\ldots \; \delta''(t), \; \delta'(t), \; \delta(t), \; h(t), \; \int_0^t h(t)\,dt, \; \ldots$$

can be written symbolically as

$$\ldots \; \frac{(-i\lambda)^2}{\sqrt{2\pi}}, \; \frac{-i\lambda}{\sqrt{2\pi}}, \; \frac{1}{\sqrt{2\pi}}, \; \frac{1}{\sqrt{2\pi}}\,\frac{1}{-i\lambda}, \; \frac{1}{\sqrt{2\pi}}\left(\frac{1}{-i\lambda}\right)^2, \; \ldots$$

Multiple Transforms

Let $f(x,y)$ be a function of the two variables (x,y), defined for $-\infty < x < \infty$, $-\infty < y < \infty$. Let the transform of $f(x,y)$ with respect to x be

$$p(\xi,y) = \frac{1}{\sqrt{2\pi}} \int_{-\infty}^{\infty} e^{i\xi x} f(x,y)\,dx$$

and let the transform of $p(\xi,y)$ with respect to y be

$$\begin{aligned} F(\xi,\eta) &= \frac{1}{\sqrt{2\pi}} \int_{-\infty}^{\infty} e^{i\eta y} p(\xi,y)\,dy \\ &= \frac{1}{2\pi} \iint_{-\infty}^{\infty} e^{i(\xi x+\eta y)} f(x,y)\,dx\,dy \end{aligned} \tag{7-80}$$

where, in the latter integral, the integration is over the entire xy plane. We confine our attention to those $f(x,y)$ for which the double integral, and hence each repeated integral, exists. The double inversion [i.e., the successive use of the inversions associated with the definitions of $p(\xi,y)$ and $F(\xi,\eta)$] of Eq. (7-80) gives

$$f(x,y) = \frac{1}{2\pi} \iint_{-\infty}^{\infty} e^{-i(\xi x+\eta y)} F(\xi,\eta)\,d\xi\,d\eta \tag{7-81}$$

Multiple transforms with respect to more than two variables and multiple sine or cosine transforms are defined analogously.

The evaluation of integrals such as those occurring in Eqs. (7-80) and (7-81) is often simplified by an appropriate change in variables. One such change is suggested by the observation that the combination $\xi x + \eta y$ is the scalar product of the two vectors (ξ,η) and (x,y). Let us introduce polar coordinates by setting $x = r\cos\theta$, $y = r\sin\theta$, and $\xi = \rho\cos\varphi$, $\eta = \rho\sin\varphi$. Then Eqs. (7-80) and (7-81) give the transform pair

$$\begin{aligned} G(\rho,\varphi) &= \frac{1}{2\pi} \int_0^{\infty} r\,dr \int_0^{2\pi} d\theta\, e^{ir\rho\cos(\theta-\varphi)} g(r,\theta) \\ g(r,\theta) &= \frac{1}{2\pi} \int_0^{\infty} \rho\,d\rho \int_0^{2\pi} d\varphi\, e^{-ir\rho\cos(\theta-\varphi)} G(\rho,\varphi) \end{aligned} \tag{7-82}$$

where $g(r,\theta)$ denotes the value of $f(x,y)$ at corresponding points. We shall subsequently find that there is a relationship between Eqs. (7-82) and the Hankel-transform theorem. In three dimensions, spherical polar coordinates could be used in a similar manner.

As an example of the use of multiple transforms, we study the heat-conduction problem for the semi-infinite plate [Eq. (7-44)] by combining a Fourier transform in x with a sine transform in y. Define

$$\Phi(\xi,\eta) = \sqrt{\frac{2}{\pi}} \int_0^\infty \sin \eta y \, dy \frac{1}{\sqrt{2\pi}} \int_{-\infty}^{\infty} e^{i\xi x} \varphi(x,y) \, dx$$

Let $A(\xi)$ be the Fourier transform of the temperature along the edge $y = 0$, so that

$$A(\xi) = \frac{1}{\sqrt{2\pi}} \int_{-\infty}^{\infty} e^{i\xi s} \varphi(s,0) \, ds \tag{7-83}$$

Then the double transform of Eq. (7-44) gives

$$-\xi^2\Phi + \sqrt{\frac{2}{\pi}} \eta \, A(\xi) - \eta^2\Phi = 0$$

This is now an algebraic equation for Φ, and we obtain

$$\Phi = \sqrt{\frac{2}{\pi}} \frac{\eta A(\xi)}{\eta^2 + \xi^2}$$

The inversion formula is

$$\varphi(x,y) = \frac{2}{\pi} \frac{1}{\sqrt{2\pi}} \int_0^\infty d\eta \int_{-\infty}^{\infty} d\xi \, e^{-i\xi x} \sin \eta y \frac{\eta A(\xi)}{\eta^2 + \xi^2}$$

Substituting for $A(\xi)$ from Eq. (7-83) and integrating first with respect to ξ and then with respect to η, we again obtain (fortunately) Eq. (7-53).

EXERCISES

17. The equation for time-dependent heat diffusion in two dimensions is

$$\alpha(\varphi_{xx} + \varphi_{yy}) = \varphi_t$$

where $\varphi(x,y,t)$ is the temperature and α the constant thermal diffusivity. Let the initial temperature in an infinite plate $(-\infty < x < \infty, -\infty < y < \infty)$ have the form

$$\varphi(x,y,0) = f(x)g(y)$$

where f and g are given functions. Use a double Fourier transform to show that $\varphi(x,y,t)$ is the product of a function of (x,t) with a function of (y,t), where each of these functions satisfies the one-dimensional diffusion equation. Verify the result by the usual separation-of-variables method.

18. Determine the steady-state temperature $\varphi(x,y)$ in a plate occupying the quarter-plane $0 \leq x < \infty$, $0 \leq y < \infty$, where $\varphi(x,0) = f(x)$ and $\varphi(0,y) = g(y)$ are prescribed. The governing equation is $\varphi_{xx} + \varphi_{yy} = 0$. Use a double sine transform to show that the general solution is

$$\begin{aligned}\varphi(x,y) &= \frac{1}{\pi}\int_0^\infty f(s)\left[\frac{y}{y^2 + (x - s)^2} - \frac{y}{y^2 + (x + s)^2}\right] ds \\ &\quad + \frac{1}{\pi}\int_0^\infty g(s)\left[\frac{x}{x^2 + (y - s)^2} - \frac{x}{x^2 + (y + s)^2}\right] ds\end{aligned}$$

What features of this formula provide checks on the answer? How could Eq. (7-53) have been used to obtain this result almost at once?

Convolution Integrals

Let $F(\lambda)$ and $G(\lambda)$ be the Fourier transforms of $f(x)$ and $g(x)$, respectively. The inverse transform of the product $F(\lambda) \cdot G(\lambda)$ is

$$\begin{aligned}\frac{1}{\sqrt{2\pi}}\int_{-\infty}^{\infty} e^{-i\lambda x}F(\lambda)G(\lambda)\, d\lambda &= \frac{1}{2\pi}\int_{-\infty}^{\infty} e^{-i\lambda x}F(\lambda)\, d\lambda \int_{-\infty}^{\infty} e^{i\lambda\xi}g(\xi)\, d\xi \\ &= \frac{1}{2\pi}\int_{-\infty}^{\infty} g(\xi)\, d\xi \int_{-\infty}^{\infty} e^{-i\lambda(x-\xi)}F(\lambda)\, d\lambda \\ &= \frac{1}{\sqrt{2\pi}}\int_{-\infty}^{\infty} f(x - \xi)g(\xi)\, d\xi \qquad (7\text{-}84)\end{aligned}$$

Alternatively, the result can be written as

$$\frac{1}{\sqrt{2\pi}}\int_{-\infty}^{\infty} e^{-i\lambda x}F(\lambda)G(\lambda)\, d\lambda = \frac{1}{\sqrt{2\pi}}\int_{-\infty}^{\infty} g(x - \xi)f(\xi)\, d\xi$$

An integral of this form is called a *convolution integral* of f and g (or a *resultant*, or a *Faltung*, of f and g). Conversely, of course, the Fourier transform of the convolution is simply $F(\lambda)G(\lambda)$.

As an example of the use of Eq. (7-84), consider Eq. (7-37), which gives the transform of $\varphi(x,t)$ as the product of the transform of $\varphi(x,0)$ and the function $\exp(-\alpha\lambda^2 t)$. Since $\exp(-\alpha\lambda^2 t)$ is the transform of $(2\alpha t)^{-1/2}$ $\exp(-x^2/4\alpha t)$, the use of Eq. (7-84) leads to

$$\varphi(x,t) = \frac{1}{2\sqrt{\pi\alpha t}}\int_{-\infty}^{\infty} \varphi(\xi,0) \exp\left[-\frac{1}{4\alpha t}(x - \xi)^2\right] d\xi \qquad (7\text{-}85)$$

in agreement with Eq. (7-65). We observe again the superposition character of the result.

Consider next the inverse transform of the product of three functions $F(\lambda)$, $G(\lambda)$, $H(\lambda)$, which are the Fourier transforms of $f(x)$, $g(x)$, $h(x)$, respectively. A calculation similar to that leading to Eq. (7-84) gives such results as

$$\frac{1}{\sqrt{2\pi}} \int_{-\infty}^{\infty} e^{-i\lambda x} F(\lambda)G(\lambda)H(\lambda)\, d\lambda = \frac{1}{2\pi} \iint_{-\infty}^{\infty} f(x - \eta - \xi) g(\eta) h(\xi)\, d\eta\, d\xi$$

$$= \frac{1}{2\pi} \iint_{-\infty}^{\infty} f(x - \eta) g(\eta - \xi) h(\xi)\, d\eta\, d\xi$$

Similar formulas are obtained for the inversion of products with more terms.

If we set $x = 0$ in Eq. (7-84), we obtain

$$\int_{-\infty}^{\infty} F(\lambda)G(\lambda)\, d\lambda = \int_{-\infty}^{\infty} f(x)g(-x)\, dx \tag{7-86}$$

An equivalent form is

$$\int_{-\infty}^{\infty} F(\lambda)G(-\lambda)\, d\lambda = \int_{-\infty}^{\infty} f(x)g(x)\, dx \tag{7-87}$$

If $g(x)$ in Eq. (7-87) is replaced by $g^*(x)$, $G(-\lambda)$ must be replaced by $G^*(\lambda)$; thus a more symmetrical form of these equations is

$$\int_{-\infty}^{\infty} F(\lambda)G^*(\lambda)\, d\lambda = \int_{-\infty}^{\infty} f(x)g^*(x)\, dx \tag{7-88}$$

For the special case $f = g$, we have

$$\int_{-\infty}^{\infty} |F(\lambda)|^2\, d\lambda = \int_{-\infty}^{\infty} |f(x)|^2\, dx \tag{7-89}$$

The similarity of Eqs. (7-87) to (7-89) to Parseval's formula for the Fourier series of Eq. (7-1),

$$\frac{1}{T} \int_{-T}^{T} f^2(t)\, dt = \tfrac{1}{2} a_0^2 + \sum_{1}^{\infty} (a_n^2 + b_n^2)$$

causes these various equations to be described as being of the Parseval type.

Equations such as (7-87) can be useful in the evaluation of certain definite integrals. For example, use of the result of Exercise 1*d* in Eq. (7-87) gives (with a, b real and positive and $a < b$)

$$\int_0^{\infty} J_0(at)J_0(bt)\, dt = \frac{2}{\pi b} \int_0^{\pi/2} \frac{d\theta}{\sqrt{1 - (a^2/b^2) \sin^2 \theta}} = \frac{2}{\pi b} K\left(\frac{a}{b}\right)$$

where K is the complete elliptic integral of the second kind.

Convolution formulas and Parseval-type formulas for sine and cosine transforms can be obtained from the corresponding results for Fourier transforms by considering even or odd functions $f(x)$ and $g(x)$. Thus, if $f(x)$

and $g(x)$ are even, Eq. (7-84) gives

$$\sqrt{\frac{2}{\pi}}\int_0^\infty \cos \lambda x\, F_c(\lambda)G_c(\lambda)\, d\lambda = \frac{1}{\sqrt{2\pi}}\int_0^\infty [f(x+\xi) + f(|x-\xi|)]g(\xi)\, d\xi$$

which, for the special case $x = 0$, reduces to

$$\int_0^\infty F_c(\lambda)G_c(\lambda)\, d\lambda = \int_0^\infty f(\xi)g(\xi)\, d\xi$$

Similar results are obtained by choosing f and g both odd or one even and the other odd in Eq. (7-84).

EXERCISES

19. Show that

$$\int_{-\infty}^{\infty} f(ax)g(bx)\, dx = \frac{1}{ab}\int_{-\infty}^{\infty} F\left(\frac{\lambda}{a}\right)G\left(-\frac{\lambda}{b}\right) d\lambda$$

20. Apply the convolution theorem to Eq. (7-47) so as to obtain the result (7-53). (The method of Exercise 9 is now seen to consist essentially in a derivation of the convolution theorem.)

21. Evaluate the following integrals, where a and b are real positive constants:

(*a*) $\displaystyle\int_0^\infty \frac{\sin ax \sin bx}{x^2}\, dx$

(*b*) $\displaystyle\int_0^\infty x^{-(\mu+\nu)}J_\mu(x)J_\nu(x)\, dx,\ \mu > 0,\ \nu > 0$

22. Show that the sine transform of

$$\int_0^x f(x-\xi)g(\xi)\, d\xi$$

is

$$\sqrt{\frac{\pi}{2}}\,[F_s(\lambda)G_c(\lambda) + F_c(\lambda)G_s(\lambda)]$$

23. Show that the inverse double Fourier transform of $F(\xi,\eta)G(\xi,\eta)$ [where F and G are the double transforms of $f(x,y)$ and $g(x,y)$] is given by

$$\frac{1}{2\pi}\iint_{-\infty}^{\infty} f(s,t)g(x-s,\ y-t)\, ds\, dt$$

and obtain some alternative forms for this expression. Use this result to obtain a general solution of the steady-state heat-conduction problem for the half-space $-\infty < x < \infty$, $-\infty < y < \infty$, $0 \le z < \infty$, where the temperature is specified on the surface $z = 0$.

24. Let the probability of finding a random variable x_1 in the range x to $x + dx$ be $p_1(x)\,dx$. In probability theory,[1] the quantity $\sqrt{2\pi}\,P_1(\lambda) = \int_{-\infty}^{\infty} e^{i\lambda\xi} p_1(\xi)\,d\xi$ is termed the *characteristic function* associated with the probability distribution $p_1(x)$; it represents the *expected value* of $e^{i\lambda x_1}$. If x_2 is a second independent random variable with the probability-distribution function $p_2(x)$, show that the formula giving the probability $p(x)\,dx$ that the sum $x_1 + x_2$ lie in the range x to $x + dx$, namely,

$$p(x)\,dx = dx \int_{-\infty}^{\infty} p_1(\xi) p_2(x - \xi)\,d\xi$$

is consistent with the theorem that the expected value of the product of two independent quantities is the product of their individual expected values.

Show that, for n independent random variables having the common distribution function $p(x) = 1$ for $0 < x < 1$, $p(x) = 0$ otherwise, the characteristic function for their mean is

$$\left(\frac{e^{i\lambda/n} - 1}{i\lambda/n}\right)^n$$

and use the method of steepest descents (set $\lambda = n\xi$) to show that the probability distribution for their mean is asymptotic to

$$\sqrt{\frac{6n}{\pi}} \exp\left[-6n\left(x - \frac{1}{2}\right)^2\right]$$

for large values of n. (This is a special case of the central-limit theorem in probability theory.)

25. The fact that the quadratic equation

$$\int [xf(t) - g(t)]^2\,dt = 0$$

where $f(t)$ and $g(t)$ are two real functions, has no real solution for x unless $g(t)$ is a constant times $f(t)$ leads to *Schwarz's inequality*

$$\left[\int f(t)g(t)\,dt\right]^2 \leq \int [f(t)]^2\,dt \int [g(t)]^2\,dt$$

with equality holding only in the special case mentioned. Use this inequality for the two functions $tf(t)$ and $f'(t)$ to show that

$$\int_{-\infty}^{\infty} t^2[f(t)]^2\,dt \int_{-\infty}^{\infty} \lambda^2\,|F(\lambda)|^2\,d\lambda \geq \tfrac{1}{4}\left\{\int_{-\infty}^{\infty} [f(t)]^2\,dt\right\}^2$$

provided that $f(t)$ vanishes sufficiently strongly at $\pm\infty$. In electrical-signal theory, this result is often described as being of the character of an *uncertainty principle;* thus, if $f(t)$ is the response function of a filter, this result gives a quantitative version of the statement that the narrower the passband the longer it takes for transients to die out.

[1] H. Cramer, "Mathematical Methods of Statistics," pp. 213ff., Princeton University Press, Princeton, N.J., 1946.

26. Let a real function $f(t)$ be defined for $-T < t < T$ and vanish identically for $t > T$ and for $t < -T$. The function

$$r(\tau) = r(-\tau) = \frac{1}{2T} \int_{-T}^{T} f(t)f(t + \tau)\, dt$$

is called the *autocorrelation* function of $f(t)$. Show that

$$R(\lambda) = \frac{\sqrt{2\pi}}{2T} |F(\lambda)|^2$$

Let the Fourier-series representation for $f(t)$ over the interval $(-T,T)$ be given by

$$f(t) = \tfrac{1}{2}a_0 + \sum_{n=1}^{\infty} \left(a_n \cos \frac{n\pi t}{T} + b_n \sin \frac{n\pi t}{T} \right)$$

Define $\nu_n = n/2T$, $\delta\nu_n = 1/2T$, and $W(\nu_n)$ by

$$W(\nu_n) \frac{1}{2T} = \frac{1}{2} (a_n^2 + b_n^2)$$

For $\lambda = \lambda_n = 2\pi\nu_n$, observe that $F(\lambda) = (T/\sqrt{2\pi})(a_n + ib_n)$. Hence

$$W(\nu_n) = \frac{2\pi}{T} |F(\lambda_n)|^2 = 2\sqrt{2\pi}\, R(\lambda_n)$$

Deduce that, as $T \to \infty$,

$$\begin{aligned} W(\nu) &= 4 \int_0^{\infty} \cos 2\pi\nu t\, r(t)\, dt \\ r(t) &= \int_0^{\infty} W(\nu) \cos 2\pi\nu t\, d\nu \end{aligned} \tag{7-90}$$

The function $W(\nu)$ is termed the *power spectral density*. Eq. (7-90) is due to Wiener.

27. Consider the problem of finding a function $\varphi(x,t)$ satisfying $\varphi_t = \alpha\varphi_{xx}$ for x in $(0,b)$, $t > 0$, where b is >0 and where the quantities $\varphi(0,t)$, $\varphi(b,t)$, and $\varphi(x,0)$ are all specified. Define a "finite" sine transform by

$$\Phi(\lambda,t) = \sqrt{\frac{2}{\pi}} \int_0^b (\sin \lambda x)\varphi(x,t)\, dx$$

so that $\Phi(\lambda,t)$ is the conventional sine transform of the function

$$\psi(x,t) = \begin{cases} \varphi(x,t) & \text{for } x \text{ in } (0,b) \\ 0 & \text{for } x > b \end{cases}$$

Multiply the equation $\varphi_t = \alpha\varphi_{xx}$ by $\sin \lambda x$, and integrate both sides (by parts if necessary) with respect to x from 0 to b so as to obtain an equation for $\Phi(\lambda,t)$. Show that this equation can be solved only if λ has one of the discrete set of values $\lambda_n = n\pi/b$, in which case $\Phi(\lambda_n,t)$ is simply a conventional Fourier coefficient.

(A finite cosine transform or a finite exponential transform could be defined similarly. Such an approach is, in general, equivalent to the use of Fourier series; however, the reader may be reminded that even here, dealing with coefficient functions rather than with the series itself—in exact analogy to the infinite-range case—can often be useful, especially if the Fourier series is not differentiable term by term.)

28. In solving a heat-conduction problem for a semi-infinite strip, one wants to find $\varphi(x,y)$ such that $\varphi_{xx} + \varphi_{yy} = 0$ for $0 \le x \le \pi$, $0 \le y < \infty$; we take the boundary conditions $\varphi(x,0) = 1$, $\varphi(0,y) = \varphi(\pi,y) = 0$. Solve for $\varphi(x,y)$ by two methods, (*a*) by writing $\varphi(x,y) = \sum_1^\infty a_n(y) \sin nx$ and solving a differential equation for $a_n(y)$ and (*b*) by first taking a sine transform in y and then summing residues. In either case, sum the resultant series to show that

$$\varphi(x,y) = \frac{2}{\pi} \arctan \frac{\sin x}{\sinh y}$$

(What assumption concerning the nature of the singularities at the two corners was necessary in order to ensure a unique solution to this problem?) Show that the quarter-plane problem $0 \le x < \infty$, $0 \le y < \infty$, with $\varphi(x,0) = 1$, $\varphi(0,y) = 0$, leads—as anticipated—to the same kind of corner singularity.

7-2 *The Application of Fourier Transforms to Boundary-value Problems*

We have already encountered a number of problems which could be solved by using Fourier-transform methods. We shall now discuss some illustrative problems which display various subtleties.

Ordinarily, the Fourier-transform method is useful in treating those partial-differential-equation problems in which at least one variable has an infinite or semi-infinite range and in which all coefficients are independent of that variable (or at most are simple polynomials in that variable). Each of the following examples is of this type:

1. A Membrane Problem

The equation governing the small deflection $\varphi(x,y)$ of an elastic membrane, subjected to a loading distribution $f(x,y)$, is[1]

$$\varphi_{xx} + \varphi_{yy} - \alpha^2\varphi = f(x,y) \tag{7-91}$$

[1] See, for example, R. Courant and D. Hilbert, "Methods of Mathematical Physics," vol. 1, p. 246, Interscience Publishers, Inc., New York, 1953.

Here α is a positive constant; physically, the term $\alpha^2\varphi$ corresponds to an elastic restraint on the membrane. Equation (7-91) is to hold for $-\infty < x < \infty$, $-\infty < y < \infty$; φ and its derivatives (as well as f) are to vanish as $x^2 + y^2 \to \infty$.

With the transform variables denoted by λ_1 and λ_2, a double Fourier transform of Eq. (7-91) leads to

$$\varphi(x,y) = -\frac{1}{2\pi}\iint_{-\infty}^{\infty} \frac{F(\lambda_1,\lambda_2)\exp(-i\lambda_1 x - i\lambda_2 y)}{\lambda_1^2 + \lambda_2^2 + \alpha^2}\, d\lambda_1\, d\lambda_2 \tag{7-92}$$

and the convolution theorem implies that

$$\varphi(x,y) = \iint_{-\infty}^{\infty} f(\xi,\eta)g(x,y;\xi,\eta)\, d\xi\, d\eta \tag{7-93}$$

where $$g(x,y;\xi,\eta) = -\frac{1}{4\pi^2}\iint_{-\infty}^{\infty} \frac{\exp[-i\lambda_1(x-\xi) - i\lambda_2(y-\eta)]}{\lambda_1^2 + \lambda_2^2 + \alpha^2}\, d\lambda_1\, d\lambda_2 \tag{7-94}$$

Changing to the polar coordinates defined by $x - \xi = \rho\cos\omega$, $y - \eta = \rho\sin\omega$, $\lambda_1 = r\cos\theta$, $\lambda_2 = r\sin\theta$, we obtain

$$g(x,y;\xi,\eta) = -\frac{1}{2\pi}\int_0^\infty \frac{rJ_0(r\rho)}{r^2 + \alpha^2}\, dr = -\frac{1}{2\pi} K_0(\alpha\rho) \tag{7-95}$$

where K_0 is the modified Bessel function defined by Eq. (5-126). $K_0(z)$ is exponentially small for large positive values of its argument z but has a logarithmic singularity at $z = 0$.

If we set $f(\xi,\eta) = \delta(\xi - \xi_0)\delta(\eta - \eta_0)$ in Eq. (7-93), that equation reduces to

$$\varphi(x,y) = \frac{1}{2\pi} K_0\left(\alpha\sqrt{(x-\xi_0)^2 + (y-\eta_0)^2}\right) \tag{7-96}$$

In the context of the membrane problem, $\varphi(x,y)$ as given by Eq. (7-96) is the deflection of a membrane which is subjected to a unit concentrated force applied at ξ_0, η_0.

Equation (7-93) is a specific illustration of the form into which the solutions of many boundary-value problems can be cast. Characteristically, if φ obeys the equation

$$L(\varphi) = f(x_i)$$

in a domain D of the independent variables $(x_1, x_2, \ldots, x_n)$ where L is some linear partial differential operator, and if the boundary conditions are homogeneous, then

$$\varphi(x_i) = \int_D f(\xi_i)g(x_i;\xi_i)\, d\xi_i$$

and $g(x_i;\xi_i)$ is a representation of the *Green's function*[1] of the boundary-value problem. Implicit in its construction are specifications of L, D, and the homogeneous boundary conditions.

We consider next the case $\alpha = 0$. If we set $\alpha = 0$ in Eq. (7-94), the integral does not exist, so that the foregoing analysis fails. One way of circumventing the difficulty is to differentiate Eq. (7-92) with respect to x, which yields (with $\alpha = 0$)

$$\varphi_x(x,y) = -\frac{1}{2\pi}\iint_{-\infty}^{\infty} \frac{-i\lambda_1 F(\lambda_1,\lambda_2)\exp(-i\lambda_1 x - i\lambda_2 y)}{\lambda_1^2 + \lambda_2^2}\, d\lambda_1\, d\lambda_2$$

The convolution theorem now gives

$$\varphi_x(x,y) = \iint_{-\infty}^{\infty} f(\xi,\eta)h(x-\xi,\, y-\eta)\, d\xi\, d\eta$$

where
$$\begin{aligned} h(x,y) &= -\frac{1}{4\pi^2}\iint_{-\infty}^{\infty} \frac{-i\lambda_1 \exp[-i(\lambda_1 x + \lambda_2 y)]}{\lambda_1^2 + \lambda_2^2}\, d\lambda_1\, d\lambda_2 \\ &= \frac{x}{2\pi(x^2+y^2)} \\ &= \frac{1}{2\pi}\frac{\partial}{\partial x}\ln(\sqrt{x^2+y^2}) \end{aligned}$$

Thus
$$\varphi(x,y) = \frac{1}{2\pi}\iint_{-\infty}^{\infty} f(\xi,\eta)\ln[(x-\xi)^2 + (y-\eta)^2]\, d\xi\, d\eta \tag{7-97}$$

provided that $f(\xi,\eta)$ is such that the $\varphi(x,y)$ defined by Eq. (7-97) vanishes as $x^2 + y^2 \to \infty$ [so that the formula for the transform of a derivative, used in deriving Eq. (7-92), is applicable]. The reader may verify that this occurs only when $\iint_{-\infty}^{\infty} f(\xi,\eta)\, d\xi\, d\eta = 0$. If this constraint on f is not met, the boundary-value problem with $\alpha = 0$, $\varphi(\infty) = 0$ has no solution.

The reader should verify that the Green's function of Eq. (7-95) or Eq. (7-97) can be obtained by treating directly the equation

$$\varphi_{xx} + \varphi_{yy} - \alpha^2\varphi = \delta(x-\xi)\delta(y-\eta) \tag{7-98}$$

Exercises

1. Let $\varphi(x,y)$ satisfy Eq. (7-91) in a simply connected finite region R, with boundary C.

[1] In honor of George Green. See "An Essay on the Application of Mathematical Analysis to the Theories of Electricity and Magnetism," 1828; facsimile reprint 1958 by Wezata-Melins AB, Göteborg, Sweden.

(*a*) Show, by use of Green's identity (Exercise 7 of Sec. 2-5) that, if $g(x,y;a,b)$ satisfies Eq. (7-98) in R, then

$$\varphi(a,b) = \int_R g(x,y;a,b)f(x,y)\,dA + \int_C \left(\varphi \frac{\partial g}{\partial n} - g\frac{\partial \varphi}{\partial n}\right) ds$$

where all integrations are with respect to (x,y). (Thus, if φ is known on C and if g satisfies the condition $g = 0$ on C, this equation provides a formal solution for φ in R. What is the appropriate boundary condition for g if $\omega\varphi + \beta\,\partial\varphi/\partial n = \gamma$ on C, where ω, β, γ are functions of position along C?)

(*b*) Use the equation satisfied by g, namely, Eq. (7-98), to show that $g(x,y;a,b) = g(a,b;x,y)$, for $g = 0$ on C. Show also that if $\partial g/\partial n = 0$ on C, the arbitrary constant in g can be adjusted so as to maintain this symmetry with respect to interchange of "source" and "effect" points.

(*c*) Let g_∞ be the Green's function for the infinite region, as given by Eq. (7-95); let g be that of Exercise 1*a*. Show that $g - g_\infty$ is nonsingular in R.

2. Consider the problem of finding a Green's function $g(x,y;\xi,\eta)$ for the infinite strip region $-\infty < x < \infty$, $0 \le y \le l$, satisfying

$$g_{xx} + g_{yy} = \delta(x-\xi)\delta(y-\eta)$$

with $g = 0$ for $y = 0$ or for $y = l$. Take a Fourier transform in x to yield

$$G = -\frac{e^{i\lambda\xi}}{\sqrt{2\pi}\,\lambda\sinh\lambda l}\begin{cases}\sinh\lambda(l-\eta)\sinh\lambda y & \text{for } y < \eta \\ \sinh\lambda(l-y)\sinh\lambda\eta & \text{for } y > \eta\end{cases}$$

Use the inversion theorem and sum residues to give

$$g = \frac{1}{4\pi}\ln\frac{\cosh(\pi/l)(\xi-x) - \cos(\pi/l)(\eta-y)}{\cosh(\pi/l)(\xi-x) - \cos(\pi/l)(\eta+y)}$$

Expand the argument of the ln function in the form of an infinite product, so as to show how this result can be interpreted in terms of the *method of images*, in which the strip is replaced by the whole plane with positive unit sources at . . . $(\xi,\ 4l+\eta)$, $(\xi,\ 2l+\eta)$, (ξ,η), $(\xi,\ -2l+\eta)$, . . . and negative unit sources at . . . $(\xi,\ 4l-\eta)$, $(\xi,\ 2l-\eta)$, $(\xi,-\eta)$, $(\xi,\ -2l-\eta)$, ... (The lines $y = l$, $y = 0$ are lines of symmetry where the effects of all sources cancel.) The final result can, of course, also be obtained by the use of a Fourier sine series in y or by conformal mapping.

3. Use Fourier transforms to show that the Green's function $g(x,y,z;\xi,\eta,\zeta)$ satisfying

$$g_{xx} + g_{yy} + g_{zz} - \alpha^2 g = \delta(x-\xi)\delta(y-\eta)\delta(z-\zeta)$$

in $-\infty < x < \infty$, $-\infty < y < \infty$, $-\infty < z < \infty$, with $g \to 0$ at ∞, is

$$g = \frac{-1}{4\pi\rho} e^{-\alpha\rho} \tag{7-99}$$

where $\rho^2 = (x - \xi)^2 + (y - \eta)^2 + (z - \zeta)^2$. [*Hint:* Use spherical polar coordinates in the inversion integral. Observe the possibility of rotating the coordinate system $(\lambda_1,\lambda_2,\lambda_3)$ into a more convenient system $(\lambda_1',\lambda_2',\lambda_3')$.]

4. Determine the Green's function for the Poisson equation $\varphi_{xx} + \varphi_{yy} = f$ in the quarter-plane $0 \leq x < \infty$, $0 \leq y < \infty$, with either φ or $\partial\varphi/\partial n$ prescribed on the boundary. Consider the two cases that arise by allowing the source point to be inside or on the boundary of the region.

2. *Some Wave-propagation Problems*

The one-dimensional wave equation

$$\varphi_{xx} - \frac{1}{c^2}\varphi_{tt} - \epsilon\varphi_t = p(x,t) \tag{7-100}$$

where c is a constant, describes the lateral deflection $\varphi(x,t)$ of an elastic string[1] subjected to a force distribution $p(x,t)$; the "wind resistance" is $\epsilon\varphi_t$, where $\epsilon > 0$ is a constant. We consider here the steady-state motion (in a plane) of an indefinitely long string when

$$p(x,t) = f(x)e^{i\omega t} \tag{7-101}$$

where $\omega > 0$ is a constant. By anticipating that φ can be written in the form

$$\varphi(x,t) = \psi(x)e^{i\omega t}$$

Eq. (7-100) becomes

$$\psi'' + \left(\frac{\omega^2}{c^2} - i\omega\epsilon\right)\psi = f(x) \tag{7-102}$$

and we take the Fourier transform of Eq. (7-102) to give

$$\left[\lambda^2 - \left(\frac{\omega^2}{c^2} - i\omega\epsilon\right)\right]\psi(\lambda) = -F(\lambda) \tag{7-103}$$

The inversion theorem yields

$$\psi(x) = \frac{1}{\sqrt{2\pi}}\int_{-\infty}^{\infty} \frac{-F(\lambda)e^{-i\lambda x}\,d\lambda}{\lambda^2 - (\omega^2/c^2 - i\omega\epsilon)} \tag{7-104}$$

Whenever $\epsilon > 0$, the path of integration is the real axis. Use of the con-

[1] See, for example, R. Courant and D. Hilbert, *op. cit.*, vol. 1, p. 245, Interscience Publishers, Inc., New York, 1953.

volution theorem gives

$$\psi(x) = \frac{ic}{2\omega} \int_{-\infty}^{\infty} f(\xi) \frac{\exp [(-i\omega/c)|\xi - x| \sqrt{1 - i\epsilon c^2/\omega}]}{\sqrt{1 - i\epsilon c^2/\omega}} \, d\xi \qquad (7\text{-}105)$$

As $\epsilon \to 0$,

$$\psi(x) \to \frac{ic}{2\omega} \int_{-\infty}^{\infty} f(\xi) e^{(-i\omega/c)|\xi - x|} \, d\xi \qquad (7\text{-}106)$$

The presence of the dissipative term $\epsilon\varphi_t$ in Eq. (7-100) means that there are no poles on the path of integration in Eq. (7-104). Even if an original problem does not contain a dissipative term, it may be convenient to introduce such a term to avoid difficulties with the path of integration; the nondissipation case can then be obtained, as here, by allowing $\epsilon \to 0$. However, it is frequently desirable to treat the nondissipative case directly, especially in problems in which the dissipative mechanism is not easily modeled or in which the dissipative contributions complicate the problem. In the present case, we are led to Eq. (7-104), now with $\epsilon = 0$. The appropriate path of integration must be so chosen that the contribution to $\psi(x)$ from each element $f(\xi)\,d\xi$ of the nonhomogeneous term provides only waves which travel outward from the segment $d\xi$. This requirement is referred to as *the Sommerfeld radiation condition;* physically, it corresponds to the absence of any source at infinity, generating inward traveling waves (for $\epsilon \neq 0$, any such waves would have no amplitude at finite values of x, so that the Sommerfeld radiation condition is then unnecessary). The only paths which meet this requirement are those which are indented below the point $\lambda_1 = -\omega/c$ and above the point $\lambda_2 = \omega/c$; the result is readily seen to agree with Eq. (7-106).

EXERCISES

5. Set $\epsilon = 0$ in Eq. (7-100), and take a Fourier transform in x to show that, if $\varphi(x,0) = \varphi_t(x,0) = 0$,

$$\varphi(x,t) = -\frac{c}{2} \int_0^t d\tau \int_{-c\tau}^{c\tau} p(x - \xi,\, t - \tau)\, d\xi$$

Why was no radiation condition required here?

6. With $\epsilon = 0$ in Eq. (7-100) and $p(x,t) = \delta(x)\delta(t)$, take a double Fourier transform to show that the displacement resulting from a concentrated impulse is a square wave. Show that the result is compatible with that of Exercise 5.

7. Using the notation of Eq. (7-78), take Fourier transforms of Eq. (7-102) (with $\epsilon = 0$) to obtain expressions for $\Psi_+(\lambda)$ and $\Psi_-(\lambda)$. Use Eq. (7-79), determining the arbitrary constants by use of the radiation condition, so as again to obtain Eq. (7-106).

We turn next to a steady-state acoustic-radiation problem, in which we wish to find the acoustic potential $\varphi(x,y,z)e^{i\omega t}$ in the half-space $-\infty < x < \infty$, $-\infty < y < \infty$, $z > 0$. The governing differential equation[1] is

$$\varphi_{xx} + \varphi_{yy} + \varphi_{zz} + k^2\varphi = 0 \tag{7-107}$$

where the constant $k = \omega/c$. As a boundary condition, we shall require $\varphi_z(x,y,0) = f(x,y)$, where f is a prescribed function (in a "piston" problem which we shall subsequently consider, f will vanish for $x^2 + y^2 > R^2$ and will be unity for $x^2 + y^2 < R^2$). Physically, this boundary condition corresponds to a prescription of the normal velocity $fe^{i\omega t}$ on the $z = 0$ plane.

A Fourier transform in x and y yields

$$\Phi_{zz} - (\lambda_1^2 + \lambda_2^2 - k^2)\Phi = 0$$

so that

$$\Phi = A(\lambda_1,\lambda_2) \exp\left(-z\sqrt{\lambda_1^2 + \lambda_2^2 - k^2}\right) \tag{7-108}$$

where the interpretation of the square root must be such that the inversion integral converges and provides a function which meets the constraints of the original problem. The boundary condition implies that

$$A(\lambda_1,\lambda_2)\sqrt{\lambda_1^2 + \lambda_2^2 - k^2} = -F(\lambda_1,\lambda_2)$$

so that, with appropriate domains of integration, we have

$$\begin{aligned}\varphi(x,y,z) &= -\frac{1}{2\pi}\iint_{-\infty}^{\infty} \frac{F(\lambda_1,\lambda_2)}{\sqrt{\lambda_1^2 + \lambda_2^2 - k^2}} \\ &\qquad \exp\left(-z\sqrt{\lambda_1^2 + \lambda_2^2 - k^2} - i\lambda_1 x - i\lambda_2 y\right) d\lambda_1\, d\lambda_2 &(7\text{-}109)\\ &= -\frac{1}{2\pi}\iint_{-\infty}^{\infty} f(\xi,\eta)q(x-\xi,\, y-\eta,\, z)\, d\xi\, d\eta &(7\text{-}110)\end{aligned}$$

where

$$q(x,y,z) = \frac{1}{2\pi}\iint_{-\infty}^{\infty} \frac{\exp\left(-z\sqrt{\lambda_1^2 + \lambda_2^2 - k^2} - i\lambda_1 x - i\lambda_2 y\right)}{\sqrt{\lambda_1^2 + \lambda_2^2 - k^2}}\, d\lambda_1\, d\lambda_2 \tag{7-111}$$

We can extend the definition of q so that $q(x,y,-z) = +q(x,y,z)$; denote the extended function by $p(x,y,z)$. Then

$$\begin{aligned}P(\lambda_1,\lambda_2,\lambda_3) &= \frac{1}{\sqrt{2\pi}}\int_0^{\infty} \frac{\exp\left(-z\sqrt{\lambda_1^2 + \lambda_2^2 - k^2} + i\lambda_3 z\right)}{\sqrt{\lambda_1^2 + \lambda_2^2 - k^2}}\, dz \\ &\qquad + \frac{1}{\sqrt{2\pi}}\int_{-\infty}^{0} \frac{\exp\left(z\sqrt{\lambda_1^2 + \lambda_2^2 - k^2} + i\lambda_3 z\right)}{\sqrt{\lambda_1^2 + \lambda_2^2 - k^2}}\, dz \\ &= \sqrt{\frac{2}{\pi}}\, \frac{1}{\lambda_1^2 + \lambda_2^2 + \lambda_3^2 - k^2}\end{aligned}$$

Thus

$$q(x,y,z) = \frac{1}{2\pi^2}\iiint_{-\infty}^{\infty} \frac{\exp(-i\lambda_1 x - i\lambda_2 y - i\lambda_3 z)}{\lambda_1^2 + \lambda_2^2 + \lambda_3^2 - k^2}\, d\lambda_1\, d\lambda_2\, d\lambda_3 \tag{7-112}$$

[1] See, for example, Lord Rayleigh, "Theory of Sound," vol. 2, chap. XI, Dover Publications, Inc., New York, 1945.

We have deliberately avoided any detailed analysis of the integration domains or the branches of $\lambda_1^2 + \lambda_2^2 - k^2$ in this transformation, but have anticipated that the appropriate domain of integration in $(\lambda_1,\lambda_2,\lambda_3)$ space will not be unduly pathological and that formal processes will lead to easily verifiable results. We continue in this way and change to spherical polar coordinates [first transforming to a λ_i' system of coordinates, rotated from the λ_i system, so that the point (x,y,z) is on the λ_3' axis] to give (with $\rho^2 = x^2 + y^2 + z^2$)

$$\begin{aligned} q(x,y,z) &= \frac{1}{2\pi^2} \int_0^{2\pi} d\varphi \int_0^\infty dr \int_0^\pi d\theta \, \frac{r^2 e^{-ir\rho\cos\theta} \sin\theta}{r^2 - k^2} \\ &= \frac{2}{\pi\rho} \int_0^\infty \frac{r \sin r\rho}{r^2 - k^2} \, dr = \frac{1}{\pi\rho} \int_{-\infty}^\infty \frac{r \sin r\rho}{r^2 - k^2} \, dr \end{aligned} \tag{7-113}$$

To obtain outgoing waves (with the time dependence $e^{i\omega t}$), we require the path of integration to pass below the point $r = -k$ and above the point $r = k$, so that

$$q(x,y,z) = \frac{1}{\rho} e^{-i\rho k} \tag{7-114}$$

Equation (7-110) then gives

$$\varphi(x,y,z) = -\frac{1}{2\pi} \iint_{-\infty}^{\infty} f(\xi,\eta) \, \frac{\exp\left(-ik\sqrt{(x-\xi)^2 + (y-\eta)^2 + z^2}\right)}{\sqrt{(x-\xi)^2 + (y-\eta)^2 + z^2}} \, d\xi \, d\eta \tag{7-115}$$

In many specific problems, the convolution form of the result is inconvenient, and frequently one can most profitably proceed from Eq. (7-109). As an illustration of this, let $f = 1$ for $x^2 + y^2 < 1$, $f = 0$ for $x^2 + y^2 > 1$ (so that the boundary is subjected to a rigid-piston motion for $x^2 + y^2 < 1$). Then

$$F(\lambda_1,\lambda_2) = \frac{1}{\rho} J_1(\rho)$$

where $\rho = \sqrt{\lambda_1^2 + \lambda_2^2}$, and Eq. (7-109) gives (with $r^2 = x^2 + y^2$)

$$\begin{aligned} \varphi(x,y,z) &= -\frac{1}{2\pi} \int_0^\infty d\rho \int_0^{2\pi} d\theta \, \frac{J_1(\rho) \exp\left(-z\sqrt{\rho^2 - k^2} - i\rho r \cos\theta\right)}{\sqrt{\rho^2 - k^2}} \\ &= -\int_0^\infty \frac{J_1(\rho) J_0(r\rho) \exp\left(-z\sqrt{\rho^2 - k^2}\right)}{\sqrt{\rho^2 - k^2}} \, d\rho \end{aligned} \tag{7-116}$$

The path is determined by the fact that, for outgoing waves to ensue, $\sqrt{\rho^2 - k^2}$ must be positive imaginary for $0 < \rho < k$ and positive for $\rho > k$. Thus, the branch line from the point $\rho = k$ lies below the path.

If $f(\rho)$ is a suitable odd function of ρ, then the reader may verify that

$$\int_\Gamma H_0^{(1)}(r\rho) f(\rho) \, d\rho = -2i \int_0^\infty J_0(r\rho) f(\rho) \, d\rho \tag{7-117}$$

where the path Γ is the real axis indented above the origin. Thus an alternative form of Eq. (7-116) is

$$\varphi(x,y,z) = \frac{1}{2i}\int_{\Gamma_1} \frac{J_1(\rho)H_0^{(1)}(r\rho)\exp\left(-z\sqrt{\rho^2-k^2}\right)}{\sqrt{\rho^2-k^2}}\,d\rho \qquad (7\text{-}118)$$

where Γ_1 consists of the real axis indented below the branch point at $-k$, above that at k, and above the origin.

EXERCISES

8. (*a*) Let $g(x,y,z,t;\xi,\eta,\xi,\tau)$ satisfy the wave equation

$$g_{xx} + g_{yy} + g_{zz} - \frac{1}{c^2}g_{tt} = \delta(x-\xi)\delta(y-\eta)\delta(z-\xi)\delta(t-\tau)$$

Use Fourier transforms to show that

$$g = -\frac{c}{4\pi\rho}\,\delta[\rho - c(t-\tau)] \qquad (7\text{-}119)$$

where $\rho^2 = (x-\xi)^2 + (y-\eta)^2 + (z-\xi)^2$.

(*b*) Let $g_1(x,y,z;\xi,\eta,\zeta)$ satisfy the Helmholtz equation

$$(g_1)_{xx} + (g_1)_{yy} + (g_1)_{zz} + \frac{\omega^2}{c^2}g_1 = \delta(x-\xi)\delta(y-\eta)\delta(z-\zeta)$$

Use Fourier transforms (and the radiation condition with time dependence $e^{i\omega t}$) to show that

$$g_1 = -\frac{1}{4\pi\rho}\,e^{-i\rho\omega/c} \qquad (7\text{-}120)$$

9. For a region V with surface S, show that if $\psi(x,y,z)$ satisfies

$$\psi_{xx} + \psi_{yy} + \psi_{zz} + \frac{\omega^2}{c^2}\psi = f(x,y,z)$$

and if $g_1(x,y,z;\xi,\eta,\zeta)$ satisfies the equation of Exercise 8*b*, then (Helmholtz)

$$\psi(\xi,\eta,\zeta) = \int_V g_1(x,y,z;\xi,\eta,\zeta)f(x,y,z)\,dV - \int_S\left(g_1\frac{\partial\psi}{\partial n} - \psi\frac{\partial g_1}{\partial n}\right)dS \qquad (7\text{-}121)$$

where all integrations are with respect to (x,y,z). Show also that, if V is a region *outside* a surface S_1 and inside a large sphere S_2 of radius R, then as a special case of the preceding theorem the surface integral over S becomes replaced by a surface integral over each of S_1 and S_2 and that, if the contribution from S_2 is to vanish as $R \to \infty$, then

$$R\left(\frac{\partial\psi}{\partial R} + i\frac{\omega}{c}\psi\right) \to 0 \qquad (7\text{-}122)$$

which is an analytical statement of Sommerfeld's radiation condition. [Had we worked throughout with a time dependence $e^{-i\omega t}$ rather than $e^{i\omega t}$, the plus sign in Eq. (7-122) would have been replaced by a minus sign.] Equation (7-121) is thus valid whether V is inside or outside S, provided that, in the latter case, the radiation condition is obeyed; it is often described as providing a mathematical statement of Huygens' principle.

10. Multiply both sides of Eq. (7-121) by $e^{i\omega t}$ so as to obtain an equation for φ at the observation point (ξ,η,ζ) and time t. Observe that, in g_1, ρ now denotes the distance between observation point and element of volume or surface. In the resulting equation, quantities such as $\psi(x,y,z)\ e^{i\omega[t-(\rho/c)]}$ can be interpreted as the value of φ at a certain time prior to t, the time difference being just that required for a signal moving with velocity c to cover the separation distance ρ from the observation point. Such a quantity is called the *retarded value* of φ and denoted by $[\varphi]_R$. Using this notation, show that the equation becomes

$$\varphi = -\frac{1}{4\pi}\int_V \frac{1}{\rho}[p]_R\,dV + \frac{1}{4\pi}\int_S \left\{\frac{1}{\rho}\left[\frac{\partial\varphi}{\partial n}\right]_R - [\varphi]_R\frac{\partial}{\partial n}\left(\frac{1}{\rho}\right) + \frac{1}{\rho c}\left[\frac{\partial\varphi}{\partial t}\right]_R\frac{\partial\rho}{\partial n}\right\} dS \qquad (7\text{-}123)$$

Since this equation does not involve ω and therefore has the same form for each frequency component of an arbitrary disturbance, it must hold generally for any function φ satisfying the wave equation (7-107). The result is known as *Kirchhoff's formula*.

11. For the special case in which S is a sphere centered on the observation point, Eq. (7-123) is called *Poisson's formula*. Show that in this case the result can be written

$$\varphi(P,t) = \frac{1}{4\pi c^2}\frac{\partial}{\partial t}\left[\frac{1}{t}\int_{r=ct}\varphi(Q,0)\,dS\right] + \frac{1}{4\pi c^2 t}\int_{r=ct}\varphi_t(Q,0)\,dS \qquad (7\text{-}124)$$

where P is the observation point and Q a point on the spherical surface of radius r surrounding P. Obtain the corresponding result for two space dimensions, and notice now that the integrations are over the entire interior of a circle, rather than just over its boundary.

12. Use the g function

$$g_2 = -\frac{1}{4\pi\rho}e^{i\rho\omega/c}$$

which differs from g_1 only in the sign of the exponent, to obtain an equation similar to Eq. (7-120) but involving *advanced* rather than retarded functions.

13. The velocity potential g corresponding to a moving acoustic source satisfies the equation

$$g_{xx} + g_{yy} + g_{zz} - \frac{1}{c^2} g_{tt} = \delta(x - Vt)\delta(y)\delta(z)$$

where V, c are constants and $-\infty < x < \infty$, $-\infty < y < \infty$, $-\infty < z < \infty$. Use Fourier transforms to show that

$$g = -\frac{1}{4\pi} \frac{1}{\{[x - V(t-\tau)]^2 + y^2 + z^2\}^{1/2} - (V/c)[x - V(t-\tau)]} \tag{7-125}$$

where τ is the time required for a signal to cover a certain distance; what is this distance? Verify the result by treating the moving impulse as a continuous superposition of impulses of the form given by Eq. (7-119).

14. A unit concentrated force is applied, for $t > 0$, at the point $x = Vt$ to an infinite stretched string. The governing equation for the deflection $y(x,t)$ is

$$Ty_{xx} - \rho y_{tt} = -\delta(x - Vt)\mu(t)$$

where T is the tension and ρ the linear density in the string, and where $\mu(t) = 1$ or 0 according to whether $t > 0$ or $t < 0$. Setting $c^2 = T/\rho$, use a transform method to show that the solution is the sum of the following three functions, and sketch the form of the solution for each of the cases $V < c$, $V > c$:

(*a*) $x > Vt$: 0

$x < Vt$: $-\dfrac{1}{T}\left(1 - \dfrac{V^2}{c^2}\right)^{-1}(Vt - x)$

(*b*) $x > ct$: 0

$x < ct$: $\dfrac{1}{2T}\left(1 - \dfrac{V}{c}\right)^{-1}(ct - x)$

(*c*) $x + ct > 0$: 0

$x + ct < 0$: $-\dfrac{1}{2T}\left(1 + \dfrac{V}{c}\right)^{-1}(ct + x)$

Why does it not matter in which way the path in the transform plane is indented about the origin?

3. *An Elasticity Problem*

As an example of the use of transforms in a problem governed by coupled differential equations, let us find the deflection pattern in an infinite elastic body, resulting from the application of a concentrated unit force at the origin. Let the coordinate axes be denoted by x_1, x_2, x_3, and denote the ith component of the displacement vector by u_i. A differentiation with respect to x_i is denoted by $(\ )_{,i}$, so that $u_{i,j}$, for example, means $\partial u_i/\partial x_j$; also, a repeated index is to be summed from 1 to 3, so that, for example, $u_{i,jj}$ means the laplacian of u_i. The governing equations[1] are

$$u_{i,jj} + \frac{1}{1-2\sigma} u_{j,ji} + \frac{2(1+\sigma)}{E} \left(\rho F_i - \rho \frac{\partial^2 u_i}{\partial t^2} \right) = 0 \qquad (7\text{-}126)$$

where σ is Poisson's ratio (usually, $0 < \sigma < \frac{1}{2}$), E is the elastic modulus, ρ is the density, and F_i is the ith component of the body force per unit mass. There are three possible choices for the "free" index i in Eq. (7-126); so Eq. (7-126) is really a set of three second-order coupled partial-differential equations.

For the present, we are interested in the case in which $\partial^2 u_i/\partial t^2 = 0$, and $\rho F_i = \delta(x_1)\delta(x_2)\delta(x_3)\delta_{is}$, where δ_{is} (the Kronecker delta) is 1 for $i = s$ and 0 otherwise; i.e., we take the concentrated force as being applied in the direction of the positive x_s axis. Taking Fourier transforms of the resultant equation, we obtain

$$-\lambda_j\lambda_j U_i - \frac{1}{1-2\sigma} \lambda_j\lambda_i U_j + \frac{2(1+\sigma)}{E} \delta_{is} \left(\frac{1}{\sqrt{2\pi}} \right)^3 = 0$$

where the three transform variables λ_1, λ_2, λ_3 correspond to x_1, x_2, x_3, respectively, and where $U_i(\lambda_1,\lambda_2,\lambda_3)$ is the triple Fourier transform of $u_i(x_1,x_2,x_3)$. At this point, it is clear that, when we solve for U_i, there will be an awkward singularity at the origin of the λ space; to circumvent this, let us introduce an "elastic-restraint" term $-\epsilon^2 u_i$ on the left-hand side of Eq. (7-126). Here ϵ is a small positive number, which $\to 0$ eventually. The transform equation then becomes

$$-(\epsilon^2 + \lambda_j\lambda_j) U_i - \frac{1}{1-2\sigma} \lambda_j\lambda_i U_j + \frac{2(1+\sigma)}{E} \delta_{is} \left(\frac{1}{\sqrt{2\pi}} \right)^3 = 0 \qquad (7\text{-}127)$$

Multiplying through by λ_i allows us to solve for $\lambda_j U_j$, whose value can then be inserted in the second term of Eq. (7-127) to give an expression for U_i. Applying the inversion theorem, we obtain

$$u_i = \frac{2(1+\sigma)}{(2\pi)^3 E} \iiint_{-\infty}^{\infty} \frac{\exp(-i\lambda_j x_j)\, d\lambda_1\, d\lambda_2\, d\lambda_3}{\epsilon^2 + \lambda_j\lambda_j} \left[\delta_{is} - \frac{\lambda_i\lambda_s}{(1-2\sigma)\epsilon^2 + (2-2\sigma)\lambda_j\lambda_j} \right] \qquad (7\text{-}128)$$

In the second term, the factor $-\lambda_i\lambda_s$ can be replaced by a differentiation of the integral with respect to each of x_i and x_s. Denote $x_j x_j$ by ρ^2, and rotate the λ_j axes so that the point x_i lies on the new λ_3 axis. Changing to spherical

[1] C. Pearson, "Theoretical Elasticity," p. 87, Harvard University Press, Cambridge, Mass., 1959. The first chapter of this reference gives a survey of index notation in vector analysis.

polar coordinates (r,θ,φ), we obtain, for the first term,

$$\frac{2(1+\sigma)}{(2\pi)^3E}\int_0^\infty dr\int_0^\pi d\theta\int_0^{2\pi} d\varphi\,\frac{e^{-ir\rho\cos\theta}r^2\sin\theta}{\epsilon^2+r^2}\,\delta_{is}$$

which $\rightarrow (1+\sigma)\delta_{is}/2\pi E\rho$ as $\epsilon\rightarrow 0$, and, for the second term,

$$\left\{\frac{2(1+\sigma)}{(2\pi)^3E}\int_0^\infty dr\int_0^\pi d\theta\int_0^{2\pi} d\varphi\,\frac{e^{-ir\rho\cos\theta}r^2\sin\theta}{(\epsilon^2+r^2)[(1-2\sigma)\epsilon^2+(2-2\sigma)r^2]}\right\}_{,is}$$

$$=\frac{1+\sigma}{2\pi\rho E\epsilon^2}\left[-e^{-\epsilon\rho}+\exp\left(-\epsilon\rho\sqrt{\frac{1-2\sigma}{2-2\sigma}}\right)\right]_{,is}$$

Upon expanding in powers of ϵ, this second term becomes

$$\frac{1+\sigma}{2\pi E}\left[\frac{1}{\epsilon}\left(1-\sqrt{\frac{1-2\sigma}{2-2\sigma}}\right)-\frac{\rho}{2}\frac{1}{2-2\sigma}+0(\epsilon)\right]_{,is}$$

The portion of this expression not dependent on s is discarded (it merely tells us that the displacement in an infinite body due to a concentrated force has an infinite component—so we measure only the relative displacement); as $\epsilon\rightarrow 0$, we now combine results to obtain the final solution as

$$u_i=\frac{1}{4\pi E}\frac{1+\sigma}{1-\sigma}\left[(2-2\sigma)\frac{\delta_{is}}{\rho}-\frac{1}{2}\rho_{,is}\right] \tag{7-129}$$

We could have performed the inversion of Eq. (7-128), without the ϵ^2 term, by observing that the inverse transform of $1/\lambda_j\lambda_j$ is $\sqrt{\pi/2}\,(1/\rho)$, and since $\rho_{,ii}=2/\rho$, the transform R of ρ is given by $-\lambda_i\lambda_iR=2\sqrt{2/\pi}/\lambda_j\lambda_j$, whence $R=-2\sqrt{2/\pi}/(\lambda_j\lambda_j)^2$, so that the second term of Eq. (7-128) is easily inverted. Of course, such a calculation is somewhat casual—however, the result could easily be verified to satisfy Eq. (7-126).

Exercises

15. Use transforms to solve the time-dependent equation (7-126) corresponding to the application of a unit impulsive force to the origin of an infinite body.

16. Show that the displacement corresponding to a unit force in the positive x_1 direction applied at the origin of the elastic half-space $x_3\leq 0$ is given by

$$u_i=\frac{1+\sigma}{2\pi Er}\left\{\delta_{1i}+\frac{x_1x_i}{r^2}+\frac{1-2\sigma}{(r-x_3)^2}\left[r(r-x_3)\delta_{1i}-x_1x_i+x_1(2x_3-r)\delta_{3i}\right]\right\}$$

(*Note:* The stress vector on the x_3 surface has components $T_i=\tau_{i3}$ where

$$\tau_{ij}=\frac{E}{2(1+\sigma)}\left(u_{i,j}+u_{j,i}+\frac{2\sigma}{1-2\sigma}u_{k,k}\delta_{ij}\right)$$

Our boundary condition requires $T_1 = \delta(x_1)\delta(x_2)$, $T_2 = T_3 = 0$ at the point (0,0,0).)

17. Maxwell's equations for the electric field E_i and magnetic field H_i resulting from a concentrated unit charge moving along the x_3 direction with velocity V are

$$\begin{aligned} E_{i,i} &= 4\pi\delta(x_1)\delta(x_2)\delta(x_3 - Vt) \\ H_{i,i} &= 0 \\ e_{ijk}E_{k,j} &= -\frac{1}{c}\frac{\partial H_i}{\partial t} \\ e_{ijk}H_{k,j} &= \frac{4\pi}{c}\,\delta_{i3}V\delta(x_1)\delta(x_2)\delta(x_3 - Vt) + \frac{1}{c}\frac{\partial E_i}{\partial t} \end{aligned}$$

where c is the velocity of light and e_{ijk} is the permutation symbol (equal to 0 unless i, j, k are a permutation of 1, 2, 3, in which case it equals 1 for an even and -1 for an odd permutation). Use transforms to find E_i and H_i; in particular, show that

$$H_i = \frac{1}{c}\left(1 - \frac{V^2}{c^2}\right)\frac{e_{i3k}Vx_k}{r^3[1 - (V^2/c^2)\sin^2\theta]^{3/2}}$$

where $\sin^2\theta = (x_1^2 + x_2^2)/r^2$ and $r^2 = x_ix_i$. The reader may verify this result by use of a Lorentz transformation applied to the electrostatic field of a stationary charge, if desired.

4. *A Viscous-flow Problem*

The following mathematical problem arises in the analysis of questions dealing with heat transfer and/or the flow of viscous fluids. We seek a function $\psi(x,y)$ which satisfies

$$\Delta\left(\Delta - \frac{\partial}{\partial x}\right)\psi = 0 \tag{7-130}$$

in the region $-\infty < x < \infty$, $y \geq 0$; here Δ is the laplacian operator $\partial^2/\partial x^2 + \partial^2/\partial y^2$. The boundary conditions are:

$$\begin{aligned} \psi(x,0) &= 0 \\ \psi_y(x,0) &= f(x) \qquad &&\text{for } x \text{ in } (a,b) \\ \psi_{yy}(x,0) &= 0 &&\text{for } x < a \text{ or for } x > b \\ \psi(x,y) &\to 0 &&\text{as } x^2 + y^2 \to \infty \end{aligned}$$

A Fourier transform in x of Eq. (7-130) leads to

$$\left(\frac{\partial^2}{\partial y^2} - \lambda^2\right)\left[\frac{\partial^2}{\partial y^2} - (\lambda^2 - i\lambda)\right]\Psi(\lambda,y) = 0$$

from which, by discarding the solutions that do not decay as $y \to \infty$, it follows that

$$\Psi = A(\lambda)e^{-|\lambda|y} + B(\lambda)\exp\left(-\sqrt{\lambda^2 - i\lambda}\,y\right)$$

The expression $\sqrt{\lambda^2 - i\lambda}$ must have a positive real part for real λ, and this can be achieved by taking its branch cut along the imaginary axis from i to $i\infty$ and, of course, by making the appropriate choice of branch. From $\psi(x,0) = 0$, we obtain $A(\lambda) = -B(\lambda)$, so that

$$\Psi = A(\lambda)[e^{-|\lambda|y} - \exp(-\sqrt{\lambda^2 - i\lambda}\, y)] \tag{7-131}$$

Denote the function $\psi_{yy}(x,0)$ by $s(x)$, so that $s(x) = 0$ for $x < a$ or $x > b$ and $s(x)$ is unknown for x in (a,b). Differentiating Eq. (7-131) twice with respect to y and setting $y = 0$, we obtain

$$i\lambda A(\lambda) = S(\lambda)$$

where $S(\lambda)$ is the (as yet unknown) transform of $s(x)$. Thus Eq. (7-131) becomes

$$\Psi = \frac{S(\lambda)}{i\lambda}[e^{-|\lambda|y} - \exp(-\sqrt{\lambda^2 - i\lambda}\, y)] \tag{7-132}$$

from which
$$\Psi_y(\lambda,0) = S(\lambda)\left(\frac{\sqrt{\lambda^2 - i\lambda} - |\lambda|}{i\lambda}\right)$$

Inverting, we now obtain

$$\psi_y(x,0) = \int_a^b s(\xi)k(x - \xi)\, d\xi \tag{7-133}$$

where we have used the fact that $s(\xi) = 0$ for ξ outside (a,b) and where

$$\begin{aligned} k(x) &= \frac{1}{2\pi}\int_{-\infty}^{\infty} e^{-i\lambda x}\left(\frac{\sqrt{\lambda^2 - i\lambda}}{i\lambda} - \frac{|\lambda|}{i\lambda}\right) d\lambda \\ &= \frac{1}{2\pi}\int_{-\infty}^{\infty} e^{-i\lambda x}\left(-\frac{1 + i\lambda}{\sqrt{\lambda^2 - i\lambda}} + \lim_{\epsilon\to 0}\frac{\epsilon + i\lambda}{\sqrt{\lambda^2 - i\epsilon\lambda}}\right) d\lambda \end{aligned}$$

Here we have replaced $|\lambda|$ by $\lim_{\epsilon\to 0}\sqrt{\lambda^2 - i\epsilon\lambda}$, $\epsilon > 0$. Both portions of this integrand have the same form; so we really need invert only one of them. We have, for example,

$$\begin{aligned} \frac{1}{2\pi}\int_{-\infty}^{\infty} e^{-i\lambda x}\left(\frac{\epsilon + i\lambda}{\sqrt{\lambda^2 - i\epsilon\lambda}}\right) d\lambda &= \left(\epsilon - \frac{\partial}{\partial x}\right)\frac{1}{2\pi}\int_{-\infty}^{\infty}\frac{e^{-i\lambda x}}{\sqrt{\lambda^2 - i\epsilon\lambda}} \\ &= \left(\epsilon - \frac{\partial}{\partial x}\right)\frac{e^{\epsilon x/2}}{2\pi}\int_{-\infty}^{\infty}\frac{e^{-i\zeta x}}{\sqrt{\zeta^2 + \frac{1}{4}\epsilon^2}}\, d\zeta \\ &= \left(\epsilon - \frac{\partial}{\partial x}\right)[e^{\epsilon x/2}K_0(\epsilon|x|/2)] \end{aligned}$$

where K_0 is the modified Bessel function of the second kind. By using the fact that $\psi_y(x,0) = f(x)$ for x in (a,b), Eq. (7-133) may now be written

$$f(x) = \int_a^b s(\xi)k(x - \xi)\, d\xi \qquad \text{for } x \text{ in } (a,b) \tag{7-134}$$

$$k(x) = -\tfrac{1}{2}e^{x/2}K_0\left(\frac{1}{2}|x|\right) + \left[\frac{1}{x} - \tfrac{1}{2}e^{x/2}(\operatorname{sgn} x)K_1\left(\frac{|x|}{2}\right)\right]$$

Equation (7-134) is an integral equation for $s(x)$, x in (a,b). Once it has been solved for $s(x)$, $S(\lambda)$ will then also be known and Ψ can be determined from Eq. (7-132). Thus, the foregoing use of the Fourier transform replaces a problem in which we must find $\Psi(x,y)$ over an infinite domain in both variables by one in which we must find $s(x)$, a function of one variable defined on the interval $a < x < b$. Integral equations of the form (7-134) are discussed in Chap. 8.

7-3 *The Laplace Transform*

Another exponential transform $F(s)$ of a function $f(t)$ which is frequently studied (sometimes without reference to the Fourier transform) is defined by

$$F(s) = \int_0^\infty e^{-st}f(t)\,dt \tag{7-135}$$

and is called the *Laplace transform*[1] of $f(t)$. Clearly, upon identifying s with $-i\lambda$ of Eq. (7-7), the Laplace transform of $f(t)$ is nothing more or less than $\sqrt{2\pi}$ times the Fourier transform of that function which coincides with $f(t)$ for $t > 0$ and which vanishes for $t < 0$.

It is useful to prove that, if the integral of Eq. (7-135) converges for $s = s_0$, then it also converges for any value of s satisfying $\operatorname{Re} s > \operatorname{Re} s_0$; thus in general there will be some *abscissa of convergence* (and therefore a half-plane of convergence). To show this, we write

$$F(s) = \lim_{T\to\infty} \int_0^T \exp\,[-(s - s_0)t]\,\exp\,(-s_0t)f(t)\,dt \tag{7-136}$$

Define $$\varphi(t) = \int_0^t \exp\,(-s_0t)f(t)\,dt$$

and observe that $\varphi(t)$ must be bounded for all t, because of the convergence hypothesis. An integration by parts of Eq. (7-136) gives

$$F(s) = \lim_{T\to\infty} \int_0^T (s - s_0)\exp\,[-(s - s_0)t]\varphi(t)\,dt$$

which clearly converges for $\operatorname{Re} s > \operatorname{Re} s_0$. Moreover, the convergence is uniform in any bounded region lying wholly within this half-plane of convergence.

For $\operatorname{Re} s > \operatorname{Re} s_0$, the application of the formula for a derivative shows that, for any fixed choice of T, the quantity $\int_0^T e^{-st}f(t)\,dt$ is an analytic function of s. Because of the uniform-convergence property just mentioned, it now follows that $F(s)$ must be analytic in the half-plane $\operatorname{Re} s > \operatorname{Re} s_0$, as well as in whatever additional region can be reached by analytic continua-

[1] We shall ordinarily use s as a Laplace-transform variable and λ as a Fourier-transform variable, so that the use of F to denote both transform functions should cause no difficulty.

tion. As a simple example of analytic continuation, we observe that the Laplace transform of $f(t) = 1$ is $1/s$, which is analytic everywhere except the origin, despite the fact that the defining integral converges only for $\operatorname{Re} s > 0$.

If we write $s = \gamma - i\lambda$, where γ and λ are real, then $F(\gamma - i\lambda) = P(\lambda)$ is the Fourier transform of the function $\sqrt{2\pi}\, p(t)$ defined by

$$p(t) = \begin{cases} e^{-\gamma t} f(t) & \text{for } t > 0 \\ 0 & \text{for } t < 0 \end{cases}$$

Consequently, the Fourier inversion theorem [Eq. (7-8)] leads to

$$\frac{1}{2\pi i} \int_{\gamma - i\infty}^{\gamma + i\infty} e^{st} F(s)\, ds = \begin{cases} 0 & \text{for } t < 0 \\ f(t) & \text{for } t > 0 \end{cases} \tag{7-137}$$

Equation (7-137) is the inversion theorem for Laplace transforms. In the foregoing derivation, γ can have any value for which s is in the half-plane of convergence; thus the path of integration in Eq. (7-137) can be any vertical line to the right of all singularities of $F(s)$. The result is then independent of the particular choice of γ. As in the case of Fourier transforms, the inversion integral yields $\frac{1}{2}[f(t+) + f(t-)]$ at a point of discontinuity of $f(t)$.

Denoting the transform of the nth derivative of $f(t)$ by $F_n(s)$ and integrating by parts, we see that

$$\begin{aligned} F_1(s) &= -f(0+) + sF(s) \\ F_2(s) &= -f'(0+) - sf(0+) + s^2 F(s) \\ F_3(s) &= -f''(0+) - sf'(0+) - s^2 f(0+) + s^3 F(s) \\ &\cdots\cdots\cdots\cdots\cdots\cdots \end{aligned} \tag{7-138}$$

where by $f(0+)$, for example, we mean the limit of $f(t)$ as $t \to 0$ with $t > 0$ [thus, if $f(t) = 1$ for $t > 0$, $f(t) = 0$ for $t < 0$, we have $f(0+) = 1$].

The convolution theorem for Laplace transforms may be proved either directly or as a special case of that for Fourier transforms. Let

$$f(t) = \int_0^t g(t - \tau) h(\tau)\, d\tau \tag{7-139}$$

where the Laplace transforms of $f(t)$, $g(t)$, $h(t)$ are $F(s)$, $G(s)$, $H(s)$; then

$$\begin{aligned} F(s) &= \int_0^\infty e^{-st}\, dt \int_0^t g(t - \tau) h(\tau)\, d\tau \\ &= \int_0^\infty d\tau \int_\tau^\infty dt\, e^{-st} g(t - \tau) h(\tau) \\ &= \int_0^\infty d\tau\, e^{-s\tau} h(\tau) \int_\tau^\infty dt\, e^{-s(t-\tau)} g(t - \tau)\, dt \\ &= H(s) G(s) \end{aligned} \tag{7-140}$$

Conversely, the inverse Laplace transform of the product $H(s) \cdot G(s)$ is given by Eq. (7-139).

Equation (7-135) implies that, as $s \to \infty$ along some ray satisfying $|\arg s| < \pi/2$, $F(s)$ must approach zero; this is a test that can often be applied to see whether or not the inverse transform of a given $F(s)$ exists—at least as a conventional function $f(t)$. This result is, of course, not valid for such improper functions as $\delta(t)$ or $\delta'(t)$; for example, the transform of $\delta(t)$ would be 1. [We interpret $\delta(t)$ as $\delta(t - \epsilon)$, where $\epsilon > 0$ is arbitrarily small, so that the range of integration in Eq. (7-135) includes the "impulse" point.] Just as in the case of Fourier transforms, the equations for transforms of derivatives, and the convolution theorem, may be used for members of the δ-function family, subject of course to our usual interpretation of the δ function as a deferred limiting process.

Tables of Laplace transforms are given in Erdélyi et al.[1] and in Doetsch, Kniess, and Voelker.[2] Among books dealing with Laplace transforms may be mentioned those by G. Doetsch[3] and by Van der Pol and Bremmer.[4]

EXERCISES

1. (*a*) Show that if $f(t) = t^\alpha$, Re $\alpha > -1$, then $F(s) = \Gamma(\alpha + 1)/s^{\alpha+1}$.
(*b*) Express the inverse function to $e^{-\alpha\sqrt{s}}/s$ in terms of the error function; here α is a positive real constant.

2. Use the generating function for the Hermite polynomials (Exercise 3 of Sec. 3-3) to show that the Laplace transform of $H_n(t)/n!$ is the coefficient of ξ^n in the expansion of $\exp(-\xi^2)/(s - 2\xi)$.

3. Take the transform of Bessel's equation for functions of order zero [noting that, if $y(t)$ has the transform Y/s, then $ty(t)$ has the transform $-Y'(s)$], and show that the transform of $J_0(t)$ is $(1 + s^2)^{-1/2}$.

Use the method of Exercise 3 to find the Laplace transform of $J_n(t)$, n a positive integer. [The resulting differential equation for the transform, $F(s)$, may be simplified by the substitutions $F(s) = G(s)(1 + s^2)^{-1/2}$ and $s = \sinh \omega$.]

4. By summing residues, invert

$$F(s) = \frac{\cosh 2x\sqrt{s}}{\sqrt{s}\sinh\sqrt{s}}$$

where x is real and lies in $(-\tfrac{1}{2},\tfrac{1}{2})$, to obtain

$$f(t) = 1 + 2\sum_{n=1}^{\infty} (-1)^n \exp(-n^2\pi^2 t)\cos 2\pi n x$$

This function $f(t)$ is denoted by $\theta(x,t)$ and is called a *theta function;* it satisfies the heat-conduction equation $\varphi_{xx} = 4\varphi_t$. The other three

[1] Erdélyi, Magnus, Oberhettinger, and Tricomi, *op. cit.*

[2] G. Doetsch, H. Kniess, and D. Voelker, "Tabellen zur Laplace Transformation," Springer-Verlag OHG, Berlin, 1947.

[3] G. Doetsch, "Handbuch der Laplace Transformation," Birkhäuser Verlag, Basel, 1955.

[4] B. Van der Pol and H. Bremmer, "Operational Calculus," Cambridge University Press, New York, 1950.

members of the theta-function family[1] and their transforms are listed here for reference.

$$\theta_1(x,t) = 2 \sum_{n=0}^{\infty} (-1)^n \exp\left[-\left(n+\frac{1}{2}\right)^2 \pi^2 t\right] \sin\,(2n+1)\pi x:$$

$$-\frac{\sinh 2x\sqrt{s}}{\sqrt{s}\cosh\sqrt{s}} \qquad \text{for } x \text{ in } \left(-\frac{1}{2},\frac{1}{2}\right)$$

$$\theta_2(x,t) = 2 \sum_{n=0}^{\infty} \exp\left[-\left(n+\frac{1}{2}\right)^2 \pi^2 t\right] \cos\,(2n+1)\pi x:$$

$$-\frac{\sinh\,(2x-1)\sqrt{s}}{\sqrt{s}\cosh\sqrt{s}} \qquad \text{for } x \text{ in } (0,1)$$

$$\theta_3(x,t) = 1 + 2 \sum_{n=1}^{\infty} \exp\,(-n^2\pi^2 t) \cos 2\pi n x:$$

$$\frac{\cosh\,(2x-1)\sqrt{s}}{\sqrt{s}\sinh\sqrt{s}} \qquad \text{for } x \text{ in } (0,1)$$

5. (*a*) Show that, if the Laplace transform of $f(t)$ is $F(s)$ and if the transform of $f(t)/t$ exists, it has the value $\int_s^\infty F(s)\,ds$. Show also that the transform of $tf(t)$ is $-F'(s)$. [It may be useful to differentiate or integrate an awkward expression for $F(s)$ before inversion, if simpler manipulations result.]

(*b*) Show that, if $sF(s)$ has an inverse Laplace transform, it will be given by $f'(t)$, where $F(s)$ is the transform of $f(t)$. Is this result compatible with Eq. (7-138)? Show also that the inverse transform of $(1/s)F(s)$ is $\int_0^t f(t)\,dt$.

6. (*a*) Let $f(t)$ be periodic, with period T. If $g(t)$ is defined to equal $f(t)$ in $(0,T)$ and to vanish elsewhere and if its transform is $G(s)$, show that the transform of $f(t)$ is given by

$$F(s) = \frac{G(s)}{1 - e^{-sT}}$$

(*b*) Find the inverse of $F(s) = (1/s^2)\tanh sT$, making use of the idea encountered in Exercise 6*a*.

7. By taking a Laplace transform in t, show that

$$\int_0^\infty \exp\left(-x^2 - \frac{t^2}{x^2}\right) dx = \frac{\sqrt{\pi}}{2}\, e^{-2t}$$

[1] Theta functions are discussed in E. Whittaker and G. Watson, "A Course of Modern Analysis," 4th ed., chap. 21, Cambridge University Press, New York, 1927. There is no standard notation for these functions.

Evaluate next the expression

$$\iint_0^\infty dt\,dx\,\exp\left(-\frac{1}{4}\frac{s^2}{x^2}-x^2t\right)f(t)$$

in two ways (i.e., with each possible order of integration) so as to show that, if $f(t)$ has the transform $F(s)$, $f(t^2)$ will have the transform[1]

$$\frac{1}{\sqrt{\pi}}\int_0^\infty \exp\left(-\frac{s^2}{4x^2}\right)F(x^2)\,dx$$

8. In the repeated convolution integral,

$$\int_0^t g(t-\tau)\,d\tau\int_0^\tau f(\tau-\xi)p(\xi)\,d\xi$$

whose transform is $G(s)F(s)P(s)$, set $p(\xi) = h(\xi)$ (the unit function; $h = 1$ for $\xi > 0$ and $h = 0$ for $\xi < 0$), and show that the integral then takes the form

$$\iint g(\tau_1)f(\tau_2)\,d\tau_1\,d\tau_2$$

where the region of integration is over all positive values of τ_1 and τ_2 for which $\tau_1 + \tau_2 \le t$. The area of a circle of radius R is given by $4\iint dx\,dy$, where the region of integration is over those positive values of x and y for which $x^2 + y^2 \le R^2$. Set $x^2 = \tau_1$ and $y^2 = \tau_2$ so as to obtain an integral of the previous form (a certain function of R), and obtain its transform; invert the transform to obtain a formula for the area of a circle. Use this same method to show that the volume of a sphere in n-dimensional space is

$$V_n = \frac{\pi^{n/2}R^n}{\Gamma(n/2+1)}$$

where $R^2 = x_1^2 + x_2^2 + \cdots + x_n^2$ is the square of the radius.

9. Use Laplace transforms to prove that

$$\int_0^t dt_1\int_0^{t_1} dt_2\cdots\int_0^{t_{m-1}} f(t_m)\,dt_m = \int_0^t \frac{(t-\tau)^{m-1}}{(m-1)!}f(\tau)\,d\tau$$

Ordinary Differential Equations

Consider a differential equation of nth order for the function $y(t)$,

$$y^{(n)} + a_{n-1}y^{(n-1)} + \cdots + a_1y' + a_0y = f(t) \tag{7-141}$$

[1] A number of rules of this character, some of them of use in electric-circuit theory, are given in chap. 11 of Van der Pol and Bremmer, *op. cit.*

where the a_i are constants, $f(t)$ is a given function of t, and $y^{(j)}$ denotes as usual the jth derivative of $y(t)$. Let the values of $y(0), y'(0), \ldots, y^{(n-1)}(0)$ be prescribed, and let it be desired to determine the solution for $t > 0$. Denote the Laplace transforms of $y(t)$ and $f(t)$ by $Y(s)$ and $F(s)$, respectively. Then, taking transforms of Eq. (7-141), we obtain

$$\begin{aligned} P(s)Y(s) = F(s) &+ y(0)(s^{n-1} + a_{n-1}s^{n-2} + \cdots + a_2 s + a_1) \\ &+ y'(0)(s^{n-2} + a_{n-1}s^{n-3} + \cdots + a_3 s + a_2) \\ &+ \cdots \\ &+ y^{(n-2)}(0)(s + a_{n-1}) \\ &+ y^{(n-1)}(0) \end{aligned} \tag{7-142}$$

where
$$P(s) = s^{n-1} + a_{n-1}s^{n-1} + \cdots + a_1 s + a_0 \tag{7-143}$$

The inversion theorem yields

$$y(t) = \frac{1}{2\pi i}\int_{\gamma-i\infty}^{\gamma+i\infty} e^{st}\frac{F(s)}{P(s)}\,ds + \frac{1}{2\pi i}\sum_{j=0}^{n-1} y^{(n-1-j)}(0)\int_{\gamma-i\infty}^{\gamma+i\infty} e^{st}\frac{P_j(s)}{P(s)}\,ds \tag{7-144}$$

where $P_j(s)$ is the polynomial of order j in s that multiplies $y^{(n-1-j)}(0)$ in Eq. (7-142). Clearly, the first term in Eq. (7-144) is a particular solution of the nonhomogeneous equation (7-141); the other terms are solutions of the corresponding homogeneous equation. Unless the polynomial $P(s)$ has multiple roots, these solutions of the homogeneous equation are sums of exponentials of the form $\exp(\alpha_i t)$, α_i constant; if $P(s)$ does have multiple roots, there will be solutions of the form $t^p e^{\alpha t}$, p integral.

More specifically, let us consider the inverse of

$$W(s) = \frac{Q(s)}{P(s)} \tag{7-145}$$

where $Q(s)$ and $P(s)$ are polynomials, with the degree of $Q(s)$ lower than that of $P(s)$. If the zeros of $P(s)$ are each of first order and are denoted by α_1, $\alpha_2, \ldots, \alpha_n$, then the inverse of Eq. (7-145) is

$$w(t) = \sum_{j=1}^{n} \frac{Q(\alpha_j)\exp(\alpha_j t)}{P'(\alpha_j)} \tag{7-146}$$

If, however, the zero α_j is of order p_j, the term in $\exp(\alpha_j t)$ is to be replaced by

$$\frac{1}{(p_j-1)!}\left\{\frac{d^{(p_j-1)}}{ds^{(p_j-1)}}\left[\frac{(s-\alpha_j)^{p_j}Q(s)e^{st}}{P(s)}\right]\right\}_{s=\alpha_j}$$

which reduces, in general, to a $(p_j - 1)$st-order polynomial in t multiplying $\exp(\alpha_j t)$. The result can, of course, also be expressed in terms of a partial-fraction expansion for $1/P(s)$. Equation (7-146), including the modification necessary for multiple zeros, is called *Heaviside's expansion theorem.*[1]

In applying the Laplace-transform method to a differential equation of order n for $y(t)$, we usually have to know the values of $y(0)$, $y'(0)$, . . . , $y^{(n-1)}(0)$; however, there may be cases where one or more of these quantities can be left as unknown constants, whose values are to be subsequently determined. Examples of such situations are given in Exercise 11 below. These examples are also ones for which the range of the independent variable is finite rather than semi-infinite; we simply assume the differential equation to hold over the entire semi-infinite range and then adjust the unknown values of some of the derivatives at the left-hand end of the interval so as to obtain the correct boundary conditions at the right-hand end.

[1] An interesting survey of Oliver Heaviside's work is given in "The Heaviside Centenary Volume," Institution of Electrical Engineers, London, 1950. Heaviside did not use transforms but rather exploited the algebraic properties of the differentiation operator $p = d/dt$. Writing the solution of $py = f$ as $y = \int_0^t f\,dt$ suggested to him that p^{-1} should be interpreted as $\int_0^t$. Then, to solve $(p - a)y = h(t)$, where the right-hand side is Heaviside's unit function, he wrote

$$\begin{aligned} y = \frac{h}{p-a} &= \frac{1}{p}\left(1 + \frac{a}{p} + \frac{a^2}{p^2} + \cdots\right)h \\ &= \frac{1}{p}\left(1 + at + \frac{a^2}{2!}t^2 + \cdots\right) \\ &= \frac{1}{p}\,e^{at} = \int_0^t e^{at}\,dt \end{aligned}$$

Thus he was led to the idea of expanding a p polynomial in partial fractions and interpreting the individual operators so obtained in terms of exponentials. If the right-hand side of the differential equation was an exponential function (e.g., an a-c signal applied to an electrical circuit), he used his "shifting theorem," by means of which

$$\frac{1}{p-a}(e^{-bt}h) = e^{-bt}\frac{1}{p-b-a}(h)$$

[Of course, only the solution for the unit function is really necessary, since any arbitrary $f(t)$ can be expressed as a continuous superposition of step functions.]

For an equation such as (7-141), with $f(t) = h(t)$, Heaviside would write

$$(p^n + a_{n-1}p^{n-1} + \cdots + a_1p + a_0)y = 1$$

Since the corresponding Laplace-transform equation is

$$(s^n + a_{n-1}s^{n-1} + \cdots + a_1s + a_0)Y = \frac{1}{s}$$

it is clear that Heaviside's method is essentially equivalent to the use of (sY) as the transform function, rather than Y itself. For that reason, some modern writers use $s\int_0^\infty e^{-st}y(t)\,dt$ as the definition of the Laplace transform of $y(t)$.

We remark also that in an equation such as (7-144), if $1/P(s)$ is expressed in partial-fraction form, then the particular integral can be written at once as a linear combination of terms of the form

$$\int_0^t f(t - \tau)e^{\alpha_i \tau}\, d\tau$$

(with an appropriate modification for the case of a multiple zero).

EXERCISES

10. Compare the D method[1] with the Laplace-transform method for the solution of the differential-equation set

$$\begin{aligned} y' + a_1 y + a_2 z &= h(t) \\ z' + b_1 z + b_2 y &= 0 \end{aligned}$$

where $y(t)$ and $z(t)$ are to be determined, a_i and b_i are constants, and $y(0)$ and $z(0)$ are specified.

11. (*a*) Let $y'' + y = f(t)$, $y(0) = y(\pi) = 0$. Use the Laplace-transform method (i.e., temporarily extending the t range to infinity) to show that this equation has a solution if and only if $\int_0^\pi f(t) \sin t\, dt = 0$. Is the solution, if it exists, unique?

(*b*) Let $y'' + \omega_0^2 y = \sin \omega t$, for $t > 0$, where ω and ω_0 are real constants and where $y(0) = y'(0) = 0$. When $\omega = \omega_0$, we expect resonance; how would this show up in the Laplace-transform method?

Solution for Large t and Small t. A Difference-equation Example

The Laplace transform of $f(t + c)$, where c is a constant, is simply related to the transform of $f(t)$; consequently, the Laplace-transform method can often be useful in dealing with certain kinds of difference equations. We choose here a simple example of a differential-difference equation, of a kind which arises in control-system analysis (in which the instantaneous rate of response of the system depends on a past value of the input function). Consider

$$y'(t) = y(t - 1) \tag{7-147}$$

which is to hold in $0 < t < \infty$. For $-1 \leq t < 0$, $y(t) = f(t)$, where $f(t)$ is some prescribed function. Given any $f(t)$, it is clear that Eq. (7-147) will then determine $y(t)$ for t in $(0, \infty)$. In fact, the Laplace transform of Eq. (7-147) is

$$-f(0) + sY(s) = e^{-s}\left[\int_{-1}^0 e^{-s\tau} f(\tau)\, d\tau + Y(s)\right] \tag{7-148}$$

[1] See, for example, F. B. Hildebrand, "Advanced Calculus for Engineers," p. 17, Prentice-Hall, Inc., Englewood Cliffs, N.J., 1949.

As an explicit example, we take $f(t) = 1$, and Eq. (7-148) becomes

$$Y(s) = \frac{1}{s} + \frac{1}{s(s - e^{-s})} \tag{7-149}$$

The path of integration for the inversion integral may be closed in the left-hand s plane, and $y(t)$ can be described by a sum of residues. It may be verified that the zeros of $s - e^{-s}$ are at $r_1 = 0.567$, $r_2 = -1.54 + i4.38$, $r_3 = -1.54 - i4.38$, $r_4 = -2.37 + i10.8$, $r_5 = -2.37 - i10.8$, . . . , in order of increasing negative real part. (Asymptotically, the zeros are given by $-\ln [(4n + 3)\pi/2] \pm i(4n + 3)\pi/2$.) The sum of the residues gives

$$y(t) = \sum_{j=1}^{\infty} \frac{\exp(r_j t)}{r_j[1 + \exp(-r_j)]} = \sum_{j=1}^{\infty} \frac{\exp(r_j t)}{r_j(1 + r_j)} \tag{7-150}$$

For large t, only the first term is important, so that $y \sim 1.125e^{0.567t}$. The choice of a different $f(t)$ would alter only the numerator, and not the denominator, of Eq. (7-149). Consequently, the exponential terms occurring in Eq. (7-150) will be the same; only the coefficients multiplying them will be different. [For example, the reader may check that, if $f(t) = \delta(t + \epsilon)$, $\epsilon > 0$ and $\epsilon \to 0$, then the leading term in Eq. (7-150) will turn out to be $0.311e^{0.567t}$.] It follows that as far as the *stability* of the control system is concerned—i.e., whether solutions of Eq. (7-147) will involve bounded or unbounded exponential functions of t—we need look only at the denominator of Eq. (7-149). In turn, this is equivalent to setting $y = e^{\alpha t}$ in Eq. (7-147) and asking whether or not there are any values of α satisfying the equation for which $\operatorname{Re} \alpha > 0$. Thus we could say at once that the modified equation $y' = -y(t - 1)$ would represent a stable control system.

There is an alternative way of inverting Eq. (7-149). The path of integration for the inversion integral in the s plane can be any vertical line to the right of r_1; let us choose a line on which $\operatorname{Re} s$ is very large. Then e^{-s} is very small compared with s, and we can write the inversion formula for Eq. (7-149) as

$$\begin{aligned} y(t) &= 1 + \frac{1}{2\pi i} \int_{\gamma - i\infty}^{\gamma + i\infty} \frac{e^{st}\, ds}{s^2[1 - (1/s)e^{-s}]} \\ &= 1 + \frac{1}{2\pi i} \int_{\gamma - i\infty}^{\gamma + i\infty} \frac{e^{st}}{s^2} \left(1 + \frac{1}{s} e^{-s} + \frac{1}{s^2} e^{-2s} + \cdots \right) ds \end{aligned} \tag{7-151}$$

Thus $y(t) = 1 + t$ for t in (0,1), $y(t) = 1 + t + (t - 1)^2/2!$ for t in (1,2), $y(t) = 1 + t + (t - 1)^2/2! + (t - 2)^3/3!$ for t in (2,3), and so on. This same result can, of course, also be obtained by direct integration of Eq. (7-147), but it is interesting to see how one obtains it from the transform expression (7-149). It is also interesting that, for all t, the result must be identical with the sum of exponential functions given by Eq. (7-150). The present result is most useful for small values of t, whereas Eq. (7-150) is most useful for large values of t.

The method used here to evaluate the inversion formula for small values of t is very similar to that used for the transmission-line problem which led to Eq. (6-60). We may remark in general that Eq. (7-135) indicates that the behavior of $F(s)$ for large values of Re s must be governed by the behavior of $f(t)$ for small values of t, and, conversely, the small-t behavior of $f(t)$ can often be determined from the large-s behavior of $F(s)$. The small-t analyses of Eqs. (6-60) and (7-151) are two illustrations of this.

If the transform function $F(s)$ can be expanded in a power series in $1/s$, convergent for $|s| > R$ [for example, $F(s)$ = ratio of polynomials], then we can readily prove that $f(t)$ possesses a power-series expansion in t, convergent everywhere. For let $F(s) = \sum_{n=1}^{\infty} a_n s^{-n}$ converge for $|s| > R$. Then there exist positive real constants M and ϵ (this latter arbitrarily small) such that $|a_n| < M(R + \epsilon)^n$ for all n. Consequently, the series

$$f(t) = \sum_{n=1}^{\infty} a_n t^{n-1}/(n-1)!$$

obtained by term-by-term inversion, converges for all t. (Moreover, $|f(t)| \leq Me^{R|t|}$.) If the $1/s$ series expansion for $F(s)$ is only asymptotic, rather than convergent, then whether or not term-by-term inversion yields an asymptotic expansion for $f(t)$ in powers of t depends on the nature of $F(s)$ and is best determined for the particular example in question (see Exercise 14).

Exercises

12. Discuss the solution of the equation

$$y''(t) = y(t-1) - y'(t-1)$$

where $y = 1$ for t in $(-1,0)$.

13. Let the Laplace-transform function $F(s)$ have the asymptotic expansion $F(s) \sim a_1/s + a_2/s^2 + \cdots$, where the a_i are constants, valid as $s \to \infty$ in the half-plane Re $s > \alpha$ (note that this is a stronger requirement than that the expansion be uniform in some sector lying wholly in the right half-plane). Show that, as $t \to 0$, $f(t) \sim a_1 + a_2(t/1!) + a_3(t^2/2!) + \cdots$.

14. Let $F(s)$ be analytic except at $s = 0$, where it has a branch point. In the neighborhood of $s = 0$, let

$$F(s) = \frac{1}{s}(a_0 + b_0 s^{1/2} + a_1 s + b_1 s^{3/2} + a_2 s^2 + b_2 s^{5/2} + \cdots)$$

Also, let $F(s)$ be such that the path of integration in the inversion integral can be replaced by a path "wrapped around" the negative half of the real axis (for $t > 0$). Show that, for large t,

$$f(t) \sim a_0 + \frac{1}{\pi\sqrt{t}}\left[b_0\Gamma\left(\frac{1}{2}\right) - b_1\frac{\Gamma(3/2)}{t} + b_2\frac{\Gamma(5/2)}{t^2} - \cdots\right]$$

Integral Equations

Laplace (or Fourier) transforms are sometimes useful in dealing with integral equations of convolution form. We shall consider a classical example first given by Abel; the problem is to find the form of a plane curve along which a particle can fall in a time which is a prescribed function of the vertical distance. Let the motion take place in the (x,y) plane, with y measured positively upward; for convenience, let the equation of the desired curve be $x = \varphi(y)$, with $\varphi(0) = 0$. A particle placed on the curve at the point (x,y) is to fall to the origin in the prescribed time $f(y)$. If g is the acceleration of gravity, the velocity at any point (ξ,η) on the curve is clearly $[2g(y - \eta)]^{1/2}$, so that the function $\varphi(y)$ must be a solution of the equation

$$\int_0^y \sqrt{\frac{[\varphi'(\eta)]^2 + 1}{2g(y - \eta)}}\, d\eta = f(y)$$

Denote the function $\{[\varphi'(\eta)]^2 + 1\}^{1/2}/(2g)^{1/2}$ by $\psi(\eta)$, so that

$$\int_0^y \psi(\eta)(y - \eta)^{-1/2}\, d\eta = f(y) \tag{7-152}$$

The Laplace transform of Eq. (7-152) gives $\Psi(s) = \pi^{-1/2}s^{1/2}F(s)$, where $\Psi(s)$ and $F(s)$ are the transforms of $\psi(y)$, $f(y)$, respectively. We want to write this result, in general form, as a convolution integral; since $s^{1/2}$ has no (conventional) inverse, we first write $\Psi(s) = \pi^{-1/2}s[(s^{-1/2})F(s)]$ to obtain

$$\begin{aligned}\psi(y) &= \frac{1}{\pi}\frac{d}{dy}\int_0^y \frac{f(\eta)}{\sqrt{y - \eta}}\, d\eta \\ &= \frac{1}{\pi}\frac{f(0)}{\sqrt{y}} + \frac{1}{\pi}\int_0^y \frac{f'(\eta)\, d\eta}{\sqrt{y - \eta}}\end{aligned}$$

In terms of the arc length S from the origin, the result can be written as

$$S = \frac{\sqrt{2g}}{\pi}\int_0^y \frac{f(\eta)\, d\eta}{\sqrt{y - \eta}}$$

(which is Abel's solution[1]). We observe that $f(y)$ cannot be given completely arbitrarily, since S must satisfy $dS/dy \geq 1$. If $f(0) = 0$, a sufficient condition that this be so is that $f'(y) \geq \frac{1}{2}\pi/\sqrt{2gy}$.

[1] N. Abel, 1832 and 1826. See "Oeuvres complètes," vol. 1, pp. 11 and 97, Christiana, 1881.

EXERCISES

15. Use Laplace transforms to solve the equation

$$\int_0^x \sin \omega(x - t)f(t)\,dt = g(x)$$

for the unknown function $f(x)$, and verify that the result is the same as that obtained by elementary means.

16. From Exercise 9, the n-fold integral of $f(t)$ is given by

$$\frac{1}{\Gamma(n)} \int_0^t (t - \tau)^{n-1} f(\tau)\,d\tau$$

This formula suggests that $\varphi(t)$, the α-fold integral of $f(t)$, where α is any real number > 0 (i.e., not necessarily integral), be defined by the same formula, with n replaced by α. Use a Laplace transform to find $f(t)$ in terms of $\varphi(t)$.

Partial-differential Equations

Frequently, in the treatment of boundary-value problems involving partial differential equations, it is convenient to use the Laplace transform. We have in fact already seen this in the analysis beginning with Eq. (7-72). In that problem, the Fourier transform with respect to time of $\varphi(x,t)$ is, except for notational convention, the Laplace transform of that function, and the reader should verify that the analysis is essentially identical to that which would have resulted from the use of a Laplace transform.

As an illustration of the use of the Laplace transform, we study the response of a finite stretched string to a forcing function $f(x,t)$. The equation of motion is

$$y_{xx} - \frac{1}{c^2} y_{tt} = f(x,t) \tag{7-153}$$

where c is as usual the constant velocity of signal propagation. The initial conditions $y(x,0)$ and $y_t(x,0)$ are prescribed. The string is fixed at $x = 0$ and $x = l$, so that $y(0,t) = y(l,t) = 0$. A Laplace transform in t leads to

$$Y_{xx} - \frac{1}{c^2}[-y_t(x,0) - sy(x,0) + s^2Y] = F(x,s)$$

whose solution is

$$Y(x,s) = A(s) \cosh \frac{s}{c} x + B(s) \sinh \frac{s}{c} x + \frac{c}{s} \int_0^x \left[F(\xi,s) - \frac{y_t(\xi,0)}{c^2} - \frac{sy(\xi,0)}{c^2} \right] \sinh \frac{s}{c}(x - \xi)\,d\xi$$

Inserting the end conditions, we obtain

$$Y(x,s) = -\frac{c \sinh (s/c)(l-x)}{s \sinh (s/c)l} \int_0^x \left[F(\xi,s) - \frac{y_t(\xi,0)}{c^2} - \frac{sy(\xi,0)}{c^2} \right] \sinh \frac{s}{c} \xi \, d\xi$$
$$- \frac{c \sinh (s/c)x}{s \sinh (s/c)l} \int_x^l \left[F(\xi,s) - \frac{y_t(\xi,0)}{c^2} - \frac{sy(\xi,0)}{c^2} \right] \sinh \frac{s}{c} (l-\xi)\, d\xi \quad (7\text{-}154)$$

where this form has been chosen in order that each of the two terms remains bounded as $s \to \infty$. These two integrals have similar forms, and we need examine only the s inversion of the first. If we temporarily denote the inverse of

$$\frac{c \sinh (s/c)(x-l) \sinh (s/c)\xi}{s \sinh (s/c)l} \quad (7\text{-}155)$$

by $w(x,\xi,t)$, then the first term of Eq. (7-154) becomes

$$\int_0^x d\xi \left[\int_0^t f(\xi,\tau) w(x,\xi,t-\tau)\, d\tau - \frac{1}{c^2} y_t(\xi,0) w(x,\xi,t) - \frac{1}{c^2} y(\xi,0) w_t(x,\xi,t) \right] \quad (7\text{-}156)$$

The next problem is to evaluate $w(x,\xi,t)$, and this we shall do in two different ways. The first method is to expand the denominator of the expression (7-155) so as to obtain, in the inversion integral,

$$w(x,\xi,t) = \frac{1}{2\pi i} \int_{\gamma - i\infty}^{\gamma + i\infty} \frac{ce^{st}}{2s} \left(e^{(s/c)(x-l)} - e^{-(s/c)(x-l)}\right)\left(e^{(s/c)\xi} - e^{-(s/c)\xi}\right)$$
$$\left(e^{-(s/c)l} + e^{-3(s/c)l} + e^{-5(s/c)l} + \cdots + ds\right) \quad (7\text{-}157)$$

As a typical term here, consider $\frac{c}{2s} e^{st+(s/c)(x-l)+(s/c)\xi-(s/c)l}$. This term has no contribution for $ct < 2l - x - \xi$; for $ct > 2l - x - \xi$, it provides a constant contribution of $c/2$. Physically, this term corresponds to a wave being initiated at ξ and being received at x after a reflection at the right-hand end. In expression (7-156), this unit-function effect is felt in the first two terms, and a delta-function effect is felt in the last term. The other terms of Eq. (7-157) may be interpreted similarly, so that the function w represents a sequence of rectangular pulses and w_t a sequence of delta functions—corresponding to all possible direct or reflection paths between ξ and x. As a second method of evaluating $w(x,\xi,t)$, we can sum the residues in the inversion integral involving expression (7-155), to give (note that there is no pole at $s = 0$)

$$w(x,\xi,t) = -\frac{2c}{\pi} \sum_{n=1}^{\infty} \frac{1}{n} \sin \frac{\pi n x}{l} \sin \frac{\pi n \xi}{l} \sin \frac{\pi n c t}{l} \quad (7\text{-}158)$$

In this form [which, as the symmetry in (x,ξ) shows, is also obtained for the inversion of the coefficient of the second term of Eq. (7-154)], what we are really doing is obtaining a solution of the problem in terms of the normal modes of the vibrating string. Incidentally, the right-hand side of Eq. (7-158) can be summed explicitly to recover the previous rectangular-wave interpretation.

In applying the Laplace-transform method, one usually has to know the values at $t = 0$ of all derivatives of order lower than the highest that occurs in the equation [cf. Eqs. (7-138)]. In some cases, however, this requirement may be avoided; we shall consider a simple example. Returning to the heat-conduction equation for the infinite rod, let us take two Laplace transforms in x of Eq. (7-34)—one for $x > 0$ and one for $x < 0$. Define

$$\Phi_+(s,t) = \int_0^\infty e^{-sx}\varphi(x,t)\,dx \qquad \Phi_-(s,t) = \int_{-\infty}^0 e^{-sx}\varphi(x,t)\,dx$$

Then, if $\varphi(x,t)$ grows no more rapidly than $e^{\alpha|x|}$ as $x \to \pm\infty$, for some real constant α, $\Phi_+(s,t)$ will exist for Re s sufficiently large positive and $\Phi_-(s,t)$ will exist for Re s sufficiently large negative. The Laplace inversion formula gives

$$\varphi(x,t) = \frac{1}{2\pi i}\int_{\gamma-i\infty}^{\gamma+i\infty} e^{sx}\Phi_+(s,t)\,ds + \frac{1}{2\pi i}\int_{-\beta-i\infty}^{-\beta+i\infty} e^{sx}\Phi_-(s,t)\,ds \qquad (7\text{-}159)$$

where the real numbers γ and β are sufficiently large positive. Multiplying Eq. (7-34) by e^{-sx} and integrating from 0 to ∞, or alternatively from $-\infty$ to 0, we obtain

$$\frac{d}{dt}\Phi_+(s,t) = \alpha[-\varphi_x(0,t) - s\varphi(0,t) + s^2\Phi_+(s,t)]$$

$$\frac{d}{dt}\Phi_-(s,t) = \alpha[\varphi_x(0,t) + s\varphi(0,t) + s^2\Phi_-(s,t)]$$

Thus,

$$\Phi_+(s,t) = -\Psi(s,t) + A(s)\exp(\alpha s^2 t)$$

$$\Phi_-(s,t) = \Psi(s,t) + B(s)\exp(\alpha s^2 t)$$

where $\quad A(s) = \int_0^\infty e^{-sx}\varphi(x,0)\,dx \qquad B(s) = \int_{-\infty}^0 e^{-sx}\varphi(x,0)\,dx$

and $\quad \Psi(s,t) = \alpha\exp(\alpha s^2 t)\int_0^t [\varphi_x(0,\tau) + s\varphi(0,\tau)]\exp(-\alpha s^2\tau)\,d\tau$

Now $\Psi(s,t)$ is analytic in s and $\to 0$ as Im $s \to \pm\infty$; thus the contribution of $\Psi(s,t)$ to the inversion formula (7-159), viz.,

$$\frac{1}{2\pi i}\int_{\gamma-i\infty}^{\gamma+i\infty}(-\Psi e^{sx})\,ds + \frac{1}{2\pi i}\int_{-\beta-i\infty}^{-\beta+i\infty}(\Psi e^{sx})\,ds$$

must vanish. We are therefore left only with

$$\varphi(x,t) = \frac{1}{2\pi i}\int_{\gamma-i\infty}^{\gamma+i\infty} e^{sx}A(s)\exp(\alpha s^2 t)\,ds + \frac{1}{2\pi i}\int_{-\beta-i\infty}^{-\beta+i\infty} e^{sx}B(s)\exp(\alpha s^2 t)\,ds$$

Inserting the formulas for $A(s)$ and $B(s)$ and inverting the order of integration, we obtain again Eq. (7-85). The difference between the previous and the present derivations of Eq. (7-85). The difference between the previous and the present derivations of Eq. (7-85) is that the latter is valid even if the temperature grows exponentially toward the two ends of the rod; this is perhaps of minor importance, since the final formula, Eq. (7-85), can always be verified to hold for such situations—irrespective of its mode of derivation.

Although the Laplace-transform method usually has as its purpose the reduction of a partial-differential equation to an ordinary differential equation, there are cases in which it is convenient to proceed in the reverse direction. Consider, for example, the problem of expanding an arbitrary function $f(x)$ in a series of eigenfunctions of the Sturm-Liouville system

$$y'' - [\lambda + q(x)]y = 0 \tag{7-160}$$

with $y(0) = y(\alpha) = 0$. Here λ is a parameter, α is real and >0, and $q(x)$ is a given function. Solutions of this system exist only for $\lambda = \lambda_n$, where the λ_n are the eigenvalues of the system; the corresponding functions $y(x,\lambda_n)$ are the eigenfunctions. We shall give a heuristic discussion of a method whereby a Laplace-transform solution of an associated partial-differential equation,

$$u_{xx} - u_t - q(x)u = 0 \tag{7-161}$$

with $u(0,t) = u(\alpha,t) = 0$, $u(x,0) = f(x)$, can provide the desired expansion; a more complete treatment will be found in Titchmarsh.[1]

Let $U(x,s)$ denote the Laplace transform of $u(x,t)$ with respect to time. Then Eq. (7-161) leads to

$$U_{xx} - [s + q(x)]U = -f(x) \tag{7-162}$$

with $U(0,s) = U(\alpha,s) = 0$. Define $w(x,\lambda)$, $v(x,\lambda)$ to be solutions of Eq. (7-160) satisfying

$$\begin{aligned} w(0,\lambda) &= 0 \qquad & v(\alpha,\lambda) &= 0 \\ w_x(0,\lambda) &= 1 \qquad & v_x(\alpha,\lambda) &= 1 \end{aligned}$$

Then the general solution of Eq. (7-162) can be written

$$U(x,s) = -\int_0^\alpha K(x,\xi,s)\, f(\xi)\, d\xi \tag{7-163}$$

where

$$K(x,\xi,s) = \frac{1}{D(s)} \begin{cases} w(x,s)v(\xi,s) & \text{for } \xi > x \\ w(\xi,s)v(x,s) & \text{for } \xi < x \end{cases} \tag{7-164}$$

$$D(s) = -v(0,s) = w(\alpha,s) \tag{7-165}$$

We now suppose that, when the inversion formula is applied to Eq. (7-163), the orders of integration may be interchanged and moreover that the s integral can be evaluated by summing residues, evaluated at the zeros of

[1] E. C. Titchmarsh, "Eigenfunction Expansions," 2d ed., pt. I, Oxford University Press, New York, 1962.

$D(s)$. From Eq. (7-165), these zeros occur at $s = \lambda_n$, so that inversion of Eq. (7-163) must lead to

$$u(x,t) = \sum a_n w(x,\lambda_n) \int_0^\alpha f(\xi) w(\xi,\lambda_n) \, d\xi \, \exp (\lambda_n t)$$

where the a_n are certain constants. If we allow $t \to 0$ and remember that $u(x,0) = f(x)$, we obtain the desired expansion.

This analysis has proceeded on the assumption that α is finite and that no point of the interval $(0,\alpha)$ is a singular point of Eq. (7-160). More general cases are considered in Titchmarsh.[1]

EXERCISES

17. Use a Laplace-transform method to solve the heat-conduction problem for the finite rod; i.e., find $\varphi(x,t)$, where $\varphi_t = \alpha\varphi_{xx}$ for $0 \leq x \leq l$, with $\varphi(0,t)$, $\varphi(l,t)$, and $\varphi(x,0)$ all prescribed.

18. As a special case of Exercise 17, let $\varphi(0,t) = 1$, $\varphi(l,t) = 0$, $\varphi(x,0) = 0$. Show that

$$y(x,t) = \operatorname{erfc} \frac{x}{2\sqrt{\alpha t}} - \operatorname{erfc} \frac{2l - x}{2\sqrt{\alpha t}} + \operatorname{erfc} \frac{2l + x}{2\sqrt{\alpha t}} - \operatorname{erfc} \frac{4l - x}{2\sqrt{\alpha t}} + \cdots$$

and observe that this form of the solution is particularly useful for small t.

19. A certain transient elastic-beam problem may be described by the equation $y_{xxxx} + y_{tt} = 0$, with $y(0,t) = t$,

$$y_x(0,t) = y(l,t) = y_x(l,t) = y(x,0) = y_t(x,0) = 0$$

Use a Laplace-transform method to find $y(x,t)$, and evaluate the inversion integral for the case $l = \infty$.

20. Use a Laplace-transform method to solve Eq. (7-100), with $\epsilon = 0$. As a special case, let Eq. (7-101) be satisfied for $t > 0$; show that (within a transient) Eq. (7-106) results. Why was no radiation condition necessary here?

21. Let ρ, θ be radial coordinates in the sector $0 \leq \rho < \infty$, $-\gamma < \theta < 0$, where $\gamma > 0$ is a given angle $< \pi/2$. Let $\varphi(\rho,\theta)$ satisfy

$$\rho^2\varphi_{\rho\rho} + \rho\varphi_\rho + \varphi_{\theta\theta} - \rho^2 k^2 \varphi = 0$$

within the sector, and the boundary conditions $\varphi_\theta - \rho\varphi = 0$ for $\theta = 0$ and $\varphi_\theta = 0$ for $\theta = -\gamma$. Here $k > 0$. Assume $\rho\varphi_\rho$ to be bounded as $\rho \to 0$ (but not necessarily φ_ρ). Take a Laplace transform [*Hint:* Rewrite the equation in terms of $(\rho\varphi_\rho)_\rho$ in order to avoid the appearance of the term $\varphi_\rho(0,\theta)$ in the transform formula; note that a term-by-term transform of the equation would not yield the correct answer] to show that the new function

$$\psi = (\sinh p)\Phi(s,\theta)$$

[1] *Ibid.*

must be harmonic in the variables p, θ, where $s = k \cosh p$. The domain of interest is now $-\gamma < \theta < 0$, $-\infty < p < \infty$. Write $\psi = \text{Re}\,[f(p + i\theta]$, and express the boundary conditions for the strip in terms of f. (Analytic continuation can then be used to find f. The method is due to Peters.)[1]

Integral Representations of Solutions of Ordinary Differential Equations

When an ordinary differential equation can be cast into a form in which its coefficients are linear in the independent variable, a Laplace-transform method can be used to obtain integral representations for its solutions. To illustrate this, we ask for those values of α for which we can find a solution of

$$(xu')' - (x - \alpha)\,u = 0 \tag{7-166}$$

which is bounded at $x = 0$ and as $x \to \infty$. The Laplace transform of Eq. (7-166) is

$$-s(sU)' + \alpha U + U' = 0$$

whence

$$U = \frac{(s-1)^{(\alpha-1)/2}}{(s+1)^{(\alpha+1)/2}}$$

Since the inversion integral,

$$u(x) = \frac{1}{2\pi i}\int_{\gamma-i\infty}^{\gamma+i\infty} e^{sx}\,\frac{(s-1)^{(\alpha-1)/2}}{(s+1)^{(\alpha+1)/2}}\,ds \tag{7-167}$$

where the path of integration passes to the right of $s = 1$, will provide contributions whose asymptotic behavior for large positive x is like e^x unless $(\alpha - 1)/2$ is a nonnegative integer, the values of α which serve our purpose are $\alpha_n = 2n + 1$ with $n = 0, 1, 2, \ldots$. The corresponding functions $u_n(x)$ are

$$u_n(x) = \frac{1}{n!}\frac{\partial^n}{\partial s^n}\left[e^{sx}(s-1)^n\right]_{s=-1}$$

Ignoring now the original question, we note that *one* of the linearly independent solutions of Eq. (7-166) is given by Eq. (7-167). We can obtain another solution by again using Eq. (7-167), taking now, however, the path of integration to be a vertical line between $s = -1$ and $s = 1$. Unless α is an odd integer, this procedure does indeed (as can be verified by direct substitution) provide a linearly independent solution of Eq. (7-166). The reader may also verify that this new solution does not exist at $x = 0$. It is clear that these two integral representations can be used to extract asymptotic and other information concerning the behavior of the two solutions.

[1] A. S. Peters, *Comm. Pure Appl. Math.*, **5:** 87 (1952).

EXERCISE

22. Find integral representations of solutions of

$$(xu')' + xu = 0$$
$$(xu')' + xu - \frac{n^2}{x}u = 0$$
$$u'' + xu = 0$$

and invent a nontrivial fourth-order equation which can be treated similarly.

Two-sided Laplace Transform. Mellin Transform

If $f(t)$ is not necessarily zero for $t < 0$, it is possible to define a two-sided Laplace transform $F_2(s)$ by

$$F_2(s) = \int_{-\infty}^{\infty} e^{-st}f(t)\,dt \tag{7-168}$$

This can be thought of as the sum of two one-sided Laplace transforms, one of which involves integration from $-\infty$ to 0 and the other from 0 to ∞. If $f(t)$ vanishes sufficiently strongly at $\pm\infty$, the two regions of convergence will overlap and the integral $F_2(s)$ will converge to an analytic function of s within some vertical strip in the s plane. In particular, let $\text{Re } s = c$ be a vertical line within this strip. Then $F_2(c - i\lambda)$ is $\sqrt{2\pi}$ times the Fourier transform of $e^{-ct}f(t)$, from which fact we can write the inversion formula for Eq. (7-168) as

$$f(t) = \frac{1}{2\pi i}\int_{c-i\infty}^{c+i\infty} e^{st}F_2(s)\,ds \tag{7-169}$$

There is, of course, no essential difference between $F_2(s)$ and a conventional Fourier transform, and we need not rewrite the various Fourier-transform formulas in terms of two-sided Laplace transforms.

However, Eq. (7-168) is a convenient departure point for the derivation of the Mellin transform. Let $e^{-t} = x$, $f(t) = g(x)$, $F_2(s) = G_M(s)$, so that Eqs. (7-168) and (7-169) become

$$G_M(s) = \int_0^{\infty} x^{s-1}g(x)\,dx \tag{7-170}$$

$$g(x) = \frac{1}{2\pi i}\int_{c-i\infty}^{c+i\infty} x^{-s}G_M(s)\,ds \tag{7-171}$$

The function $G_M(s)$ is called the *Mellin transform* of $g(x)$. There is a certain similarity between Eq. (7-170) and the formula for the Laplace transform; in particular, values of $g(x)$ occur only for $x > 0$. As with our previous transform formulas, $g(x)$ in Eq. (7-171) should be replaced by $\frac{1}{2}[g(x+) + g(x-)]$ at a point where $g(x)$ is discontinuous.

Again, there is no essential difference between a Mellin transform and a

Fourier transform. However, the somewhat different form of the kernel (a power of x rather than an exponential function) is sometimes convenient. For reference, we list some Mellin-transform formulas which may be readily derived by the reader:

1. If the transform of $f(x)$ is $F_M(s)$, then that of $f'(x)$ is $-(s-1)F_M(s-1)$, provided that $[x^{s-1}f(x)]_0^\infty = 0$.

2. If the transforms of $f(x)$ and $g(x)$ are $F_M(s)$ and $G_M(s)$, respectively, then

(*a*) The inverse of $F_M(s)G_M(s)$ is $\int_0^\infty f\left(\frac{x}{\xi}\right) g(\xi)\,\frac{d\xi}{\xi}$.

(*b*) $\frac{1}{2\pi i}\int_{c-i\infty}^{c+i\infty} F_M(s)G_M(s)\,ds = \int_0^\infty f\left(\frac{1}{\xi}\right) g(\xi)\,\frac{d\xi}{\xi}$.

(*c*) The transform of $f(x)g(x)$ is $\frac{1}{2\pi i}\int_{c-i\infty}^{c+i\infty} F_M(s-\tau)G_M(\tau)\,d\tau$.

(*d*) $\int_0^\infty f(x)g(x)\,dx = \frac{1}{2\pi i}\int_{c-i\infty}^{c+i\infty} F_M(s)G_M(1-s)\,ds$ (Parseval formula for Mellin transforms).

(*e*) $\int_0^\infty [f(x)]^2\,dx = \frac{1}{2\pi}\int_{-\infty}^{\infty} \left|F_M\left(\frac{1}{2}+i\lambda\right)\right|^2 d\lambda$.

EXERCISES

23. Find the Mellin transform of each of e^{-x}, $(2+x^2)^{-1}$, $J_0(x)$, and verify the inversion formulas directly.

24. Relate the Mellin transforms of $f(px)$, $x^q f(x)$, $f(1/x)$, and $f(x^m)$ to that of $f(x)$. Discuss the range of allowed values for the constants p, q, m. Find also the transform of $[x(d/dx)]^n f(x)$.

25. One way of inventing new transforms is to try to find kernel functions $k(\lambda,x)$ and $h(\lambda,x)$ such that, for any function $f(x)$, the "transform function" $g(\lambda)$ defined by

$$g(\lambda) = \int_a^b k(\lambda,x)f(x)\,dx$$

may be used to recover $f(x)$ via

$$f(x) = \int_a^b h(\lambda,x)g(\lambda)\,d\lambda$$

For the case in which k and h are functions of the product λx alone and in which the limits of integration are 0 to ∞, show that a necessary condition for such a relationship to hold is that $K_M(s)H_M(1-s) = 1$. Verify that this relation is satisfied by the Fourier cosine transform kernel. In deriving a relation between transforms of k and h, why were Mellin transforms more directly useful than other kinds of transforms?

7-4 Hankel Transforms

In Eqs. (7-82), we have already noticed the usefulness of polar coordinates in dealing with Fourier transforms in two space variables. For axially symmetric problems, the resulting formulas lead to the idea of the Hankel transform. Equations (7-82) relate a function $g(r,\theta)$ to its transform $G(\rho,\varphi)$, where (r,θ) and (ρ,φ) are, respectively, the original and image sets of polar coordinates; if g does not depend on θ, then it is clear from the first of Eqs. (7-82) that G does not depend on φ, so that we obtain

$$\begin{aligned} G(\rho) &= \frac{1}{2\pi}\int_0^\infty g(r)r\,dr\int_0^{2\pi} e^{ir\rho\cos\theta}\,d\theta \\ &= \int_0^\infty rJ_0(r\rho)g(r)\,dr \end{aligned} \tag{7-172}$$

and similarly,

$$g(r) = \int_0^\infty \rho J_0(r\rho)G(\rho)\,d\rho \tag{7-173}$$

We shall call $G(\rho)$ the Hankel transform (of order zero) of $g(r)$. As in previous Fourier-transform formulas, $g(r)$ in Eq. (7-173) should be replaced by $\frac{1}{2}[g(r+) + g(r-)]$ at a point of discontinuity.

If we replace $g(r)$ in Eq. (7-172) by $g'(r)$ and integrate by parts in the usual manner, it soon becomes clear that we cannot obtain a simple formula for the transform of a derivative in terms of the transform of the parent function. However, since our Hankel transform is that special case of a double Fourier transform which arises in axially symmetric situations, we can anticipate that the Hankel transform of the laplacian of $g(r)$—the simplest and most common isotropic differential operator—will be expressible in terms of $G(\rho)$. In fact, if the Fourier-transform variables corresponding to cartesian coordinates x and y are denoted by ξ and η, the transform of $g_{xx} + g_{yy}$ is $-(\xi^2 + \eta^2)$ times the transform of g; but since $\xi^2 + \eta^2 = \rho^2$, we have the result that the Hankel transform of

$$\Delta g = g_{rr} + \frac{1}{r}g_r$$

is $-\rho^2 G(\rho)$ (provided, of course, that g and its derivatives vanish sufficiently rapidly at ∞ that the integrations-by-parts formulas used in the Fourier-transform derivation are valid. Otherwise, the δ-function family must be permitted).

We now see that several of our previous examples of Green's functions (Sec. 7-2) essentially involved Hankel transforms. For example, the Hankel transform of a function g satisfying

$$g_{xx} + g_{yy} - \alpha^2 g = \delta(x)\delta(y)$$

is seen at once to satisfy the equation

$$-\rho^2 G - \alpha^2 G = \frac{1}{2\pi}$$

from which
$$G = -\frac{1}{2\pi}\frac{1}{\rho^2+\alpha^2}$$

so that
$$g(r) = -\frac{1}{2\pi}\int_0^\infty \rho \frac{J_0(r\rho)}{\rho^2+\alpha^2}\,d\rho$$
$$= -\frac{1}{2\pi}K_0(\alpha r)$$

in agreement with Eq. (7-95).

EXERCISES

1. By considering triple Fourier transforms for a three-dimensional spherically symmetric function, show that J_0 in Eqs. (7-172) and (7-173) can be replaced by $J_{1/2}$.

2. (*a*) Derive the following pairs of Hankel transform (of order zero):

(1) $g(r) = 1/r,\ G(\rho) = 1/\rho$

(2) $g(r) = \begin{cases}(a^2 - r^2)^{-1/2} & \text{for } r < a\\ 0 & \text{for } r > a\end{cases},\ G(\rho) = \frac{1}{\rho}\sin a\rho$

(3) $g(r) = (1 - e^{-r})/r^2,\ G(\rho) = \text{arcsinh}\,(1/\rho)$

(*b*) Prove Parseval's theorem for Hankel transforms (of order zero),

$$\int_0^\infty rf(r)g(r)\,dr = \int_0^\infty \rho F(\rho)G(\rho)\,d\rho$$

3. Let $V(x,r)$ be an axially symmetric solution of Laplace's equation $V_{rr} + (1/r)V_r + V_{xx} = 0$, in the cylindrical coordinates (r,x), valid for $x > 0$, $0 \le r < \infty$, and vanishing sufficiently strongly as $x^2 + r^2 \to \infty$ in this region. Take a Hankel transform in r to show that

$$V(x,r) = \int_0^\infty \rho A(\rho)J_0(r\rho)e^{-\rho x}\,d\rho$$

where the function $A(\rho)$ is unknown. Set $r = 0$ to show that $\rho A(\rho)$ is, however, the inverse Laplace transform of $V(x,0)$, substitute the inversion formula for $\rho A(\rho)$, and interchange the order of integration (wrap a contour around a branch cut) to derive

$$V(x,r) = \frac{1}{\pi}\int_0^\pi V(x + ir\cos\theta,\ 0)\,d\theta$$

provided that the function $V(x,0)$ is analytic in x. This equation determines $V(x,r)$ in terms of its values along the axis; it can be expected to be valid somewhat more generally than under the conditions hypothesized. Derive a similar result[1] for Helmholtz's equation $V_{rr} + (1/r)V_r + V_{xx} + k^2V = 0$.

Fourier-Bessel Series

In Sec. 7-1, we obtained the Fourier-transform theorem as a limiting case of a Fourier-series expansion; there is a similar relationship between the Hankel expansion theorem and the *Fourier-Bessel series*. Consider a function $f(x)$ defined in (0,1), and let j_n be the nth positive zero of the Bessel function $J_\nu(x)$. Then, if $f(x)$ satisfies conditions analogous to those required for a Fourier-series expansion, it is possible to find constants c_n such that

$$f(x) = \sum_{n=1}^{\infty} c_n J_\nu(j_n x) \tag{7-174}$$

for x in (0,1) (where, at a discontinuity of $f(x)$, we must interpret the left-hand side as $\frac{1}{2}[f(x+) + f(x-)]$). Such a series, with $\nu = 0$, was first used by Fourier in connection with a problem in radial heat conduction; the generalization to arbitrary real values of $\nu > -1$ was made by Lommel. Since the Fourier-Bessel series is perhaps not so familiar to the reader as the Fourier series, we shall begin with a brief discussion of its properties.

Consider the ordinary differential equation

$$(xy_x)_x + \left(k^2x - \frac{\nu^2}{x}\right)y = 0 \tag{7-175}$$

where ν is real and > -1, k is real and positive, and a subscript denotes differentiation. A solution of Eq. (7-175) is any multiple of $J_\nu(kx)$; if a prime is used to indicate differentiation of a function with respect to its argument, we have $[J_\nu(kx)]_x = kJ'_\nu(kx)$. Writing Eq. (7-175) for two different choices of parameter, say k and l, we obtain

$$[kxJ'_\nu(kx)]_x + \left(k^2x - \frac{\nu^2}{x}\right)J_\nu(kx) = 0$$

$$[lxJ'_\nu(lx)]_x + \left(l^2x - \frac{\nu^2}{x}\right)J_\nu(lx) = 0$$

Multiplying the first of these equations by $J_\nu(lx)$, the second by $J_\nu(kx)$, subtracting, and integrating by parts from 0 to 1, we obtain

$$\int_0^1 xJ_\nu(kx)J_\nu(lx)\,dx = \frac{1}{k^2 - l^2}\,[lJ'_\nu(l)J_\nu(k) - kJ'_\nu(k)J_\nu(l)] \tag{7-176}$$

So far, k and l are any real positive numbers. As a special case, let k and l be two different zeros of $J_\nu(x)$, say j_m and j_n with $m \neq n$; then the right-hand

[1] A discussion of formulas of this character is given by P. Henrici, *Comm. Math. Helv.*, *27*: 235 (1953).

side of Eq. (7-176) vanishes, so that the functions $J_\nu(j_n x)$ are orthogonal over the unit interval with respect to the weight function x. Second, let l be a chosen zero of $J_\nu(x)$, and let $k \to l$; then by L'Hospital's rule the right-hand side of Eq. (7-176) becomes [by using $J_\nu(l) = 0$]

$$\frac{1}{2l}\,[lJ'_\nu(l) \cdot J'_\nu(l)] = \frac{1}{2}\,[J_{\nu+1}(l)]^2$$

Thus,
$$\int_0^1 xJ_\nu(j_m x)J_\nu(j_n x) = \begin{cases} 0 & \text{if } m \neq n \\ \tfrac{1}{2}[J_{\nu+1}(j_m)]^2 & \text{if } m = n \end{cases} \tag{7-177}$$

If the validity of the expansion (7-174) is assumed, the coefficients c_n can be found by means of Eq. (7-177).

We shall next demonstrate the validity of Eq. (7-174). In order to be able to give a simple proof, we shall restrict ourselves to functions $f(x)$ which are differentiable and such that $f(1) = 0$; the more general case is dealt with in Titchmarsh.[1] Define $\varphi_n(x)$ to be a multiple of $J_\nu(j_n x)$, the factor to be so chosen that $\int_0^1 x\varphi_n^2\,dx = 1$. Let $\alpha_n = \int_0^1 x\varphi_n f\,dx$. We shall show that the weighted error E, defined by

$$E^2 = \int_0^1 x\left(f - \sum_{n=1}^{N} \alpha_n\varphi_n\right)^2 dx \tag{7-178}$$

approaches zero as $N \to \infty$; this is the sense in which Eq. (7-174) will thereby be established.

Consider now the problem of minimizing

$$I = \int_0^1 \left[x(g')^2 + \frac{\nu^2}{x}\,g^2\right] dx$$

for functions $g(x)$ satisfying $g(1) = 0$, $\int_0^1 xg^2\,dx = 1$, $\int_0^1 xg\varphi_p\,dx = 0$ for $p = 1, 2, \ldots, N$. By straightforward variational procedures, this minimum is attained for $g(x) = \varphi_{N+1}(x)$, and the value of the minimum of I is simply j_{N+1}^2. Consequently, for any function $g(x)$ satisfying the above requirements,

$$\int_0^1 \left[x(g')^2 + \frac{\nu^2}{x}\,g^2\right] dx \geq j_{N+1}^2 \tag{7-179}$$

One such function is

$$g(x) = \frac{1}{E}\left(f - \sum_{n=1}^{N} \alpha_n\varphi_n\right)$$

[1] E. C. Titchmarsh, "Eigenfunction Expansions," 2d ed., pt. I, p. 81, Oxford University Press, New York, 1962.

and applying Eq. (7-179), we obtain

$$\frac{1}{E^2}\int_0^1 \left\{x(f')^2 - 2xf'\left(\sum \alpha_n\varphi_n'\right) + x\left(\sum \alpha_n\varphi_n'\right)\left(\sum \alpha_n\varphi_n'\right) + \frac{\nu^2}{x}\left[f^2 - 2f\left(\sum \alpha_n\varphi_n\right) + \left(\sum \alpha_n\varphi_n\right)^2\right]\right\} dx \geq j_{N+1}^2$$

or, after integration by parts and use of the differential equation satisfied by φ_n,

$$\frac{1}{j_{N+1}^2}\int_0^1 \left[x(f')^2 + \frac{\nu^2}{x}f^2 - \sum_{n=1}^{N} j_n^2\alpha_n^2\right] dx \geq E^2$$

Since $j_{N+1} \to \infty$ [consider the asymptotic expansion of $J_\nu(x)$], and since f is a fixed function, it does indeed follow that $E \to 0$, so that the theorem is proved.

We turn now to the Hankel transform of order ν. If we use the interval $(0,L)$ rather than $(0,1)$ and substitute the values of the coefficients as obtained with the help of Eq. (7-177), Eq. (7-174) becomes

$$f(x) = \sum_{n=1}^{\infty} \left\{\frac{2}{L^2[J_{\nu+1}(j_n)]^2} J_\nu\left(\frac{j_n}{L}x\right)\right\} \int_0^L \xi f(\xi) J_\nu\left(\frac{j_n}{L}\xi\right) d\xi$$

Upon introducing the variable $\rho = j_n/L$, it is clear that the process $L \to \infty$ will generate the generalized Hankel-transform theorem,

$$f(x) = \int_0^\infty \rho J_\nu(\rho x)\, d\rho \int_0^\infty \xi J_\nu(\rho\xi) f(\xi)\, d\xi$$

provided that

$$\frac{2}{L^2[J_{\nu+1}(j_n)]^2}$$

can be interpreted as $\rho\, d\rho$ (at least for large n, since, as $L \to \infty$, the integration region corresponding to small values of n becomes of negligible extent). That is, we want to show that

$$\frac{2}{L^2[J_{\nu+1}(j_n)]^2} \to \frac{j_n}{L}\,\frac{j_{n+1} - j_n}{L}$$

as $n \to \infty$, and an examination of the asymptotic behavior of J_ν and $J_{\nu+1}$ shows that this is indeed the case. We have now derived the general Hankel-transform formulas:

$$G_\nu(\rho) = \int_0^\infty rJ_\nu(r\rho)g(r)\, dr \tag{7-180}$$

$$g(r) = \int_0^\infty \rho J_\nu(r\rho)G_\nu(\rho)\, d\rho \tag{7-181}$$

The reader should prove the Parseval relation

$$\int_0^\infty rf(r)g(r)\, dr = \int_0^\infty \rho F_\nu(\rho)G_\nu(\rho)\, d\rho \tag{7-182}$$

(a source of useful integral identities) and the following formulas for the transforms $G_\nu^{(1)}(\rho)$ and $G_\nu^{(2)}(\rho)$ of the derivatives $g'(r)$ and $g''(r)$,

$$G_\nu^{(1)}(\rho) = \frac{\rho}{2\nu}[(\nu - 1)G_{\nu+1}(\rho) - (\nu + 1)G_{\nu-1}(\rho)] \tag{7-183}$$

[provided that $rg(r) \to 0$ as $r \to 0$ or as $r \to \infty$]

$$G_\nu^{(2)}(\rho) = \frac{\rho^2}{4}\left[\frac{\nu - 1}{\nu + 1}G_{\nu+2}(\rho) - 2\frac{\nu^2 - 3}{\nu^2 - 1}G_\nu(\rho) + \frac{\nu + 1}{\nu - 1}G_{\nu-2}(\rho)\right] \tag{7-184}$$

[provided that $rg(r)$ and $rg'(r) \to 0$ as $r \to 0$ or as $r \to \infty$].
Also, if $rg(r)$ and $rg'(r)$ vanish at $0, \infty$, we have

$$\int_0^\infty rJ_\nu(r\rho)\left(g'' + \frac{1}{r}g' - \frac{\nu^2}{r^2}g\right)dr = -\rho^2 G_\nu(\rho) \tag{7-185}$$

The utility of the Hankel transform of order ν is primarily a consequence of this equation.

Some Examples

An interesting classical problem, examined by Bernoulli and Euler, concerns the free vibrations of a hanging chain. Consider first a finite chain, of length L; let the arc length from the suspension point be s, and let the (x,y) coordinates of any point on the chain (as measured from the point of suspension; y horizontally and x vertically downward) be functions of s and t. If T is the tension force in the chain, g the acceleration of gravity, and ρ the density per unit length, the governing equations are

$$(Ty_s)_s = \rho y_{tt} \qquad (Tx_s)_s + \rho g = \rho x_{tt} \qquad x_s^2 + y_s^2 = 1$$

For small displacements from the vertical equilibrium position, it is reasonable to set $x \cong s$, so that the second equation gives $T \cong \rho g(L - x)$; the first equation is then approximated by

$$g(L - x)y_{xx} - gy_x = y_{tt} \tag{7-186}$$

where y is now $y(x,t)$. The boundary conditions require $y(0,t) = 0$ and that y be well behaved at $x = L$. As initial conditions, the values of $y(x,0)$ and $y_t(x,0)$ must be specified.

One way of solving Eq. (7-186) is to use a Laplace transform in t so as to reduce the equation to an ordinary differential equation. This equation may be solved in terms of Bessel functions, and the inversion formula then leads to a sum of residues, which may be recognized as a Fourier-Bessel series. As has previously been pointed out in connection with trigonometric series, it is also possible to begin directly with a series expansion in the space variable. In general, it is desirable to deal with the coefficients rather than with the series itself (this sometimes avoids convergence difficulties arising from series differentiation). We first make the change in variable $\xi = L - x$ to put Eq. (7-186) into Bessel form,

$$g(\xi y_{\xi\xi} + y_\xi) = y_{tt} \tag{7-187}$$

A Fourier-Bessel series is clearly indicated, but since we anticipate that the "finite" Hankel transform gives a simple result for the differential operator $y_{\xi\xi} + (1/\xi)y_\xi$, we introduce a new independent variable $\zeta = \zeta\sqrt{\xi/L}$. Then

$$y_{\zeta\zeta} + \frac{1}{\zeta}y_\zeta = \frac{4L}{g}y_{tt} \tag{7-188}$$

where ζ ranges from 0 to 1. Write

$$y = \sum_{n=1}^{\infty} c_m(t)J_0(j_n\zeta) \tag{7-189}$$

where $$c_n(t) = \frac{2}{[J_1(j_n)]^2}\int_0^1 \xi J_0(j_n\xi)y(\xi,t)\,d\xi$$

Multiplying both sides of Eq. (7-188) by $\zeta J_0(j_n\zeta)$ and integrating by parts from 0 to 1, we obtain

$$-j_n^2c_n = \frac{4L}{g}c_n''$$

whence

$$c_n(t) = c_n(0)\cos\left(\tfrac{1}{2}j_n\sqrt{\frac{g}{L}}\,t\right) + \frac{2}{j_n}\sqrt{\frac{L}{g}}\,c_n'(0)\sin\left(\tfrac{1}{2}j_n\sqrt{\frac{g}{L}}\,t\right) \tag{7-190}$$

and since $c_n(0)$ and $c_n'(0)$ can be determined from the initial conditions, the solution is complete. In terms of the original x variable, we have

$$y(x,t) = \sum_{n=1}^{\infty} c_n(t)J_0\left(j_n\sqrt{\frac{L-x}{L}}\right)$$

where $c_n(t)$ is given by Eq. (7-190), with

$$c_n(0) = \frac{1}{L[J_1(j_n)]^2}\int_0^L J_0\left(j_n\sqrt{\frac{L-x}{L}}\right)y(x,0)\,dx$$

and a similar formula for $c_n'(0)$, with $y(x,0)$ replaced by $y_t(x,0)$.

As a modification of the preceding example, consider a very long (infinite) chain subjected to a lateral force $f(\xi)$, where ξ denotes the vertical distance measured upward from the lower end of the chain. Assuming as before that the tension T is affected negligibly by lateral motion, we obtain in place of Eq. (7-187)

$$g(\xi y_{\xi\xi} + y_\xi) + \frac{1}{\rho}f = y_{tt} \tag{7-191}$$

The change in variables $\xi = r^2$ gives

$$\frac{g}{4}\left(y_{rr} + \frac{1}{r}y_r\right) + \frac{1}{\rho}f(r^2,t) = y_{tt}$$

and a Hankel transform gives

$$-\frac{g\lambda^2}{4}Y(\lambda,t) + \frac{1}{\rho}\int_0^\infty rf(r^2,t)J_0(\lambda r)\,dr = Y_{tt}(\lambda,t)$$

As a special case, let f represent a concentrated force $h(t)$ applied just above the lower end (which means, incidentally, that there is some "boundary-layer" region at the lower end within which our equations are not accurate). Then $f(\xi,t) = h(t)\delta[\xi - (0+)]$, and the previous equation becomes

$$-\frac{g\lambda^2}{4}Y + \frac{1}{2\rho}h(t) = Y_{tt}$$

If the chain is at rest in its equilibrium position at $t = 0$, the solution of this equation is

$$Y = \frac{1}{\rho\sqrt{g}\lambda}\int_0^t h(\tau)\sin\left[\lambda\frac{\sqrt{g}}{2}(t-\tau)\right]d\tau$$

Inversion gives

$$y = \frac{1}{\rho\sqrt{g}}\int_0^t h(\tau)\,d\tau\int_0^\infty J_0(\lambda r)\sin\left[\lambda\frac{\sqrt{g}}{2}(t-\tau)\right]d\lambda$$

or

$$y(\xi,t) = \frac{1}{\rho\sqrt{g}}\int_{2\sqrt{\xi/g}}^t h(t-\tau)\left(\frac{g}{4}\tau^2 - \xi\right)^{-1/2}d\tau$$

As an example involving Hankel transforms of order higher than zero, let us examine a set of slow viscous-flow equations encountered by Haskell[1] in a geological problem,

$$\frac{1}{r}(ru_r)_r - \frac{1}{r^2}u + u_{zz} = f_r$$

$$\frac{1}{r}(rv_r)_r + v_{zz} = f_z$$

$$\frac{1}{r}(ru)_r + v_z = 0$$

where $0 < r < \infty$, $f(r,z)$ is a prescribed function, and where it is desired to find those functions u and v which satisfy these equations, subject to certain boundary conditions, which we shall not describe in detail. Defining $U(\rho,z)$ as the Hankel transform of order 1 of u and $V(\rho,z)$ as the Hankel

[1] N. A. Haskell, *Physics*, **6**: 265 (1935).

transform of order 0 of v, we obtain

$$-\rho^2 U + U_{zz} = \int_0^\infty rJ_1(\rho r)f_r\,dr = -\rho F(\rho,z)$$
$$-\rho^2 V + V_{zz} = F_z$$
$$\rho U + V_z = 0$$

where F is the zero-order Hankel transform of f. Appropriate manipulation of these equations leads to $[(d^2/dz^2) - \rho^2]V = 0$, etc.; we shall not carry the example further.

EXERCISES

4. The equation of axially symmetric heat conduction in an infinitely long cylinder, for which there are no axial variations in temperature, is $\varphi_t = \alpha[\varphi_{rr} + (1/r)\varphi_r]$, where α is a constant and $\varphi(r,t)$ is the temperature. Let $\varphi(r,0) = f(r)$, and impose the radiation boundary condition $\varphi_r(a,t) + \beta\varphi(a,t) = 0$, where a is the radius of the cylinder and β is a constant. Solve this problem by a series expansion of the form $\varphi = \Sigma c_i(t)J_0(\omega_i t)$, for an appropriate choice of ω_i. Discuss the orthogonality of the terms in the series solution. Such a series, a generalization of the Fourier-Bessel series, is called a *Dini series;* in the general form, the series expansion is in terms of J_ν rather than J_0, and the ω_i satisfy $\omega_i J_\nu'(\omega_i) + \beta J_\nu(\omega_i) = 0$. (The special case $\nu = 0$, considered here, was first given by Fourier.)

5. (*a*) Show that the usual Fourier-Bessel series arises naturally from the separation-of-variables procedure for the equation of heat conduction in a cylinder (the equation of Exercise 4), with boundary condition $\varphi = 0$ at $r = 1$.

(*b*) Generalize (*a*) to the case of heat conduction in a cylinder containing an inner concentric hole; let the inner radius be a, the outer radius b. Show that

$$\varphi = \Sigma c_i(t)[Y_0(\omega_i a)J_0(\omega_i r) - J_0(\omega_i a)Y_0(\omega_i r)]$$

where ω_i satisfies

$$J_0(\omega_i a)Y_0(\omega_i b) = J_0(\omega_i b)Y_0(\omega_i a)$$

Is it feasible to obtain an integral transform theorem by the process $b \to \infty$?

6. The equation for radially symmetric heat conduction in a sphere is $\varphi_t = (\alpha/r^2)(r^2\varphi_r)_r$. Since this can be written $(r\varphi)_t = \alpha(r\varphi)_{rr}$, the function $(r\varphi)$ can be expanded in a sine series for a finite sphere, or a sine transform can be used for an infinite sphere. Is each of a Fourier-Bessel series and a Hankel transform feasible? If so, are they equivalent? Can the problem involving a finite or infinite sphere containing a hole be treated similarly?

7. The equation governing the displacement $w(r,t)$ in the radially symmetric vibration problem for a very large plate is

$$\left(\frac{\partial^2}{\partial r^2} + \frac{1}{r}\frac{\partial}{\partial r}\right)^2 w + \alpha^2 w_{tt} = 0$$

where α is a real constant. If $w(r,0)$ is given and $w_t(r,0) = 0$, use a Hankel transform in r to find w. Second, solve the same problem (*a*) by a Laplace transform in t and (*b*) by taking both transforms simultaneously.

8
special techniques

8-1 The Wiener-Hopf Method

There is an extensive array of important problems whose solution by Fourier-transform methods requires the use of a ingenious technique whose modern usage is due to Wiener and Hopf[1] but which was actually invented by Carleman (cf. Sec. 8-6). The technique is most conveniently displayed in connection with an illustrative example chosen to minimize algebraic complication; accordingly, we study the boundary-value problem[2]

$$\Delta\varphi - \varphi_x = 0 \qquad \text{in } D \tag{8-1}$$

with $\qquad \varphi(x,y) \to 0 \qquad$ as $x^2 + y^2 \to \infty$

and $\qquad \varphi = e^{-ax} \qquad$ on $y = 0,\ x \geq 0 \qquad$ (8-2)

Here $a > 0$, and D is the domain exterior to the half-line $y = 0$, $x \geq 0$. We require that φ be bounded everywhere and that φ_y be integrable everywhere.

[1] N. Wiener and E. Hopf, *S.B. Preuss. Akad. Wiss.*, 1931, p. 696.

[2] This problem arises in connection with the flow of heat from a semi-infinite plate to a fluid moving past the plate with uniform velocity.

When $y \neq 0$, the Fourier transform in x of Eq. (8-1) is

$$\Phi_{yy} - (\lambda^2 - i\lambda)\Phi = 0 \tag{8-3}$$

(Throughout this chapter, we continue to use the convention whereby the Fourier transform of a function is denoted by the corresponding capital letter.) The transform of Eq. (8-2) is

$$\Phi(\lambda,0) = \frac{1}{\sqrt{2\pi}} \frac{1}{a - i\lambda} + U(\lambda) \tag{8-4}$$

where $u(x)$ is defined by

$$u(x) = \begin{cases} 0 & \text{for } x > 0 \\ \varphi(x,0) & \text{for } x < 0 \end{cases} \tag{8-5}$$

The solution of Eq. (8-3) is

$$\begin{aligned} \Phi &= A(\lambda) \exp\left(-|y|\sqrt{\lambda^2 - i\lambda}\right) \\ &= \left[\frac{1}{\sqrt{2\pi}} \frac{1}{a - i\lambda} + U(\lambda)\right] \exp\left(-|y|\sqrt{\lambda^2 - i\lambda}\right) \end{aligned} \tag{8-6}$$

The branch of $\sqrt{\lambda^2 - i\lambda}$ in Eq. (8-6) must be so chosen that $\sqrt{\lambda^2 - i\lambda} \to +\infty$ as $\lambda \to \pm\infty$. Thus the inversion path will pass between $\lambda = 0$ and $\lambda = i$. We define[1]

$$F(\lambda) \equiv \Phi_y(\lambda,0+) - \Phi_y(\lambda,0-) = -2\left[\frac{1}{\sqrt{2\pi}} \frac{1}{a - i\lambda} + U(\lambda)\right]\sqrt{\lambda^2 - i\lambda} \tag{8-7}$$

and we note that, since continuity within D requires that

$$\varphi_y(x,0+) = \varphi_y(x,0-)$$

for $x < 0$, we must have $f(x) = 0$ for $x < 0$.

Thus, Eq. (8-7) is an equation from which we must determine each of the half-known functions $f(x)$ and $u(x)$. We note that, if $f(x)e^{-\beta x} \to 0$ as $x \to \infty$, with β real, then

$$F(\lambda) = \frac{1}{\sqrt{2\pi}} \int_0^\infty e^{i\lambda x} f(x)\, dx$$

describes a function which exists and is analytic at each point of $\operatorname{Im} \lambda > \beta$. Similarly, if $u(x)e^{-\alpha x} \to 0$ as $x \to -\infty$, with α real, $U(\lambda)$ is analytic at each point of $\operatorname{Im} \lambda < \alpha$.

We now *hope* that $f(x)e^{-\beta x}$ does vanish as $x \to \infty$ for some β and that $u(x)e^{-\alpha x}$ does vanish as $x \to -\infty$ *for some* $\alpha > \beta$, and we proceed as though we were assured of these two facts. We denote any function G analytic in

[1] By $\Phi_y(\lambda,0+)$, for example, we mean $\lim_{\epsilon \to 0} \Phi_y(\lambda,\epsilon)$, where $\epsilon > 0$.

Im $\lambda > \beta$ by G_+ and any function H analytic is Im $\lambda < \alpha$ by H_-. Equation (8-7) becomes

$$F_+(\lambda) = -2\left[\left(\frac{1}{\sqrt{2\pi}}\frac{1}{a - i\lambda}\right)_+ + U_-(\lambda)\right]\sqrt{\lambda^2 - i\lambda} \tag{8-8}$$

If Eq. (8-8) is to be at all meaningful, there must be a common domain, containing the inversion path, in which F_+ and U_- are defined; so our optimistic expectation that $\alpha > \beta$ is entirely consistent. The foregoing remarks imply the situation depicted in Fig. 8-1. Consistent with Fig. 8-1, we can write Eq. (8-8) in the form

$$\frac{-F_+(\lambda)}{2\sqrt{\lambda}} = \frac{\sqrt{\lambda - i}}{\sqrt{2\pi}\,(a - i\lambda)} + \sqrt{\lambda - i}\;U_-(\lambda) \tag{8-9}$$

Furthermore, the second term of Eq. (8-9) can be written

$$\frac{\sqrt{\lambda - i}}{\sqrt{2\pi}\,(a - i\lambda)} = \left[\frac{\sqrt{\lambda - i} - \sqrt{-ia - i}}{\sqrt{2\pi}\,(a - i\lambda)}\right]_- + \left[\frac{\sqrt{-ia - i}}{\sqrt{2\pi}\,(a - i\lambda)}\right]_+ \tag{8-10}$$

Accordingly,

$$\left[\frac{-F(\lambda)}{2\sqrt{\lambda}} - \frac{\sqrt{-ia - i}}{\sqrt{2\pi}\,(a - i\lambda)}\right]_+ = \left[\frac{\sqrt{\lambda - i} - \sqrt{-ia - i}}{\sqrt{2\pi}\,(a - i\lambda)} + \sqrt{\lambda - i}\;U(\lambda)\right]_- \tag{8-11}$$

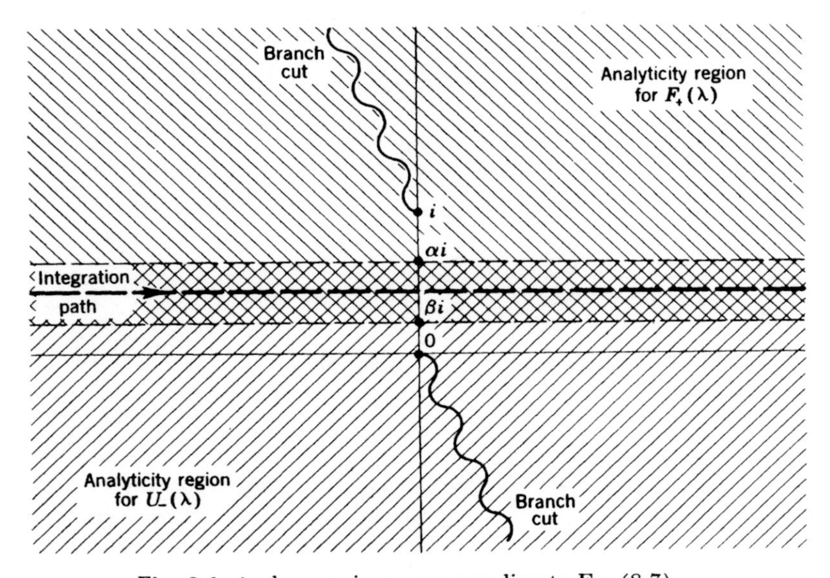

Fig. 8-1 λ-plane regions corresponding to Eq. (8-7).

The function on the left side of Eq. (8-11) is the analytic continuation of the function on the right side (they are equal on a dense set of points), and, together, they define an entire function $E(\lambda)$. Since $\varphi_y(x,0)$ is integrable at $x = 0$, $F_+ \to 0$ as $|\lambda| \to \infty$ in $\text{Im}\ \lambda > \beta$, and since $\varphi(x,0)$ is bounded at $x = 0$, $\lambda U_-(\lambda)$ is bounded as $|\lambda| \to \infty$ in $\text{Im}\ \lambda < \alpha$. Thus, each side of Eq. (8-11) tends to zero as $|\lambda| \to \infty$ in its domain of definition, and, by Liouville's theorem, $E(\lambda)$ is identically zero.

We now have

$$F(\lambda) = \frac{-2\sqrt{\lambda}\sqrt{-ia - i}}{\sqrt{2\pi}\,(a - i\lambda)} \tag{8-12}$$

and

$$\Phi(\lambda,y) = \frac{\sqrt{1 + a}\, e^{-i\pi/4}}{\sqrt{2\pi}\,(a - i\lambda)\sqrt{\lambda - i}} \exp\left(-|y|\sqrt{\lambda^2 - i\lambda}\right)$$

Fourier inversion now yields $\varphi(x,y)$. In particular,

$$\varphi_y(x,0+) = \begin{cases} 0 & \text{for } x < 0 \\ \sqrt{1 + a}\left(\dfrac{-1}{\sqrt{\pi x}} - i\sqrt{a}\, e^{-ax} \operatorname{erf}(i\sqrt{ax})\right) & \text{for } x > 0 \end{cases} \tag{8-13}$$

For $a = 0$, we obtain $\varphi(x,y) = \operatorname{erfc} \eta$, where $\eta = \text{Im}\sqrt{x + iy}$, with $0 \le \arg(x + iy) \le 2\pi$. The reader should verify that this result can also be obtained by writing Eq. (8-1) in terms of the independent variables ξ, η, with $\xi + i\eta = \sqrt{x + iy}$, and asking whether or not solutions depending only on η satisfy all constraints.

The novel item in the foregoing problem arose in connection with the boundary condition on the half-line $y = 0$, $x > 0$. Typically, in problems with mixed boundary conditions on the line $y = 0$, $-\infty < x < \infty$, or with boundary conditions on a half-line, one is led via the Fourier-transform technique to a functional equation such as Eq. (8-8). From that equation one must identify two analytic functions with different domains of analyticity. Ordinarily, that functional equation can be cast in the form

$$F_+(\lambda)K(\lambda) = A(\lambda) + U_-(\lambda) \tag{8-14}$$

The Wiener-Hopf procedure then requires that $K(\lambda)$ be factored into the form

$$K(\lambda) = \frac{L_+(\lambda)}{M_-(\lambda)} \tag{8-15}$$

so that

$$F_+(\lambda)L_+(\lambda) = A(\lambda)M_-(\lambda) + U_-(\lambda)M_-(\lambda)$$

The only "mixed" function is $A(\lambda)M_-(\lambda)$, and this must be decomposed into the sum

$$A(\lambda)M_-(\lambda) = P_-(\lambda) + Q_+(\lambda) \tag{8-16}$$

One can then use the analytic-continuation arguments and note that

$$F_+(\lambda)L_+(\lambda) - Q_+(\lambda) = P_-(\lambda) + U_-(\lambda)M_-(\lambda) = E(\lambda) \qquad (8\text{-}17)$$

where E is entire. One must then find $E(\lambda)$; ordinarily, this is done by using the constraints at $x = 0$ on $u(x)$ and $f(x)$ to identify the behavior of $E(\lambda)$ as $|\lambda| \to \infty$.

Another problem, encountered in connection with the flow of a viscous fluid,[1] is defined by

$$\Delta\Delta\psi - \Delta\psi_x = 0 \qquad \text{in } D \qquad (8\text{-}18)$$

where D is again the xy plane exterior to the half-line $y = 0$, $x > 0$ and where

$$\psi \text{ is odd in } y,\ \psi(x,0) = 0,\ \psi_y(x,0) = \begin{cases} ? & x < 0 \\ 1 & x > 0 \end{cases}$$

$$\psi \to 0, \text{ as } x^2 + y^2 \to \infty, \text{ in any sector excluding the line } y = 0,\ x > 0$$

Furthermore, ψ and grad ψ must be bounded for all x, y.

The transform of Eq. (8-18) is

$$\left(\frac{\partial^2}{\partial y^2} - \lambda^2\right)\left(\frac{\partial^2}{\partial y^2} - \lambda^2 + i\lambda\right)\Psi = 0 \qquad (8\text{-}19)$$

and, taking account of the boundary conditions $\Psi(x,0) = 0$ and

$$\Psi(\lambda,y) = -\Psi(\lambda,-y)$$

we infer that

$$\Psi(\lambda,y) = \frac{|y|}{y} A(\lambda)\left[\exp\left(-|y|\sqrt{\lambda^2}\right) - \exp\left(-|y|\sqrt{\lambda^2 - i\lambda}\right)\right] \qquad (8\text{-}20)$$

The function $\sqrt{\lambda^2}$ is that branch of the multivalued analytic function for which

$$\sqrt{\lambda^2} = \begin{cases} \lambda & \operatorname{Re}\lambda > 0 \\ -\lambda & \operatorname{Re}\lambda < 0 \end{cases} \qquad (8\text{-}21)$$

Among many other possibilities, it can be represented as the limit

$$\sqrt{\lambda^2} = \lim_{\substack{\epsilon > 0 \\ \epsilon \to 0}} \sqrt{\lambda^2 - i\epsilon\lambda} \qquad (8\text{-}22)$$

and we shall see later that this is a most convenient interpretation.

With $u(x)$ defined as in the previous example, the requirement on $\psi_y(x,0)$ implies that

$$\begin{aligned} \Psi_y(\lambda,0) &= A(\lambda)\left(\sqrt{\lambda^2 - i\lambda} - \sqrt{\lambda^2}\right) \\ &= U_-(\lambda) + \left(\frac{1}{\sqrt{2\pi}}\frac{1}{-i\lambda}\right)_+ \end{aligned} \qquad (8\text{-}23)$$

We also note that the Fourier transform of $f(x) \equiv \psi_{yy}(x,0+) - \psi_{yy}(x,0-)$ is

$$F(\lambda) = 2i\lambda A(\lambda) \qquad (8\text{-}24)$$

[1] J. A. Lewis and G. F. Carrier, *Quart. Appl. Math.*, **7**: 228 (1949).

Since ψ obeys Eq. (8-18) on $y = 0$, $x < 0$, $f(x)$ vanishes for $x < 0$ and $F(\lambda)$ is $F_+(\lambda)$.

Equations (8-23) and (8-24) now imply that

$$U_-(\lambda) - \left(\frac{1}{\sqrt{2\pi}}\frac{1}{i\lambda}\right)_+ = F_+(\lambda)\frac{\sqrt{\lambda^2 - i\lambda} - \sqrt{\lambda^2}}{2i\lambda}$$
$$= \frac{-F_+(\lambda)}{2(\sqrt{\lambda^2 - i\lambda} + \sqrt{\lambda^2})} \tag{8-25}$$

The role of $K(\lambda)$ in Eq. (8-14) is played in Eq. (8-25) by $(\sqrt{\lambda^2 - i\lambda} + \sqrt{\lambda^2})^{-1}$. Unfortunately, there is no horizontal strip, not even a horizontal line, in the λ plane on which $\sqrt{\lambda^2 - i\lambda} + \sqrt{\lambda^2}$ is single-valued and analytic.[1] Accordingly, either we must find more powerful theorems than those we invoked in the foregoing arguments, or we must find some simple device which eliminates the difficulty. We like the device wherein we replace $\sqrt{\lambda^2}$ by $\sqrt{\lambda^2 - i\epsilon\lambda}$ and where we defer the limiting process of Eq. (8-22) until a convenient point in the analysis. Replacing $\sqrt{\lambda^2}$ by $\sqrt{\lambda^2 - i\epsilon\lambda}$ in Eq. (8-25), we can write

$$\sqrt{\lambda^2 - i\lambda} + \sqrt{\lambda^2 - i\epsilon\lambda} = (\sqrt{\lambda})_+(\sqrt{\lambda - i} + \sqrt{\lambda - i\epsilon})_- \tag{8-26}$$

Clearly, the strip in which both factors are analytic is that lying in $0 < \operatorname{Im}\lambda < \epsilon$.

Equation (8-25) can now be written in the form

$$U_-(\lambda)(\sqrt{\lambda - i} + \sqrt{\lambda - i\epsilon})_- - \left(\frac{\sqrt{\lambda - i} + \sqrt{\lambda - i\epsilon} - \sqrt{-i} - \sqrt{-i\epsilon}}{\sqrt{2\pi}\, i\lambda}\right)_-$$
$$= \left(\frac{-F(\lambda)}{2\sqrt{\lambda}} + \frac{\sqrt{-i} + \sqrt{-i\epsilon}}{\sqrt{2\pi}\, i\lambda}\right)_+ = E(\lambda) \tag{8-27}$$

and, once again, as the reader may verify, $E(\lambda) \equiv 0$ because of the boundedness of ψ and grad ψ. In the limit as $\epsilon \to 0$,

$$F(\lambda) = \sqrt{\frac{2}{\pi}}\, e^{-i\frac{3}{4}\pi}\frac{1}{\sqrt{\lambda}} \tag{8-28}$$

and the solution can be completed by using Eqs. (8-24) and (8-20) and the Fourier inversion formula. The reader should find it an interesting exercise to fill in the details.

In view of the artificial limit process we have introduced into the problem, it is also the better part of valor to show directly that the final result satisfies all the constraints of the problem. As in the previous example, the use of the variables $\xi + i\eta = \sqrt{x + iy}$ would lead to the same result by a more elementary route; however, it should be noted that, unlike the Wiener-Hopf method, such a device will fail unless the boundary condition on $y = 0$, $x > 0$ is very special indeed.

[1] The reader should verify that alternative interpretations of $\sqrt{\lambda^2}$ and $\sqrt{\lambda^2 - i\lambda}$ which admit such a horizontal strip do not lead to a solution of the problem.

EXERCISES

Solve the boundary-value problems:

1. $\Delta\varphi + k^2\varphi = 0$ in D

with $\varphi_y = e^{i\beta x}$ on $y = 0,\ x \geq 0$

and $r(\varphi_y + ik\varphi) \to 0$ as $r^2 \equiv x^2 + y^2 \to \infty$

D is the domain exterior to the half-line $y = 0,\ x \geq 0$; k is real and positive; $0 < \beta < k$.

2. $\Delta\varphi - k^2\varphi = 0$ in D

$\varphi_y = e^{i\beta x}$ on $y = 0,\ x \geq 0$

$\varphi \to 0$ as $r \to \infty$

D is the domain described in Exercise 1. Contrast the answer and the problem specification with those of Exercise 1.

3. $\Delta\Delta\psi - \Delta\psi_x = 0$ in D

(where D is as in Exercise 1. ψ is odd in y, $\psi_y(x,0) = 0$ on $x < 0$. $\psi_y(x,0) = ?$ on $x > 0$. $\psi_{yy}(x,0) = 0$ on $x < 0$, $\psi_{yy}(x,0\pm) = \pm e^{-\beta x}$ on $x > 0$. $\psi \to 0$ as $\eta \to \infty$, where $\xi + i\eta = (x + iy)^{1/2}$.

8-2 *The Kernel Decomposition*

Ordinarily, the only unpleasant task encountered in the use of the Wiener-Hopf method is the identification of the functions $L_+(\lambda)$ and $M_-(\lambda)$ of Eq. (8-15). The factors $L_+(\lambda)$ and $M_-(\lambda)$ such that

$$K(\lambda) = \frac{L_+(\lambda)}{M_-(\lambda)} \tag{8-29}$$

are usually not available by inspection [at least to most persons; e.g., try $K(\lambda) = a^2 + \sqrt{1 + \lambda^2}$]. We must therefore devise a constructive method for obtaining L_+ and M_-.

Let the strip in which $K(\lambda)$ is analytic contain a substrip $\beta < \operatorname{Im}\lambda < \alpha$ in which $K(\lambda)$ has no zeros. We write

$$\ln K(\lambda) = \ln L_+(\lambda) - \ln M_-(\lambda) \tag{8-30}$$

and observe that, provided that L_+ and M_- have no zeros in their half-planes of analyticity, each of $\ln L_+$ and $\ln M_-$ is itself analytic in the corresponding domain. Thus the decomposition of $K(\lambda)$ as a quotient is equivalent to the decomposition of $\ln K$ as a sum.

We now recall that, whenever Γ encloses λ but does not enclose any singularities of an analytic function $G(z)$.

$$G(\lambda) = \frac{1}{2\pi i}\int_\Gamma \frac{G(z)}{z - \lambda}\,dz \tag{8-31}$$

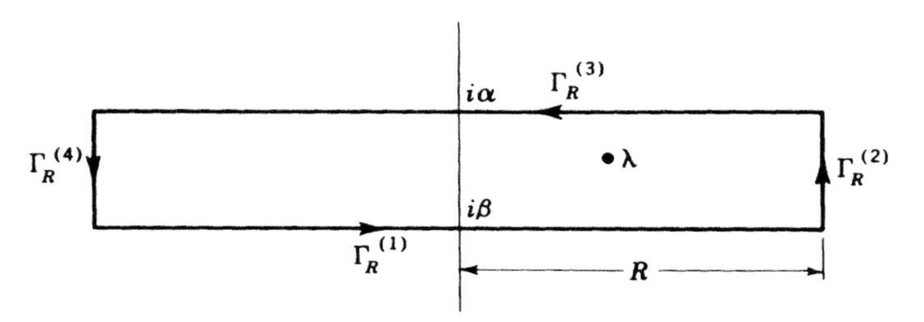

Fig. 8-2 Contour for Eq. (8-31).

Accordingly, we choose Γ_R to be the contour of Fig. 8-2 and identify segments $\Gamma_R^{(j)}$ of the contour as shown in that figure. We consider only functions $G(z)$ for which G is analytic at each point inside Γ_R for all R, and we limit our attention to those $G(z)$ for which the integrals on $\Gamma_\infty^{(1)}$ and $\Gamma_\infty^{(3)}$ exist and for which $\Gamma_\infty^{(2)} + \Gamma_\infty^{(4)} = 0$.

The integral over $\Gamma_\infty^{(1)}$ defines a function which is analytic everywhere above $\Gamma_\infty^{(1)}$. Correspondingly, the integral along $\Gamma_\infty^{(3)}$ defines a function which is analytic everywhere below $\Gamma_\infty^{(3)}$. Thus we have

$$G(\lambda) = R_+(\lambda) - S_-(\lambda)$$

with

$$R_+(\lambda) = \frac{1}{2\pi i}\int_{\Gamma_\infty^{(1)}} \frac{G(z)}{z-\lambda}\,dz \tag{8-32}$$

and

$$S_-(\lambda) = \frac{1}{2\pi i}\int_{-\Gamma_\infty^{(3)}} \frac{G(z)}{z-\lambda}\,dz \tag{8-33}$$

If we now identify our $G(z)$ with $\ln K(z)$, we see that the desired splitting is obtained by setting $\ln L_+ = R_+$ and $\ln M_- = S_-$.

Consider the example $K(\lambda) = (a^2 + \sqrt{1+\lambda^2})/\sqrt{1+\lambda^2}$, with $a > 0$. From Eq. (8-32),

$$R_+(\lambda) = \frac{1}{2\pi i}\int_{-\infty}^{\infty} \frac{\ln(a^2+\sqrt{1+z^2}) - \ln\sqrt{1+z^2}}{z-\lambda}\,dz$$

and

$$\begin{aligned} R_+'(\lambda) &= \frac{1}{2\pi i}\int_{-\infty}^{\infty} \frac{\ln(a^2+\sqrt{1+z^2}) - \ln\sqrt{1+z^2}}{(z-\lambda)^2}\,dz \\ &= \frac{1}{2\pi i}\int_{-\infty}^{\infty} \frac{-a^2 z\,dz}{(z^2+1)(z-\lambda)(a^2+\sqrt{z^2+1})} \end{aligned} \tag{8-34}$$

The reader can complete the calculation [a direct method is to set $z = \frac{1}{2}(x - 1/x)$ and use partial fractions; a little ingenuity will reduce the labor] to obtain an expression for $R_+'(\lambda)$, and similarly for $S_-'(\lambda)$, in terms of logarithms or arccosines. The reader should identify carefully which branches of these functions are the appropriate ones, and he should check this identification by verifying that his $R_+(\lambda)$ and his $S_-(\lambda)$ are such that

$$R_+' - S_-' = (\ln K)'$$

Returning to Eq. (8-29), we can do no better than to write

$$
\begin{aligned}
L_+(\lambda) &= \exp\left[\int^\lambda R'_+(\lambda)\,d\lambda\right] \\
M_-(\lambda) &= \exp\left[\int^\lambda S'_-(\lambda)\,d\lambda\right]
\end{aligned}
\tag{8-35}
$$

It is evident that $L_+(\lambda)$ and $M_-(\lambda)$ are not particularly simple functions. It is also evident that many problems of the Wiener-Hopf type which can be answered in principle by the foregoing method lead to formulas which are exceedingly cumbersome to interpret. We shall deal with this difficulty in Sec. 8-B.

The reader should note that:

1. A technique for the additive decomposition represented in Eq. (8-16) has automatically been included in the foregoing discussion.

2. When $A(\lambda)$ in Eq. (8-16) has only poles, the decomposition of $A(\lambda)M_-(\lambda)$ is straightforward. For example, let a and γ be constants, with

$$A(\lambda) = \left[\frac{a}{(\lambda - i\gamma)}\right]_+$$

then $Q_+(\lambda)$ is given by

$$Q_+(\lambda) = \frac{aM_-(i\gamma)}{\lambda - i\gamma} \tag{8-36}$$

3. For a function $K_1(\lambda)$ such as $a^2 + \sqrt{1 + \lambda^2}$, whose logarithm does not qualify under the constraints preceding Eq. (8-32), it may be that we need only divide by an easily decomposed function (here $\sqrt{1 + \lambda^2}$, say) to obtain a new function for which the constraints *are* satisfied. Thus, for $K_1(\lambda)$, we obtain

$$K_1(\lambda) = \frac{a^2 + \sqrt{1 + \lambda^2}}{\sqrt{1 + \lambda^2}}\sqrt{1 + \lambda^2} = \frac{[L_+(\lambda)\sqrt{\lambda + i}]_+}{[M_-(\lambda)/\sqrt{\lambda - i}]_-}$$

where $L_+(\lambda)$ and $M_-(\lambda)$ are defined by Eqs. (8-35).

If $K(\lambda)$ in Eq. (8-29) involves an entire function, then it may be convenient to use an infinite-product expression for that entire function, so as to obtain at once its two appropriate factors. Some general remarks on the choice of associated convergence factors and the determination of the asymptotic behavior at ∞ have been given by Carlson and Heins[1] in connection with the reflection of a plane wave by an infinite set of plates. Where trigonometric functions are encountered, such gamma-function formulas as Eq. (5-9) can be useful.

[1] J. F. Carlson and A. E. Heins, *Quart. Appl. Math.*, *4:* 313 (1947).

A large collection of problems requiring a factoring process of the Wiener-Hopf type will be found in the book by Noble.[1]

EXERCISES

1. (*a*) Obtain the analog of Eq. (8-36) for the case in which $A(\lambda)$ has a number of poles, some of which may be multiple (but no other singularities).

(*b*) Obtain an additive splitting for the function $\sqrt{1 + \lambda^2}$ [cf. Eq. (1-14)].

2. Let $\Delta\varphi = 0$ in $y < 0$, with $\varphi = 1$ on $y = 0, x < 0$ and $\varphi_y + \varphi = 0$ on $y = 0$, $x > 0$. Find $\varphi(x,y)$.

3. Solve the problem of the propagation of a wave down a two-dimensional duct, with a strip across it, described by the following equations:

(*a*) $\Delta\varphi + k^2\varphi = 0$

for $-\infty < x < \infty, 0 < y < l$, except on the strip $x = 0, 0 < y < l/2$.

(*b*) $\partial\varphi/\partial n = 0$ on $y = 0$, $y = l$; $\partial\varphi/\partial n = -ik$ on $x = 0$, $0 < y < l/2$, where $\partial/\partial n$ indicates the normal derivative.

(*c*) $\varphi \to 0$ as $|x| \to \infty$.

[*Hint:* Consider the combinations $\Phi_+(\lambda,y) + \Phi_+(-\lambda,y)$, $\Phi_-(\lambda,y) + \Phi_-(-\lambda,y)$.]

4. The acoustic radiation from a semi-infinite pipe of circular cross section has been treated by Levine and Schwinger.[2] Postulating circular symmetry, let the cylindrical coordinates be (r,z), and let the pipe occupy the surface $r = a, z \leq 0$. A wave e^{ikz} is incident from $z = -\infty$, inside the pipe. (The time factor is $e^{-i\omega t}$. For convenience, k may be considered to have a small positive imaginary part.) The acoustic potential φ is to satisfy the wave equation $\Delta\varphi + k^2\varphi = 0$, with $\varphi_r(a,z) = 0$ for $z \leq 0$, the radiation condition at ∞, and a "derivative-integrability" condition at the end of the pipe. Taking Fourier transforms of the functions φ (for $r > a$) and $\varphi - e^{ikz}$ (for $r < a$), use the Wiener-Hopf method to determine φ. (Simplify the integrals which occur as much as you can.)

Occasionally, the additive decomposition of Eq. (8-16) is cumbersome. Suppose, for example, that one wishes to solve a linear problem identical with Exercise 3 of Sec. 8-1 except that $\psi_{yy}(x,0+) - \psi_{yy}(x,0-)$ is given not as $2e^{-\beta x}$ but as $(x + a)^{-1/2}$. The decomposition of Eq. (8-16) is "messy," and it leads to a $P_-(\lambda)$ and a $Q_+(\lambda)$ which are most inconvenient for subsequent manipulation. A very convenient procedure is as follows: Denote the solu-

[1] B. Noble, "Methods Based on the Wiener-Hopf Technique," Pergamon Press, New York, 1958.

[2] H. Levine and J. Schwinger, *Phys. Rev.*, *73:* 383 (1948).

tion of the original problem of that exercise by $\psi_3(x,y,\beta)$. Clearly, since

$$\frac{1}{\sqrt{x+\alpha}} = \int_0^\infty \frac{e^{-(x+\alpha)\beta}}{\sqrt{\pi\beta}}\, d\beta \tag{8-37}$$

the solution of the new problem is given by

$$\psi(x,y) = \int_0^\infty \psi_3(x,y,\beta) \frac{e^{-\alpha\beta}}{\sqrt{\pi\beta}}\, d\beta \tag{8-38}$$

8-3 Integral Equations with Displacement Kernels

Wiener and Hopf made their contribution[1] to the foregoing technique in conjunction with the treatment of the integral equation

$$f(x) = \tfrac{1}{2} \int_0^\infty Ei(|x-t|) f(t)\, dt \qquad x > 0 \tag{8-39}$$

where $Ei(x)$ is the exponential integral

$$Ei(x) = \int_x^\infty \frac{e^{-s}}{s}\, ds \tag{8-40}$$

and where $f(x)$ is to be found.

We consider, more generally, the problem of determining that function $f(x)$ for which, in $x > 0$,

$$\int_0^\infty k(x-t) f(t)\, dt = g(x) \tag{8-41}$$

where $k(x)$ and $g(x)$ are given functions; we assume that k, f, and g are sufficiently well behaved that the transform operations we are about to perform will lead to a meaningful result.

Without loss of generality, we can set $f(t) = 0$ for $t < 0$, so that Eq. (8-41) becomes, for $x > 0$,

$$\int_{-\infty}^\infty k(x-t) f(t)\, dt = g(x) \tag{8-42}$$

and the Fourier transform of the left side can now conveniently be calculated. However, Eq. (8-42) is not correct for $x < 0$; therefore, in order to make the problem accessible to Fourier-transform techniques, we rewrite Eq. (8-42) in the form (valid in $-\infty < x < \infty$)

$$\int_{-\infty}^\infty k(x-t) f(t)\, dt = m(x) + h(x) \tag{8-43}$$

[1] *Loc. cit.*

where
$$f(x) = \begin{cases} 0 & \text{for } x < 0 \\ ? & \text{for } x > 0 \end{cases}$$
$$m(x) = \begin{cases} 0 & \text{for } x < 0 \\ g(x) & \text{for } x > 0 \end{cases}$$
$$h(x) = \begin{cases} ? & \text{for } x < 0 \\ 0 & \text{for } x > 0 \end{cases}$$

Note that, in $x > 0$, Eq. (8-43) is identical with Eq. (8-41), so that the solution of Eq. (8-43) is precisely the solution of Eq. (8-41). Note also that—again—we have a problem in which there are two half-known, half-unknown functions.

The Fourier transform of Eq. (8-43) is

$$K(\lambda)F(\lambda) = M(\lambda) + H(\lambda) \tag{8-44}$$

and since $F(\lambda)$ and $H(\lambda)$ can be written as $F_+(\lambda)$ and $H_-(\lambda)$, respectively, and since $K(\lambda)$ and $M(\lambda)$ are known, we have again a Wiener-Hopf type of problem. In fact, Eq. (8-44) is identical in form with Eq. (8-14).

As an example, let

$$k(x) = -\frac{1}{\sqrt{2\pi}} \exp\left(\tfrac{1}{2}x\right)K_0\left(\tfrac{1}{2}|x|\right)$$

where K_0 is the Hankel function of imaginary argument, as defined in Eq. (5-126). Then the transform integral defining $K(\lambda)$ converges for $0 < \operatorname{Im}\lambda < 1$, and Eq. (8-44) becomes

$$-\frac{1}{2\sqrt{\lambda^2 - \lambda i}}\,F_+(\lambda) = \frac{1}{\sqrt{2\pi}}\,\frac{1}{a - i\lambda} + H_-(\lambda) \tag{8-45}$$

We need proceed no further, since Eq. (8-45) is the same as Eq. (8-8), which we have previously solved. Conversely, we could have obtained Eq. (8-41) from Eq. (8-8), by Fourier inversion.

It is clear that any partial-differential-equation boundary-value problem which leads to a functional equation of the Wiener-Hopf type can always be recast as an integral equation—simply by taking the inverse transform of that equation, using the convolution theorem. The kernel function $k(x - t)$ that results from this process is the appropriate Green's function for the original problem. [In fact, when that Green's function is not already known, the simplest way of finding it is to employ the analysis leading to Eq. (8-8) and follow that by Fourier inversion, using the convolution theorem.] Once one has obtained the integral equation, one proceeds through the analysis that leads to Eqs. (8-43) and (8-44)—i.e., one backs up part of the way—and then continues exactly as if the integral equation had never been found. Thus it is ordinarily more time-consuming than useful to identify an integral equation with a boundary-value problem on which, ultimately, the Wiener-

Hopf technique is to be used. However, we shall see that the *existence* of an equivalent integral equation provides the motivation for the powerful approximation techniques studied in the next section.

EXERCISES

1. We have seen that Eq. (8-8) leads to an integral equation of the form (8-41) when $f(x)$ (for $x > 0$) in Eq. (8-8) is treated as the unknown. Alternatively, we could treat $u(x)$ (for $x < 0$) as the unknown; show that the resulting integral equation can be written

$$\frac{d^2}{dx^2}\int_{-\infty}^{0} u(t)k(x-t)\,dt = m(x) \qquad \text{for } x < 0$$

and identify $k(x)$ and $m(x)$.

2. Solve the integral equations:

(*a*) $\int_0^\infty K_0(|x-t|)u(t)\,dt = 1$ in $x > 0$

(*b*) $\int_0^\infty e^{-\alpha|x-t|}u(t)\,dt = e^{-\beta x}$ in $x > 0$

(*c*) $\int_0^\infty (1 + k|x-t|)e^{-\alpha|x-t|}u(t)\,dt = e^{-\beta x}$ in $x > 0$

where α, β, k are all > 0 and where K_0 is the usual modified Bessel function. Note that the solutions to (*b*) and (*c*) may involve δ functions. Determine also the ordinary differential equations to which (*b*) and (*c*) correspond, and invent a boundary-value problem involving partial-differential equations which is equivalent to (*a*).

3. Recast the problems described in Exercises 1, 2, and 3 of Sec. 8-1, and Exercises 3 and 4 of Sec. 8-2, into integral-equation form.

The Homogeneous Problem

Before dealing with the homogeneous Wiener-Hopf problem associated with the integration range $(0, \infty)$, it is useful to study the homogeneous integral equation

$$\int_{-\infty}^{\infty} k(x-t)f(t)\,dt = \sigma f(x) \tag{8-46}$$

in which the integration range is $(-\infty, \infty)$. Here σ is a constant. To begin with, we proceed formally. A Fourier transform yields

$$\left[K(\lambda) - \frac{\sigma}{\sqrt{2\pi}}\right]F(\lambda) = 0 \tag{8-47}$$

so that $F(\lambda)$ can be different from zero only at the roots $\lambda = \alpha_j$ of

$$K(\lambda) - \frac{\sigma}{\sqrt{2\pi}} = 0 \tag{8-48}$$

In order to obtain a nontrivial result, we therefore try $F(\lambda) = \sum_j A_j\delta(\lambda - \alpha_j)$, which upon inversion yields

$$f(x) = \sum \frac{1}{\sqrt{2\pi}} A_j \exp(-i\alpha_j x) \tag{8-49}$$

where the A_j are arbitrary constants. If Eq. (8-48) had a double root at α_1, say, we might also include a term $\delta'(\lambda - \alpha_1)$ in $F(\lambda)$ and so be led to an added term of the form $x \exp(-i\alpha_1 x)$ in Eq. (8-49).

As we have said, such an analysis—although often useful—is purely formal and is difficult to justify directly. We therefore sketch an alternative procedure, based on Eqs. (7-78) and (7-79). Using the notation of these two equations, we employ Eq. (7-79) to substitute for $f(x)$ and $f(t)$ in Eq. (8-46),

$$\int_{-\infty}^{\infty} k(x-t)\left[\frac{1}{\sqrt{2\pi}}\int_{-\infty+ib_1}^{\infty+ib_1} e^{-i\lambda t}F_+(\lambda)\,d\lambda + \frac{1}{\sqrt{2\pi}}\int_{-\infty-ic_1}^{\infty-ic_1} e^{-i\lambda t}F_-(\lambda)\,d\lambda\right]dt$$
$$= \frac{\sigma}{\sqrt{2\pi}}\int_{-\infty+ib_1}^{\infty+ib_1} e^{-i\lambda x}F_+(\lambda)\,d\lambda + \frac{\sigma}{\sqrt{2\pi}}\int_{-\infty-ic_1}^{\infty-ic_1} e^{-i\lambda x}F_-(\lambda)\,d\lambda \tag{8-50}$$

Equation (8-50) is valid even if $f(x)$ has exponential growth as $|x| \to \infty$. We now change the order of integration and replace the integration variable t by $\xi = x - t$, to give

$$\int_{-\infty+ib_1}^{\infty+ib_1}\left[K(\lambda) - \frac{\sigma}{\sqrt{2\pi}}\right]F_+(\lambda)e^{-i\lambda x}\,d\lambda$$
$$+ \int_{-\infty-ic_1}^{\infty-ic_1}\left[K(\lambda) - \frac{\sigma}{\sqrt{2\pi}}\right]F_-(\lambda)e^{-i\lambda x}\,d\lambda = 0 \tag{8-51}$$

for all x, where we assume $K(\lambda)$ to exist for a strip which includes the lines $\operatorname{Im}\lambda = b_1$, $\operatorname{Im}\lambda = -c_1$. Multiply Eq. (8-51) by $e^{ix\omega}$, where $\operatorname{Im}\omega > b$, integrate with respect to x over $(0, \infty)$, and move the line of integration in the first integral to $\operatorname{Im}\lambda = b_2$, where $b_2 > \operatorname{Im}\omega$. We obtain

$$\int_{-\infty+ib_2}^{\infty+ib_2}\left[K(\lambda) - \frac{\sigma}{\sqrt{2\pi}}\right]F_+(\lambda)\frac{d\lambda}{\lambda-\omega} + \int_{-\infty-ic_1}^{\infty-ic_1}\left[K(\lambda)\right.$$
$$\left.- \frac{\sigma}{\sqrt{2\pi}}\right]F_-(\lambda)\frac{d\lambda}{\lambda-\omega} = -2\pi i\left[K(\omega) - \frac{\sigma}{\sqrt{2\pi}}\right]F_+(\omega) \tag{8-52}$$

The left-hand side of Eq. (8-52) is analytic for $c_1 < \operatorname{Im}\omega < b_2$, and so consequently is the right-hand side. Similarly,

$$\int_{-\infty+ib_1}^{\infty+ib_1}\left[K(\lambda) - \frac{\sigma}{\sqrt{2\pi}}\right]F_+(\lambda)\frac{d\lambda}{\lambda-\omega} + \int_{-\infty-ic_2}^{\infty-ic_2}\left[K(\lambda)\right.$$
$$\left.- \frac{\sigma}{\sqrt{2\pi}}\right]F_+(\lambda)\frac{d\lambda}{\lambda-\omega} = 2\pi i\left[K(\omega) - \frac{\sigma}{\sqrt{2\pi}}\right]F_-(\omega) \tag{8-53}$$

where $-c_2 < \text{Im}\,\omega$. The left-hand side, and so the right-hand side, of Eq. (8-53) is analytic for $b_1 > \text{Im}\,\omega > -c_2$. But if $-c_1 < \text{Im}\,\omega < b_1$, the left-hand sides of Eqs. (8-52) and (8-53) are clearly equal, so that, throughout a strip which includes the lines $\text{Im}\,\omega = b_1$, $\text{Im}\,\omega = -c_1$, we have

$$\left[K(\omega) - \frac{\sigma}{\sqrt{2\pi}}\right] F_+(\omega) = -\left[K(\omega) - \frac{\sigma}{\sqrt{2\pi}}\right] F_-(\omega) = \text{analytic} \tag{8-54}$$

Thus the singularities of $F_+(\omega) = -F_-(\omega)$ in this strip can be at worst poles, whose locations and orders coincide with those of the zeros of $K(\omega) - \sigma/\sqrt{2\pi}$. Since $F_+(\omega) = -F_-(\omega)$, Eq. (7-79) takes the form of a contour integral and we obtain again Eq. (8-49)—with now the added result that if $K(\omega) - \sigma/\sqrt{2\pi}$ has an nth-order zero at α_j, then terms $x^p \exp(-i\alpha_j x)$, for all $p < n$, must be included in Eq. (8-49).

As an example, let $k(x) = K_0(|x|)$, where K_0 is the modified Bessel function. Then $\alpha_1 = -\alpha_2 = \sqrt{(\pi^2/\sigma^2) - 1}$, and

$$f(x) = A_1 \exp(i\alpha_1 x) + A_2 \exp(i\alpha_2 x)$$

for all σ for which the integral in Eq. (8-46) exists.

We now turn our attention to the equation

$$\int_0^\infty k(x - t) f(t)\, dt = \sigma f(x) \tag{8-55}$$

for $x > 0$, in which the range of integration is $(0, \infty)$. We write

$$\int_{-\infty}^\infty k(x - t) f(t)\, dt = \sigma f(x) + h(x) \tag{8-56}$$

for x in $(-\infty, \infty)$, where $f(x) = 0$ for $x < 0$, $h(x) = 0$ for $x > 0$. The transform of Eq. (8-56) is

$$\sqrt{2\pi}\, K(\lambda) F_+(\lambda) = \sigma F_+(\lambda) + H_-(\lambda)$$

The splitting $\sqrt{2\pi}\, K(\lambda) - \sigma = G_+(\lambda)/G_-(\lambda)$ leads to

$$G_+(\lambda) F_+(\lambda) = G_-(\lambda) H_-(\lambda) = E(\lambda)$$

where $E(\lambda)$ is entire. In each individual problem, one must identify $E(\lambda)$; typically, $E(\lambda) \equiv \text{const.}$

To illustrate, let $k(x) = e^{-\alpha|x|}$, where $\alpha > 0$. Also, let $\sigma > 0$. Then

$$\frac{2\alpha - \sigma\alpha^2 - \alpha\lambda^2}{(\lambda + i\alpha)(\lambda - i\alpha)} F_+(\lambda) = H_-(\lambda)$$

and
$$\left(\frac{2\alpha - \sigma\alpha^2 - \sigma\lambda^2}{\lambda + i\alpha}\right)_+ F_+(\lambda) = (\lambda - i\alpha) H_-(\lambda) \tag{8-57}$$

Since f and h are assumed bounded, we deduce that each side of Eq. (8-57) is a constant, A. Thus

$$F_+(\lambda) = A \frac{\lambda + i\alpha}{2\alpha - \sigma\alpha^2 - \sigma\lambda^2}$$

$$H_-(\lambda) = \frac{A}{\lambda - i\alpha}$$

Note that the inversion contour must pass above both poles of the denominator in $F_+(\lambda)$. The solution is

$$f(x) = A_1\left(\cos x\sqrt{\frac{2\alpha}{\sigma} - \alpha^2} + \frac{\alpha}{\sqrt{2\alpha/\sigma - \alpha^2}}\sin x\sqrt{\frac{2\alpha}{\sigma} - \alpha^2}\right)$$

We observe that there is only one arbitrary constant, rather than the two which would appear if the integration limits were $(-\infty, \infty)$. In other words, we are here not free to adjust the "phase" of the solution arbitrarily.

A second example is afforded by the Milne[1] problem of Eq. (8-39). We consider

$$\tfrac{1}{2}\int_0^\infty Ei(|x - t|)f(t)\,dt = \sigma f(x)$$

for $x > 0$, where $\sigma > 1$ and $|\sigma - 1| \ll 1$. We are then led to the Wiener-Hopf problem

$$\left(\frac{\arctan\lambda}{\lambda} - \sigma\right)F_+(\lambda) = H_-(\lambda) \tag{8-58}$$

where the function $\arctan\lambda = (1/2i)[\ln(1 + i\lambda) - \ln(1 - i\lambda)]$ has branch points at $\lambda = \pm i$, with the branch lines drawn away from the origin. Denote the two roots of $\arctan\lambda = \sigma\lambda$, near the origin, by $\pm\lambda_0$; for $|1 - \sigma| \ll 1$, $\lambda_0 \cong i\sqrt{3(\sigma - 1)}$. To use the general splitting method on $\lambda^{-1}\arctan\lambda - \sigma$, we must first modify this function so that it has no zeros and so that it $\rightarrow 1$ as $\text{Re}\,\lambda \rightarrow \pm\infty$ in the strip. We can do this by multiplying through, for example, by $-(\lambda^2 + 1)/[(\lambda^2 - \lambda_0^2)\sigma]$, before performing the splitting. If we write

$$\frac{-(\lambda^2 + 1)}{\rho(\lambda^2 - \lambda_0^2)}\left(\frac{\arctan\lambda}{\lambda} - \rho\right) = \frac{G_+(\lambda)}{G_-(\lambda)}$$

we obtain

$$(\lambda + i)^{-1}(\lambda^2 - \lambda_0^2)G_+(\lambda)F_+(\lambda) = -\frac{\lambda - i}{\sigma}H_-(\lambda)G_-(\lambda)$$

from which

$$F_+(\lambda) = \frac{\text{const}}{G_+(\lambda)(\lambda^2 - \lambda_0^2)}(\lambda + i)$$

[1] See E. Milne, "Handbuch der Astrophysik," vol. 3, p. 1, Springer Verlog, Berlin, 1930, or E. Hopf, Mathematical Problems of Radiative Equilibrium, *Cambridge Univ. Tract* 31, Cambridge Univ. Press, 1934.

Inversion yields

$$f(x) = A\left[\frac{(\lambda_0 + i)\exp(-i\lambda_0 x)}{2\lambda_0 G_+(\lambda_0)} - \frac{(-\lambda_0 + i)\exp(i\lambda_0 x)}{2\lambda_0 G_+(-\lambda_0)}\right] + q(x)$$

where A is an arbitrary constant and where $q(x)$ represents the contribution from the integral around the branch line of $G_+(\lambda)$, and so is $0(e^{-x})$ Thus

$$f(x) = A_1 \sin(\lambda_0 x + \nu) + 0(e^{-x})$$

where the phase ν is fixed. Comparing with the result for Eq. (8-46), we see that [if the kernel in Eq. (8-46) were the same] the solutions are very similar, except that (1) the phase is now not arbitrary and (2) we must supplement the answer by a contribution which is appreciable only near $x = 0$.

It is typical of most positive kernels $k(x)$ that the solution of Eq. (8-55) should consist of a trigonometric function whose phase is fixed and whose argument is real when $\sigma < \int_{-\infty}^{\infty} k(x)\,dx$, plus an edge correction which is appreciable in a neighborhood of the origin, the size of this neighborhood being of order $\left[\int_{-\infty}^{\infty} x^2\, k(x)\,dx \Big/ \int_{-\infty}^{\infty} k(x)\,dx\right]^{1/2}$ (cf. Exercise 4).

EXERCISES

4. Verify the statements in the last preceding paragraph. [*Hint:* Expand $G_+(\lambda)$ in powers of λ.]

5. Solve the integral equation (8-55), where

(*a*) $k(x) = \begin{cases} e^{-x}, & x > 0 \\ e^{\beta x}, & x < 0 \end{cases}$

where $\beta > 0$

(*b*) $k(x) = e^{-|x|} + Ae^{-R|x|}$, R and $A > 0$, with $R \gg 1$.

Study the result as a function of A, R, σ. Determine also the ordinary differential equation and boundary conditions to which each of (*a*) and (*b*) corresponds.

6. Solve

$$\int_0^{\infty} K_0(|x - t|) f(t)\,dt = \sigma f(x) \qquad x > 0$$

Identify the boundary-value problem, in two independent variables, to which this equation is equivalent. Determine the behavior of $f(x)$ near $x = 0$, (*a*) from the Wiener-Hopf solution and (*b*) from a local expansion in $x^2 + y^2$ about the origin, substituted into the partial-differential equation.

7. Complete the solution of Eq. (8-39).

8-4 *The Use of Approximate Kernels*

The objective of many scientific investigations is a reasonably accurate description of the macroscopic features of the phenomenon under study and, particularly, their dependence on the parameters of the problem. Ordinarily, in such an investigation, the ease of interpretability of the answers is more important than meticulous numerical accuracy, and the useful approximations are those which lead to simple descriptions at the expense of, at most, uninformative numerical detail. One group of problems of this type is that in which the Fourier transform and the Wiener-Hopf technique are the tools by which the solution is obtained. In a great many of these problems the description of the answer is not simple, usually because the required factors of the transform of the kernel are very complicated functions.

As we have seen, the Wiener-Hopf functional equation is equivalent to an integral equation, and the kernel associated with that equation is the function whose transform must be factored. We anticipate that a replacement of that kernel by another which duplicates its important features (singularities in the integration domain, area, first few moments) will lead to a solution whose dominant features duplicate those of the solution of the original problem. If the substitute kernel is also one whose factors are elementary functions, the answer will gain considerably in simplicity.

In this section we treat several problems by this kernel-substitution method and indicate its effectiveness, when possible, by comparison with the more laboriously obtained solutions of the original problems. We find[1] that the foregoing objective is attained when the substitute kernel has the same singularity, the same area, and the same first moment as the original kernel. When further moments are also retained, greater accuracy can be achieved; however, even in problems in which a kernel whose second moment does not exist is replaced by one with unit second moment, the results can be extremely useful.

As a first example, consider a problem in which an inviscid incompressible fluid moves past a flat plate; at time zero, the temperature of the plate is suddenly raised. In (dimensionless) mathematical form, the problem becomes that of finding $\theta(x,y,t)$ such that

$$\Delta\theta - \theta_x - \theta_t = 0 \qquad \text{in } t > 0 \text{ and in } D \tag{8-59}$$

(where D is the xy plane exterior to the half-line $y = 0$, $x > 0$). For $t < 0$, $\theta = 0$ everywhere; for $t > 0$, $\theta \to 0$ as $\operatorname{Im}(x + iy)^{1/2} \to \infty$. On the half-line, $\theta(x,0,t) = h(t)$ for $x > 0$, where $h(t)$ is the unit step function.

[1] See W. T. Koiter, *Koninkl. Ned. Akad. Wetenschap. Proc.*, (B)*57:* 558 (1954), and G. F. Carrier, *J. Appl. Phys.*, *30:* 1769 (1959).

The use of a Fourier transform in x and a Laplace transform in t leads to

$$\Theta(\xi,y,s) = A(\xi,s) \exp\left(-|y| \sqrt{\xi^2 - i\xi + s}\right) \tag{8-60}$$

where A is an unknown function, and where the branch choice is clear. Define

$$f(x,t) = \theta_y(x,0+,t) - \theta_y(x,0-,t) \tag{8-61}$$

(which vanishes for $t < 0$), and observe that $F(\xi,s)$ can be written as $F_+(\xi,s)$. Define also $v(x,t)$ by

$$\theta(x,0,t) = h(x)h(t) + v(x,t)$$

so that $v(x,t)$ is nonzero only for $x < 0$ and $t > 0$ and $V(\xi,s)$ is therefore a $V_-(\xi,s)$ function. The subscripts $+$ and $-$ refer of course to the ξ plane.

The Wiener-Hopf problem becomes

$$-\frac{F_+(\xi,s)}{2\sqrt{\xi^2 - i\xi + s}} = -\frac{1}{i\sqrt{2\pi}\, s\xi} + V_-(\xi,s) \tag{8-62}$$

The inversion contour in the ξ plane is to pass above the origin and below the point $\xi = ir_+$, where

$$\begin{aligned} r_+ &= \tfrac{1}{2} + \sqrt{\tfrac{1}{4} + s} \\ r_- &= \tfrac{1}{2} - \sqrt{\tfrac{1}{4} + s} \end{aligned} \tag{8-63}$$

This problem was deliberately chosen because $(\xi^2 - i\xi + s)^{-\frac{1}{2}}$ has an elementary decomposition; we are led to

$$-\frac{F_+(\xi,s)}{2\sqrt{\xi - ir_-}} + \frac{\sqrt{-ir_+}}{i\sqrt{2\pi}\, s\xi} = \frac{\sqrt{-ir_+} - \sqrt{\xi - ir_+}}{i\sqrt{2\pi}\, s\xi} + \sqrt{\xi - ir_+}\, V_-(\xi,s) \tag{8-64}$$

Thus

$$F_+(\xi,s) = \frac{2\sqrt{-ir_+}\sqrt{\xi - ir_-}}{i\sqrt{2\pi}\, s\xi} \tag{8-65}$$

The dimensionless heat-transfer rate $\theta_y(x,0+,t)$ is $\frac{1}{2}f(x,t)$ and is given by

$$\theta_y(x,0+,t) = -\frac{1}{\sqrt{\pi t}} + \frac{1}{2\pi i\sqrt{\pi}} \int_x^\infty \left(\frac{1}{\sqrt{u}} - \frac{1}{\sqrt{x}}\right) du \int_{c-i\infty}^{c+i\infty} \exp\left[st - \left(\sqrt{s + \frac{1}{4}} - \frac{1}{2}\right)u\right] ds \tag{8-66}$$

For all values of $x > 0$, the asymptotic behavior as $t \to \infty$ is

$$\theta_y(x,0+,t) \sim -\frac{1}{2\sqrt{\pi t}} \operatorname{erfc}\frac{t-x}{2\sqrt{t}} - \frac{1}{2\sqrt{\pi x}} \operatorname{erfc}\frac{x-t}{2\sqrt{t}} \tag{8-67}$$

Thus $\theta_y(x,0+,t) \sim -(\pi t)^{-\frac{1}{2}}$ if $x - t \gg 2\sqrt{t}$, and $\theta_y(x,0+,t) \sim -(\pi x)^{-\frac{1}{2}}$ if $t - x \gg 2\sqrt{t}$.

We now turn our attention to the approximation for whose introduction this problem was chosen. From Sec. 8-3, our problem may be put into the form of an integral equation for $\Phi(x,s)$, the Laplace transform of $f(x,t)$,

$$\frac{1}{s} = \int_0^\infty k(x - x',s)\Phi(x',s)\,dx' \tag{8-68}$$

for $x > 0$, where

$$k(x,s) = -\frac{1}{4\pi}\int_{-\infty}^{\infty} \frac{e^{-i\xi x}\,d\xi}{\sqrt{\xi^2 - i\xi + s}} \tag{8-69}$$

It is clear that $\Phi(x,s)$ as specified by Eq. (8-68) would be different from that which would arise were some other kernel $k_1(x,s)$ substituted for $k(x,s)$. However, it is also clear that if the area under $k_1(x,s)$, its singularity at $x = 0$, and the first few moments

$$\int_{-\infty}^{\infty} x^n k_1(x,s)\,dx$$

are identical with those for $k(x,s)$, then the solution of the substitute problem might be very close to that of the original problem. If such moments of k_1 are close to those of k only for small s (hence large time), then $f_1(x,t)$ (the solution of the substitute problem) will lead to a good asymptotic estimate in time of $f(x,t)$.

Since the area and moments of $k(x,s)$ are proportional, respectively, to $K(0,s)$, $K_\xi(0,s)$, $K_{\xi\xi}(0,s)$, etc., it is a simple matter to select a $k_1(x,s)$ which is close to $k(x,s)$ in the sense used above. Ordinarily, this selection should be made in such a way that K_1 can be factored with much greater ease than can K. In this case, since K can already be factored trivially, we choose K_1 so that $\Phi_1(x,s)$ can be easily inverted. We choose

$$K_1(\xi,s) = -\frac{1}{2\sqrt{2\pi}}(\xi^2 - i\xi + s + i\xi s)^{-1/2}$$

Note that the ratios of the areas and first two moments of K_1 to K are, respectively, 1, $1 - s$, $(1 + \frac{4}{3}s - s^2)/(1 + \frac{4}{3}s)$. The singularities of $k_1(x,s)$ and $k(x,s)$ are identical.[1] The exact solution for $f_1(x,s)$ is now

$$f_1(x,s) = \begin{cases} -2(\pi t)^{-1/2} & \text{for } x > t \\ -2(\pi x)^{-1/2} & \text{for } t > x \end{cases}$$

which agrees with the large t behavior of the original function

$$f(x,s) = 2\theta_y(x,0+,t)$$

It is interesting to observe that we would have obtained this same $f_1(x,s)$ (but not the same θ) had we formulated the original problem so as to ignore horizontal diffusion—that is, $\theta_{yy} - \theta_x - \theta_t = 0$.

[1] To show this, wrap the integration contour around an appropriate branch cut, and observe that the singularity behavior for small $|x|$ depends on the integrand behavior for large $|\xi|$.

Another example of the effectiveness of the kernel substitution is associated with the problem

$$1 = \int_0^\infty k(x - y)(g(y)\, dy \tag{8-70}$$

for $x > 0$, where

$$k(x) = \frac{1}{\pi} \operatorname{Re}\, [e^{ix} Ei(ix)] \tag{8-71}$$

with $Ei(x)$ defined by Eq. (8-40). The transform of $k(x)$ is $\sqrt{1/2\pi}\,(1 + |\xi|)^{-1}$. The kernel $k(x)$ has a logarithmic singularity at $x = 0$, unit area, and vanishing first moment; its second moment fails to exist. We are led to the Wiener-Hopf problem

$$N_-(\xi) + \frac{1}{\sqrt{2\pi}} \frac{1}{-i\xi} = (1 + |\xi|)^{-1} G_+(\xi)$$

where N_- is the transform of an unknown function $n(x)$, with $n(x) = 0$ for $x < 0$, and where $G_+(\xi)$ is the transform of a function vanishing for $x < 0$ and coinciding with $g(x)$ for $x > 0$. We can replace $|\xi|$ by $\sqrt{\xi^2 + \epsilon^2}$, with ϵ small and positive, and carry out the factoring process and the Wiener-Hopf argument. In the limit as $\epsilon \to 0$, the factors become

$$\frac{1}{1 + |\xi|} = \left[\frac{1}{\sqrt{1 + \xi}} \exp\left(\frac{1}{\pi i} \int_0^\xi \frac{\ln z\, dz}{1 - z^2}\right)\right] \left[\frac{1}{\sqrt{1 + \xi}} \exp\left(-\frac{1}{\pi i} \int_0^\xi \frac{\ln z}{1 - z^2}\, dz\right)\right] \tag{8-72}$$

where the first factor is analytic in the upper half-plane $(-\pi/2 < \arg z < 3\pi/2)$ and the second in the lower half-plane $(-3\pi/2 < \arg z < \pi/2)$. If we label these two factors by $C_+(\xi)$ and $C_-(\xi)$, respectively, we obtain

$$G_+(\xi) = \frac{1}{-i\sqrt{2\pi}\,\xi} \frac{1}{C_+(\xi)} \frac{1}{C_-(0)} \tag{8-73}$$

where the inversion path is to be indented above the origin. It now follows that, for large $x > 0$, $g(x) \sim 1 + 0(1/x)$, and, for small $x > 0$, $g(x) \sim 1/\sqrt{\pi x}$.

We now approximate the original kernel $k(x)$, as given by Eq. (8-71), by

$$k_1(x) = \frac{1}{\pi} K_0(|x|) \tag{8-74}$$

where $K_0(x)$ is the modified Bessel function of order zero. This modified kernel has the same singularity at the origin and the same first moment as the original kernel $k(x)$, but its second moment is unity instead of being nonexistent. The solution obtained by replacing $k(x)$ by $k_1(x)$ should therefore provide a useful estimate of the importance of maintaining the second moment in the substitute kernel. Denoting by $g_1(x)$ the solution of Eq. (8-70) with k replaced by k_1, we obtain without further approximation

$$g_1(x) = \frac{e^{-x}}{\sqrt{\pi x}} + \operatorname{erf} \sqrt{x} \tag{8-75}$$

Thus, for large x, $g_1(x) \sim 1 + 0(e^{-x}/\sqrt{x})$, and, for small x, $g_1(x) \sim 1/\sqrt{\pi x}$. The dominant contributions to the answer are identical with those of the original problem, and it would appear that the singularity of the kernel at $x = 0$ is a more important measure of the "width" of the kernel than its second moment. However, as we shall see when we treat the homogeneous problem, the retention of both the singularity and the second moment can improve the accuracy substantially. One should also note, however, that the algebraic $(1/x)$ approach to the asymptotic value of $g(x)$ is dictated by the unbounded character of the second moment of $k(x)$; the kernel $k_1(x)$ with its unit second moment provides an exponential decay to the asymptotic value.

As a final example, consider the homogeneous problem

$$\lambda \int_0^\infty k(x - t)f(t)\,dt = f(x) \tag{8-76}$$

where (1) $k(x) = (1/\pi)K_0(|x|)$ (the modified Bessel function) and (2) $k(x) = \frac{1}{2}Ei(|x|)$ [cf. Eq. (8-40)]. We obtain, as the transform function to be factored,

$$\begin{aligned} &(1)\ \frac{\lambda}{\sqrt{\xi^2 + 1}} - 1 \\ &(2)\ \lambda\,\frac{\arctan \xi}{\xi} - 1 \end{aligned} \tag{8-77}$$

A first approximation for (1) would be

$$\begin{aligned} \frac{\lambda}{\sqrt{\xi^2 + 1}} - 1 &= \frac{\lambda^2 - 1 - \xi^2}{\sqrt{\xi^2 + 1}\,(\lambda + \sqrt{1 + \xi^2})} \\ &\cong \frac{\lambda^2 - 1 - \xi^2}{\sqrt{\xi^2 + 1}\,\sqrt{\xi^2 + (1 + \lambda)^2}} \end{aligned} \tag{8-78}$$

This approximation retains the area and singularity of the original kernel. A better approximation is

$$\frac{\lambda}{\sqrt{\xi^2 + 1}} - 1 \cong \frac{(\lambda^2 - 1 - \xi^2)(\xi^2 + \beta^2)}{\sqrt{\xi^2 + 1}\,\sqrt{\xi^2 + \alpha^2}\,(\xi^2 + \gamma^2)} \tag{8-79}$$

where α, β, γ are to be chosen so that the area and second moment are retained; this leaves the choice of one parameter free to assure that no zeros or singularities lie close to the real axis. It is clearly desirable to admit spurious singularities only in locations where they cannot affect the major features of the description of the eigenfunction. The choice $\alpha = 1$ is convenient, and we obtain

$$\frac{\lambda}{\sqrt{\xi^2 + 1}} - 1 \cong \frac{(\lambda^2 - 1 - \xi^2)(\xi^2 + 2)}{(\xi^2 + 1)[\xi^2 + 2(\lambda + 1)]} \tag{8-80}$$

The solution of Eq. (8-76), by using either the correct kernel or one of the approximations, is

$$f(x) = \sin p(x + q) + 0(e^{-x}) \tag{8-81}$$

where p and q are constants which depend on λ. In fact, $p = \sqrt{\lambda^2 - 1}$, and this is rigorously correct; q is the only other macroscopically important parameter. Table 1 compares the values of q obtained from the exact calculation, those obtained using Eq. (8-78), and those obtained using Eq. (8-80).

Table 1

λ	k	q (exact)	q [Eq. (8-78)]	q [Eq. (8-80)]
1	0	0.817	0.75	0.793
1.02	0.2	0.814	0.742	0.78
1.12	0.5	0.797	0.697	0.720

The same type of approximation can be used in case 2. We write

$$\lambda \frac{\arctan \xi}{\xi} - 1 \cong \frac{(p^2 - \xi^2)(\xi^2 + \beta^2)}{(\xi^2 + \alpha^2)(\xi^2 + \gamma^2)} \tag{8-82}$$

where $p^{-1} \arctan p = 1/\lambda$. To preserve the area and second moment, we can choose $\alpha^2 = 1$, $\beta^2 = (1 - \lambda R)p/(R - \frac{1}{5})$, $\gamma^2 = (1 - \lambda R)/[\lambda(R - \frac{1}{5})]$, $R = \frac{1}{3} - \frac{1}{5}p^2$. The values of q in Eq. (8-81), corresponding to the exact solution and also to the use of Eq. (8-82), are listed in Table 2.

Table 2

λ	p	q (exact)	q [Eq. (8-82)]
1	0	0.710	0.67
1.013	0.2	0.701	0.67
1.053	0.4	0.676	0.664

Exercises

1. Solve Eq. (8-76) for (*a*) the exponential integral kernel and (*b*) the kernel $K_0(|x|)$, using substitute kernels of the form $\Sigma a_n|x|^n$ exp

$(-\alpha_n|x|)$. Compare the approximate and exact solutions, especially near $x = 0$.

2. Solve Eq. (8-76) with

$$k(x) = \frac{1}{\pi} K_0(|x|) + \frac{A}{\pi\epsilon} K_0 \frac{|x|}{\epsilon}$$

both exactly, and using an approximate kernel.

3. Let $\Delta\Delta\psi - \Delta\psi_x - \Delta\psi_t = 0$ in $x^2 + y^2 < \infty$, except on $y = 0$, $x > 0$, and let $\psi(x,y,t) = -\psi(x,-y,t)$, $\psi(x,y,t) = 0$ in $t < 0$,

$$\psi_y(x,0,t) = h(t)$$

[where $h(t) = 1$ for $t > 0$, 0 for $t < 0$] on $x > 0$; let $\psi(x,0,t) = 0$. Find $\psi(x,y,t)$ using an approximate kernel. Then compare your results with those obtained for this problem by Greenspan and Carrier.[1]

8-5 *Dual Integral Equations*

A number of problems, usually similar in nature to those which we have been discussing, can be phrased in terms of dual integral equations. As a first example, we consider the much-discussed Sommerfeld diffraction problem.

As shown in Fig. 8-3, let the plane wave

$$\varphi_i = e^{-ik(x\cos\theta + y\sin\theta)} \tag{8-83}$$

be incident from an angle θ onto a sheet occupying the half-plane $x < 0$, $y = 0$, $-\infty < z < \infty$. (A time factor $e^{-i\omega t}$ is to be understood, and k is a positive constant.) The presence of the sheet alters the field by an amount $\varphi(x,y)$, so that the total field φ_T is given by

$$\varphi_T = \varphi_i + \varphi \tag{8-84}$$

where

$$\Delta\varphi + k^2\varphi = 0 \tag{8-85}$$

[1] H. P. Greenspan and G. F. Carrier, *J. Fluid Mech.*, 7: 22 (1959).

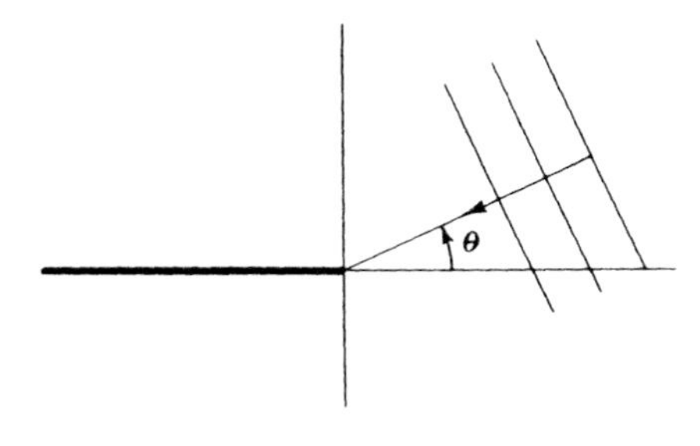

Fig. 8-3 Sommerfeld diffraction problem.

The particular physical problem we consider (e.g., diffraction by a rigid plate of an acoustic wave whose potential is φ_i) is such that $\partial\varphi_T/\partial_y$ must vanish on the sheet, so that

$$\varphi_y = ik(\sin\theta)e^{-ikx\cos\theta} \tag{8-86}$$

for $x < 0$. The statement of the mathematical problem is completed by requiring φ_y to be continuous across the line $y = 0$, by requiring φ to satisfy the radiation condition as $x^2 + y^2 \to \infty$, and by requiring any singularities of $(\varphi_x)^2$ and $(\varphi_y)^2$ near the origin to be no worse than $(x^2 + y^2)^{-1+\epsilon}$, where $\epsilon > 0$ (in the acoustic problem mentioned, the total energy in the diffraction field is then integrable).

The Fourier transform in x of Eq. (8-85) has the solution

$$\Phi(\lambda,y) = \begin{cases} A(\lambda)\exp\left(-y\sqrt{\lambda^2 - k^2}\right) & y > 0 \\ -A(\lambda)\exp\left(y\sqrt{\lambda^2 - k^2}\right) & y < 0 \end{cases} \tag{8-87}$$

where we have used the continuity of φ_y, and hence of Φ_y, across the line $y = 0$. When we take the inverse transform of Eq. (8-87), the inversion contour must be indented above the branch point at $\lambda = -k$ and below that at $\lambda = k$, and $\sqrt{\lambda^2 - k^2}$ is to be positive for λ real and $>k$. [The reader may prefer to move the branch points to $-k - i\epsilon$, $k + i\epsilon$, respectively, where $\epsilon > 0$, and to permit the contour to be any horizontal line lying between these two branch points. As we have seen in Chap. 7, this is equivalent to including a small amount of damping in Eq. (8-85).]

From Eq. (8-86), we have

$$-\sqrt{\lambda^2 - k^2}\,A(\lambda) = \frac{1}{\sqrt{2\pi}}\frac{k\sin\theta}{\lambda - k\cos\theta} + \frac{1}{\sqrt{2\pi}}\int_0^\infty \varphi_y(x,0)e^{i\lambda x}\,dx \tag{8-88}$$

[We note that the inversion contour must pass below the pole at $\lambda = k\cos\theta$. This may be seen either by use of the radiation condition or by introducing a decay term $e^{\epsilon x}$, $\epsilon > 0$, into the right-hand side of Eq. (8-86).] Denote the limiting value of $\varphi(x,y)$ as $y \to 0$ from above or below by $\varphi(x,0+)$ and $\varphi(x,0-)$, respectively. For $x > 0$, we have $\varphi(x,0+) = \varphi(x,0-)$. Then from Eq. (8-87)

$$\begin{aligned} A(\lambda) &= \frac{1}{\sqrt{2\pi}}\int_{-\infty}^0 e^{i\lambda x}\varphi(x,0+)\,dx + \frac{1}{\sqrt{2\pi}}\int_0^\infty e^{i\lambda x}\varphi(x,0)\,dx \\ -A(\lambda) &= \frac{1}{\sqrt{2\pi}}\int_{-\infty}^0 e^{i\lambda x}\varphi(x,0-)\,dx + \frac{1}{\sqrt{2\pi}}\int_0^\infty e^{i\lambda x}\varphi(x,0)\,dx \end{aligned} \tag{8-89}$$

These equations imply[1]

[1] We note that Eqs. (8-89) also imply the interesting result that $\varphi(x,0) = 0$ for $x > 0$; thus, $\varphi(x,0+) = -\varphi(x,0-)$.

$$2A(\lambda) = \frac{1}{\sqrt{2\pi}} \int_{-\infty}^{0} e^{i\lambda x}[\varphi(x,0+) - \varphi(x,0-)]\, dx \tag{8-90}$$

and elimination of $A(\lambda)$ from Eqs. (8-88) and (8-90) leads to a Wiener-Hopf problem, from whose solution we deduce that

$$\varphi = \mp \frac{\sqrt{2k}\sin(\theta/2)}{2\pi} \int_{-\infty}^{\infty} \frac{\exp(-i\lambda x \mp y\sqrt{\lambda^2 - k^2})}{(\lambda - k\cos\theta)\sqrt{\lambda - k}}\, d\lambda \tag{8-91}$$

where the upper and lower signs correspond to $y > 0$ and $y < 0$, respectively. As we have previously noted, the integration path passes below the points $\lambda = k \cos \theta$ and $\lambda = k$ and above the point $\lambda = -k$; also, $\sqrt{\lambda^2 - k^2}$ is positive for λ real, with $\lambda > k$.

Our interest here, however, is not to use the Wiener-Hopf method but rather to examine the consequences of describing the problem in terms of dual integral equations. The integral equations in question are easily derived; in fact, Eq. (8-90) tells us that $A(\lambda)$ is the transform of a function which is zero for $x > 0$, so that

$$\frac{1}{\sqrt{2\pi}} \int_{-\infty}^{\infty} e^{-i\lambda x} A(\lambda)\, d\lambda = 0 \qquad \text{for } x > 0 \tag{8-92}$$

and similarly, from Eq. (8-88),

$$-\frac{1}{\sqrt{2\pi}} \int_{-\infty}^{\infty} e^{-i\lambda x} \sqrt{\lambda^2 - k^2}\, A(\lambda)\, d\lambda = ik(\sin\theta)e^{-ikx\cos\theta} \qquad \text{for } x < 0 \tag{8-93}$$

In each of Eqs. (8-92) and (8-93), the integration path is indented above $\lambda = -k$ and below $\lambda = k$. Our problem is now to determine $A(\lambda)$ from Eqs. (8-92) and (8-93). One method is obvious—we could simply take a Fourier transform of each of these two equations, a process which will require the introduction of two unknown functions [i.e., the right-hand sides of Eqs. (8-92) and (8-93) for $x < 0$ and $x > 0$, respectively]; elimination of $A(\lambda)$ then gives a Wiener-Hopf problem identical to that previously obtained.

A different but related method has been used by Clemmow[1] and others. In describing this method, it is convenient to think of the branch points at $\lambda = -k$ and $\lambda = k$ as being moved slightly off the real axis, so as to lie below and above it, respectively. The path of integration is then any horizontal line between the new branch points. From Eq. (8-92), $A(\lambda)$ is the transform of a function which is zero for $x > 0$, so that $A(\lambda)$ is analytic in some lower half-plane; moreover, $|A(\lambda)| \to 0$ as $|\lambda| \to \infty$ in the lower

[1] P. C. Clemmow, *Proc. Roy. Soc. (London)*, (A)**205**: 286 (1951).

half-plane. We assume that we can take the path of integration as the upper boundary of this region. Next, let $P(\lambda)$ be an as yet undetermined function, analytic above the path of integration, with $|P(\lambda)/\lambda| \to 0$ as $|\lambda| \to \infty$ in the upper half-plane. Contour integration then shows that (8-93) would be satisfied if

$$-\frac{1}{\sqrt{2\pi}}\sqrt{\lambda^2 - k^2}\,A(\lambda) = \frac{k \sin \theta}{2\pi}\,\frac{1}{\lambda - k \cos \theta}\,\frac{P(\lambda)}{P(k \cos \theta)} \tag{8-94}$$

on the path of integration, provided that this path of integration passes below the pole at $\lambda = k \cos \theta$. Thus, if we can find appropriate functions $A(\lambda)$ analytic below the path and $P(\lambda)$ analytic above the path, such that Eq. (8-94) is satisfied on the path, then we shall have found a solution of Eqs. (8-92) and (8-93). Equation (8-94) may be written

$$-\frac{1}{\sqrt{2\pi}}\,A(\lambda) = \frac{k \sin \theta}{2\pi}\,\frac{1}{P(k \cos \theta)(\lambda - k \cos \theta)}\,\frac{P(\lambda)}{\sqrt{\lambda^2 - k^2}} \tag{8-95}$$

and an obvious choice for $P(\lambda)$ is $\sqrt{\lambda + k}$, for then $P(\lambda)$ is analytic above the path and $A(\lambda)$ is analytic below the path. Moreover, the behavior of $A(\lambda]$ and $P(\lambda)$ as $|\lambda| \to \infty$ allows us to verify, a posteriori, that Eqs. (8-92) and (8-93) are satisfied. The reader may check that the formula for $A(\lambda)$ as given by Eq. (8-95), with $P(\lambda) = \sqrt{\lambda + k}$, leads again to Eq. (8-91). In the general case (cf. Exercise 1), there is of course no guarantee that we can obtain an equation analogous to Eq. (8-95), nor that we can solve that equation. However, in those special cases where the method does work, it avoids one of the splittings required by the Wiener-Hopf method.

A more general method of solving a pair of coupled-integral equations like (8-92) and (8-93), is based on a property of the convolution theorem. Let $a(x)$ and $n(x)$ be two functions defined in $-\infty < x < \infty$. Define

$$I(x) = \int_{-\infty}^{\infty} N(\lambda)A(\lambda)e^{-i\lambda x}\,d\lambda = \int_{-\infty}^{\infty} n(\xi)a(x - \xi)\,d\xi$$

It follows that if $n(x) = 0$ for $x < 0$ and if $n(x)$ is prescribed for $x > 0$ [we shall denote the transform of such a function by $N_+(\lambda)$, the $+$ sign indicating analyticity in some upper-half-plane region], then $I(x)$ for $x < 0$ depends only on values of $a(x)$ for $x < 0$. Similarly, if $n(x) = 0$ for $x > 0$ [we denote its transform by $N_-(\lambda)$] and if $n(x)$ is prescribed for $x < 0$, then $I(x)$ for $x > 0$ depends only on values of $a(x)$ for $x > 0$.

To illustrate the use of this result, we consider again Eqs. (8-92) and (8-93). Using as yet undetermined functions[1] $N_+(\lambda)$ and $N_-(\lambda)$ of the

[1] An alternative version of this process is described in Noble, *op. cit.*, p. 58; see also p. 222.

type just described, we obtain

$$\frac{1}{\sqrt{2\pi}} \int_{-\infty}^{\infty} N_-(\lambda) A(\lambda) e^{-i\lambda x}\, d\lambda = 0 \qquad \text{for } x > 0$$

$$\frac{1}{\sqrt{2\pi}} \int_{-\infty}^{\infty} N_+(\lambda)[\sqrt{\lambda^2 - k^2}\, A(\lambda)] e^{-i\lambda x}\, d\lambda = \frac{-ik \sin\theta}{\sqrt{2}} \int_0^{\infty} n_+(\xi) e^{-ik(x-\xi)\cos\theta}\, d\xi \qquad \text{for } x < 0 \quad (8\text{-}96)$$

where $n_+(\xi)$ is the inverse transform of $N_+(\lambda)$. If we could determine N_+ and N_- so that

$$N_-(\lambda) = N_+(\lambda)\sqrt{\lambda^2 - k^2} \tag{8-97}$$

then the left-hand sides of the two equations (8-96) would be identical and Fourier inversion would permit us to solve the problem. However, this is not possible here, since we also have the requirement that N_+ and N_- be the transforms of functions zero for $x < 0$ and $x > 0$, respectively, and so vanishing as $\text{Im}\,\lambda \to +\infty$ and $-\infty$, respectively. We therefore choose

$$N_+(\lambda) = (\lambda + k)^{-1/2} \qquad N_-(\lambda) = (\lambda - k)^{-1/2}$$

and then obtain the same integrands by multiplying the first of Eqs. (8-96) by e^{ikx} and differentiating that equation with respect to x. We thus obtain

$$\frac{1}{\sqrt{2\pi}} \int_{-\infty}^{\infty} \sqrt{\lambda - k}\, A(\lambda) e^{-i\lambda x}\, d\lambda = \begin{cases} 0 & \text{for } x > 0 \\ -\dfrac{ik \sin\theta}{2\pi} \displaystyle\int_0^{\infty} n_+(\xi) e^{-ik(x-\xi)\cos\theta}\, d\xi & \text{for } x < 0 \end{cases} \tag{8-98}$$

where

$$n_+(x) = \frac{1}{\sqrt{2\pi}} \int_{-\infty}^{\infty} e^{-i\lambda x} \frac{d\lambda}{\sqrt{\lambda + k}}$$

The reader should verify that all conditions of the problem are satisfied and that Eq. (8-98) again leads to Eq. (8-91). A feature of this method is that general formulas can sometimes be obtained for the solution of a Wiener-Hopf type of problem in which a boundary condition is given in terms of an arbitrary function. Of course, as we have seen in Sec. 8-2, a convolution method can also give a general result from the Wiener-Hopf solution of a specific problem; cf. also Sec. 8-6, where the Carleman method is used for a similar problem.

Exercises

1. Examine the applicability of the above two methods to the general problem

$$\begin{aligned} \int_{-\infty}^{\infty} A(\lambda) e^{-i\lambda x} &= f(x) \qquad x > 0 \\ \int_{-\infty}^{\infty} K(\lambda) A(\lambda) e^{-i\lambda x} &= g(x) \qquad x < 0 \end{aligned} \tag{8-99}$$

[*Hint:* Consider, for example, functions $K(\lambda)$ for which $|K(\lambda)| \sim |\lambda|$ as $|\lambda| \to \infty$.] Show also that this pair of equations may be replaced by a single integral equation.

2. Carry through the details of the procedure leading to Eq. (8-91); cf. Exercise 1 of Sec. 8-1. Use a saddle-point method [*Hint:* Set $y = \alpha x$, α a constant, and let $x \to \pm\infty$] to determine the nature of φ_T in various regions of Fig. 8-1 and interpret these results in terms of reflected, transmitted, or diffracted waves. Also, express Eq. (8-91) in terms of the complex Fresnel integral $\int_z^\infty \exp(-it^2)\,dt$. Finally, derive the single integral equation to which the original problem is equivalent.

3. Discuss, in terms of coupled integral equations and also by a conventional Wiener-Hopf method, the problem of diffraction of a plane wave by the two staggered half-planes $y = b$, $x < 0$ and $y = 0$, $x < c$. The incident wave is again given by Eq. (8-83), and the boundary condition (8-86) is to be satisfied on each half-plane.

In problems involving polar coordinates, one sometimes encounters dual integral equations in which the kernel is a Bessel function. A well-known example arises in the problem of finding the potential field $\varphi(r,z)$ surrounding an electrified disk whose center is at the origin and whose axis lies along the z axis, the potential $\varphi(r,0)$ on the disk being specified. The mathematical problem is defined by

$$\varphi_{rr} + \frac{1}{r}\varphi_r + \varphi_{zz} = 0 \tag{8-100}$$

in $z > 0$, $0 < r < \infty$, subject to $\varphi(r,0) = f(r)$ (a specified function) for $0 < r < 1$, $\varphi_z(r,0) = 0$ for $r > 1$, and $\varphi \to 0$ as $r^2 + z^2 \to \infty$. Defining the Hankel transform

$$\Phi(\rho,z) = \int_0^\infty rJ_0(r\rho)\varphi(r,z)\,dr$$

we obtain

$$\Phi = A(\rho)e^{-\rho z}$$

for $z > 0$, and insertion of the boundary conditions leads to

$$\begin{aligned} &\int_0^\infty b(\rho)J_0(r\rho)\,d\rho = f(r) \qquad 0 < r < 1 \\ &\int_0^\infty \rho b(\rho)J_0(r\rho)\,d\rho = 0 \qquad r > 1 \end{aligned} \tag{8-101}$$

where $b(\rho) = \rho A(\rho)$.

These equations may be put into a different form by use of the Parseval relation (Sec. 7-3) for Mellin transforms; we obtain

$$\begin{aligned} &\frac{1}{2\pi i}\int_{c-i\infty}^{c+i\infty} B_M(s)\,\frac{2^{-s}\Gamma(\tfrac{1}{2} - \tfrac{1}{2}s)}{\Gamma(\tfrac{1}{2} + \tfrac{1}{2}s)}\,r^{s-1}\,ds = f(r) \qquad 0 < r < 1 \\ &\frac{1}{2\pi i}\int_{c-i\infty}^{c+i\infty} B_M(s)\,\frac{2^{1-s}\Gamma(1 - \tfrac{1}{2}s)}{\Gamma(\tfrac{1}{2}s)}\,r^{s-2}\,ds = 0 \qquad r > 1 \end{aligned} \tag{8-102}$$

where we take $0 < c < 1$. The usefulness of this transformation is that the variable r has been freed from its imprisonment in the argument of the Bessel function. If we now set $r = e^t$, we are back on familiar ground and our previous methods—including the Wiener-Hopf method—may be used. However, we choose instead to describe an alternative procedure due to Titchmarsh.[1]

Set

$$\frac{2^{-s}\Gamma(\tfrac{1}{2} - \tfrac{1}{2}s)}{\Gamma(\tfrac{1}{2}s)} B_M(s) = H(s) \tag{8-103}$$

so that Eqs. (8-102) become

$$\begin{aligned} \frac{1}{2\pi i}\int_{c-i\infty}^{c+i\infty} H(s)\,\frac{\Gamma(\tfrac{1}{2}s)}{\Gamma(\tfrac{1}{2} + \tfrac{1}{2}s)}\, r^{s-1}\,ds &= f(r) \qquad 0 < r < 1 \\ \frac{1}{2\pi i}\int_{c-i\infty}^{c+i\infty} H(s)\,\frac{\Gamma(1 - \tfrac{1}{2}s)}{\Gamma(\tfrac{1}{2} - \tfrac{1}{2}s)}\, r^{s-1}\,ds &= 0 \qquad r > 1 \end{aligned} \tag{8-104}$$

The Γ-function factor in the first of Eqs. (8-104) has no poles or zeros for Re $s > 0$, and that in the second of Eqs. (8-104) has no poles or zeros for Re $s < 1$. Multiply the first equation by $r^{-\omega}$, where Re $\omega < c$, and integrate over (0,1) to obtain

$$\frac{1}{2\pi i}\int_{c-i\infty}^{c+i\infty} (Hs)\,\frac{\Gamma(\tfrac{1}{2}s)}{\Gamma(\tfrac{1}{2} + \tfrac{1}{2}s)}\,\frac{ds}{s - \omega} = \int_0^1 f(r)r^{-\omega}\,dr \tag{8-105}$$

where the right-hand side is presumed known. Move the line of integration to Re $s = c'$, with $c' <$ Re ω, to give

$$\frac{1}{2\pi i}\int_{c'-i\infty}^{c'+i\infty} H(s)\,\frac{\Gamma(\tfrac{1}{2}s)}{\Gamma(\tfrac{1}{2} + \tfrac{1}{2}s)}\,\frac{ds}{s - \omega} = \int_0^1 f(r)r^{-\omega}\,dr - H(\omega)\,\frac{\Gamma(\tfrac{1}{2}\omega)}{\Gamma(\tfrac{1}{2} + \tfrac{1}{2}\omega)} \tag{8-106}$$

The left-hand side of Eq. (8-106) is analytic for Re $\omega > c'$; since c' may be taken arbitrarily close to zero, it is analytic for Re $\omega > 0$. So also then is the right-hand side, and so also is

$$H(\omega) - \frac{\Gamma(\tfrac{1}{2} + \tfrac{1}{2}\omega)}{\Gamma(\tfrac{1}{2}\omega)}\int_0^1 f(r)r^{-\omega}\,dr$$

Thus, if the behavior as $|\omega| \to \infty$ is appropriate,

$$\frac{1}{2\pi i}\int_{k-i\infty}^{k+i\infty}\left[H(s) - \frac{\Gamma(\tfrac{1}{2} + \tfrac{1}{2}s)}{\Gamma(\tfrac{1}{2}s)}\int_0^1 f(r)r^{-s}\,dr\right]\frac{ds}{s - \omega} = 0 \tag{8-107}$$

where Re $\omega < k$ and $k > 0$.

Now multiply the second of Eqs. (8-104) by $r^{-\omega}$, with Re $\omega > c$, and

[1] E. C. Titchmarsh, "Theory of Fourier Integrals," 2d ed., p. 335, Oxford University Press, New York, 1948.

integrate from 1 to ∞ to give

$$\frac{1}{2\pi i}\int_{c-i\infty}^{c+i\infty} H(s)\,\frac{\Gamma(1-\frac{1}{2}s)}{\Gamma(\frac{1}{2}-\frac{1}{2}s)}\,\frac{ds}{s-\omega} = 0$$

Move the line of integration to Re $s = c''$, with $c'' >$ Re ω, and proceed as before to deduce that $H(\omega)$ is analytic for Re $\omega < 1$. Thus

$$\frac{1}{2\pi i}\int_{k-i\infty}^{k+i\infty} H(s)\,\frac{ds}{s-\omega} = H(\omega) \tag{8-108}$$

for Re $\omega < k$ and $k < 1$. We now see that $\int_{k-i\infty}^{k+i\infty} H(s)[ds/(s-\omega)]$ can be eliminated from Eqs. (8-107) and (8-108) to yield a formula for $H(\omega)$; using Eq. (8-103) and an inverse Mellin transform, we then obtain $b(\rho)$ and so $A(\rho)$.

Using a similar method, Titchmarsh obtained also the solution of the integral-equation pair

$$\begin{aligned} &\int_0^\infty y^\alpha f(y)J_\nu(xy)\,dy = g(x) \qquad 0 < x < 1 \\ &\int_0^\infty f(y)J_\nu(xy)\,dy = 0 \qquad x > 1 \end{aligned} \tag{8-109}$$

where $\alpha > 0$, in the form

$$f(x) = \frac{(2x)^{1-\frac{1}{2}\alpha}}{\Gamma(\frac{1}{2}\alpha)}\int_0^1 \mu^{1+\frac{1}{2}\alpha}J_{\nu+\frac{1}{2}\alpha}(\mu x)\,d\mu \int_0^1 g(\rho\mu)\rho^{\nu+1}(1-\rho^2)^{\frac{1}{2}\alpha-1}\,d\rho \tag{8-110}$$

For $\alpha > -2$, Bushbridge[1] has derived the alternative form

$$\begin{aligned} f(x) = \frac{2^{-\frac{1}{2}\alpha}x^{-\alpha}}{\Gamma(1+\frac{1}{2}\alpha)}\Big[&x^{1+\frac{1}{2}\alpha}J_{\nu+\frac{1}{2}\alpha}(x)\int_0^1 \mu^{\nu+1}(1-\mu^2)^{\frac{1}{2}\alpha}g(\mu)\,d\mu \\ &+ \int_0^1 \mu^{\nu+1}(1-\mu^2)^{\frac{1}{2}\alpha}\,d\mu \int_0^1 g(\rho\mu)(x\rho)^{2+\frac{1}{2}\alpha}J_{\nu+1+\alpha/2}(x\rho)\,d\rho\Big] \end{aligned} \tag{8-111}$$

EXERCISES

4. Use Titchmarsh's method to solve the equation pair (8-99).

5. The problem of the electrified disk, which led to Eqs. (8-101), can also be approached as the limiting case of the problem[2] of finding the potential field surrounding the cone $0 < r < 1$, $\theta = \theta_0$, where r and θ are spherical polar coordinates and where there is symmetry around the axis of the cone. Let the cone have the uniform potential 1. Take a Mellin transform of the governing equation

$$(r^2\varphi_r)_r + \frac{1}{\sin\theta}[(\sin\theta)\varphi_\theta]_\theta = 0$$

[1] I. W. Bushbridge, *Proc. London Math. Soc.*, **44**: 115 (1938).

[2] S. N. Karp, *Comm. Pure Appl. Math.*, **3**: 411 (1950).

to obtain

$$\Phi(s) = \begin{cases} A(s)P_{s-1}(\cos\theta) & 0 < \theta < \theta_0 \\ A(s)\dfrac{P_{s-1}(\cos\theta_0)}{P_{s-1}(-\cos\theta_0)}P_{s-1}(-\cos\theta) & \theta_0 < \theta < \pi \end{cases}$$

Introduce

$$\Phi_+(s,\theta) = \int_0^1 \varphi(r,\theta)r^{s-1}\,dr \qquad \Phi_-(s,\theta) = \int_1^\infty \varphi(r,0)r^{s-1}\,dr$$

which are analytic in right and left half-planes, respectively. Consider partial derivatives with respect to θ, and obtain the Wiener-Hopf equation for this problem. Note that

$$P_{s-1}(-\cos\theta_0)P'_{s-1}(\cos\theta_0) - P_{s-1}(\cos\theta_0)P'_{s-1}(\cos\theta_0) = \frac{2\sin\pi s}{\pi\sin\theta_0}$$

Show that, in general, the splitting can be obtained in terms of an infinite product. Complete the details for the case of a disk, and show that the result coincides with that obtained by setting $f = 1$ in Eqs. (8-101).

6. Another approach to the electrified-disk problem has been given by Heins.[1] From Exercise 3 of Sec. 7-4, the solution of Eq. (8-100) may be written

$$\varphi(r,z) = \frac{1}{\pi}\int_0^\pi \varphi(0,\, z + ir\cos\theta)\,d\theta$$

With $f(r) = 1$ for $r < 1$, this equation is equivalent to

$$\frac{1}{\pi}\int_0^r \frac{\varphi(0,it) + \varphi(0,-it)}{\sqrt{r^2 - t^2}}\,dt = 1 \qquad \text{for } r < 1$$

Solve this equation (note that the substitutions $r^2 = \alpha$, $t^2 = \beta$ put it into convolution form) to show that Re $\varphi(0,it) = 1$, $0 < t < 1$.

If $\sigma(r)$ is proportional to the electrostatic-charge density on the disk at radius r, then it may also be shown that

$$\varphi(0,z) = \frac{1}{2}\int_0^1 \frac{\sigma(t)t\,dt}{\sqrt{t^2 + z^2}}$$

Replace z by $\rho + i\tau$, and let $\rho \to 0+$, to obtain

$$\varphi(0,i\tau) = \frac{1}{2}\left[-i\int_0^\tau \frac{\sigma(t)t\,dt}{\sqrt{\tau^2 - t^2}} + \int_\tau^1 \frac{\sigma(t)t\,dt}{\sqrt{t^2 - \tau^2}}\right]$$

[1] A. E. Heins, "Symposium on Electromagnetic Waves," R. E. Langer (ed.), p. 99, University of Wisconsin Press, Madison, Wis., 1962. See also A. E. Heins and R. C. MacCamy, *Z. angew. Math. Phys.*, ***11*:** 249 (1960). In both these papers, similar methods are used to obtain a Fredholm equation for the problem of scattering of waves by a disk; a similar equation has also been obtained by D. S. Jones, *Comm. Pure Appl. Math.*, ***9*:** 713, 1956.

and thus obtain another formula for Re $\varphi(0,it)$. Using the previous result that Re $\varphi(0,it) = 1$, obtain an integral equation for $\sigma(t)$, and solve it to show that $\sigma(t) = 4/(\pi\sqrt{1-t^2})$ for $t < 1$.

8-6 Singular Integral Equations

Most of the integral equations we have so far discussed have been of convolution form, with limits of integration such that transform methods were applicable. We now turn our attention to convolution integrals involving finite limits of integration; our particular interest will continue to lie in the use of function theoretic methods. However, before proceeding, it is useful to recapitulate briefly some of the general theory of integral equations.

An equation of the form

$$\varphi(x)f(x) = g(x) + \lambda \int_a^b k(x,t)f(t)\,dt \tag{8-112}$$

valid for x in (a,b), in which $\varphi(x)$, $k(x,t)$, $g(x)$ are prescribed functions and a, b prescribed constants, is said to be a *linear integral equation* for the unknown function $f(x)$. The function $k(x,t)$ is called the *kernel*. If $\varphi(x) \equiv 0$, the integral equation is said to be of the *first kind;* if $\varphi(x)$ vanishes nowhere in (a,b) (so that we can divide through by it), the equation is of *second kind*. If $k(x,t) = 0$ for $t > x$, so that b can be replaced by x in Eq. (8-112), the equation is of the *Volterra* type; otherwise it is of the *Fredholm* type. The parameter λ arises naturally in eigenvalue problems; its presence is more generally useful for series-expansion purposes.

Although Abel's solution of his integral equation [Eq. (7-148)] was given in 1823 and Liouville used successive substitutions to solve another integral equation in 1837, the modern interest in integral equations dates largely from about 1900, when Fredholm[1] began his extensive study of the equation

$$f(x) = g(x) + \lambda \int_a^b k(x,t)f(t)\,dt \tag{8-113}$$

and its relationship to existence proofs for solutions of the Dirichlet problem. Fredholm's theory was motivated by the idea of approximating the integral by a sum and replacing Eq. (8-112) by a set of linear algebraic equations. It turns out that, when $k(x,t)$ is continuous in the square $a \le x \le b$, $a \le t \le b$ (or continuous in the triangle $a \le t \le x \le b$ and zero for $t > x$) and when λ does not have one of a certain set of exceptional values, there is a unique continuous solution of Eq. (8-112) given by

$$f(x) = g(x) + \int_a^b \frac{D(x,t;\lambda)}{D(\lambda)}\,g(t)\,dt \tag{8-114}$$

[1] E. Fredholm, *Acta Math.*, *27:* 365 (1903).

where $D(x,t;\lambda)$ and $D(\lambda)$ are entire functions of the complex variable λ. The exceptional values of λ are those for which $D(\lambda) = 0$; since $D(\lambda)$ is entire and not identically zero, these exceptional λ values must be isolated points in the λ plane, and only a finite number of them can lie in any bounded region. From Eq. (8-114), the homogeneous equation

$$f(x) = \lambda \int_a^b k(x,t)f(t)\, dt \tag{8-115}$$

can have no nontrivial solution if $D(\lambda) \neq 0$; on the other hand, it may be shown that, if $D(\lambda) = 0$, Eq. (8-115) does indeed have a solution. Such special values of λ and the corresponding solutions of Eq. (8-115) are called *eigenvalues* and *eigenfunctions*, respectively; a theorem due to Schmidt states that, if $k(x,t)$ is symmetric, then $D(\lambda)$ has at least one zero. There is an extensive literature (stemming largely from Hilbert) dealing with expansion possibilities of certain functions (and kernels) in terms of eigenfunctions. It has been shown by Carleman and others that most of the Fredholm theory can be generalized so as to hold for the case in which $k(x,t)$ is singular, with $\int_a^b \int_a^b |k^2(x,t)|\, dx\, dt$ finite; the importance of this result lies in the fact that many practical problems lead to such singular kernels.

We shall not discuss Fredholm theory[1] here, since its main implications relate to eigenvalue problems rather than to function theory. For our purposes, it is enough to be able to construct (in principle, at least) solutions to Eq. (8-113) by the *Liouville-Neumann method* of successive substitutions. In Exercises 1 and 2, the reader is asked to prove that—even if $k(x,t)$ has integrable logarithmic or algebraic singularities—the process of successively substituting the right-hand side of Eq. (8-113) for $f(t)$ under the integral sign will yield a power series in λ, which for sufficiently small $|\lambda|$ converges to the solution of the equation. For larger values of $|\lambda|$, we represent the kernel in the form

$$k(x,t) = k_1(x,t) + \sum_{j=1}^{n} \alpha_j(x)\beta_j(t) \tag{8-116}$$

where the functions $\alpha_j(x)$ and $\beta_j(t)$ are so chosen that $\int_a^b |k_1(x,t)|^2$ is sufficiently small for the successive-substitution process to be valid. [The possibility of a double Fourier-series expansion of $k(x,t)$ suggests that the separation (8-116) is usually feasible.] Incidentally, if $k_1 \equiv 0$ in Eq. (8-116) for some finite value of n, the kernel is termed *degenerate*.

[1] See S. G. Mikhlin, "Integral Equations," Pergamon Press, New York, 1957, or F. G. Tricomi, "Integral Equations," Interscience Publishers, Inc., 1957, or F. Smithies, "Integral Equations," Cambridge University Press, New York, 1958.

Integral equations of the first kind,

$$\int_a^b k(x,t)f(t)\,dt = g(x) \tag{8-117}$$

are of a very different character from Eq. (8-113). In fact, for well-behaved $k(x,t)$, Eq. (8-117) often does not have a solution; this is at once clear for the simple case of a degenerate kernel, for if $k_1 \equiv 0$ in Eq. (8-116), Eq. (8-117) can have a solution only if $g(x)$ is a linear combination of the n functions $\alpha_j(x)$ (and it will then have an infinite number of solutions). We therefore expect that those equations of the form (8-117) which arise in applications will involve kernels which are singular in some way or other (an infinite interval of integration, as in Fourier-transform formulas, is sometimes also considered to constitute a singularity).

Our main purpose in this section is to discuss the use of function-theoretic methods to attack certain singular integral equations. They will usually be of the form of Eq. (8-117), with $k(x,t)$ exhibiting a logarithmic, algebraic, or "principal-value" type of singularity.

In Eq. (7-148), we have already encountered one such equation, that of Abel,

$$\int_0^x f(t)(x-t)^{-\alpha}\,dt = g(x) \tag{8-118}$$

with $0 < \alpha < 1$. Just as for the special case $\alpha = \frac{1}{2}$ involved in Eq. (7-148), Eq. (8-118) is amenable to a transform method (Exercise 3). A less tractable equation arises in airfoil theory,[1]

$$(P)\int_0^1 \frac{f(t)\,dt}{x-t} = g(x) \tag{8-119}$$

The symbol (P) indicates that the principal value of the integral is to be taken, in the sense

$$(P)\int_0^1 = \lim_{\epsilon\to 0}\int_0^{x-\epsilon} + \int_{x+\epsilon}^1$$

As another example, the calculation of the heat transfer from a finite heated strip into a uniformly moving fluid can be reduced to the problem of solving

$$\int_0^1 f(t)e^{\alpha(x-t)}K_0(|x-t|)\,dt = g(x) \tag{8-120}$$

where the modified Bessel function K_0 has a logarithmic singularity at $x = t$.

It will turn out that there is a close relationship between certain of these singular integral equations and what is called the *Riemann-Hilbert* problem, in which one seeks an analytic function whose limiting values on the two sides of a cut bear a linear relation to one another. The method we shall

[1] L. Prandtl and O. G. Tietjens, "Applied Hydro and Aeromechanics," p. 199, Dover reprint, 1957.

use for such problems involves product and sum-splitting operations identical in nature to those encountered in the Wiener-Hopf method. The method was developed by Carleman[1]; in fact, the Wiener-Hopf method might with justice be renamed the WHC method.

EXERCISES

1. Schwarz's inequality

$$\left[\int_a^b f(t)g(t)\,dt\right]^2 \leq \int_a^b f^2(t)\,dt \int_a^b g^2(t)\,dt \qquad (8\text{-}121)$$

follows easily from the requirement that the quadratic equation in x

$$\int_a^b [f(t) - xg(t)]^2\,dt = 0$$

can have no real root (unless f is a constant times g).

Define the nth iterated kernel $k_n(x,t)$ by

$$k_n(x,t) = \int_a^b k_{n-1}(x,\tau)k(\tau,t)\,d\tau \qquad (8\text{-}122)$$

with $k_1(x,t) = k(x,t)$; then successive substitutions of the right-hand side of Eq. (8-113) for $f(t)$ under the integral sign leads to the formal series

$$f(x) = g(x) + \lambda \int_a^b k_1(x,t)g(t)\,dt + \lambda^2 \int_a^b k_2(x,t)g(t)\,dt + \cdots \qquad (8\text{-}123)$$

For the case in which each of $\iint_a^b k^2(x,t)\,dx\,dt$ and $\int_a^b k^2(x,t)\,dt$ is bounded by constants S and M, respectively, use Schwarz's inequality to obtain an inequality for the bound of $k_n^2(x,t)$, and integrate this inequality over (a,b) to obtain a bound for $\int_a^b k_n^2(x,t)\,dt$. Hence show that the series (8-123) converges to the unique solution of Eq. (8-113) for $|\lambda| < S^{-1/2}$. What kind of singularity of $k(x,t)$ is permitted by this proof? Discuss also the use of Eq. (8-117) for larger values of λ; how would this calculation be affected by setting λ equal to one of the eigenvalues?

2. If the process of successive substitutions described in Exercise 1 is terminated after $n - 1$ iterations, we obtain the new integral equation

$$f(x) = \left[g(x) + \lambda \int_a^b k(x,t)g(t)\,dt + \cdots + \lambda^{n-1} \int_a^b k_{n-1}(x,t)g(t)\,dt\right] + \lambda^n \int_a^b k_n(x,t)f(t)\,dt \qquad (8\text{-}124)$$

[1] T. Carleman, *Arkiv Math., Astron., Fysik*, **16**: 141 (1922), *Math. Z.*, **15**: 161 (1922).

It is clear that, if λ does not have one of its critical values, any solution of Eq. (8-113) is also a solution of Eq. (8-124). Show conversely that, if $f(x)$ is a solution of Eq. (8-124), so is $g(x) + \lambda \int_a^b k(x,t)f(t)\,dt$, which must therefore coincide with $f(x)$ because of uniqueness; hence $f(x)$ is also a solution of Eq. (8-113). Thus Eqs. (8-113) and (8-124) are equivalent. Using this fact, we can extend the proof of Exercise 1 so as to hold for kernels $k(x,t)$ which do not satisfy the conditions of Exercise 1 but which are such that some iterated kernel $k_n(x,t)$ does satisfy the conditions. Prove that this will be the case for the kernel

$$k(x,t) = \frac{m(x,t)}{|x - t|^\alpha}$$

where $0 < \alpha < 1$ and where $m(x,t)$ is bounded in the square region $a \leq x \leq b$, $a \leq t \leq b$.

3. Use a transform method to show that the solution of Eq. (8-118) is given by

$$f(x) = \frac{\sin \alpha\pi}{\pi} \frac{d}{dx} \int_0^x g(t)(x - t)^{\alpha-1}\,dt \tag{8-125}$$

Carry out the differentiation so as to obtain an alternative form for this result; observe that $g(0)$ need not be zero.

4. (*a*) Abel devised two methods for solving Eq. (8-118). One of these consisted in multiplying both sides by $(z - x)^{\alpha-1}$ and integrating with respect to x from 0 to z. Carry out this process; change the order of integration, and use the substitution $(z - x)/(x - t) = y$ to obtain an integral whose evaluation leads to Eq. (8-125).

(*b*) Abel's second method was to write $f(t) = f_1'(t)$ and to expand $f_1(t)$ in the series $f_1(t) = \sum_0^\infty a_n t^{n+\alpha}$. A corresponding power-series expansion for $g(x)$ then permits a term-by-term comparison; sum the resulting series for $f(x)$ to recover Eq. (8-125).

5. Devise a method to solve the integral equation

$$\int_0^x \frac{p(x - t)}{(x - t)^\alpha} f(t)\,dt = g(x) \tag{8-126}$$

where $0 < \alpha < 1$ and $p(x - t)$ is a polynomial in $x - t$. Remark on the case $p(x - t) = x - t$.

Cauchy Integrals

We begin our study of singular integral equations by establishing some properties of the Cauchy integral,

$$F(z) = \frac{1}{2\pi i} \int_C \frac{f(t)\,dt}{t - z} \tag{8-127}$$

where the path of integration C is some curve in the complex t plane, $f(t)$ is a complex-valued function prescribed on C, and z is a point not on C. The curve C may be an arc or a closed contour (or more generally a collection of such arcs and closed contours). For suitable curves C and functions $f(t)$, $F(z)$ will clearly be an analytic function of z. Denote by t_0 a point on C, other than an end point (we consider the end-point case in Exercise 7). Our main purpose is to examine the behavior of $F(z)$ as $z \to t_0$.

If we orient ourselves at the point t_0 on C so as to be facing along the direction of integration, then the limits of $F(t)$ as $z \to t_0$ from the left and from the right (if these limits exist) will be denoted by $F_+(t_0)$ and $F_-(t_0)$, respectively. If C is a closed contour, then since our standard sense of traversal is counterclockwise, the convention just adopted means that $F_+(t_0)$ and $F_-(t_0)$ are the limits as $z \to t_0$ from the inside and outside, respectively. We also define the principal-value integral

$$\begin{aligned} F_p(t_0) &= \frac{1}{2\pi i} (P) \int_C \frac{f(t)\,dt}{t - t_0} \\ &= \frac{1}{2\pi i} \lim_{\epsilon \to 0} \int_{C - C_\epsilon} \frac{f(t)\,dt}{t - t_0} \end{aligned} \tag{8-128}$$

where C_ϵ is the portion of the curve C contained within a small circle of radius ϵ, centered on t_0. We want to obtain relations between $F_+(t_0)$, $F_-(t_0)$, and $F_p(t_0)$.

The desired relations are readily obtained if $f(t)$ is analytic at the point t_0 (and continuous elsewhere). If this is the case, there is a small circle around t_0 such that $f(t)$ is analytic within this circle (Fig. 8-4). By Cauchy's theorem, C may be indented around t_0 by use of an (approximate) semicircle of radius ϵ lying within the circle of analyticity, as shown. We may then carry out the limiting process in which a point z inside C approaches

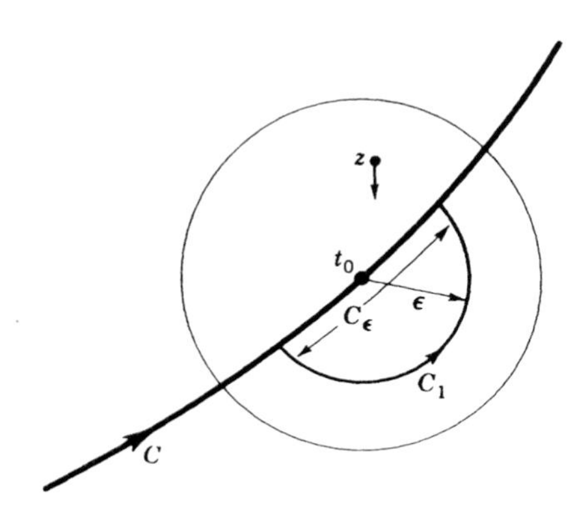

Fig. 8-4 Neighborhood of point t_0 on C.

t_0, and we obtain

$$F_+(t_0) = \frac{1}{2\pi i} \int_{C-C_\epsilon} \frac{f(t)\,dt}{t - t_0} + \frac{1}{2\pi i} \int_{C_1} \frac{f(t)\,dt}{t - t_0} \tag{8-129}$$

where C_ϵ denotes the omitted portion of the original contour C and where C_1 denotes the semicircle contour. As $\epsilon \to 0$, Eq. (8-129) tends to

$$F_+(t_0) = \tfrac{1}{2}f(t_0) + F_p(t_0) \tag{8-130}$$

[we have here assumed that C does not have a corner at t_0; cf. Eq. (8-133)]. Similarly,

$$F_-(t_0) = -\tfrac{1}{2}f(t_0) + F_p(t_0) \tag{8-131}$$

The reader may readily prove that $F_+(t_0)$ and $F_-(t_0)$ are, locally at least, continuous functions of t_0.

Equations (8-130) and (8-131) are known as the *Plemelj formulas*. Although our derivation requires $f(t)$ to be analytic at t_0, we shall see in Exercise 6 that weaker conditions[1] on $f(t)$ are possible; it is sufficient to require $f(t)$ to be continuous on C and to satisfy the Lipschitz condition

$$|f(t_1) - f(t_0)| < A|t_1 - t_0|^\alpha \tag{8-132}$$

for all points t_1 on C in some neighborhood of t_0, where A and α are constants, with $0 < \alpha \leq 1$. If C is an arc, then we shall in general also permit $f(t)$ to have a logarithmic or integrable algebraic singularity at each end of the arc; from Exercise 7, this means that $F(z)$ can grow no faster than some power $\beta > -1$ of $|z - t_e|$ as $z \to t_e$, where t_e is an end point. Our reason for permitting $f(t)$ to have these end-point singularities on an arc is that certain integral equations of interest are solved only by functions of this character.

As a further generalization of the Plemelj formulas, we can permit the curve C to have a corner at t_0, as in Fig. 8-5. If the left-hand angle is θ, then, by a calculation analogous to that leading to Eqs. (8-130) and (8-131), the reader may show that

$$F_+(t_0) = \left(1 - \frac{\theta}{2\pi}\right) f(t_0) + F_p(t_0) \tag{8-133}$$

$$F_-(t_0) = -\frac{\theta}{2\pi} f(t_0) + F_p(t_0) \tag{8-134}$$

We note that addition or subtraction of the Plemelj formulas leads to

$$f(t_0) = F_+(t_0) - F_-(t_0) \tag{8-135}$$

$$F_p(t_0) = \tfrac{1}{2}[F_+(t_0) + F_-(t_0)] \tag{8-136}$$

[1] An extensive discussion of the conditions for validity of Eqs. (8-130) and (8-131) is given in N. I. Muskhelishvili, "Singular Integral Equations," Erven P. Nordhoff, NV, Gronigen, Netherlands, 1953.

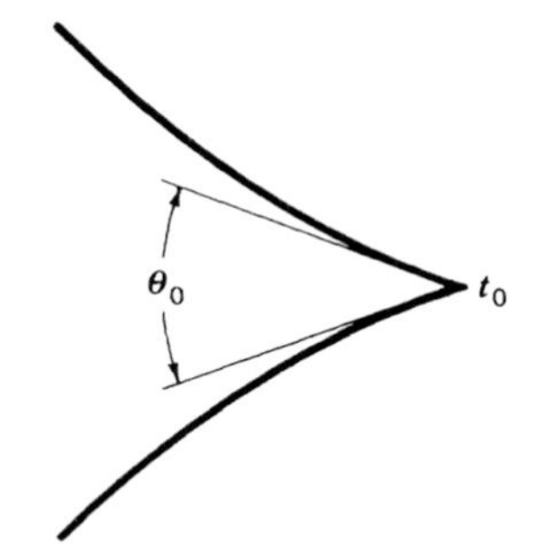

Fig. 8-5 Corner at t_0.

A problem which will often arise in the sequel is that of determining an analytic function possessing a jump discontinuity $f(t)$ across a certain arc; Eq. (8-135) suggests that one solution to this problem is given by $F(z)$ as defined by Eq. (8-127). To find the general solution, we proceed as follows:

Let $f(t)$ be prescribed along a curve C (an arc or a closed contour). It is required to find a function $W(z)$ analytic for all points z not on C, with limiting values as $z \to t_0$ denoted by $W_+(t_0)$ and $W_-(t_0)$ (from the left and right, respectively), such that (1) $W_+(t_0) - W_-(t_0) = f(t_0)$ for a point t_0 on C (not an end point); (2) $W(z)/z^n \to 0$ as $z \to \infty$, for some integer $n \geq 0$; and (3) if C is an arc, then as $z \to t_e$, where t_e is an end point of the arc, $|W(z)| < B|z - t_e|^\beta$ for some constants B and β, with $\beta > -1$.

One solution is certainly $F(z)$, since $F(z) \to 0$ as $z \to \infty$ and since condition (3) is satisfied for the kinds of functions $f(t)$ that we are permitting. Consider the difference function $W(z) - F(z)$; this difference function is analytic everywhere except possibly at the end points of C if C is an arc, but a consideration of the growth rate permitted in each of $F(z)$ and $W(z)$ near the end points shows that any singularities there must be removable. Thus $W(z) - F(z)$ is entire, so that the general solution is

$$W(z) = F(z) + p_{n-1}(z) \tag{8-137}$$

where $p_{n-1}(z)$ is an arbitrary polynomial of order $n - 1$. For ease of reference, we shall call this result the *discontinuity theorem*.

As an example of the use of this result, let us evaluate the principal-value integral

$$I(x) = (P) \int_{-1}^{1} \frac{(1 - t)^{\alpha-1}\, dt}{(1 + t)^\alpha (t - x)} \tag{8-138}$$

where $0 < \alpha < 1$ and $-1 < x < 1$. We consider the function

$$G(z) = \frac{(z - 1)^{\alpha-1}}{(z + 1)^\alpha} \tag{8-139}$$

with the branch cut $(-1,1)$ and with the branch chosen so as to correspond to principal values for z real and >1. Then, for t_0 in $(-1,1)$, it follows from Eq. (8-139) that

$$\begin{aligned} G_+(t_0) &= (1-t_0)^{\alpha-1}(1+t_0)^{-\alpha}e^{i(\alpha-1)\pi} \\ G_-(t_0) &= (1-t_0)^{\alpha-1}(1+t_0)^{-\alpha}e^{-i(\alpha-1)\pi} \end{aligned}$$

Thus $G_+(t_0) - G_-(t_0) = (1-t_0)^{\alpha-1}(1+t_0)^{-\alpha}\,(-2i \sin \alpha\pi)$, and from the discontinuity theorem it follows that

$$G(z) = (-2i \sin \alpha\pi)\,\frac{1}{2\pi i}\int_{-1}^{1} \frac{(1-t)^{\alpha-1}\,dt}{(1+t)^{\alpha}(t-z)}$$

Using Eq. (8-136), we now obtain

$$I(x) = \pi \cot \alpha\pi\,(1-x)^{\alpha-1}(1+x)^{-\alpha} \tag{8-140}$$

(Notice the interesting special case $a = \frac{1}{2}$.)

Exercises

6. Let $f(t)$ be continuous and satisfy Eq. (8-132). Writing $f(t) = [f(t) - f(t_0)] + f(t_0)$, show that $F_p(t_0)$ exists and that the Plemelj formulas hold.

7. Let the arc C be the line segment $(0,1)$. Investigate the behavior of $F(z)$ as $z \to 0$ for the cases $f(t) = 1$, $f(t) = \ln t$, $f(t) = t^{\alpha}$, with $-1 < \alpha < 0$, and explain how the results can be made to carry over to the case of an arbitrary arc. An intuitive way in which to think of the effect of an integrable singularity of $f(t)$ at any end point t_e is that it acts like a concentrated impulse at t_e [which would give $F(z) \sim \text{const}/(t_e - z)$ in Eq. (8-127)]—except that it is weaker.

8. Consider a simple closed contour C, enclosing a region S_+; denote the region exterior to C by S_-. If a function $f(t)$ is prescribed on C, show that the necessary and sufficient condition that a function $w(z)$ exist, analytic in S_+ and satisfying $w(z) \to f(t)$ as $z \to t$ from the interior, is that $F(z) \equiv 0$ for z in S_-. State an equivalent criterion involving $F_p(t)$.

9. Prove Eq. (8-140) by a direct contour-integration method. Use any method to show that

$$(P)\int_{-1}^{1} \frac{(1-t^2)^{1/2} \ln (1+t)\,dt}{t-x} = \pi\left[x \ln 2 - 1 + (1-x^2)^{1/2}\left(\frac{\pi}{2} - \arcsin x\right)\right]$$

10. Let C be a simple closed contour. Let $g(t)$ be a given function, and consider the integral equation

$$\frac{1}{\pi i}\,(P)\int_C \frac{f(t)\,dt}{t-t_0} = g(t_0) \tag{8-141}$$

Observe that the left-hand side can be written as $F_+(t_0) + F_-(t_0)$; define a function $W(z)$ such that $W(z) = F(z)$ for z inside C and $W(t) = -F(z)$ for z outside C. Proceeding in this way, show that the solution of Eq. (8-141) is given by the symmetrical formula

$$f(t_0) = \frac{1}{\pi i} (P) \int_C \frac{g(t)\, dt}{t - t_0} \tag{8-142}$$

11. By considering the special case of Exercise 10 in which C is a circle, show that if $f(\theta)$ is a periodic function of θ, with period 2π, and such that $\int_0^{2\pi} f(\theta)\, d\theta = 0$, then the function $g(\theta)$ defined by

$$g(\theta) = \frac{1}{2\pi} (P) \int_0^{2\pi} f(t) \cot \frac{1}{2} (\theta - t)\, dt \tag{8-143}$$

is also periodic, with the property that $\int_0^{2\pi} g(\theta)\, d\theta = 0$; moreover $f(\theta)$ can be recovered via

$$f(\theta) = -\frac{1}{2\pi} (P) \int_0^{2\pi} g(t) \cot \frac{1}{2} (\theta - t)\, dt \tag{8-144}$$

The function $g(\theta)$ is said to be *Hilbert transform* of $f(\theta)$; it follows that $f(\theta)$ in turn is the Hilbert transform of $-g(\theta)$. Show also that $f(\theta) + ig(\theta)$ is the boundary value of a function analytic inside the circle, so that in a sense f and g are conjugate to each other. Finally, show that, if $f(\theta) = \sum_1^\infty (a_n \cos n\theta + b_n \sin n\theta)$, then $g(\theta) = \sum_1^\infty (a_n \sin n\theta - b_n \cos n\theta)$.

12. Deduce from Eq. (8-142) that, for a closed contour C,

$$f(t_0) = \frac{1}{\pi i} (P) \int_C \frac{dt}{t - t_0} \frac{1}{\pi i} (P) \int_C \frac{f(\tau)\, d\tau}{\tau - t} \tag{8-145}$$

which is called the *Poincaré-Bertrand* formula. Use this formula to show that the solution of

$$af(t_0) + \frac{b}{\pi i} (P) \int_C \frac{f(t)\, dt}{t - t_0} = g(t_0) \tag{8-146}$$

(where a, b are constants, with $a^2 \neq b^2$) is

$$f(t_0) = \frac{a}{a^2 - b^2} g(t_0) - \frac{b}{(a^2 - b^2)\pi i} (P) \int_C \frac{g(t)\, dt}{t - t_0}$$

Discuss the homogeneous equation obtained by setting $g(t_0) = 0$ in Eq. (8-146).

13. In Exercise 8, a criterion was obtained for a function prescribed on a circle to be the boundary value of a function analytic inside the circle. Examine the possibility of obtaining a similar criterion for a function to exist which is analytic outside an arc, vanishes at ∞, and takes on prescribed boundary values on the arc. [*Hint:* Consider conformal mapping onto the exterior of a circle.]

Riemann-Hilbert Problem

In his thesis, Riemann posed the problem of determining a function which was analytic inside a closed contour C, when a prescribed linear relation between the boundary values of its real and imaginary parts was to be satisfied on C. By conformal mapping, we can take C as a circle. Denote the desired analytic function by $W_+(z)$; let its real and imaginary parts be u, v, respectively, and suppose that the prescribed relation is

$$\alpha(t)u(t) + \beta(t)v(t) = \gamma(t) \tag{8-147}$$

for t on C, where $\alpha(t)$, $\beta(t)$, $\gamma(t)$ are given real functions. A function which is analytic outside the circle is

$$W_-(z) = W_+^*\left(\frac{1}{z^*}\right) \tag{8-148}$$

As $z \to t$ on C, we have $W_-^*(t) = W_+(t) = u(t) + iv(t)$, so that Eq. (8-147) can be rewritten

$$\frac{\alpha(t) - i\beta(t)}{2} W_+(t) + \frac{\alpha(t) + i\beta(t)}{2} W_-(t) = \gamma(t) \tag{8-149}$$

Consequently, Riemann's problem can be rephrased so as to require us to find two functions $W_+(z)$ and $W_-(z)$, analytic inside and outside the circle, respectively, such that their boundary values on the circle satisfy the linear relation (8-149). [The behavior of $W_-(z)$ at ∞ must also be prescribed for completeness; from Eq. (8-148), we require $W_-(z) \to W_+(0)$ as $z \to \infty$.]

As a generalization of Riemann's problem, Hilbert posed the problem of determining a function $W(z)$ analytic for all points z not lying on a curve C, such that, for t on C,

$$W_+(t) = g(t)W_-(t) + f(t) \tag{8-150}$$

where $g(t)$ and $f(t)$ are given complex-valued functions.[1] Although Hilbert dealt only with the case of a closed contour, the general problem (8-150) with C either an arc or a closed contour (or more generally a collection of areas and contours) has come to be called the *Riemann-Hilbert problem.*

[1] If C is a closed contour, then we have, not one function $W(z)$, but two functions $W_+(z)$ and $W_-(z)$; nevertheless we shall loosely speak of the two functions as constituting a single function. Also we shall, for example, use the notation $W_+(z)$ to denote an analytic function for z inside C and the notation $W_+(t)$ to designate its limiting value as $z \to t$ on C.

For completeness, the behavior of $W(z)$ at ∞ must be specified, as must also the behavior of $W(z)$ near the ends of C if C is an arc.

Carleman[1] encountered this same problem in his work on singular integral equations (see next section) and devised an effective means of attack. The method is to first find a function $L(z)$, such that

$$L_+(t) = g(t)L_-(t) \tag{8-151}$$

with $L_+(t)$ and $L_-(t)$, as well as $L(z)$, nonzero. Substitution of Eq. (8-151) into Eq. (8-150) then gives

$$\frac{W_+(t)}{L_+(t)} - \frac{W_-(t)}{L_-(t)} = \frac{f(t)}{L_+(t)} \tag{8-152}$$

The function $W(z)/L(z)$ is analytic for z not on C [since $L(z) \neq 0$], so that the conditions of the discontinuity theorem are satisfied, and the most general function $W(z)/L(z)$ satisfying Eq. (8-152) is given by that theorem. Thus, since $L(z)$ is known, $W(z)$ has been determined.

Before discussing the problem of determining $L(z)$, we may remark that the use of the discontinuity theorem in solving Eq. (8-152) is equivalent to defining

$$M(z) = \frac{1}{2\pi i}\int_C \frac{f(t)/L_+(t)}{t - z}\,dt$$

and writing

$$\frac{f(t)}{L_+(t)} = M_+(t) - M_-(t)$$

Equation (8-152) then becomes

$$\frac{W_+(t)}{L_+(t)} - M_+(t) = \frac{W_-(t)}{L_-(t)} - M_-(t)$$

so that the function

$$\frac{W(z)}{L(z)} - M(z)$$

has the same boundary values on each of the two sides of C and so is entire. The function $W(z)$ is thereby determined, within an arbitrary polynomial compatible with the prescribed finite order at ∞. For the special case in which C is an infinite straight line parallel to the real axis, the method just described is identical to the Wiener-Hopf method.

We turn now to the problem of finding a nonvanishing function $L(z)$ to satisfy Eq. (8-151). Assuming $g(t)$ to be nonzero on C, we can take logarithms of Eq. (8-151) to give

$$\ln L_+(t) - \ln L_-(t) = \ln g(t) \tag{8-153}$$

which could be solved by use of the discontinuity theorem—at least for the

[1] *loc. cit.*

case in which C is an arc, for then we do not have to worry about multiple-valuedness of $\ln g(t)$. For the moment, then, let C be an arc. We also assume that $g(t)$ is continuous onto the ends z_1, z_2 of the arc, with end-point values $g(z_1)$, $g(z_2)$. Then a particular solution of Eq. (8-153) is given by

$$\begin{aligned} \ln L(z) &= \frac{1}{2\pi i} \int_C \frac{\ln g(t)}{t - z}\, dt \\ &= Q(z), \text{ say} \end{aligned} \tag{8-154}$$

Since $L(z) = e^{Q(z)}$, $L(z)$ is indeed nonzero, and, further,

$$\frac{L_+(t)}{L_-(t)} = \exp\,[Q_+(t) - Q_-(t)] = e^{\ln g(t)} = g(t)$$

so that Eq. (8-151) is satisfied. At ∞, $L(z) \to 1$. However, the behavior of $L(z)$ as $z \to z_1, z_2$ may not be appropriate for the application of the discontinuity theorem to Eq. (8-152); fortunately, we can adjust this behavior by incorporating an integral power of $z - z_1$ or $z - z_2$ into L. We know from Exercise 7, for example, that

$$Q(z) \sim -\frac{1}{2\pi i} \ln g(z_1) \ln (z_1 - z)$$

for z near the initial point z_1 of the arc. Thus

$$\begin{aligned} L(z) &\sim (z - z_1)^{-1/2\pi i \ln g(z_1)} \\ &\sim (z - z_1)^{a+ib} \end{aligned}$$

where a and b are real; we would then normally revise $L(z)$ by incorporating a factor $(z - z_1)^{k_1}$, where k_1 is a real integer, with $-1 < a + k_1 < 0$. A similar factor would be inserted in order to produce the desired behavior at the other end.

If C is a closed contour, Eq. (8-153) is not generally useful, since $\ln g(t)$ will in general not return to its initial value after a complete transversal of C; thus the function $\ln g(t)$ occurring in the integral defining $Q(z)$ in Eq. (8-154) has a discontinuity, and the Plemelj formulas are not valid. Let $\ln g(z)$ increase by $2\pi in$, n an integer, during a transversal of C. Then we can avoid the multiple-valuedness difficulty by defining

$$g_0(t) = (t - z_0)^{-n} g(t)$$

where z_0 is a point within C_0. If we define

$$N(z) = \begin{cases} L(z) & \text{for } z \text{ inside } C \\ (z - z_0)^n L(z) & \text{for } z \text{ outside } C \end{cases} \tag{8-155}$$

our problem is now to solve

$$N_+(t) = g_0(t) N_-(t)$$

where $g_0(t)$ is single-valued. The same procedure that was used for the arc may now be applied.

As a simple example of the method, consider the frequently occurring problem of determining a function $W(z)$ which satisfies

$$W_+(t) + W_-(t) = f(t) \tag{8-156}$$

for t on an arc C, with $W(z)$ to be of finite degree at ∞ and to have singularities near the ends which are no worse than algebraic with degree > -1. The function $f(t)$ may have an integrable singularity at the end points of C, denoted by z_1 and z_2. Equation (8-156) is of the form of Eq. (8-150), with $g(t) = -1$, so we begin by solving Eq. (8-151) with $g(t) = -1$. We need only a particular solution; one such is obtained from Eq. (8-154) as

$$\ln L(z) = \frac{1}{2\pi i} \int_{z_1}^{z_2} \frac{i\pi \, dt}{t - z}$$

$$L(z) = \sqrt{\frac{z - z_2}{z - z_1}} \tag{8-157}$$

It is easily verified that Eq. (8-151) is satisfied. However, in order that $W(z)/L(z)$ not grow too fast as $z \to z_1$ or z_2, it is desirable to make $L(z)$ grow algebraically (with exponent > -1) as $z \to z_1$, z_2; we therefore use for $L(z)$ a function obtained by multiplying the right-hand side of Eq. (8-157) by $1/(z - z_2)$. Thus our choice for $L(z)$ is

$$L(z) = \frac{1}{\sqrt{(z - z_1)(z - z_2)}} \tag{8-158}$$

and for definiteness we choose a branch cut along the arc, with $L(z) \sim z^{-1}$ for large z. Then, for t on C, $L_+(t)$ and $L_-(t)$ are easily calculated from Eq. (8-158); for example, if C is the line segment $(-1,1)$ of the real axis, then

$$L_+(t) = \frac{-i}{\sqrt{1 - t^2}} \quad \text{and} \quad L_-(t) = \frac{i}{\sqrt{1 - t^2}} \tag{8-159}$$

In any event, Eq. (8-152) now leads to

$$\frac{W(z)}{L(z)} = \frac{1}{2\pi i} \int_C \frac{f(t) \, dt}{L_+(t)(t - z)} + p_n(z) \tag{8-160}$$

(where p_n is an arbitrary polynomial of degree n) as the most general solution for which $W(z)/z^n \to 0$ as $z \to \infty$.

Exercises

14. Let the arc C of Eq. (8-156) be the segment $(-1,1)$ of the real axis, and let $f(t) = 1$. Evaluate Eq. (8-160) to obtain the general solution as

$$W(z) = \frac{1}{2} - \frac{z}{2\sqrt{z^2 - 1}} + \frac{p_n(z)}{\sqrt{z^2 - 1}}$$

where $\sqrt{z^2 - 1} \sim z$ for large z. [Thus it is not possible to make $W(z)$ vanish faster than $1/z^2$ as $z \to \infty$.]

15. Use the Carleman method to determine a function $F(z)$, analytic outside the real axis segment (0,1) and of finite degree at ∞, such that if $F(z) = U + iV$, U and V real, then $U_+(t) = f(t)$ and $U_-(t) = g(t)$, where $f(t)$ and $g(t)$ are prescribed for t in (0,1). It is also required that $F(z)$ have no worse than an algebraic singularity near the points 0, 1, with degree > -1. Determine the conditions under which a solution exists. [*Hint:* Let $G(z) = F^*(z^*)$. Then the boundary condition requires that

$$[F_+(t) + G_+(t)] + [F_-(t) + G_-(t)] = 2[f(t) + g(t)]$$
$$[F_+(t) - G_+(t)] - [F_-(t) - G_-(t)] = 2[f(t) - g(t)]$$

and then treat $F + G$, $F - G$ as the two functions to be determined.] This problem can, of course, be solved by conformal mapping; however, the present method carries over to the case in which the $f(t)$ and $g(t)$ are prescribed on several line segments on the real axis. It is clear that similar methods can be used for a variety of problems involving harmonic functions (or biharmonic functions, via the Goursat formula), where certain boundary conditions are prescribed on the real axis (or the unit circle). A number of examples are exhaustively treated by Muskhelishvili.[1]

16. It was remarked above that the general splitting procedures were the same in each of the Wiener-Hopf and Carleman methods. Each method provides a formal mechanism for carrying out the splitting process; show that these formal mechanisms are also identical.

Integral Equations with Cauchy Kernels

As an example of such an equation, we consider

$$a(x)f(x) = \lambda(P) \int_{-1}^{1} \frac{f(t)\,dt}{t - x} + g(x) \tag{8-161}$$

for x in $(-1,1)$, where λ is real and > 0[2] and where $a(x)$, $g(x)$ are prescribed real functions.

Defining

$$F(z) = \frac{1}{2\pi i} \int_{-1}^{1} \frac{f(t)\,dt}{t - z} \tag{8-162}$$

[where we shall permit $F(z)$ to have an algebraic singularity, of degree > -1, at the points -1, 1], we have

$$[a(x) - \lambda\pi i]F_+(x) = [a(x) + \lambda\pi i]F_-(x) + g(x) \tag{8-163}$$

[1] *Loc. cit.*

[2] For $\lambda < 0$, simply reverse the signs of $a(x)$ and $g(x)$.

which is of the form discussed in the preceding section. We seek first a nonzero function $L(z)$ such that

$$\frac{L_+(x)}{L_-(x)} = \frac{a(x) + \lambda\pi i}{a(x) - \lambda\pi i}$$

A suitable $L(z)$ is given by

$$L(z) = \frac{1}{z - 1} e^{Q(z)} \tag{8-164}$$

where

$$Q(z) = \frac{1}{2\pi i} \int_{-1}^{1} \frac{dt}{t - z} \ln \frac{a(t) + \lambda\pi i}{a(t) - \lambda\pi i} \tag{8-165}$$

We take

$$\frac{1}{2\pi i} \ln \frac{a(t) + \lambda\pi i}{a(t) - \lambda\pi i}$$

(which is purely real) to lie in the range (0,1). The factor $(z - 1)^{-1}$ in Eq. (8-164) has been inserted to ensure that $L(z)$ grows algebraically, with index between -1 and 0, as z approaches either point -1, 1.

Equation (8-163) becomes [for x in $(-1,1)$]

$$\frac{F_+(x)}{L_+(x)} - \frac{F_-(x)}{L_-(x)} = \frac{g(x)}{L_+(x)[a(x) - \lambda\pi i]} \tag{8-166}$$

where

$$\begin{aligned} L_+(x) &= \frac{1}{x - 1} \sqrt{\frac{a(x) + \lambda\pi i}{a(x) - \lambda\pi i}}\, e^{\omega(x)} \\ L_-(x) &= \frac{1}{x - 1} \sqrt{\frac{a(x) - \lambda\pi i}{a(x) + \lambda\pi i}}\, e^{\omega(x)} \end{aligned} \tag{8-167}$$

$$\omega(x) = \frac{1}{2\pi i} (P) \int_{-1}^{1} \frac{dt}{t - x} \ln \frac{a(t) + \lambda\pi i}{a(t) - \lambda\pi i} \tag{8-168}$$

Using the discontinuity theorem and taking account of the behavior of $F(z)$ and $L(z)$ as $z \to \infty$, we obtain the most general solution of Eq. (8-166) as

$$\frac{F(z)}{L(z)} = \frac{1}{2\pi i} \int_{-1}^{1} \frac{g(t)\, dt}{(t - z)L_+(t)[a(t) - \lambda\pi i]} + k_1$$

where k_1 is an arbitrary constant. From $f(x) = F_+(x) - F_-(x)$ and Eqs. (8-167), we now obtain the general solution of Eq. (8-161) as

$$\begin{aligned} f(x) = \frac{a(x)g(x)}{a^2(x) + \lambda^2\pi^2} &+ \frac{\lambda e^{\omega(x)}}{\sqrt{a^2(x) + \lambda^2\pi^2}} (P) \int_{-1}^{1} \frac{g(t)e^{-\omega(t)}\, dt}{(t - x)\sqrt{a^2(t) + \lambda^2\pi^2}} \\ &+ \frac{k e^{\omega(x)}}{(1 - x)\sqrt{a^2(x) + \lambda^2\pi^2}} \end{aligned} \tag{8-169}$$

where k is a new arbitrary constant and where $\omega(t)$ is given by Eq. (8-168). [In verifying this result, note that an integrand factor of $t - 1$ can be written as $(t - x) + (x - 1)$.] The $(1 - x)^{-1}$ singularity at $x = 1$ in the last term of Eq. (8-169) is weakened by the factor $e^{\omega(x)}$, so that the last term is integrable. We observe that the homogeneous equation, obtained by setting $g(x) = 0$ in Eq. (8-161), has a solution for all λ—that is, the spectrum is continuous.

For the special case $a(x) = 0$, $g(x) = -\lambda h(x)$, we obtain the solution of

$$(P) \int_{-1}^{1} \frac{f(t)\,dt}{t - x} = h(x) \tag{8-170}$$

as
$$f(x) = \frac{-1}{\pi^2} \sqrt{\frac{1 - x}{1 + x}} (P) \int_{-1}^{1} \frac{h(t)\sqrt{1 + t}}{\sqrt{1 - t}\,(t - x)}\,dt + \frac{k}{\sqrt{1 - x^2}} \tag{8-171}$$

If $h(x) = 1$ in Eq. (8-170), the solution is

$$f(x) = -\frac{1}{\pi}\sqrt{\frac{1 - x}{1 + x}} + \frac{k}{\sqrt{1 - x^2}} \tag{8-172}$$

Exercises

17. (*a*) Obtain the general solution of

$$(P) \int_{-1}^{1} \frac{1}{t - x} [1 + \alpha(t - x)] f(t)\,dt = 1$$

where α is a constant, by treating $\int_{-1}^{1} f(t)\,dt$ as a constant to be determined a posteriori. Discuss the special case $\alpha = 1$; is the solution unique?

(*b*) Let $p(x)$ be some function of x. By placing the term

$$\int_{-1}^{1} p(t - x) f(t)\,dt$$

on the right-hand side of the equation and using Eq. (8-171), show that

$$(P) \int_{-1}^{1} \left[\frac{1}{t - x} + p(t - x)\right] f(t)\,dt = h(x)$$

is transformed into a Fredholm equation for x; what is the character of the kernel?

(*c*) Discuss the use of the method of Exercise 17*a* for the problem of Exercise 17*b*, for the case in which $p(x)$ is a polynomial.

(*d*) Discuss the extent to which the methods of this exercise can be used for the equation

$$a(x)f(x) = \lambda(P) \int_{-1}^{1} f(t) \left[\frac{1}{t - x} + p(t - x)\right] dt + g(x)$$

18. Obtain the solution of

$$a(x)f(x) = \lambda(P) \int_{C} \frac{f(t)\,dt}{t - x} + g(x) \tag{8-173}$$

where C is a simple closed contour, x is any point on that contour, and all functions may be complex-valued. What accessory condition, if any, must λ, $a(x)$, and $g(x)$ satisfy for a solution to exist?

19. Solve

$$(P) \int_0^1 \frac{f(t)\,dt}{t - x} + (P) \int_2^3 \frac{f(t)\,dt}{t - x} = x$$

where this equation is to hold for x in either (0,1) or (2,3).

20. Discuss

$$a(x)f(x) + \lambda(P) \int_0^1 \frac{f(t) - f(x)}{t - x}\,dt = g(x)$$

where all quantities are real.

21. Show that the general solution of

$$f'(x) + \lambda(P) \int_C \frac{f(t)\,dt}{t - x} = 1$$

[where C is a simple closed contour, λ a constant, and $f(x)$ a differentiable function of position x on C] is given by

$$f(x) = \frac{1}{\pi i \lambda} + ke^{-\pi i \lambda x}$$

where k is an arbitrary constant. Generalize the result to the case of an arbitrary function $g(x)$ on the right-hand side, where $g(x)$ is analytic for x inside C.

22. Let $f(z)$ be a given function, analytic inside a simple closed curve C. Show that it is generally possible to find a *real* function $u(t)$ such that, for z inside C,

$$f(z) = \frac{1}{2\pi i} \int_C \frac{u(t)}{t - z}\,dt + ik$$

where k is some real constant. [*Hint:* Define $w(t) = u(t) - f_+(t)$; then $w(t)$ is the negative of the boundary value of a function analytic outside C and vanishing at ∞.]

23. Show that the solution of

$$(P) \int_C \left[\frac{1}{t - t_0} + P(t - t_0)\right] f(t)\,dt = g(t_0)$$

where C is a simple closed curve and $P(t)$ a given entire function of t, is

$$f(t) = -\frac{1}{\pi^2}(P) \int_C \frac{g(\tau)\,d\tau}{\tau - t} - \frac{1}{\pi^2} \int_C g(\tau)P(\tau - t)\,d\tau$$

Integral Equations with Algebraic Kernels

As an example of an equation whose kernel has an algebraic singularity, we consider an Abel-type integral equation with finite limits

$$\int_0^1 \frac{f(t)\,dt}{|x-t|^\alpha} = g(x) \tag{8-174}$$

for x in (0,1), where α is real, with $0 < \alpha < 1$. This equation is not of the Cauchy type; nevertheless, as shown by Carleman[1] it is still useful to introduce an analytic function related to $f(t)$ in a way analogous to that used for Cauchy integrals. Define

$$F(z) = \int_0^1 \frac{f(t)\,dt}{(z-t)^\alpha} \tag{8-175}$$

for z anywhere in the complex plane cut along the real axis from $-\infty$ to 1. For z real and $z > 1$, we use principal values in Eq. (8-175). It follows easily that, for x in (0,1),

$$F_+(x) = \int_0^x \frac{f(t)\,dt}{(x-t)^\alpha} + e^{-i\alpha\pi} \int_x^1 \frac{f(t)\,dt}{(t-x)^\alpha} \tag{8-176}$$

$$F_-(x) = \int_0^x \frac{f(t)\,dt}{(x-t)^\alpha} + e^{i\alpha\pi} \int_x^1 \frac{f(t)\,dt}{(t-x)^\alpha} \tag{8-177}$$

where, as before, $F_+(x)$ and $F_-(x)$ denote the limits of $F(z)$ as $z \to x$ from above or below, respectively. Equations (8-176) and (8-177) may be thought of as the appropriate replacements for the Plemelj formulas for Cauchy integrals. Since

$$e^{i\alpha\pi}F_+(x) - e^{-i\alpha\pi}F_-(x) = 2i \sin \alpha\pi \int_0^x \frac{f(t)\,dt}{(x-t)^\alpha} \tag{8-178}$$

the function $f(x)$ can be determined from a knowledge of $F_+(x)$ and $F_-(x)$, via the solution of a conventional Abel equation.

We can solve Eqs. (8-176) and (8-177) for each of

$$\int_0^x \frac{f(t)\,dt}{(x-t)^\alpha} \qquad \int_x^1 \frac{f(t)\,dt}{(t-x)^\alpha}$$

in terms of $F_+(x)$, $F_-(x)$, and then use Eq. (8-174) to obtain

$$F_+(x) = -e^{-i\alpha\pi}F_-(x) + (1 + e^{-i\alpha\pi})g(x) \tag{8-179}$$

for x in (0,1). For x in $(-\infty,0)$, Eq. (8-175) gives

$$F_+(x) = e^{-2i\alpha\pi}F_-(x) \tag{8-180}$$

[1] *Loc. cit.*

This is again a Riemann-Hilbert problem involving the two arcs $(-\infty,0)$ and $(0,1)$. Our previous formal method for solving Eqs. (8-179) and (8-180) is not applicable, because it leads to a nonconvergent integral over $(-\infty,0)$. However, the coefficients in Eqs. (8-179) and (8-180) are simply constants, and, by experimentation with a factor of the form $z^\mu(z-1)^\lambda$, we are led to define the new function

$$W(z) = z^{(\alpha-1)/2}(z-1)^{(\alpha-1)/2}F(z)$$

Equation (8-180) then becomes

$$W_+(x) = W_-(x) \tag{8-181}$$

for x in $(-\infty,0)$, and Eq. (8-179) becomes

$$W_+(x) = W_-(x) - 2i\cos\frac{\alpha\pi}{2}\, x^{(\alpha-1)/2}(1-x)^{(\alpha-1)/2}g(x) \tag{8-182}$$

for x in $(0,1)$, with

$$\begin{aligned} W_+(x) &= x^{(\alpha-1)/2}(1-x)^{(\alpha-1)/2}e^{i\pi(\alpha-1)/2}F_+(x) \\ W_-(x) &= x^{(\alpha-1)/2}(1-x)^{(\alpha-1)/2}e^{-i\pi(\alpha-1)/2}F_-(x) \end{aligned} \tag{8-183}$$

for x in $(0,1)$.

The solution of Eq. (8-182) is

$$W(x) = -\frac{1}{\pi}\cos\frac{\alpha\pi}{2}\int_0^1 \frac{[t(1-t)]^{(\alpha-1)/2}g(t)}{t-z}\,dt \tag{8-184}$$

where we have permitted $F(z)$ to have algebraic singularities near the points 0, 1, of order not greater than $-\frac{1}{2}(1+\alpha)$; this corresponds, in general, to permitting $f(t)$ to have nothing worse than an algebraic integrable singularity at each end. Computing $W_+(x)$ and $W_-(x)$ and using Eqs. (8-183), (8-178), and (8-125), we obtain

$$\begin{aligned} f(x) = \frac{\sin\alpha\pi}{2\pi}\frac{d}{dx}\int_0^x \frac{g(t)\,dt}{(x-t)^{1-\alpha}} - \frac{\cos^2(\alpha\pi/2)}{\pi^2} \\ \frac{d}{dx}\int_0^x \frac{[t(1-t)]^{(1-\alpha)/2}}{(x-t)^{1-\alpha}}\,dt\,(P)\int_0^1 \frac{g(\tau)[\tau(1-\tau)]^{(\alpha-1)/2}}{\tau-t}\,d\tau \end{aligned} \tag{8-185}$$

Exercises

24. Show that the solution of Eq. (8-174), with $g(x) = 1$, is given by

$$f(x) = \frac{\cos(\alpha\pi/2)}{\pi}(x-x^2)^{(\alpha-1)/2} \tag{8-186}$$

25. Solve

$$\int_0^x \frac{f(t)\,dt}{\sqrt{x-t}} + A\int_x^1 \frac{f(t)\,dt}{\sqrt{t-x}} = 1$$

where A is a real positive constant. Outline briefly the appropriate method if A is a function of x.

26. Show that the equation

$$\int_0^1 \frac{P(x-t)}{|x-t|^\alpha} f(t)\,dt = g(x)$$

(where P is a polynomial) can be solved by a method analogous to that used for Eq. (8-174).

Integral Equations with Logarithmic Kernels

We consider first

$$\int_{-1}^{1} \ln|x-t| f(t)\,dt = g(x) \tag{8-187}$$

for x in $(-1,1)$. Since this equation is one way of formulating the Dirichlet problem for a plane region exterior to a slit, its solution can be found, indirectly, by conformal mapping. To solve it by a direct method, we define

$$F(z) = \int_{-1}^{1} \ln(z-t) f(t)\,dt \tag{8-188}$$

and observe that, for $x < -1$,

$$\begin{aligned} F_+(x) &= \int_{-1}^{1} \ln|x-t| f(t)\,dt + i\pi \int_{-1}^{1} f(t)\,dt \\ F_-(x) &= \int_{-1}^{1} \ln|x-t| f(t)\,dt - i\pi \int_{-1}^{1} f(t)\,dt \end{aligned} \tag{8-189}$$

whereas, for x in $(-1,1)$,

$$\begin{aligned} F_+(x) &= \int_{-1}^{1} \ln|x-t| f(t)\,dt + i\pi \int_{x}^{1} f(t)\,dt \\ F_-(x) &= \int_{-1}^{1} \ln|x-t| f(t)\,dt - i\pi \int_{x}^{1} f(t)\,dt \end{aligned} \tag{8-190}$$

The discontinuity for $x < -1$ can be avoided by using $F'(z)$ instead of $F(z)$; for x in $(-1,1)$, we have[1]

$$F'_+(x) + F'_-(x) = 2g'(x)$$

In terms of

$$W(z) = F'(z)\sqrt{z^2-1}$$

this equation becomes

$$W_+(x) - W_-(x) = 2i\sqrt{1-x^2}\,g'(x)$$

for x in $(-1,1)$. The solution is

$$W(z) = \frac{1}{\pi}\int_{-1}^{1} \frac{\sqrt{1-t^2}\,g'(t)}{t-z}\,dt + \int_{-1}^{1} f(t)\,dt$$

[in considering the behavior of $W(z)$ near -1, $+1$, we permit $f(t)$ to have an integrable algebraic singularity at each end point]. Thus, since

[1] The reader should show that, for example, $[F(x+i0)]' = F'(x+i0)$.

$F'_+(x) - F'_-(x) = -2\pi i f(x)$, we obtain

$$f(x) = \frac{1}{\sqrt{1-x^2}}\left[\frac{1}{\pi^2}(P)\int_{-1}^{1}\frac{\sqrt{1-t^2}\,g'(t)\,dt}{t-x} + \frac{1}{\pi}\int_{-1}^{1} f(t)\,dt\right] \quad (8\text{-}191)$$

To evaluate the second term, we notice first that, if $g(t) \equiv 1$, then Eq. (8-191) shows that

$$\int_{-1}^{1}\frac{\ln|x-t|\,dt}{\sqrt{1-t^2}}$$

is a constant, whose value is easily found to be $-\pi \ln 2$. We now multiply Eq. (8-187) by $(1-x^2)^{-1/2}$ and integrate with respect to x from -1 to 1 to give a formula for $\int_{-1}^{1} f(t)\,dt$; using this result in Eq. (8-191), we obtain Carleman's formula

$$f(x) = \frac{1}{\pi^2\sqrt{1-x^2}}\left[(P)\int_{-1}^{1}\frac{\sqrt{1-t^2}\,g'(t)\,dt}{t-x} - \frac{1}{\ln 2}\int_{-1}^{1}\frac{g(t)\,dt}{\sqrt{1-t^2}}\right] \quad (8\text{-}192)$$

For the more general equation

$$\int_{-1}^{1}[\ln|x-t|p(x-t) + q(x-t)]f(t)\,dt = g(x) \quad (8\text{-}193)$$

for x in $(-1,1)$, where $p(x)$ and $q(x)$ are arbitrary polynomials, we proceed by defining

$$F(z) = \frac{1}{\sqrt{z^2-1}}\int_{-1}^{1}\left[p(z-t)\ln\frac{z-t}{z+1} + q(z-t)\right]f(t)\,dt \quad (8\text{-}194)$$

which is single-valued in the plane cut along the real axis from -1 to 1. For x in $(-1,1)$, we have

$$\begin{aligned} F_+(x) &= \frac{-i}{\sqrt{1-x^2}}\left[g(x) - \ln(1+x)\int_{-1}^{1}p(x-t)f(t)\,dt + i\pi\int_{x}^{1}p(x-t)f(t)\,dt\right] \\ F_-(x) &= \frac{i}{\sqrt{1-x^2}}\left[g(x) - \ln(1+x)\int_{-1}^{1}p(x-t)f(t)\,dt - i\pi\int_{x}^{1}p(x-t)f(t)\,dt\right] \end{aligned} \quad (8\text{-}195)$$

Forming $F_+(x) - F_-(x)$ and examining the behavior of $F(z)$ at ∞ and near the points -1, 1, we find

$$F(z) = -\frac{1}{\pi}\int_{-1}^{1}\frac{1}{\sqrt{1-t^2}}\left[g(t) - \ln(1+t)\int_{-1}^{1}p(t-\tau)f(\tau)\,d\tau\right]\frac{dt}{t-z} + R(z) \quad (8\text{-}196)$$

where $R(z)$ is the "entire" part of $F(z)$—i.e., that part of the Laurent series for $F(z)$, in the region outside the unit circle, which does not involve negative powers of z. From Eq. (8-194), $R(z)$ may be expressed in terms of a finite number of unknown constants c_n defined by

$$c_n = \int_{-1}^{1} t^n f(t)\, dt \tag{8-197}$$

(n integral and ≥ 0). These same c_n occur also in the term arising from $\int_{-1}^{1} P(t-\tau)f(\tau)\, d\tau$ in Eq. (8-196); thus, except for a finite number of these c_n, $F(z)$ is known. We next form

$$F_+(x) + F_-(x) = \frac{2\pi}{\sqrt{1-x^2}} \int_x^1 p(x-t)f(t)\, dt \tag{8-198}$$

and solve Eq. (8-198) by Laplace transforms to give

$$f(x) = \frac{d}{dx} \int_x^1 M(t-x)\{2\pi \sqrt{1-t^2}\, [F_+(t) + F_-(t)]\}'\, dt \tag{8-199}$$

where $M(t)$ is the inverse transform of $[s^2 P_1(s)]^{-1}$, $P_1(s)$ being the transform of $p(-t)$. Since

$$F_+(x) + F_-(x)$$
$$= -\frac{2}{\pi}(P) \int_{-1}^{1} \frac{1}{\sqrt{1-t^2}} \left[g(t) - \ln(1+t) \int_{-1}^{1} p(t-\tau)f(\tau)\, d\tau \right] \frac{dt}{t-x} + 2R(x)$$

the solution is complete, except for the evaluation of the constants c_n. However, a set of linear algebraic equations for the c_n may be obtained by multiplying Eq. (8-199) by appropriate powers of t and integrating over $(-1,1)$; the coefficients of the c_n in these equations may be explicitly evaluated in terms of Bessel functions.[1] For the special case $p(t) = 1$, $q(t) = 0$, the result is

$$f(x) = \frac{d}{dx}\left[\frac{1}{\pi^2} \sqrt{1-x^2}\, (P) \int_{-1}^{1} \frac{g(t) - c_0 \ln(1+t)}{\sqrt{1-t^2}\,(t-x)}\, dt \right] \tag{8-200}$$

where c_0 may be determined from the condition that the $F(z)$, as given by Eq. (8-196), have no term of order $1/z$ as $z \to \infty$ [cf. Eq. (8-194) for $p = 1$, $q = 0$]; this gives

$$c_0 = -\frac{1}{\pi \ln 2} \int_{-1}^{1} \frac{q(t)\, dt}{\sqrt{1-t^2}}$$

as before. The reader may show that Eqs. (8-200) and (8-192) are equivalent.

It is sometimes possible to apply the Carleman method directly to an

[1] C. Pearson, *Quart. Appl. Math.*, *15:* 203 (1957).

integral equation of Wiener-Hopf form, without the necessity of first taking transforms. As an example of this, consider[1]

$$\int_0^\infty H_0^{(1)}(|x - t|)f(t)\,dt = g(x) \tag{8-201}$$

for $x > 0$. Define

$$F(z) = \int_0^\infty H_0^{(1)}(t - z)f(t)\,dt \tag{8-202}$$

where the positive half of the real axis is to be a branch cut; we take $\arg(t - z) = 0$ for z real and < 0. Then $|\arg(t - z)| < \pi$ in the cut plane, and

$$H_0^{(1)}(t - z) \sim \sqrt{\frac{2}{\pi(t - z)}}\, e^{i(t-z-\frac{1}{4}\pi)}$$

for large $|z|$. It follows that the function $e^{iz}F(z)$ will be bounded as $|z| \to \infty$ in the cut plane. We define next

$$\varphi(z) = \frac{1}{\sqrt{z}}\, e^{iz}F(z) \tag{8-203}$$

with $-2\pi < \arg z < 0$) and obtain, for x in $(0, \infty)$,

$$\varphi_+(x) - \varphi_-(x) = -\frac{2}{\sqrt{x}}\, e^{ix}g(x) \tag{8-204}$$

$$\varphi_+(x) + \varphi_-(x) = -\frac{4}{\sqrt{x}}\, e^{ix} \int_0^x J_0(t - x)f(t)\,dt \tag{8-205}$$

The appropriate solution of Eq. (8-204) is

$$\varphi(z) = \frac{1}{2\pi i}\int_0^\infty \frac{-2e^{it}g(t)}{\sqrt{t}\,(t - z)}\,dt \tag{8-206}$$

and we can then solve Eq. (8-205), by Laplace transforms, to give

$$f(t) = \left(\frac{d^2}{dt^2} + 1\right)\int_0^x J_0(x - t)\left(\frac{\sqrt{t}\,e^{-it}}{2\pi i}\,(P)\int_0^\infty \frac{e^{i\tau}g(\tau)\,d\tau}{\sqrt{\tau}\,(\tau - t)}\right) \tag{8-207}$$

EXERCISES

27. Solve

$$\int_{-1}^{1} [1 + k(x - t)] \ln |x - t| f(t)\,dt = 1$$

where k is a constant. Show in particular that

$$c_0 = -\frac{I_0(k)}{(\ln 2)I_0(k) + kI_1(k)} \qquad c_1 = -\frac{(1 - \ln 2)I_1(k)}{(\ln 2)I_0(k) + kI_1(k)}$$

Obtain also an approximate form of the solution valid for small k.

[1] A related discussion of this example, using a Riemann surface viewpoint, is given by A. E. Heins and R. C. MacCamy, *Quart. J. Math. Oxford*, **9**: 132 (1958).

28. Show that the constants c_r in Eq. (8-197) are easily determined from a knowledge of the solutions of the special equation

$$\int_{-1}^{1} [\ln|x - t|p(t - x) + q(t - x)]w_r(t)\, dt = x^r$$

29. Show rigorously that the result of differentiating Eq. (8-187) once is

$$(P)\int_{-1}^{1} \frac{f(t)\, dt}{x - t} = g'(x)$$

and verify that Eqs. (8-171) and (8-192) are consistent.

Obtain also a formula for the second derivative of Eq. (8-187), involving the term $f(x) - f(t)$ in the integrand numerator.

30. An interesting method of attacking certain singular integral equations, due to Latta,[1] makes use of the differential equation satisfied by the kernel function. For the case $k(x - t) = \ln|x - t|$, the appropriate differential equation is $xk'(x) = 1$. Define

$$\Gamma(f) = \int_{-1}^{1} \ln|x - t|f(t)\, dt$$

and show that

$$\Gamma'(xf) = x\Gamma'(f) - \int_{-1}^{1} f(t)\, dt \tag{8-208}$$

Show also that, if $y(x)$ is such that $y(-1) = y(1) = 0$, then $\Gamma'(y) = \Gamma(y')$. Consider now the equation $\Gamma(f) = 1$ for x in $(-1,1)$, and use Eq. (8-208) to evaluate $\Gamma(xf)$ and $\Gamma(x^2 f)$. Choosing $y = (1 - x^2)f$, show that $\Gamma(y' + xf) = 0$; assuming that $\Gamma(h) = 0 \Rightarrow h = 0$, solve the resulting differential equation for $f(x)$. Discuss the possibility of solving $\Gamma(f) = g$ for other right-hand sides than unity. [Latta discusses also the case of the Bessel kernel $K_0(x - t)$ and for certain right-hand sides is able to obtain a rather complicated differential-equation eigenvalue problem describing f.]

31. For what values of $b > 0$ does

$$\int_{-b}^{b} \ln|x - t|f(t)\, dt = 1$$

not have a solution?

[1] G. Latta, *J. Rational Mech. Anal.*, ***5:*** 821 (1956).

index